Preface

Lecithins are complex mixtures of polar lipids, mainly phospholipids, obtained by physical procedures from a variety of vegetable and animal sources.

Does that make it sound as if there is anything phenomenal about lecithin? Is the title of this book justified? The answer is definitely YES. Because what makes lecithin so phenomenal are its remarkable properties and its history.

Phospholipids are natural components of all animal and vegetable cell membranes, where they have an essential function. That explains the physiological benefits of lecithin. Because of the amphiphilic structure of the phospholipids, which gives them their surfactant properties, it may be said that lecithin is the most important of all the natural emulsifiers. It also means that from the techno-functional point of view, lecithin is universally usable and practically indispensable for certain applications.

Although considerable literature has become available in recent years referring to the specific properties and benefits of phospholipids, the last books devoted to the scientific and technological principles of the production, processing and use of lecithins in general were – as far as we know – published back in the 1990s. So, the editors, who work with lecithin day in, day out, felt it was important to publish an updated overview of the subject.

The history, technical development and current status of lecithin as a product are very closely bound up with the market situation and changes in our society. Precisely this is the topic of the first chapter of the book. The theoretical principles – the chemical, physiological and physical properties of phospholipids that constitute the basis for understanding the book – are explained in the second part. This is followed by the practical section which deals in considerable detail with the origins, production, processing and modification of lecithin. A very important element of the production process is quality control by chemical, physical and further analyses which are covered in the next chapter. The subsequent descriptions of the numerous applications of lecithin in the food, nutrition, feed, pharmaceutical, cosmetic and technical industries make up the core of the work.

The book "Phenomenon of Lecithin – Science, Technology and Applications" is intended as a source of information not only for both scientists and technologists involved in production, research and development, and quality assurance but also for everyone who works with lecithin or wishes to do so.

I wish to thank the numerous renowned and internationally acknowledged experts in the fields of lecithin research and lecithin applications who agreed to contribute chapters to this book. My thanks also go to my co-editors, Parris Kidd and Rüdiger Ziegelitz, for their valuable cooperation.

In particular, I wish to thank Volkmar Wywiol who initiated the book project and gave it his active support.

Waldemar Buxmann
Hamburg, April 2021

Foreword

I have been closely associated with lecithin for over 65 years. First as a young commercial clerk, then as an entrepreneur, I devoted my time, energy, creativity and passion to this unrivalled nutrient. Lecithin became my mission in life, one that has taken me across all the continents and into all the spheres of application. Whereas worldwide consumption of lecithin was only about 50,000 tonnes in the 1960s, the past 60 years have seen a twelvefold increase in demand to over 550,000 – 600,000 tonnes annually.

Besides soy lecithins, sunflower and rapeseed lecithins and also lecithins from egg, milk and marine species are now on the advance. The increasing significance of this outstanding natural substance has made it necessary to update our knowledge base. The diversity of its applications in foods, animal nutrition, food supplements, pharmaceuticals, cosmetics and the technical industries is unique.

Our book "Phenomenon of Lecithin" sprang from a need of mine to offer experts around the globe a forum for expressing their views and passing on their experience to the younger generation. I thank all 23 authors for their willingness to contribute and for taking time to document their knowledge in this way. My special appreciation and thanks go to Rüdiger Ziegelitz – a legend in the history of lecithin – and to Dr Michael Schneider, both of whom helped me with their many suggestions and encouraged me to complete this project. Dr Parris Kidd took on the laborious task of standardizing the structure and language of the contributions, for which I thank him.

But I also especially wish to thank Dr Waldemar Buxmann, who compiled this outstanding work with untiring diligence, working at night in addition to his function as scientific head of the lecithin company Sternchemie. A shooting star in the lecithin community.

May the book "Phenomenon of Lecithin – Science, Technology and Applications" – give chemists, biologists, pharmacists, technologists and also entrepreneurs a deeper insight into this miraculous natural substance. And may it inspire you to continue with intensive research work. For "Lecithin is Life"!

Volkmar Wywiol
Hamburg, April 2021

Phenomenon of Lecithin
Science | Technology | Applications

An initiative of Sternchemie on the occasion of the 40th Stern-Wywiol Gruppe anniversary.
www.sternchemie.com

Editors Waldemar Buxmann, Paris Kidd, Volkmar Wywiol, Rüdiger Ziegelitz
English translation Marion Braase, Wieners+Wieners GmbH, Hamburg
Proofreading Sternchemie, Paris Kidd, Wieners+Wieners GmbH, Hamburg
Layout & Design sachdesign, Hamburg
Photos & Illustrations Sternchemie, sachdesign
Project coordination Tina Höhne, Marketing Sternchemie

Publisher
Robert Wenzel
An der Alster 81, 20099 Hamburg, Germany
www.robert-wenzel.de

Production & Print
AlsterWerk MedienService GmbH
www.alsterwerk.com

1st edition 2021

ISBN 978-3-9820751-3-6

Library of Congress Control Number: 2021939002

Distributor
Taylor & Francis Group, LLC
Suite 300, 6000 Broken Sound Parkway, NW
Boca Raton, Florida 33487, United States
www.taylorandfrancis.com

CRC Press
Taylor & Francis Group

Contents

4. Modification and Fractionation of Lecithin 194

1 Lecithin: Past – Present – Future

- 1846 **Discovery of lecithin**
- 1850 – 1884 Elucidation of the structure of phosphorus-containing lipids
- 1900 Isolation of individual phospholipids
- 1920 First industrial production of lecithin from soybeans
- 1930 Lecithin becoming increasingly popular
- 1934 Start of production of lecithin in the United States
- 1946 Lecithin again marketed in Europe
- 1955 Lecithin fractions and modifications
- 1960 Lecithin used for parenteral nutrition
- 1990 New generations of lecithin
- 1992 Central Soya and Sternchemie establish a joint venture in lecithin
- 1996 Genetic engineering transforms the world of lecithin
- 2000 Restructuring of the lecithin market
- 2003 Beginning of IP lecithin
- 2010 Upswing in sunflower and rapeseed lecithin
- 2015 New oil mill groups increase the options
- 2020 Lecithin market today
- **Outlook for the future**

1.1 History of lecithin

Volkmar Wywiol, Michael Schneider, Rüdiger Ziegelitz

The term 'lecithin' stands for a complex mixture of different, highly specialized ingredients. Its many useful properties derive mainly from a group of substances known as phospholipids, each of which has quite specific technical or physiological characteristics (for more details see Chapter 2).

For decades, lecithin has been an irreplaceable ingredient of chocolate, sweets, instantized cocoa, bakery products, margarine and many other foods. Even a small percentage achieves excellent emulsifying and dispersing effects, thus enhancing the quality of the products and prolonging their shelf life. But it is not only their technical properties that make the lecithins so valuable.

Used in medicines, they offer a wide range of therapeutic benefits, for example in the treatment of liver disorders and for lowering elevated cholesterol levels. In the form of food supplements, certain ingredients act as an aid to memory and alleviate the symptoms of stress. They also have important uses in modern medicinal formulations – the liposomal dosage forms – and in products for intravenous nutrition.

And, not least, manufacturers of animal feed are making increasing use of the emulsifying and nutritional benefits.

1846
The discovery of lecithin

Lecithin was discovered more than 170 years ago. After the French pharmacist L.N. Vauquelin had extracted a substance containing phosphorus from brain tissue as long ago as 1812, the chemist and pharmacologist Nicolas-Théodore Gobley isolated a phosphorus-containing substance from egg yolk for the first time in 1846/1847. In 1850 he gave it the name 'lecithin', after the Greek word *lekithos* for egg yolk. Gobley was also the first to isolate glycerophosphoric acid and a nitrogen compound, which he was unable to identify, from a natural substance. The nitrogen compound was not identified until 15 years later, by the chemists C. Diakanow and Adolph Strecker. In 1862 Strecker, who was then private assistant to Justus Liebig, found in bile a compound he called choline, based on the Greek word *chole* for bile. The two scientists also discovered that two fatty acids were joined by the glycerophosphate skeleton. They came to the conclusion, for the first time, that phosphoric acid serves as a link between the diglyceride moiety of the lecithin and the choline.

1850–1884
Elucidation of the structure of the phosphorus-containing lipids

At about the same time as Gobley, Diakonow and Strecker, the German chemist Johann Ludwig Wilhelm Thudichum made an intensive study of the chemistry of the brain. Under the influence of Justus Liebig, he first explored the chemical structure of the brain at the University of Giessen. In 1853 he moved to London, where he ran a private research laboratory. With his *Treatise on the Chemical Constitution of the Brain*, published in 1884, he laid the foundations of phospholipid chemistry. Between 1850 and 1882 he discovered the structure of numerous phospholipid compounds in the brain; these originally became known under the names cephalin, colamine lecithin and sphingomyelin.

The works of Adolph Strecker, his successor Felix Hoppe-Seyler and Johann Thudichum served as a stimulus to numerous scientists. They were followed by a flood of publications. In the course of time it emerged that phospholipids, and therefore lecithin, occur in all cells – of plants, animals and microorganisms.

1900
Isolation of individual phospholipids – elucidation of their structure

In the meantime, however, lecithin research had come up against its limits. It proved very difficult to produce the individual ingredients of lecithin in a chemically pure form. In the early 20th century, Levine and West reported that 30 recrystallization procedures were often necessary in order to obtain substances with sufficient purity. No appreciable progress was achieved until new analytical methods had been developed. In the 1920s, for instance, chromatography by means of adsorption to aluminium oxide or silica was introduced by Thannhauser and Setz and used by many researchers for concentrating and purifying the different ingredients of lecithin.

At first, little attention was given to the diversity of the lecithin ingredients. Only gradually did it become clear that Gobley's term 'lecithin' was really a collective name for a large number of individual lipids containing phosphorus.

Even the number of fatty acids and their types – saturated or unsaturated – that can occur in the phospholipids (bound to the common glycerol skeleton) is extremely large. The diversity becomes even greater when two different fatty acids are bound to the glycerol. Further variants occur through molecules like choline, ethanolamine and inositol bound to the phosphate.

From 1900
Development of the first dietetic lecithin preparations

From 1900 there was great interest in researching and exploiting the health benefits attributed to lecithin in the form of medicines and food supplements. The idea stemmed mainly from the observation that vital human and animal organs contained large amounts of phospholipids. On the basis of these findings, scientists tried to extract lecithin for pharmaceutical products from animal sources. Lecithin attracted the interest of researchers for dietetic purposes, too, and in this field the physician Dr Heinrich Buer played an important role. With his "Dr. Buer Neura Lecithin" he launched one of the first dietetic lecithins on the German market, a product that later became known as Buer-lecithin and is sold to this day.

1920
The first industrial production of lecithin from soybeans

The main source for pharmaceutical lecithin preparations was egg yolk, but this was unsuitable for industrial-scale production. For one thing, lecithin from egg was too expensive; for another, the substance was not available in sufficient amounts. It was then that soybeans came on the scene as a cheap and practically inexhaustible source.

Large amounts of soybeans were already being imported to Germany from Manchuria at the beginning of the 20th century. From the 1930s, soybeans were then grown successfully as a food plant in the USA, with rapidly increasing annual quantities. After 85 years, the US crop of 2016 reached a record-breaking 119 million tonnes. In that year, 328 million tonnes of soybeans were harvested worldwide.

Around 1920, Hamburg became the scene of the first large-scale industrial production of lecithin from soybeans. The foundations were laid by Hermann Bollmann, manager of the Hansa-Mühle oil mill established in 1911 (now called Ölmühle Hamburg and owned by the ADM group). His patented extraction method permitted the economically viable production of lecithin from soybeans for the first time. Dr Bruno Rewald, Bollmann's assistant, worked consistently on improving the quality of the lecithin. He sought and tested more and more new applications. The many patents from Hansa-Mühle are proof of the tireless research work carried out there. By 1933, the company had applied for more than 65 patents worldwide.

1927
The United States imports the first batches of lecithin from Germany

Whereas the large-scale industrial processing of soybeans was already in full swing in Hamburg, no lecithin was yet being extracted from soybeans in the USA. It was not until the American Joseph Eichberg heard of the activities in Europe and visited the Hansa-Mühle in 1927. After that, lecithin began to be used in food production in the United States – mainly in chocolate. Subsequently, in 1930, the American Lecithin Company (ALC) was established with the participation of Hansa-Mühle and took on the distribution of the lecithin from Hamburg in the USA. At the same time Eichberg ensured that lecithin was produced in the United States, too. USA is now one of the biggest lecithin producers worldwide.

From 1930
Lecithin becoming increasingly popular as an additive for various industries

Soy lecithin was initially used mainly as an emulsifier in the production of margarine, in order to create a cheap product similar to butter. The chocolate industry started using more and more lecithin, too. It improved the dispersibility of the sugar in the chocolate mass and saved a large proportion of the expensive cocoa butter without impairing the flow properties and processability. Moreover, lecithin came to be used in cosmetic articles like skin creams, soaps and other toiletries. It was also used as a sizing agent and dyeing auxiliary in technical products from the rubber, leather and textile industries. Lecithin production increased continuously as a result of the rising demand. By 1927, Hansa-Mühle had increased its annual capacity from the initial 150 tonnes to 800 tonnes. Other mills now also started to produce lecithin from the soybean oil.

1934
ADM – the first company to produce lecithin in the United States

In 1934, Archer-Daniels-Midland (ADM) in Chicago built the first US American factory to produce lecithin. Although the focus of all oil mills was not on the production of lecithin (only 0.6% of the content of the soybean) but on extracting soybean oil (20%) and soy flour or soy protein, the separation and sale of lecithin proved to be an attractive additional benefit. Shortly after ADM, the Glidden company built a similar factory in the USA, which was later taken over by Central Soya. In the USA, too, lecithin became an important additive used for a wide range of purposes, mainly in the food industry.

1935
The search for a substitute for egg lecithin for therapeutic applications

Besides Dr Heinrich Buer, the chemist Dr Hans Eikermann from Cologne was another scientist to devote himself to this fascinating substance. In the 1930s, his research initially focused on egg lecithin. But shortly after the outbreak of the Second World War there was a shortage of the raw material egg yolk powder, which was imported from China. Eikermann looked for a substitute, and found it.

Hansa-Mühle sent him soy lecithin. In his first laboratory tests he concentrated on purifying the soy lecithin so far as to obtain a quality comparable to that of egg lecithin. In doing so he found that the fatty-acid composition of the two substances differed considerably. A comparison of the iodine values, an expression of the degree of unsaturation of a fat, showed the value to be 60 – 64 for egg lecithin and 84 – 87 for soy lecithin. It also emerged that the concentrate produced by him from soy lecithin contained a higher percentage of choline phospholipids – with a surprisingly high proportion of unsaturated fatty acids.

When the war ended, Eikermann resumed his research work. The Institute of Physiological Chemistry of the University of Cologne succeeded in determining the structure of this 'new' lecithin. It was a lecithin with over 70% phosphatidylcholine and a high percentage of essential linoleic acid (dilinoleoyl phosphatidylcholine).

1946
Lecithin again marketed in Europe

After the war, in 1946, the Hamburg company Lucas Meyer started standardizing and processing lecithin. The soy mills concentrated on extracting oil and protein; in many cases, lecithin was a troublesome by-product that occurred in greatly varying qualities and quantities. It was from this starting point that the company (now Cargill) developed its successful business model: the processing of lecithin. In 1948 the chemical engineer Rüdiger Ziegelitz and, in 1953, the future businessman Volkmar Wywiol joined the lecithin team. Along with comprehensive applications consultancy, knowledge of the use of lecithin for over a hundred different applications from food, aquaculture farming, pet food and pharmaceuticals to auxiliaries for textiles and leather made its way from Germany around the globe.

The situation in America developed differently. The great advantage there was that top-quality, locally grown soybeans could be used. Moreover, the mills had not been damaged in the war and produced a good, light-coloured lecithin. In some cases it was marketed by the oil mills themselves, in others through specialist distributors like the American Lecithin Company. Central Soya (now Solae/DuPont) was the leading American lecithin producer alongside ADM. As with Lucas Meyer, the focus at Central Soya was on research and development work with lecithin, and it contributed to the high status lecithin now enjoys in the producing industries worldwide.

From 1955
Milestones in research: Lecithin fractions and modifications

Acetylation and enzymatic hydrolysis
Work on enhancing the properties of lecithin continued steadily. Crude lecithin was mixed with hydrogen peroxide to provide antibacterial treatment and lighten the colour. Developments started with the aim of creating new forms with special or intensified properties from natural lecithin. Acetylation and enzymatic hydrolysis improved the emulsifying properties of the lecithin, for example, and increased its temperature stability.

Deoiling, Fractionation
There followed the industrial-scale deoiling of lecithin with acetone in order to transform the viscose, oily raw lecithin into oil-free, powdered or granulated lecithin. Alcohol extraction, column chromatography and hydrogenation

resulted in highly specialized products such as pure phosphatidylcholine for various applications in pharmacy and cosmetics. Dry-spraying and cold-spraying methods brought further improvements in the use of lecithins in foods. In 1955, the German dairy cooperative Dahlenburg patented the first spray-dried lecithin.

Margarine
Unilever, the world's biggest margarine manufacturer, carried out continuous development work in the fields of fat processing and the use of lecithin. Following the switch from discontinuous to continuous margarine production in the 1960s, normal commercial lecithin was no longer suitable for stabilizing the emulsion. The Unilever research group under Hermann Pardun developed new emulsifiers for margarine on the basis of lecithins produced by alcoholic fractionation, with a phosphatidylcholine content of about 33%.

Pharmacy
Under Dr Eikermann, the company A. Nattermann & Cie carried out extensive pharmacological and clinical investigations that showed phosphatidylcholine to have a significant effect on lipid metabolic disorders and liver disease. The company – now Sanofi, after various changes of ownership and demergers – put a number of pharmaceutical products on the market.

In order to produce the new phosphatidylcholine fractions, suitable processes had to be developed; in the 1970s, this led to the construction of the world's biggest fractionation plant, now Phospholipid GmbH. The new compound with about 70% phosphatidylcholine was called EPL substance (essential phospholipids). The EPL was later purified up to a content of over 90%. It is also known as PPC (polyenylphosphatidylcholine). The scientists Dr B.R. Günther, Dr H. Betzing, Dr D. Lekim and Dr M. Ghyczy from Nattermann in Cologne also made a major contribution to these developments.

Liposomes
In 1964, rather by chance, Bangham and Horne observed under the electron microscope that lecithin spontaneously forms tiny enclosed globules in an aqueous dispersion. A two-molecule-thick bilayer phospholipid membrane enclosed a small amount of the added water. Since this discovery, continuous research into the physical and chemical properties of liposomes (Greek: liposoma = fat body) and work on their possible applications and effects has been carried out.

Increasing scientific interest in liposomes was also shown by biochemists and the medical profession. It was realized, for example, that these microscopic vesicles offer an interesting possibility of transporting medications enclosed within them to their required site of action. In the years that followed, liposomes were also discussed as a new dosage form and mechanism of action for skincare products. When it emerged that they can be produced on a large scale and worked into already marketable skincare preparations, the biotechnological innovation was used successfully in 1987 with Dior's patent for 'Capture' (slogan: 'The triumph of science over ageing').

1960
Lecithin used for parenteral nutrition

From the late 1950s, detailed work was carried out on the development and use of fat emulsions for parenteral nutrition of seriously ill patients receiving intensive care. Initially, fractions from soy lecithin were used as an emulsifier, but the considerable side effects made it necessary to take the product off the market after only a short time. In 1962, Intralipid from Kabi-Vitrum was approved in Sweden as the first fat emulsion based on egg lecithin for intravenous application; it is used clinically to this day.

The undesirable side effects of the soy lecithin fractions stimulated the development of egg lecithin as the physiologically better-tolerated emulsifier. In the late 1970s, the German company Lipoid began to make a close study of egg lecithin for parenteral nutrition under Dr H. Rebmann. The company is now the international market leader in this segment.

Initially, the main reason for administering fat emulsions was simply to supply energy, but in the following years they also came to be used as carriers for fat-soluble medications such as the anaesthetic Propofol. The more recent use of highly unsaturated oils, such as fish oil, offers further health benefits in addition to adequate caloric intake.

1982
Sternchemie – a new lecithin specialist captures the markets

In the meantime, the global market for lecithin had increased to over 200,000 tonnes annually. Besides uniform quality, all efforts were directed towards ensuring an adequate supply of this natural substance. Brazil and Argentina were producing larger and larger quantities, but transportation in drums or 20-tonns containers became too expensive. With the construction of a large stainless-steel tank farm in the Dutch town of Ouderkerk (Rotterdam) in 1986, Volkmar Wywiol, the owner of Sternchemie, created the conditions necessary for taking in shipments of 500 to 1,000 tonnes of crude lecithin and processing them for the European food industry.

1992
Central Soya and Sternchemie establish a joint venture in lecithin

The increasing demand, the wide diversity of applications and the worldwide marketing of lecithins almost automatically led to closer cooperation between suppliers. Competitors became partners. Jim McCarthy, Vice-President of Central Soya, and Volkmar Wywiol, the owner of Sternchemie, pooled their European interests and in 1992 established the joint venture Stern-Lecithin & Soja. Within a few years the company became the world's biggest supplier of lecithin. Central Soya was later acquired by Bunge and their lecithin and protein businesses were merged with DuPont to form the Solae Company, which became part of DuPont's Nutrition and Biosciences Group that was recently acquired by International Flavors and Fragrances (IFF). It is now managed under the name IFF Nourish Division.

From the 1990s
New generations of lecithin

Non-GM vegetable lecithin
A fundamental change of attitude on the part of food consumers began with the introduction of genetically modified oilseeds, in particular soybeans, largely under the influence of Monsanto. Over the years, the European countries came to demand – and use – only non-GM lecithin as an ingredient of foods. A further factor was the obligation to declare allergens in labelling, recently introduced in Europe; it resulted in an increasing demand for alternative vegetable lecithins, for example from sunflower or rape seeds, in addition to soy (which is subject to allergen declaration).

Synthesis of phospholipids
As a result of the increasing interest of the cosmetics and pharmaceutical industries in highly specialized lecithins and phospholipids, more and more attention was given to hydrogenated and therefore oxidation-resistant vegetable lecithins (trans-

formation of unsaturated fatty acids into saturated fatty acids). Even complete or partial synthesis of phospholipids was not considered too expensive for meeting the demand for 100% defined and reproducible substances for highly specialized pharmaceutical preparations. Enzyme technology now offered an attractive way of producing new phospholipids, or simply such with a higher concentration compared to natural sources. In this way it was easily possible not only to modify the fatty-acid composition but even to transform one phospholipid into another.

Phosphatidylserine (PS)
Surprisingly, it was BSE ('mad cow disease') that triggered a new era in the development and use of lecithin in the middle of the1990s. A special extract from the brain and spinal marrow of cattle, a substance rich in a specific phospholipid (phosphatidylserine (PS)), had been used successfully to treat memory disorders until about 1994, but the outbreak of BSE made it impossible to continue such an application. The company Lucas Meyer then had the idea of producing this PS, which is not present in soy lecithin, from other soy phospholipids by an enzymatic process. In a number of clinical studies, this product showed itself to be practically identical to the original substance derived from bovine brain in respect of its effects, and it achieved considerable international success as a food supplement.

Lecithins from animal raw materials
Innovations in alternative raw materials for lecithins were largely in the field of lecithins from animal raw materials. The hype surrounding the health benefits of omega-3 fatty acids was reflected in the growing number of lecithins from marine sources, for example from krill, fishmeal or fish roe. Lecithins were even produced from squid skin.

The health benefits were said to lie in a modified and therefore more positive ratio of omega-3 to omega-6 fatty acids in the human body, with effects on fat metabolism and a number of other conditions detrimental to health.

Milk lecithins
Lecithin from milk could prove to be enormously important, since its composition is completely different from that of other known lecithins. Not only do milk lecithins contain very specific phospholipids. They also contain phosphorus-free lipids, so-called polar lipids, like the highly effective gangliosides and ceramides.

Manufacturers of baby food are extremely interested in this new product, since its addition to milk preparations could bring these even closer to the composition of breast milk. But this is not the only incentive: the clinically proven effects are extraordinarily diverse. Milk lecithins prevent the infestation of the digestive tract with pathogenic bacteria and viruses, which is common in babies and may lead to severe diarrhoea. Milk lecithin has been shown to contribute to better protection and better development of the entire digestive system and is even said to promote brain development.

Whether the preventive effect of milk lecithin on bowel cancer, demonstrated in animal experiments, can be applied to human beings remains to be seen.

We have summarized the developments of the past 25 years in a separate section under the heading 'Lecithin market'.

1.2 Lecithin market
Significant developments in the last 25 years

Volkmar Wywiol

1996
Genetic engineering transforms the world of lecithin

The introduction of genetically modified oilseeds – in particular soybeans (GM soy) – by the American Monsanto group marks the beginning of a new era in the history of lecithin.

A totally new soy product: in 1997, American soybeans containing minimal amounts of non-declared GM soy from experimental farms are exported to Europe for the first time. The lecithin extracted from them – about 0.5% – 0.7% of the soybean – is analysed for genetically modified DNA by PCR testing and found to be positive.

Practically overnight, the market splits into GM and non-GM soy lecithins. Rapeseed and sunflower lecithin are discussed as potential alternatives for the first time. But only to a marginal extent, largely because the qualities cannot yet compete with those of the established soy lecithin products.

2000
Restructuring of the lecithin market

The start of the new millennium is marked by turbulence, the supplier structure for lecithins changes abruptly. The Hamburg companies Lucas Meyer and Stern-Lecithin & Soja, until then the two main players, find themselves with new owners.

Lucas Meyer – 'The Lecithin People' – sell their company to SKW/Degussa. Central Soya acquires the remaining 50% of the shares of Stern-Lecithin & Soja. Later Central Soya was taken over by Bunge and merged into Solae, a Joint Venture between DuPont and Bunge. After that it becomes part of DuPont Nutrition & Health and is now merged with IFF. In 2005 Sternchemie was reactivated as lecithin specialist. With ADM, a further oil mill group extends its own direct marketing of its lecithins into Europe.

Specialized European suppliers
Besides the two established oil mill groups and DuPont/Solae (today IFF), it is mainly specialized European suppliers that assert themselves in the global market for lecithin. They include the German companies Sternchemie and Lipoid and also the Spanish firm Lasenor, in particular with processed and modified special lecithins.

As opposed to the raw and standard lecithins offered by the oil mills, the three specialists focus on further purification steps and a greater breadth and depth of applications. Lipoid makes a name for itself chiefly with lecithin fractions and egg lecithin. Sternchemie concentrates on increasing the depth of its lecithin applications, mainly in the food sector, with customized formulations based on soybean, rape and sunflower seeds.

2003
Beginning of IP lecithin

New food regulations bring fundamental changes to the lecithin market.

Regulation (EC) No. 1829/2003 of the European Parliament on genetically modified food and feed and Regulation (EC) No. 1830/2003 concerning the traceability and labelling of genetically modified organisms and the traceability of food and feed products produced from these provide the basis for IP (Identity Preserved) soy lecithin. The regulations ensure continuous traceability from the bean to the lecithin.

The dividing lines, which were somewhat woolly in the past, are drawn more clearly by these regulations. In the European region, at least, this means the end, in the food sector, of so-called PCR-negative lecithins that no longer contain any DNA because of filtration and are therefore negative in the PCR test although they might originate from a GM soybean. This marks the beginning of IP soy lecithins, which guarantee traceability and may contain no more than 0.9% genetically modified DNA if this is technically unavoidable, purely by chance, and a GMO variety approved in the European Union.

These regulations apply to the food and feed market. Foods of animal origin (e.g. milk, eggs, meat) obtained from animals fed with GM soy do not have to be labelled in respect of GMOs. British food manufacturers, especially, discover this gap in the labelling requirements as a means of differentiation. They insist on meal from conventionally grown soybeans for their livestock breeding and thus – temporarily – create a lucrative market that preserves the growing, processing and segregation of conventional soybeans in Brazil.

Potential allergens must be declared in labelling
In 2005 and 2006 respectively, certain allergens as food ingredients become subject to declaration under the German food labelling regulation (LMKV) and through publication of the LOAEL (Lowest Observed Adverse Effect Level) by the FDA. They include soy and soy lecithin. These changes in the regulations bring about further shifts in the lecithin markets.

2008/09 – IP lecithins from Brazil become scarce and expensive

At this time, the largest volumes of non-GM soy lecithin traditionally come from Brazil. But the cultivation of GM soybeans is very much on the advance in that country, with the result that considerably less non-GM soy lecithin is available. A further problem is the risk of cross-contamination with other seeds during the local transportation chain. The new situation necessitates a large amount of work for documentation and proof of IP status. Many consignments have to be rejected and downgraded to GM soy. This is reflected in appreciable extra cost for non-GM soybeans. For the oil mills, lecithin is only a by-product (approximately 0.5 – 0.7% of the soybean), which means that the main product streams – oil, flour and in some cases protein fractions – have to justify and compensate for the extra cost of IP lecithin. Moreover, the demand for non-GM flour is declining rapidly because of a change in the marketing strategy of European supermarket chains. For the most part segregation is abandoned, and in effect conventional beans become GM soybeans.

The result is a shortage of IP soy lecithins, which become much more expensive compared to GM soy lecithin.

2008/09 – India comes into focus for non-GM soy

New suppliers and other countries of origin begin to participate in the market; from 2008/2009 India, especially, moves into the centre of attention. Cultivation of GM soybeans is prohibited there, as are imports of soybeans. These are ideal conditions for non-GM soy lecithins. But there are other obstacles to be overcome, for example the pattern of very small farms, complex warehousing and transport chains, substantial overcapacity of the oil mills and very short seasonal processing. Major differences in storage and processing conditions lead to vastly fluctuating qualities. A further problem is that although no GM soybeans may be imported at this time, imports of GM soybean oil and GM lecithin compete with locally produced lecithin. Moreover, other GM plants such as cotton are grown in India and their seeds are crushed, which involves a risk of contamination in the transport chains and during processing.

In particular the European lecithin specialists take advantage of these geographic changes. With affiliates of their own, Indian partner companies and above all a transfer of know-how, they establish a reliable supply chain in India. In this way they create alternatives to the increasingly scarce Brazilian IP lecithins.

2010 – Lecithin specialists organize supply chains for rapeseed and sunflower products

In the first years from 2010, supply bottlenecks and rising prices for non-GM soy greatly enhance the climate for investments. In close cooperation with the oil mills, the lecithin specialists create new processing capacities and logistics systems for sunflower and rapeseed lecithin. Substantially improved qualities and greater availability – together with the obligation to declare allergens and GMOs in labelling – make real alternatives to IP soy feasible.

Biodiesel affects the quality of lecithins
Not all lecithins are the same. Industrial production processes have a very marked effect on their functionality as well as their quality.

High subsidies cause a meteoric increase in biodiesel production. Among other things, biodiesel requires very pure vegetable oils. This means that the oil mill industry has to make changes to its processes. The classic method involves degumming the raw oil with water in order to extract the lecithin – which, though valuable, is a disruptive element in the oil. In order to produce an oil with a much lower phosphorus content, which is optimally suitable for biodiesel production, some oil mills convert their processes to degumming with acid or by enzymatic treatment.

This causes a substantial fall in the availability of classic lecithin, since the material derived from soy oil production for biodiesel may have a different functionality as well as a different composition. Moreover, lecithin produced in this way no longer conforms to the purity criteria. The production methods for lecithins therefore constitute a threat to availability and quality.

At the same time, the growing demand for vegetable oil from the biodiesel industry leads to rising oil prices and an increased demand for raw lecithins as an alternative energy source in animal feed.

The decade from 2010
An upswing in sunflower lecithin

The tighter regulations on the declaration of allergens in labelling in effect from 2011 in conjunction with a shorter supply of IP soy lecithins give powerful momentum to the development of sunflower lecithin. Broad applications research, the continuous improvement of production processes and enlarged capacities make sunflower lecithin an attractive alternative.

Oil mills in the Black Sea region, in southern and eastern Europe and also Argentina provide further capacity for the production of sunflower lecithin. Once a product for which there was little demand, often chosen as a cheap alternative in the feed sector, sunflower lecithin experiences a tremendous upswing.

In the middle of the decade the vigorous demand, great variations in quality and the conflict between Russia and Ukraine make sunflower lecithin a scarce product for a time. Attractive prices lead further oil mills to invest in the drying of sunflower lecithin or optimize their existing production processes.

From 2015
New oil mill groups increase the options

Besides the changing market conditions on the product side, the choice of suppliers widens, too. Bunge – once Solae – places its lecithin business on an independent footing. The Singapore agribusiness group Wilmar takes a similar step together with its Indian partner Adani, as does the French agricultural enterprise Groupe Avril. All begin to push ahead with marketing their lecithins from different oilseeds on an independent basis.

From 2017 – organic lecithins

The growing demand for organic foods gives rise to an increasing need for organic lecithins. After a transition period of several years, the European regulation 2016/673 requires that as of 1 January 2019, only organic lecithin may be used in the production of organic foods.

This presents the industry with new challenges, since the conventional production process for most vegetable oils consists not only of pressing but also extraction with solvents. Without additional solvent extraction the lecithin yield from the process is much smaller. Moreover, organic oil mills have a much lower throughput than conventional oil mills, which leads to a further reduction of production capacities. And the acquisition of seeds and beans in organic quality is a further constant challenge to organic oil mills.

Extensive monitoring coupled with comprehensive documentation and an unbroken chain of certifications are indispensable for ensuring that customers can rely on the suitability and qualification of the organic lecithin.

From 2020
Lecithin market today

The uniqueness of the natural substance lecithin and its triad of effects – as an emulsifier, dispersant and physiological agent – is a constant stimulus to the sciences and technology. It becomes the mainspring of extensive research and the development of new applications in many countries around the globe, on the part of both suppliers and the downstream industries.

This interest more than doubles the volume of lecithin used during the first two decades of the new millennium.

An ever-growing demand for expertise
On the one hand, global flows of trade and goods ensure availability, but on the other hand they present major challenges in respect of quality, traceability and also reliable functionality in the various regions in which the products are used. More than 80% of the worldwide flow of lecithin crosses borders between continents. So besides knowing about the GMOs that are permitted and used in the various regions and the different production methods, it is essential to have a knowledge of local laws, the use of pesticides and herbicides, and any pollution, for example with bacterial toxins. Analysis and documentation become more and more extensive and automatically result in more work and higher costs.

Production and demand increase continuously
The production of vegetable lecithins increases rapidly. More stringent requirements in respect of oil, flour and proteins and also tax advantages for exports provide important motivation for investment. At the same time the demand for vegetable lecithins increases steadily. Especially in the field of clean-label products and for avoiding artificial emulsifiers, lecithin plays an increasingly important role because of its broad functionality and good reputation. Again and again new benefits are discovered and further applications opened up.

Estimated volumes according to industry
We estimate global lecithin processing at a volume of about 600,000 tonnes.

- The biggest buyer is currently the feed industry, at approximately 220,000 tonnes. Especially the aquafeed segment has increased substantially in recent years. The greatest amount is used purely as a source of energy – e.g. as a substitute for fat or oil – and only about 30% as a functional agent.
- The second-largest market for lecithins is the food industry, with approximately 210,000 tonnes. In this industry, as in the applications below, lecithin is used exclusively as a functional ingredient.
- This is followed by the technical industry with a volume of about 110,000 tonnes.
- In the pharmaceutical and food supplements sector and the cosmetics industry we estimate the demand for processing at about 60,000 tonnes.

Estimated volumes according to origin
In terms of quantity, lecithins derived from soybeans constitute the largest share of the vegetable lecithins, with a volume of over 400,000 tonnes.

- Most of this quantity consists of GM lecithin, with over 300,000 tonnes.
- IP soy lecithin may vary from 60,000 to 90,000, due to climatic and political influences of soyabeans, respectively.
- The production of sunflower lecithin is increasing and currently has a volume of over 40,000 tonnes.
- The production of rapeseed lecithin is also expanding and has reached a volume of approximately 35,000 tonnes. It must be remembered, however, that part of this is obtained from GM seeds.

The extraction method is decisive for quality

When choosing a lecithin quality it is especially important to consider the different extraction methods – using water, acid or enzymatic degumming. The different processes result in completely different product qualities, some of which have poor functionalities and also fail to meet the purity criteria for lecithin.

Other sources of lecithin and phospholipids

Besides oilseeds as the classic raw material, use is made of other lecithin sources for specific functions in the food industry and also for food supplements and products of the pharmaceutical and cosmetics industries. In such cases, special lecithins and lecithin fractions from egg and milk play an important role, but fish and krill are used, too. Again and again new variants come onto the market from other vegetable sources such as rice, maize or oats. From the point of view of quantity, these alternatives are fairly insignificant at present and are estimated at 1,000 to 1,500 tonnes.

New applications in medical treatment

It is already foreseeable that medical treatment will make substantial advances, not least through the use of phospholipids – whether for liposomal preparations or in the form of so-called prodrugs (coupling active ingredients to a phospholipid to achieve better bioavailability) or simply because of their emulsifying properties and safety in improving the bioavailability of substances that have poor uptake. Here, especially, there is increasing potential since most of the recently developed active substances can be allocated to this category.

The outstanding importance of phospholipids in almost all areas of life is also expressed in the composition of the revolutionary mRNA vaccines against SARS-CoV-2 (COVID-19) infections. The extremely labile mRNA contained therein is encapsulated in a membrane (liposomes) consisting in substantial part of phospholipids. The membrane of this cell-like construct consists of double layers of special phospholipids that protect the messenger RNA, the blueprint for a virus-specific surface protein. The purpose of the vaccine is to induce an immune response in the human body against this protein and thus against the virus itself.

Other ingredients of the vaccine serve to protect the phospholipid membrane from rapid degradation or preserve the vaccine and adjust to the correct acidity.

Outlook for the future

'Lecithin is a miracle', said Rüdiger Ziegelitz, one of the greatest lecithin technologists in the last 70 years. His words have been confirmed by numerous scientific papers and books such as "Vegetable Lecithins" by Hermann Pardun, "Lecithin – the Unrivalled Nutrient" by Werner Schäfer and Volkmar Wywiol and monographs from international congresses on the subject of lecithin, including "Lecithins: Sources, Manufacture & Uses" by Bernard F. Szuhaj, to mention just a few examples. Moreover, countless scientific publications have contributed to repeated innovations concerning the use and optimization of lecithin. There is no other natural ingredient with such a wide diversity of applications.

The volumes sold as a functional ingredient or alternative energy source in animal feed will continue to increase. But lecithin has always been, and will remain, a by-product of the oil mill industry. Price, quality and availability are not just the result of supply and demand; often, too, they are a reflection of external factors such as the demand for biodiesel, the price of vegetable oils, or economic and geopolitical developments.

With global sourcing and increasingly stringent requirements in respect of quality and functionality, the extraction, fractionation and processing of optimally functional lecithins will remain a challenge for the future, too.

2 Theoretical **Principles**

2.1 Physical effects of lecithin

Eckhard Flöter, Tabea Masiak

2.1.1 Introduction

Lecithins are naturally occurring, surfactant substances with numerous applications. Since they are important constituents of the cell membrane systems of plants, animals and microorganisms, they can be isolated from a wide range of natural sources such as soybeans, rapeseeds, other plants, eggs, and fish or other marine organisms (1). Because of their natural origin and pronounced functionality they are used in many industrial settings to facilitate manufacturing processes and enhance the quality of the products.

There is no uniform definition of 'lecithin' in the current literature. In industry, lecithin is generally taken to be a complex mixture of polar phospholipids and glycolipids, non-polar triglycerides and small amounts of carbohydrates, water and other companion substances like sterols, tocopherols and free fatty acids, derived from natural sources (2). According to European Directive 96/77/EC, lecithins are 'mixtures or fractions of phosphatides derived from animal or vegetable foods by physical methods'. The content of acetone-insoluble polar substances must be at least 60% in lecithin and at least 56% in hydrolysed lecithins. Because of their functionality these substances, which are also termed 'raw lecithins', are frequently used for various purposes in the food industry as an additive designated as E322. Deoiled lecithins are widely used, too. Besides these, modified lecithins are used; they have specific properties and can therefore be adjusted for use in the relevant processes.

In biochemistry and medicine, the term 'lecithin' has sometimes been synonymous with 1,2-diacyl-sn-3-phosphatidylcholine, which is just one of many phospholipid molecular species found in commercially produced lecithin. This usage is progressively less common (3).

Most lecithin is derived from soybeans; this main source accounts for about 90% of the annual global production of commercial lecithin. But early lecithins were produced from chicken eggs and currently lecithin can also be commercially produced from sunflower, rapeseed and maize. Against the problematic background of the widespread growing of GM soy, which makes access to non-GM lecithins more difficult, lecithins from alternative sources are becoming increasingly important (4).

2.1.2 Composition of lecithin

Being complex mixtures, and varying considerably in the methods used to produce them, commercial lecithins have no strictly uniform chemical composition. Their exact composition depends, above all, on the source of the raw material, but is also subject to considerable influence by the extraction method used and by the purification process (5). The average composition of liquid standard raw lecithin of vegetable origin by weight is shown in Table 1. Vegetable lecithins can be produced as a by-product during the degumming stage of refining plant oils. The raw lecithin thus obtained contains residual triglycerides, which are not completely removed in the degumming process. In some cases oil is subsequently added to the raw lecithins in order to improve their flow properties and obtain a standardized product.

Table 1.

Average compositions of raw lecithins.

Components	by wt %
Phospholipids	47
Triglycerides	36
Water	1
Glycolipids	5–7
Carbohydrates	5
Companion substances (sterols, tocopherols)	1–3

With a share of some 36%, the triglycerides in the raw lecithin constitute its acetone-soluble content, in addition to the about 1% water that is also present because of raw lecithin's hygroscopic properties. The almost complete removal of the triglycerides results in deoiled lecithins in powder form, with their phospholipid functional ingredients concentrated. This is reflected chiefly in an increase in the phospholipid content, whereas the ratio of the individual phospholipid fractions remains unchanged by deoiling (6).

The acetone-insoluble content of the lecithins (approximately 60%) consists of phospholipids, glycolipids, carbohydrates and various companion substances such as sterols and tocopherols. The share of the glycolipids in raw lecithin is about 5–7%. These compounds contain a lipid unit, which often consists of sphingosine or glycerol, to which a monosaccharide or oligosaccharide group is bound. Within each molecule is a polar and a non-polar part, and therefore glycolipids have an amphiphile character. Unlike the phospholipids, they belong to the category of phosphate-free lipids.

Besides containing about 5% carbohydrates like saccharose, raffinose or stachyose, lecithins contain small amounts of free fatty acids, pigments, sterols, sterol glycosides, esters and tocopherols. The largest fraction of lecithin, about 47%, consists of the phospholipids. Not only are they the largest fraction in terms of quantity, they are also the main functional constituents of the commercial lecithins (7).

2.1.3 Chemical structure of phospholipids

The molecular core of most phospholipids is the 3-carbon molecule glycerol, esterified with a fatty acid at positions sn-1 and sn-2. The hydroxyl group of the glycerol at position sn-3 is esterified with phosphoric acid, to which mostly either choline, ethanolamine, serine or inositol is bound, so that the sn-3 position amounts to a phosphorylated group or 'head group' specific to that phospholipid class. Depending on the head group, phospholipids are assigned to different fractions or molecular classes, including phosphatidylcholines, phosphatidylethanolamines, phosphatidylserines and phosphatidylinositols. Fig. 1 shows the structural formulae of the most important phospholipids occurring in lecithins. The fatty acid chains designated as R1 and R2 may differ according to the origin of the lecithin (2). The characteristic feature of the individual fractions is the chemical structure of the specific phosphorylated groups. Whereas in the molecular structure of phosphatidic acid only a single hydrogen atom is bound to the phosphate group, in phosphatidylcholine and phosphatidylethanolamine the phosphate group is bound to either of the amino alcohols choline or ethanolamine. In the phosphatidylinositol molecule, on the other hand, the cyclic alcohol inositol is bound as a specific phosphate group. Some lecithins contain small amounts of phosphatidylserine, in which the amino acid serine is bound to the phosphorylated group. Raw lecithins always contain a mixture of different phospholipid fractions (1).

The exact proportion of the various phospholipid fractions in lecithins depends mainly on the source of the raw material. Lecithin composition is also influenced by the growing and storage conditions of the raw material and the extraction method used. In other words: each lecithin has its own specific composition in respect of the percentages of the different phospholipid fractions. Table 2 compares some lecithins from different sources in respect of the average percentage of the phospholipid fractions they contain, each in relation to raw lecithin as a whole, so that the wt % totals add up to less than 100%.

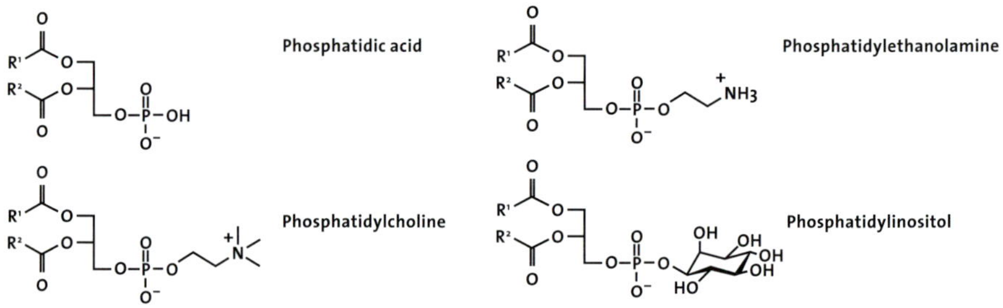

Figure 1.

Structural formulas of different phospholipid fractions (8).

Table 2.

Average phospholipid composition of different lecithins (9,10).

Phospholipid fraction	Soy wt %	Sunflower wt %	Rapeseed wt %	Egg yolk wt %
Phosphatidylcholine (PC)	15	16	17	68
Phosphatidylethanolamine (PE)	11	8	9	11
Phosphatidylinositol (PI)	10	14	10	–
Phosphatidic acid (PA)	4	3	4	–
Other phospholipids	7	6	6	6

The differences between the lecithins of vegetable origin are smaller in comparison to animal lecithins, which show a very divergent composition. All raw lecithins have phosphatidylcholine as their single most common phospholipid class (fraction), irrespective of the source of the raw material. This fraction also has an important influence on the functionality of the lecithins, and for that reason it is often separated off by complex procedures and used industrially as a purified fraction. At 68% the proportion of phosphatidylcholine is especially high in egg yolk lecithin, whereas in vegetable lecithins it is in the range of about 15–17%. The phosphatidylethanolamines and phosphatidylinositols are present in lesser amounts, followed by phosphatidic acid and occasionally also phosphatidylserine (typically far less than 1%). On average, vegetable lecithins contain less than 5% phosphatidic acid, and in some animal lecithins it is not present at all. In addition, lysophospholipids (phospholipids lacking acyl tails at the sn-1 and/or sn-2 positions) may be present in small amounts in raw lecithins, and these can also be produced by modification. Thus the overall phospholipid content of lecithins may vary considerably and depends largely on the raw material source used.

Besides their differing percentages of phospholipid fractions, lecithins also differ in respect of their fatty acid compositions. The main differences exist between the various raw material classes. But since lecithin is a natural product, deviations also occur within a lecithin class. Table 3 shows the average fatty acid composition of different lecithins.

Table 3.
Average fatty acid compositions of different lecithins (1,10).

Fatty acid	Soy wt %	Sunflower wt %	Rapeseed wt %	Egg yolk wt %
16:0	16	11	7	30
18:0	4	4	1	16
18:1	17	18	56	29
18:2 omega-6	55	63	25	14
18:3 omega-3	7	0	6	1
Others	1	4	5	8

Lecithins of vegetable origin contain large amounts of the unsaturated fatty acids linoleic acid and oleic acid, whereas the maximum saturated fatty acid content is 20% (soy lecithin). The oils from which vegetable lecithins are obtained show a similar composition to that of the specific lecithins. No significant concentration of specific fatty acids in the phospholipids is therefore to be seen, which means that the fatty acid composition can be attributed largely to the source of the raw material. Animal lecithin obtained from egg yolk contains much higher proportions of saturated fatty acids (about 50%) and therefore differs considerably from the lecithins of vegetable origin in respect to specific fatty acid composition as well as the distribution of the phospholipid fractions.

Besides the differing percentages of fatty acids in the various lecithins, the fatty acid distributions in the phospholipids of the lecithins also vary. The individual phospholipid fractions are therefore also mixtures of individual components with different distributions of the fatty acid chains bound to the glycerol. As a rule, the fatty acid chain in the sn-1 position has a higher degree of saturation than the fatty acid in the sn-2 position (8).

The phospholipid content of lecithin therefore consists of a complex mixture of different phospholipids, with each head group type having numerous possible combinations of fatty acid acyl groups. In total, these have a wide range of potential influences on the chemical and physical properties of the lecithins and ultimately determine their technological applications.

2.1.3.1 Possibilities of modification

The chemical structure and composition of the lecithins and therefore their pattern of properties can be further adjusted specifically through modification. For example deoiled, fractionated, hydrolysed, hydrated, hydroxylated or acetylated lecithins can be produced. Deoiled lecithins, that is, lecithins from which the bulk of the non-polar triglycerides have been removed, are in wide use. Deoiling can be carried out by various methods such as solvent extraction with acetone, supercritical CO_2 extraction, near-critical propane extraction or the use of membrane technology. Since the deoiling process predictably concentrates the remaining fractions, deoiled lecithins have a higher phospholipid content (10). Lysolecithins obtained from raw lecithin via a hydrolysis step are also in wide use. Hydrolysis causes one of the fatty acid chains to be split off the phospholipid molecule, thus reducing the size of its non-polar region. Lysophospholipids therefore have stronger polarity than unhydrolysed phospholipids and are thus more readily soluble in water. For more details see Chapter 4.

2.1.4 Physical properties

A uniform definition of the physical properties of lecithins is not possible because of their complex composition. Clearly, however, the phospholipid fraction has the greatest influence on the overall properties and functional profiles of the lecithins. For example, no characteristic melting point for lecithin can be determined, since this is influenced by the total of the components present – especially the fatty acid composition of the lipids. As the latter can vary widely, a distinction must be made between the different molecules occurring within the phospholipid fraction. Anhydrous dipalmitoylphosphatidylcholine, for example, melts in a range of about 80 – 100°C. Above this temperature range the fatty acid chains are much more mobile, but the polar head groups and the glycerol still retain their orderly structure. Further heating subsequently results in chemical degradation processes within the structure of the molecules, especially in the region of the polar head groups. For phosphatidylcholine, the temperature for decomposition is about 230°C and thus much higher than the temperature at which the fatty acid chains are found to melt. Phosphatidylethanolamine, on the other hand, already decomposes at a temperature of about 196°C, and phosphatidic acid at 71°C (1). The addition of water to the phospholipids further lowers the melting point. In the case of dipalmitoylphosphatidylcholine the melting point falls to around 50°C when water is added (11). Thus each individual phospholipid fraction has a specific melting point range that depends on the hydrophobic head group and also the fatty acid composition.

2.1.4.1 Self-organization of lecithin

Phospholipids have both polar and non-polar domains in their structure, and like the glycolipids they are therefore classified as amphiphiles. The fatty acid chains at sn-1 and sn-2 of the glycerol molecule form the non-polar domain, and the specific head group at sn-3 forms the polar domain (12). The presence of these opposing domains within the same molecule gives a phospholipid the unique ability to create self-organizing structures. Self-organization is the term for a reversible process in which the molecules of a component spontaneously associate with each other in a solvent. The nature of the aggregates formed in this way can be altered by changes in system parameters such as concentration or temperature. In their aggregated form, the molecules are held together by intermolecular interactions such as bridge-type hydrogen bonds or van der Waals forces (12).

2.1.4.2 Hydrophobic effect

Self-organization requires a driving force that brings about an arrangement in the form of aggregates. In an aqueous environment the hydrophobic effect plays an important role in the aggregation of the molecules. The term 'hydrophobic effect' describes the ability of hydrophobic ('water-avoiding') particles to associate preferably with each other in an aqueous environment in such a way as to avoid contact with water, and as a result of which self-organizing structures are formed. When water molecules make contact with hydrophobic, non-polar sections of a hydrocarbon, their hydrogen bridges become initially directed away from the non-polar section and this causes the water to take on a quasi-crystalline structure, which consequently restricts the relative mobility of the water molecules. Therefore water molecules in the vicinity of a hydrocarbon chain show a higher degree of order. This process is accompanied by a thermodynamically unfavourable decrease in entropy. But then congregation of the hydrophobic groups causes the water molecules' mobility to increase again; the order breaks down, and water molecules are released into the surrounding medium. This causes the entropy to increase again. As a result, when in aqueous solutions amphiphiles like the phospholipids have a propensity for minimizing the area of contact between the non-polar side chains and the water for entropic reasons; this leads to the formation of aggregates and a concentration at the interface. This process can be expressed by the equation

$$\Delta G = \Delta H - T\Delta S$$

ΔG being Gibbs free energy, T the temperature, ΔH the enthalpy change and ΔS the entropy change. As described above, an association of hydrophobic particles in an aqueous environment leads to a reduction in the number of states of low entropy in the water. The contribution of entropy to hydrophobic interaction is therefore positive. On the other hand, the effect of the enthalpy change on the release of non-polar particles into water is rather slight. Because of the positive contribution of entropy to the hydrophobic interaction, the Gibbs free energy becomes negative, and self-aggregation with $\Delta G < 0$ is therefore a spontaneous, energy-efficient process (13).

2.1.4.3 Formation of liquid-crystalline structures

Because of their hydrophobic domains, single phospholipid molecules have very poor solubility in water. Since congregation of the amphiphilic molecules causes a decrease in the hydrophobic interface, aggregates form in water even at a low lipid concentration. The nature of the associates formed depends largely on the geometry of the monomer molecules. If the critical concentration for micelle formation is exceeded, phospholipids in an aqueous environment form micelles, vesicles or reverse micelles. Higher phospholipid concentrations cause these aggregates to arrange into regular superstructures, the liquid-crystalline phases (14).

The term liquid crystal is used for substances that have both the attributes of a liquid and the properties of a crystal. The liquid-crystalline state therefore unites the properties of the liquid and crystalline states of matter. Within liquid-crystalline structures the mobility of the molecules is restricted, and a long-range directional and positional order exists which causes some physical properties to show directional dependence. This attribute is termed anisotropy.

Lyotropic liquid crystals are formed by amphiphilic substances when the latter are dissolved in a suitable solvent. As already explained, amphiphiles like the phospholipids are made up of incompatible molecular domains which are covalently linked. Because of these differences the incompatible molecular domains seek to separate, but this is not possible because of the covalent bonds. Instead of separation, different areas emerge through regular arrangement of the molecules, this being facilitated by the hydrophobic effect. The shape of the aggregates formed sometimes depends on the size and geometry of the amphiphile, since these influence the curvature of the interfaces of the separated phases (15).

Phospholipids with similarly large hydrophilic and hydrophobic regions tend to form the lamellar L_α phase, which is made up of amphiphilic bilayers. Here, the non-polar regions form the inside of the bilayer, while the polar head groups are directed outwards. Because of their molecular structure, phosphatidylcholine and phosphatidylinositol tend to arrange themselves into such planar or slightly curved lipid bilayers. In biological membranes, which consist largely of phospholipids, lipid bilayers are an important structural component. Another liquid-crystalline phase that occurs is the inverse hexagonal (H II) phase, characterized above all by the presence of cylindrical aggregates. The polar head groups of the phospholipids are directed towards the inside, where they enclose the aqueous phase. This is made up mainly of phosphatidylethanolamine and phosphatidic acid. These phospholipid fractions are characterized by the presence of a small polar head group and a large non-polar region, which means that their molecules are shaped like the frustum of a cone. Because of this geometry, they form aggregates with a negative curvature and therefore tend to arrange themselves as the inverse hexagonal H II phase or the cubic phase, with polar head groups directed inwards. The cubic phase is a phase with a bicontinuous structure in which the water and the molecules of the amphiphile form extensive, continuous areas which penetrate each other. The polar groups are oriented towards the water. Cubic phases are often situated in an area between lamellar and hexagonal phases (16). Furthermore, the hexagonal (H 1) phase is also seen to occur among the phospholipids; this is characterized by the presence of cylindrical structures with polar head groups directed outwards. Lysophospholipids have only one fatty acid chain and therefore a small non-polar and a large polar domain. Their molecules are shaped like an inverted cone, which means that they tend to aggregate into structures with a positive curvature like the hexagonal (H 1) phase in which the polar regions are directed outwards. See also Fig. 4 in Chapter 8.2.

Since the space required by the polar and non-polar domains of the molecules determines the shape of the aggregates formed, the fatty acid distribution in the phospholipid molecules also plays an important role in determining the organization of the aggregate. Different aggregate structures can therefore come about within a lipid fraction, influenced by the chain length and degree of saturation of the fatty acids. Because of the elongated shape of their hydrocarbon chain, saturated fatty acids require less space than unsaturated fatty acids, which have rigid kinks in the chain because of their double bonds and therefore take up more space. So unsaturated fatty acids in the phospholipid molecules enlarge their non-polar regions. The non-polar region also becomes larger as the chain length of the fatty acids increases, and consequently tends to induce formation of the inverse hexagonal H II phase.

Since the distribution of the fatty acid chains at the glycerol segment also varies within a fraction, the formation of different aggregates is possible. For example, phosphatidylethanolamine with a high percentage of unsaturated fatty acids mainly forms the inverse hexagonal H II phase. Saturated phosphatidylethanolamine, on the other hand, prefers to form the lamellar phase, since a closer packing of the molecules is possible (17,18).

The shape of the phospholipid aggregates in an aqueous environment is therefore influenced by the specific phosphate group at the glycerol on the one hand, and on the other by the chain length and degree of saturation of the fatty acids. But the temperature and the concentration of the solvent in a water/phospholipid system also exercise considerable influence on the structure of the liquid crystal phase formed. In different temperature and concentration ranges, phospholipids may form different liquid-crystalline phases through interaction with water or some other polar solvent. The ability to vary the structure when certain parameters change is termed lyotropic mesomorphism, and can be observed in a number of amphiphiles. Because of this characteristic, phospholipids are able to form various mesophases if the relevant parameters in the system are changed. Besides the temperature and lipid concentration, their phase behaviour depends on the specific polar head groups present, as well as the chain length of the fatty acids and their degree of saturation, and can be further influenced by changes in the pressure and pH parameters.

2.1.4.4 Lyotropic mesomorphism

Phospholipids differ greatly in respect of their lyotropic phase behaviour. Fig. 2 shows a possible binary phase diagram of a lysolecithin (in this case lysophosphatidylcholine). In different temperature and concentration ranges a lamellar mesophase, a cubic mesophase and a hexagonal H I phase occur. The hexagonal H I phase can be observed chiefly in less hydrophobic phospholipids like the lyso-forms.

Moreover, in wide ranges of temperatures and concentrations, lysolecithin occurs in the form of micelles with polar head groups directed outwards. At high lipid concentrations and low temperatures, hydrated crystals in the form of monohydrates may be seen to occur because of the hygroscopic nature of the lecithins. The phase diagram for pure phosphatidylcholine (in this case egg yolk lecithin) (Fig. 3) differs considerably from that of the lysolecithins in Fig. 2. Over a wide temperature and concentration range, the dominant phase in phosphatidylcholine/water mixtures is the lamellar L_α phase. At a higher percentage of water, the lamellar L_α phase and an aqueous solution exist side by side. In this range, liposomes may form if energy is introduced. From a temperature of about 220°C, restructuring of the molecules into micelles takes place and an isotropic solution forms. Low percentages of water (less than 10%) lead to the formation of further mesophases such as a lamellar L_β phase, a cubic phase and, at high temperatures, a hexagonal H II phase.

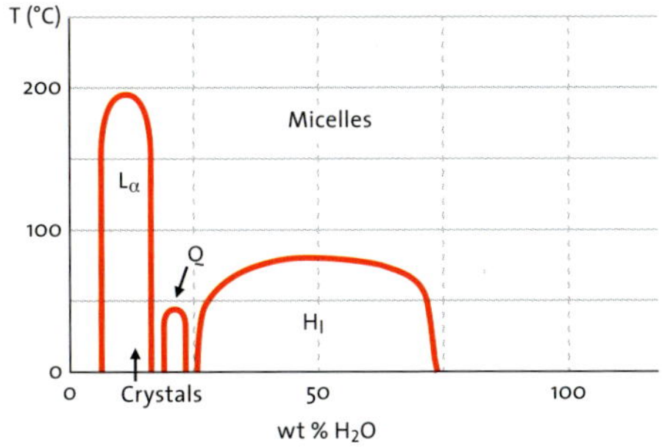

Figure 2.
Binary phase diagram of lysolecithin (18,19).

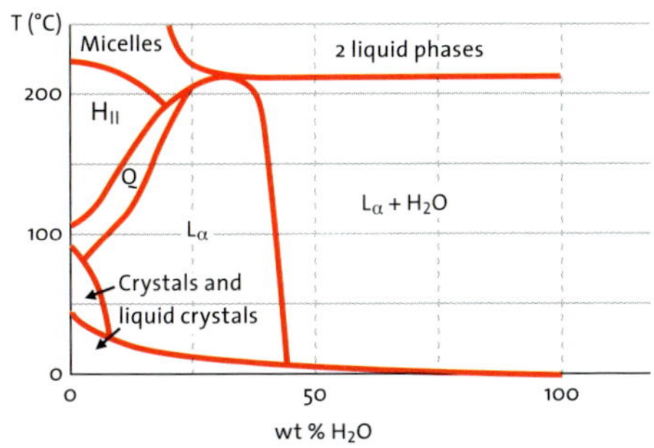

Figure 3.
Binary phase diagram of pure phosphatidylcholine from egg yolk lecithin (12,20).

The phase diagram changes significantly when the specific phosphate group is replaced. In the case of zwitterionic phospholipids like phosphatidylethanolamine, the phase transitions are very much determined by temperature changes. The L_α phase is only stable within a narrow temperature range. At higher temperatures, on the other hand, formation of the hexagonal H II phase can be observed, irrespective of the amount of water. Moreover, the phase behaviour changes through reduction of the chain length of the fatty acids at the glycerol segment, and various cubic phases occur between the lamellar and the inverse hexagonal phase.

The formation of liquid-crystalline structures of phospholipids is not only observed in water; it also occurs in some other solvents, for instance through the addition of phosphatidylcholine to formamide, glycerol or ethylene glycol at an appropriate concentration and temperature (12).

2.1.4.5 Formation of liposomes

Besides the various liquid crystal mesophases, liposomes, microemulsions and nanoemulsions can be produced with the aid of phospholipids. In products for the pharmaceutical and cosmetic industries especially, encapsulation of ingredients in liposomes plays an important role. Liposomes are made up of concentric, closed bilayers of phospholipids and make it possible to envelop valuable substances (see also Fig. 8 in Chapter 3.2). The formation of liposomes can be attributed to the disappearance of energetically inefficient peripheral areas of the lipid bilayers through fusion of the latter into closed double layers. Liposomes can be produced by methods such as extrusion, high-pressure homogenization or ultrasound treatment. They are usually made from pure phosphatidylcholine since its geometry causes it to form lipid bilayers in an aqueous environment, and these can transform themselves into liposomes through the input of energy. The optimum range for the formation of liposomes lies in the concentration/temperature range in which the L_α phase and an aqueous phase exist simultaneously. If other phospholipid fractions are present, liposome formation is disturbed by a change in the radius of curvature that unfavourably changes the geometry of the aggregates. This means that, as a rule, only individual phosphatidylcholine fractions with high purity can be used (12,21).

2.1.4.6 Formation of wormlike micelles in organic solvents

In non-polar organic solvents, lecithins usually form spherical reverse micelles. In the presence of small amounts of water, these micellar solutions can be transformed into transparent, viscoelastic and thermodynamically stable systems. When pure phosphatidylcholine is used, this behaviour can be observed in numerous organic solvents. The phosphatidylcholine molecules initially form spherical reverse micelles. The addition of small amounts of water then causes polar elements to enter the system, and hydrogen bridge bonds form between the water molecules and the phosphate groups of neighbouring phosphatidylcholine molecules, thus reducing the curvature of the aggregates. Rearrangement of the amphiphilic molecules and non-axial growth of the aggregates induce the formation of a chain-like structure. In this way cylindrical reverse micelles, so-called wormlike micelles, with a polymer-like structure are formed. The polar head regions are directed towards the inside of the micelles, whereas the non-polar regions project into the solvent. The feature essential for the formation of these micelles is that they have a stronger curvature at their end caps than along the sides, so that growth results in minimization of the free energy in the system because of the reduction in the number of end caps. Branching of the wormlike micelles into a three-dimensional network leads to immobilization of the organic solvent; as a result, the viscosity of the solution increases spontaneously and a system with viscoelastic properties comes about. In this process the amphiphilic molecules undergo a dynamic exchange with the solvent phase, so that positions are changed and the aggregate chains can be broken off and recombined (22,23).

Gelation can be observed in linear and cyclic alkanes in particular and in amines and esters of fatty acids. Formation of the gel network is impaired even by very minor impurities in the system. The presence of phosphatidylethanolamine, for example, that occurs naturally in raw lecithins, prevents the formation of a transparent gel. This is due on the one hand to the geometric differences between the phospholipids and on the other to the functional phosphate group, which has a decisive effect on the formation of hydrogen bridge bonds (3,24).

2.1.5 Formation of lecithin-based emulsions

Because of their surfactant properties, lecithins are used as emulsifiers in many applications and processes. Due to the diversity of the lecithins, specific choices can be made to permit the production of both water-in-oil (W/O) and oil-in-water (O/W) emulsions. Since lecithins derived from natural sources are complex mixtures of phospholipids and other substances, their individual surfactant characteristics are outcomes of interaction between all the surface-active substances they contain. Of all these substances the phospholipids have a decisive influence on the emulsifying properties of the lecithins due to their amphiphilic character.

The presence of hydrophilic and hydrophobic regions within the phospholipid molecules leads them to attach to oil-and-water interfaces, where they reduce the surface tension and are able to stabilize emulsions. The various phospholipid fractions show divergent behaviour in respect of their emulsifying properties, which means that the phospholipid compositions of lecithins play a decisive role in their uses as emulsifiers (5,6).

When individual phospholipid fractions are used as emulsifiers, the stabilizing effect is often attributed to the formation of a liquid-crystalline phase around the droplets of the emulsion. This results in a local increase in viscosity, leading to reduced coalescence and thus improving the stability of the emulsion (25,26). The structure of the liquid-crystalline phases at the oil/water interface depends on the geometry of the phospholipids. Phosphatidylcholine and phosphatidylinositol form lamellar structures because of their cylindrical geometry; phosphatidylethanolamine and phosphatidic acid usually aggregate to form an inverse hexagonal phase, while lysolecithins tend to form a hexagonal phase at the interface. Whether an O/W or a W/O emulsion forms depends to a large extent on the shape of the phases. Less hydrophilic phospholipids with small polar head groups, like phosphatidylethanolamine and phosphatidylinositol, are suitable for the production of W/O emulsions, while O/W emulsions can be stabilized better with more strongly hydrophilic phospholipids with a large polar head group, like phosphatidylcholine or the lysophospholipids (14).

Table 4.

HLB values of different lecithins (27,28).

	HLB
Raw soy lecithin	3
Deoiled soy lecithin	4
Soy lysolecithin	8–11
Acetylated and hydrolysed soy lecithin	9–10
Phosphatidylcholine	7

The balance between the hydrophilic and hydrophobic properties of surface-active substances can be described with the aid of the hydrophilic-lipophilic balance (HLB) value. But because of the differences in the composition of natural lecithins, especially in respect of their phospholipid composition, it is not possible to establish a standard HLB value. Moreover, modification or enrichment of lecithins with individual phospholipid fractions can change their emulsifying properties to a large extent, which also causes a change in the HLB value. Table 4 shows the HLB values of different soy lecithins. Compared with raw lecithin (HLB 3) and deoiled lecithin (HLB 4), lysolecithin has a much higher HLB value of about 8 to 11, depending on the degree of hydrolysis. Hydrolysis increases the proportion of hydrophilic regions and thus the polarity of the phospholipids, which explains the superior emulsifying qualities.

Moreover, the hydrophilic head groups of the phospholipids carry differing electric charges, and this too has an effect on the stability of the emulsion.

Phosphatidylcholine and phosphatidylethanolamine display zwitterionic properties, and under physiological conditions they are electrically neutral since the negative charge at the phosphate group and the positive charge at the nitrogen atom of the choline or ethanolamine cancel each other out. Phosphatidylinositol and phosphatidic acid, on the other hand, are anionic molecules under physiological conditions, since only the phosphate group contributes to the charge of the molecule, whereas the polar head group is uncharged. Because of this anionic chemical structure these phospholipids contribute negative charge to an emulsion. This means that due to the additional repulsive electrostatic forces between the droplets of the emulsion, the emulsion's stability can be improved by using such anionic phospholipids (5,6). On the other hand, zwitterionic substances like phosphatidylcholine and phosphatidylethanolamine have no influence on the electric charge of the droplets of the emulsion, so that such emulsions have more of a tendency towards coalescence because of the absence of an electric barrier (29).

In all lecithins the emulsifying properties are influenced primarily by the chemical structure and electric charge of the phospholipids. But the fatty acid chains of the phospholipid molecular species also contribute to the surface activity of the individual fractions. Phospholipids with a higher degree of saturation of their fatty acids have greater surface activity than those with fewer saturated fatty acids. A high degree of saturation as in the case of phosphatidylinositol, for example, contributes to increased surface activity when compared to lipids like phosphatidylcholine, which usually have a low degree of saturation (30).

2.1.6 Solubilities

Lecithins are soluble or dispersible in most solvents. Above all it is their acetone-insoluble content that influences solubility – in particular their phospholipid portion largely determines their overall solubility. The individual phospholipid fractions differ in their solubility because of their differing chemical structure. Most phospholipids – with the notable exception of the lysophospholipids – have hydrophobic fatty acid chains, and therefore are insoluble in water, but they can be dispersed. Successful dispersion relies on their amphiphilic character and the resulting predisposition to form aggregates on account of their hydrophobic effect (5).

The solubility of the lecithins in organic solvents depends largely on the fatty acid composition of the phospholipids and the structure of their polar head

industries, oleogelation constitutes a forward-looking alternative to the conventional structuring of oils and organic solvents and is the subject of much current research (39).

References

1. Bueschelberger, H.G., Tirok, S., Stoffels, I., Schoeppe, A., 2014. Lecithins. Emulsifiers in Food Technology. Chichester In: Viggo Norn (ed.):, UK: John Wiley & Sons, Ltd, 21–60.
2. Scholfield, C.R., 1981. Composition of soybean lecithin. *J. Am. Oil Chem. Soc.* 58 (10), 889–892.
3. Shchipunov, Y.A, 2001. Lecithin organogel. A micellar system with unique properties. *Colloids and Surfaces A: Physicochemical and Engineering Aspects* 183-185, pp. 541–554.
4. Schneider, M., 1992. Lecithine – Gewinnung, Eigenschaften und Bedeutung für die industrielle Anwendung. *Fett Wiss. Technol.* 94 (S1), 524–533.
5. Somasundaran, P. (ed.), 2006. Encyclopedia of Surface and Colloid Science. Lecithin. 2nd Ed. New York: Taylor & Francis.
6. Xu, Q., Nakajima, M., Liu, Z., Shii, T., 2011. Soybean-based surfactants and their applications. In: Tzi-Bun Ng (ed.): Soybean - Applications and Technology: InTech.
7. Koolman, J., Röhm, K-H., Wirth, J., 2003. Taschenatlas der Biochemie. 3., revised version: Thieme.
8. Gunstone, F.D., 2008. Oils and Fats in the Food Industry. Oxford, Ames, Iowa: Wiley-Blackwell Pub (Food Industry Briefing Series).
9. Gunstone, F.D., 2011. Vegetable Oils in Food Technology. Composition, Properties and Uses, 2nd Ed. Somerset: Wiley.
10. van Nieuwenhuyzen, W., Tomás, M.C., 2008. Update on vegetable lecithin and phospholipid technologies. *Eur. J. Lipid Sci. Technol.* 110 (5), 472–486.
11. Löffler, G., Petrides, P.E., Heinrich, P.C., 2007. Biochemie und Pathobiochemie. 8., fully revised edition, Heidelberg: Springer (Springer-Lehrbuch).
12. Shchipunov, Y.A., 1997. Self-organising structures of lecithin. *Russ. Chem.* Rev. 66 (4), 301–322.
13. Adam, G., Läuger, P., Stark, G., 2009. Physikalische Chemie und Biophysik. Mit 51 Tabellen. Revised edition, Berlin, Heidelberg: Springer (Springer-Lehrbuch).
14. McClements, D.J., 2016. Food Emulsions. Principles, Practices, and Techniques, 3rd Edition. Boca Raton: CRC Press, Taylor & Francis Group.
15. Collings, P.J., Hird, M., 1997. Introduction to Liquid Crystals Chemistry and Physics. London, Bristol, PA: Taylor & Francis (The Liquid Crystals Book Series).
16. Tolédano, P., Figueiredo Neto, A.M., 1998. Phase Transitions in Complex Fluids: World Scientific.
17. Jouhet, J., 2013. Importance of the hexagonal lipid phase in biological membrane organization. *Frontiers in Plant Science* 4, 494.
18. Seddon, J.M., 1990. Structure of the inverted hexagonal (HII) phase, and non-lamellar phase transitions of lipids. *Biochimica et Biophysica Acta (BBA) - Reviews on Biomembranes* 1031 (1), 1–69.
19. Belitz, H-D., Grosch, W., Schieberle, P., 2008. Lehrbuch der Lebensmittelchemie. Mit 634 Tabellen. Substantially revised edition, Berlin, Heidelberg: Springer (Springer-Lehrbuch).
20. Small, D.M., 1967. Phase equilibria and structure of dry and hydrated egg lecithin. *Journal of Lipid Research* 8 (6), 551–557.
21. Knauf, K., 2007. Physikochemische Charakterisierung homologer kationischer Tenside und deren Wechselwirkung mit Modellmembranen bestehend aus DMPC und DMPG. Dissertation.
22. Cummings, S., Trickett, K., Enick, R., Eastoe, J., 2011. CO2. A wild solvent, tamed. *Phys. Chem. Chem. Phys.* 13 (4), 1276–1289.
23. Dreiss, C.A., 2007. Wormlike micelles: where do we stand? Recent developments, linear rheology and scattering techniques. *Soft Matter* 3 (8), 956.
24. Shchipunov, Y.A., Hoffmann, H., 1998. Growth, branching, and local ordering of lecithin polymer-like micelles. *Langmuir* 14 (22), 6350–6360.
25. Hasenhuettl, G.L., Hartel, R.W., 2008. Food Emulsifiers and Their Applications, 2nd Edition. New York, NY: Springer Science+Business Media, LLC.
26. Friberg, S.E., Solans, C., 1986. Surfactant association structures and the stability of emulsions and foams. *Langmuir* 2 (2), 121–126.

27. Gunstone, F.D., Padley, F.B., 1997. Lipid Technologies and Applications. New York: Marcel Dekker.
28. Belitz, H-D., Grosch, W., Schieberle, P., 2009. Food Chemistry. 4th Rev. and Extended Ed. Berlin: Springer-Verlag.
29. Tadros, T.F., 2016. Emulsions. Formation, Stability, Industrial Applications. Berlin, Boston: De Gruyter (De Gruyter Textbook).
30. Wu, Y., Wang, T., 2003. Soybean lecithin fractionation and functionality. *J. Am. Oil Chem. Soc.* 80 (4), 319–326.
31. Liu, D., Ma, F., 2011. Soybean Phospholipids. In: Krezhova, D (ed.): Recent Trends for Enhancing the Diversity and Quality of Soybean Products: InTech.
32. BeMiller, J.N., Whistler, R.L., 2009. Starch. Chemistry and Technology, 3rd Ed. London: Academic (Food Science and Technology).
33. Nakamura, A., Shimizu, R., Kôno, T., Funahashi, S., 2014. Nature of lysolecithin in rice grains. *Bulletin of the Agricultural Chemical Society of Japan* 22 (5), 324–329.
34. Tiwari, B.K., Singh, N., 2012. Pulse Chemistry and Technology. Cambridge: Royal Society of Chemistry.
35. Schäfer, W., 2006. Future of Flour. A Compendium of Flour Improvement. Ed. from Lutz Popper. Bergen/Dumme: AgriMedia.
36. Nikiforidis, C.V., Scholten, E., 2014. Self-assemblies of lecithin and α-tocopherol as gelators of lipid material. *RSC Adv* 4 (5), 2466–2473.
37. Pernetti, M., Mallsen, K., van; Kalnin, D., Flöter, E., 2007. Structuring edible oil with lecithin and sorbitan tri-stearate. *Food Hydrocolloids* 21 (5-6), 855–861.
38. Han, L-J, Li, Lin, Zhao, L., Li, B., Liu, G.Q., Liu, X.Q., Wang, X-D., 2013. Rheological properties of organogels developed by sitosterol and lecithin. *Food Research International* 53 (1), 42–48.
39. Rogers, M.A., Wright, A.J., Marangoni, A.G., 2009. Oil organogels: the fat of the future? *Soft Matter* 5 (8), 1594.

2.2 Physiological properties of lecithin phospholipids

Parris M. Kidd

The classic textbook of Guyton and Hall defines physiology as 'the science that seeks to explain the physical and chemical mechanisms that are responsible for the origin, development and progression of life' (page 3) (1). Human physiology attempts to explain the life processes that make the human body a living being. Physiology by necessity integrates the functioning of individual cells, tissues and organs of the body into a living whole. But physiology is founded even more fundamentally on the biological molecules that make life happen. This is dramatically evident in the contributions phospholipids make to the functioning of the human body at every level of physiology.

The phospholipid (PL) substances that are found concentrated in lecithins are all fundamentally required both for the physiology of plants and microorganisms and for virtually all of human physiology. This chapter reviews the numerous contributions that lecithin's phospholipids make to human physiology. For practical reasons it emphasizes those contributions that are the best established and most illustrative of the molecular character of phospholipids. This chapter is also a prelude to Chapter 6.6, which reviews the uses of lecithins as dietary supplements.

2.2.1 Lecithin phospholipids – fundamental to life

The amphiphilic properties of PL molecules – charged at one end, uncharged at the other, as illustrated in Fig. 1 – are fundamental to their central importance for life as we know it. This amphiphilic molecular character is also the key to the huge commercial success of lecithin preparations. Commercially, lecithins are most concisely defined by the European Union E322 regulation: mixtures of phospholipids extracted from foods of vegetable or animal origin with a minimum of 60% acetone-insoluble substances (3). The acetone-insoluble substances in lecithins are mostly the phospholipids (PL). Besides their content of PL, lecithins typically contain lesser and varying amounts of sphingolipids, glycolipids, triglycerides, sterols and carbohydrates (2,3).

Three molecular classes of PL predominate in commercial lecithins, being consistently present at levels higher than one per cent by weight (3). These are the phosphatidylcholine class (PC), the phosphatidylethanolamines (PE) and the phosphatidylinositols (PI). Another two PL classes – the phosphatidylserines (PS) and phosphatidic acid (PA) – are not as consistently present in commercial lecithins but are also covered in this chapter since they are important in human physiology (4,5). These five PL classes – PC, PE, PI, PS and PA – are the major focus of this review.

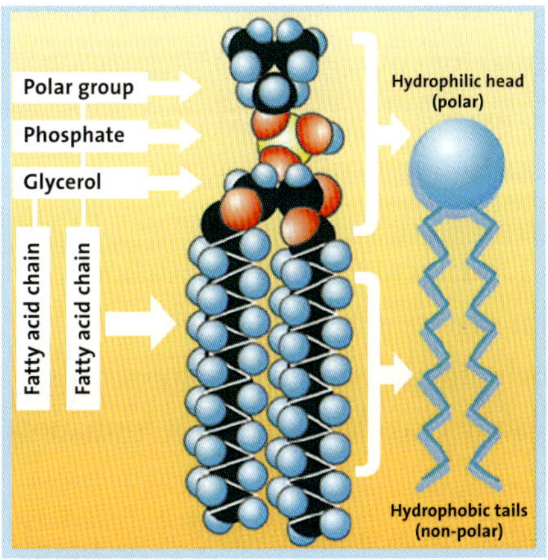

Left: The distinct polar group, the phosphate group and the glycerol molecule together make the 'head group'. Two fatty acid chains are attached to the glycerol molecule and are termed 'tail groups' or simply 'tails'.

Right: The PL molecule is amphiphilic, with its polar (charged) head being hydrophilic ('water-loving') and its two non-polar (uncharged) tails being hydrophobic ('water-avoiding').

Figure 1.

Molecular organization of the glycerophospholipids (GPL), often called phospholipids (PL). From Schneider and Schmitt, 2013 by permission (2).

To understand the broad importance of these five PL classes in human physiology, one must first comprehend their relatively complex molecular organization (see Fig. 1) (2). The glycerol molecular structure forms the backbone (^{1}CH^{2}OH-^{2}CHOH-^{3}CH2OH). The third carbon (C3) of the glycerol backbone bears the molecule's 'head group': C3 is esterified with phosphoric acid, which at its other end is esterified either with choline (for PC), ethanolamine (for PE), inositol (for PI), serine (for PS) or a hydrogen atom (for PA). C1 and C2 of the glycerol backbone most often are esterified with fatty acids that make the two 'tail groups' (6). This molecular organization generates the amphiphilic molecule: the charged head group makes a water-loving or hydrophilic head at one end of the molecule, while the uncharged tails make the other end hydrophobic.

The five PL classes with their amphiphilic properties provide the bases for much of human physiology: (1,6–24)

- Forming the cell membranes that manage many life processes.
- Essential for most of the body's energy generation by mitochondria.
- Fundamental to cell regulation by signal transduction.
- Sources of messenger substances for integrating cell, tissue and organ functions.
- Enabling electrical transmission in the nervous system.
- Building blocks for HDL and other circulating lipoproteins.
- Providing dietary choline, broadly essential for epigenetic DNA regulation.
- Serving as surfactants ('surface-active agents') for lung gas exchange.
- Reinforcing the barrier function of the gastrointestinal epithelium.
- Lubricating the joint epithelial linings.
- Important for cell volume regulation, a fundamental survival mechanism.
- Providing PC to make bile, an important digestive fluid.
- Providing PS for numerous processes including disposal of unwanted cells.
- Providing PE, to complement PS for healthy coagulation and clot formation.

These physiological roles of PL are detailed in the sections that follow.

2.2.2 Phospholipids – molecular building blocks for cell membranes

The human body has many trillions of cells, each with networks of cell membranes that manage many of their life processes (Fig. 2) (6,25). An outer membrane (OM), sometimes called the *plasma membrane* and occasionally the *plasmalemma,* encapsulates the entire cell, and other membranes are distributed within the cell's interior. Some of these create structures called organelles that have specialized functions. The OM and the various internal membranes have very similar molecular layouts and at places have continuity with each other, making a *cell membrane system*. The entire membrane system relies on PL molecules for its structural integrity and diverse functions (4,6–11).

All cell membranes are organized on a lipid bilayer plan. This is a double molecular layer composed mostly of PL but with some cholesterol molecules present (Fig. 2) (8). The amphipathic nature of PL molecules drives them thermodynamically to assemble into lipid bilayers when in the presence of water (8). The hydrophobic tail groups avoid contact with water molecules by facing towards each other directly and at close range, thus creating an inner hydrophobic zone. The hydrophilic head groups are drawn to water molecules, and so face away from the inner zone into the water medium. This lipid bilayer, very thin but continuous and semi-fluid, is the foundation of the working cell membrane (4,6–11).

Figure 2.

Diagram of a generalized animal cell. The outer membrane encloses the cytoplasm, in which are suspended smaller structures or organelles. These include the nucleus (blue) and mitochondria (brown) with their double membranes. The endoplasmic reticulum (reddish-brown, wrapped around the nucleus) and Golgi apparatus (purplish) are single-membrane networks.

In addition to its complement of PL, the bilayer houses cholesterol molecules that serve as a kind of 'glue'. Cholesterol is an essential membrane constituent and adds side-to-side cohesiveness to the PL molecules, thereby regulating the bilayer's fluidity between fluid and solid (6). Adding further cohesion to the bilayer is a substance related to phospholipids, the sphingolipid known as sphingomyelin (SM). This membrane constituent is found in some lecithins.

Then there are the myriad types of membrane proteins, applied into or onto the lipid bilayer and responsible for a range of enzymic, transporter and other catalytic activities that power the membrane's physiological functions (6,8). Membrane proteins fall into three broad categories (8). Integral membrane proteins have amphipathic molecular zones that allow them to penetrate deeply into the hydrophobic zone of the lipid bilayer, as well as hydrophilic zones that line up with the hydrophilic outer zone of the bilayer and even may protrude from the bilayer into the surrounding water. Peripheral membrane proteins attach to the bilayer, and may partly penetrate it but do not extend into the inner hydrophobic zone. These proteins help to link the membrane's activities into the activities within the cytoplasm or the water phase that surrounds the cell (6,8–11).

The membrane-associated proteins contribute to membrane functions though not necessarily directly connected with the lipid bilayer (8). Some serve as receptors on the outer cell surface, to link other membrane proteins with messenger substances arriving from elsewhere. Others are located on the cytoplasmic side of the OM or other membranes and serve as internal scaffolding, to enable shape changes or movement for example. Still others act as attachment points to allow water-phase proteins to hook up with the membrane (8,10,11).

The present culmination of decades of intensive research on cell membrane organization is the revised Fluid Mosaic Membrane Model (rFMMM), illustrated in Fig. 3 (8). The initial fluid mosaic membrane model emerged in the early 1970s (for an illustration refer to Figure 1 in Nicolson (8)) and became progressively more sophisticated as it incorporated a deeper understanding of how the lipids, proteins and other cell membrane constituents fit together.

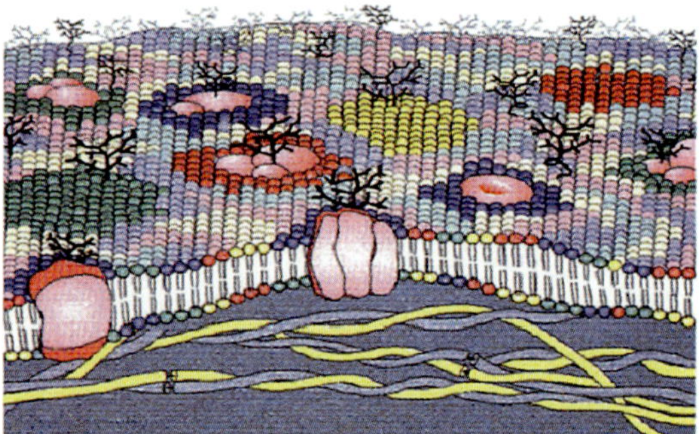

Figure 3.

The revised Fluid Mosaic Membrane Model (rFMMM). Note the variable thickness of the lipid bilayer and the groupings ('domains') of phospholipids portrayed in various colours and clustered around membrane proteins (pink). The branching black 'trees' represent carbohydrate components on the non-cytoplasmic surface. The rope-like cables represent membrane-associated 'scaffolding' proteins on the cytoplasmic membrane surface. From Nicolson (8).

Fluidity was recognized as a central character of cell membranes from early on, but the rFMMM reflects a major advance in the concept of membrane fluidity (8). Initially it was believed that the lipid bilayer was a relatively uniform, fluid 'sea of lipid' with the different PL classes more or less randomly distributed, and with the integral proteins having considerable freedom of mobility in this lipid ocean. Now the rFMMM depicts the bilayer as a mosaic of 'lipid domains' each with its own PL composition, fluidity and other characteristics (8,11).

In this rFMMM membrane model, micro-variations in the physical and chemical character of the PL domains result in specific proteins being preferentially localized to specific domains. Fig. 3 illustrates the extreme mosaic character of this model, with the further addition of 'skeletal' or 'scaffolding' membrane-associated proteins able to link with peripheral or integral membrane proteins and further restrict their movement within the bilayer (8,10,11).

2.2.2.1 Phospholipid properties strongly influence membrane behaviour

The functioning of membrane proteins is closely linked to the lipid characteristics of the membranes in which they are located (4,6–11). The numerous PL molecular species of membranes (numbering well over 100) act in conjunction with cholesterol and SM to influence the proteins, whether at close range between neighbouring molecules or via 'bulk' physical or chemical effects that act from a greater distance. Of these bulk lipid effects, one of the most studied is membrane fluidity.

The early studies suggested cell membranes were semi-fluid – microviscosity and other experimental data measures suggested they were not fully liquid and not fully solid (8). Cholesterol has a consistent solidifying influence, as does the SM. The membrane's PL molecules have individually variable effects on fluidity, depending on the total numbers of double bonds in their fatty acid tails (8,25,26). This is highly relevant to the physiological effects of PL, especially since the tail group compositions are determined by the composition of the individual's diet.

For a membrane PL molecule, the more carbon – carbon double bonds in its tail groups, the greater its fluidizing effect, and vice versa (25,26). Fatty acid tails that are saturated have no double bonds and therefore little if any fluidizing influence. Tails with just one double bond (monounsaturated, including omega-7 or -9 fatty acids) have mild fluidizing effects. Tails with multiple double bonds (polyunsaturated, including omega-6 and omega-3) have relatively substantial fluidizing effects, in direct proportion to the total double bonds they contain. Thus lecithin constituents of the diet (eggs, meat, milk, for example) or dietary supplement regimens can be tailored to deliver more or less fluidizing power.

Simplistically, higher membrane fluidity should better support membrane efficiency by enabling greater freedom of motion for the membrane's proteins (25,26). This correlation may not fully apply in the so-called raft zones of membranes.

The rFMMM model incorporates 'lipid rafts', zones within membranes that are less fluid than the surrounding zones, thereby giving an impression of 'rafts floating in a sea' of membrane. Often the lipid bilayer is also thicker in the raft zones (8,10,11). The proteins in lipid rafts are less mobile than those in the surrounding zones. This can be due to higher densities of cholesterol and/or SM molecules in the raft zone, or to linkages between the raft zone and membrane-associated scaffolding proteins (8). Lower mobility of a protein within a raft zone does not necessarily equate with less activity, however: certain transporter proteins may be more active when associated with specific lipids within a raft (8,11).

The impacts of specific fatty acid profiles on the health benefits of the various commercially available lecithin preparations have become very important research objectives and are covered from a clinical perspective in Chapter 6.6. (3,7,26).

Other membrane physical characteristics besides fluidity can strongly influence cell behaviour and function. Some of these are curvature, deformation, compression and expansion, all influenced by the physical character of individual PL in the membrane (8,11). Convex curvature, for example, can support membrane-to-membrane fusion for purposes of secretion, and concave curvature can support internalization of small segments of the OM for remodelling or other purposes. The lyso-PC (LPC) molecule, which is PC with just one tail group, is cone-shaped and facilitates positive curvature. The PE molecule has an inverted cone shape and encourages negative curvature (11).

2.2.2.2 Specific phospholipids activate important membrane proteins

The ongoing intensive research on PL and cell membranes is increasingly being focused on the very close-range interactions between specific PL and specific membrane proteins (27). Most of this focus is on integral membrane proteins (IMP), which are said to be 'drug targets' in possibly 50% of the current pharmacology research (28).

It has been estimated that as many as one-third of all cell proteins are involved in membrane transport and signalling (29), and most are integral (IMP). Researchers are finding that important membrane proteins have molecular sites that bind specifically with selected membrane PL and depend on such bonding for optimal function (30–32).

Integral membrane proteins can be highly responsive – and vulnerable – to fluctuations in the membrane's lipid profile. Perhaps the best understood of these is the sodium/potassium ATPase, also called Na,K-ATPase or simply NKA. This enzyme is very expensive for our cells to run, consuming as much as 25% of the body's total energy and a remarkable 50–70% of the energy used by the brain and the kidneys (25).

NKA is an active transport enzyme: it uses the energy of ATP (adenosine triphosphate), the body's 'energy currency' molecule, to move sodium ions out of the cell while simultaneously moving potassium ions into the cell (6). NKA's activity creates an electrical potential difference across the OM, an inside-negative millivolt charge that is vital for cell survival. NKA's activity is markedly influenced by at least three different PLs (30).

The functional NKA unit actually is a molecular complex, made up of several protein molecules, and extends completely across the OM (6,30). This complex has at least three different sites that preferentially bind with PL, and the binding at each site has a different effect on the functioning of the complex (30). At one site PS has a stabilizing effect; at another, PE with polyunsaturated tails stimulates the transport activity; at the third site, PC with saturated tails inhibits the activity. These characteristics likely make the NKA highly sensitive to changes in the PL content of its membrane environment. The reliance of this energetically expensive enzyme system on three of lecithin's PLs underscores the relevance of lecithins to human health.

Another IMP known to be highly sensitive to membrane PL is the nicotinic acetylcholine receptor (NAR). It is part of the cell's sensing system for acetylcholine, which is distributed not only in the brain but has messenger functions throughout the body. The NAR unit is made from five protein molecules and requires a lipid bilayer that contains predominantly PC but also with a relatively high amount of PA (phosphatidic acid) (31).

Yet another integral protein with specific PL preferences is cytochrome c oxidase, a fundamental player in energy production by the mitochondria. Each molecule of this protein requires tightly bound PC (one molecule), PE (three molecules), PG (four molecules) and CL (cardiolipin, two molecules) (32).

2.2.2.3 Cell membranes manage much of cell physiology

The body's many different cell populations use their membrane systems to perform their essential 'housekeeping' functions, most of which relate to homeostasis – the maintenance of nearly constant conditions in their internal environment (1).

Homeostasis at the cell level requires membrane networks to generate energy, import nutrients into the cell, discharge waste products from the cell, manage cell growth and proliferation, execute shape changes and movement, and enable controlled releases of newly made hormones or other regulatory substances to the outside ('secretion') (1,6). In addition, most human cells have to maintain their specialized functions, which invariably require their membrane networks to be up and running at all times (1,4,6,9–11).

Of the various cell membrane networks, the OM is the best understood. The OM is the definitive barrier that demarcates the cell's living contents from the non-living surroundings (see Fig. 2) (6). But this is no passive wall; the OM is a major contributor to cell homeostasis by dynamically transporting substances into and out of the cell, 'on demand'. The OM has numerous proteins that are channels, gates or ATP-consuming active transporters (such as NKA) to finely regulate the movements of vitamins, other metabolic factors, water or waste products across its width (10,28,33).

The OM operates in synchrony with the cell's other membrane networks. It makes physical connections with the endoplasmic reticulum (ER), which is the cell's main factory for making both proteins and PL (10). The ER further connects with the Golgi network, which prepares proteins and other substances for export from the cell. The ER also makes connections with the double membrane systems of the nuclear envelope and the mitochondria. Essentially, all the cell's membrane networks interconnect and work as an integrated whole (1,6).

The cell's membrane networks can readily interconnect precisely because they share structural continuity. All are organized on the same lipid bilayer plan, with PL as their main building blocks (6,8–10). However, there are subtle differences from one network to the next, one being patterns of PL asymmetry between the two halves of the bilayer. The exact nature of this so-called membrane asymmetry is related to each membrane network's functional priorities.

Membrane asymmetry is most evident in the cell's outer membrane. PC, which is consistently the most abundant membrane PL, is most concentrated in the outer half of the OM bilayer, the membrane half that faces away from the cytoplasm (34,35). SM shows a similar distribution. By contrast, most of the outer membrane's PE, PS, PI and PA are positioned in its inner bilayer half, the side that borders the cytoplasm. These patterns are enforced by enzymes in the membrane (10,34,35).

Membrane asymmetry is under homeostatic control and requires energy. Thermodynamic forces cause the PL molecules to intrinsically 'flip' from one membrane half to the other in order to randomize their distribution, but 'floppase' membrane enzymes constantly push them back to their earlier position to maintain the membrane's asymmetric nature (8,34). Such active (meaning energy-dependent) maintenance of membrane asymmetry seems essential to life, because experiments that disturb asymmetry result in the severe impairment or death of the cell (10,34,35).

The membrane asymmetries of mitochondria particularly illustrate the harmonization of membrane form and function. The mitochondria are the cell's energy organelles, powerhouses that can number 100 or more per cell and generate at least 90% of all its energy as ATP (6,37). They are typically tubular in shape, average about 100 per cell, and have a double membrane structure (see Fig. 2). The mitochondrion's outer membrane has a PL distribution that resembles the OM and the cell's other membrane networks, and makes contacts with the ER. But the inner mitochondrial membrane is very different from the cell's other membranes. It houses most of the energy-generating proteins and has a high concentration of cardiolipin (CL), a fluidizing phospholipid. Most of the CL is in the inner half of this inner membrane, the side that faces the inner watery matrix of this organelle (36). The composition and unique asymmetry of the mitochondrial inner membrane makes it less similar to the cell's membrane networks and more similar to the primitive one-celled organisms from which the mitochondria most likely derived (6,37).

2.2.3 Phospholipids sustain mitochondrial energetics

The inner mitochondrial membrane is unique not just for its high CL content but for its extremely dense population of proteins, perhaps as many as three proteins for each phospholipid molecule present (38). The vast majority of these participate in energy generation and they occur packed together into functionally cooperative molecular complexes (named Complexes 1, 2, 3, 4 and 5) (6,38,39). These complexes in turn are closely packed into higher-order protein assemblies or supercomplexes, which are the functional ATP-generating units of the mitochondria. These functional units seem to require a high membrane content of the fluidizing CL and a low content of the rigidifying SM and cholesterol, which allows them relatively high mobility within the membrane (38).

These energy-generating proteins of the inner membrane complexes are highly vulnerable to alterations in the membrane's PL profile, and even mild membrane abnormalities can result in disease (9,38,39). Complexes 3 and 4 need PE. PG (phosphatidylglycerol) is highly concentrated in Complex 4 (39). Cardiolipin is crucial for all the energy-generating complexes (for details refer to Figure 1 in Paradies and collaborators (39)). CL abnormalities are linked to an array of problems including the life-threatening Barth syndrome and other disease states (9,39,40).

Cardiolipin is composed of two molecules of phosphatidic acid linked together by a glycerol molecule and has four fatty acid tails, all of which tend to be highly unsaturated (38). CL is necessary for the energy proteins to properly assemble into the supercomplexes, for optimal energy generation performance. CL also is used for another function: mitochondria that are dying move CL to their outer surface, the outer half of their outer membrane, and this causes the cell to destroy them. Going further, a high degree of this 'CL externalization' by its mitochondria can drive the cell to apoptosis, a kind of cell suicide (9,38,39).

Barth syndrome is a genetic disorder linked to abnormalities in the amounts and types of CL molecules within the inner mitochondrial membrane (9,40). It affects males and features growth retardation, heart problems, impaired immunity, neuromuscular problems, exercise intolerance, fatigue and defective mitochondrial energetics (40). Barth syndrome is an instructive example of how a membrane lipid abnormality can result in a severe health outcome (9).

The fundamental importance of the mitochondria for the cell's continued survival rests on their capacity to generate ATP, which drives practically all the cell's life processes. The ATP generation capacity of their membrane proteins relies not just on CL but also on PC, PE and PG (38–40). Further research into mitochondrial phospholipids and their interactions with the energy-generating protein complexes should help us better understand energy dysfunctions that compromise the physiology of the heart, brain and other organs (41).

2.2.4 Phospholipids are pivotal in signal transduction

Signal transduction (ST) is the term applied to a complex web of processes that convert signals coming to the cell into functional changes within the cell (6,11). Signal transduction is primarily a membrane phenomenon and directly involves membrane PL. The roles of membrane PL in ST are currently subjects of intensive ongoing research (11,42,43). The literature on this topic is so voluminous that this section can cover only the best-documented contributions of lecithin PL to ST processes.

Most of the signals that cells transduce originate outside the cell and must traverse the cell's outer membrane (OM) to reach the cell's interior (8,11). The outer face of the OM is studded with a variety of protein receptors, some with carbohydrates or lipids attached, that extend out from the membrane into the surrounding environment. Once a receptor 'senses' the signal, which is often a chemical messenger substance, it responds by activating a number of other proteins within the membrane and 'downstream' from it. Each of these in turn may activate a number of other membrane proteins further downstream. The result is a cascading wave of protein activation that amplifies the signal and conveys it into the cell interior (42,43).

At the end of this chain of linked membrane proteins are enzymes that convert certain membrane PL into smaller molecules that readily diffuse out of the membrane and into the cell cytoplasm (11). These small molecules are termed 'second messengers', able to convey the ST cascade into the cell interior. Second messengers can activate enzymes and other proteins located in the cytoplasm's watery matrix as well as the ER, mitochondria, nucleus and practically all the cell's other membrane networks. Second messengers are the hallmark of ST (8,11,42).

By far the major source of second messengers for signal transduction is PI (phosphatidylinositol), one of the lecithin phospholipids. PI makes up 10–20% of the total cell membrane PL (42). PI is distributed across practically all the cell membrane networks, asymmetrically concentrated in the membrane halves that face the cytoplasm (42). It has structural roles, such as anchoring certain cytoplasmic proteins to the membrane. However, PI molecules also are available to ST-related membrane enzymes, which attach phosphate groups onto them. This activity generates a variety of PI molecules with varying numbers of phosphate groups attached, known as polyphosphoinositides (PPI). At least seven PPIs are known, and these are the main sources of second messengers for ST (43).

Wherever PI molecules are located in a cell membrane network, PPI can be generated from them (43). PPI are so important for ST that one researcher could state they 'control almost all aspects of a cell's life and death' (42). In the OM, PPI regulate a variety of membrane ion channels, proteins with watery pores that span the membrane and facilitate the entry of sodium, potassium and calcium into the cell (42). These contribute to cell activation from ST. The various PPI are also distributed in the ER and Golgi membranes, in the membranes of the cell's nucleus, even in the most minor internal membrane networks (43). The PPI second messengers are the main signal bridges between membranes and the cytoplasm.

Like their parent PI molecules, the PPI are located mostly in the membrane's cytoplasmic half. As the signal cascade arrives in this zone, it activates enzymes that convert PPI molecules to smaller inositol phosphates (IP), which are hydrophilic ('water-loving') and therefore miscible with water (43). These IPs diffuse out of the membrane and easily mix into the watery medium of the cytoplasm. There they activate water-phase proteins, some of which can have their own 'downstream' effects on other water-phase proteins, so that the ST activation cascade now spreads through the cell cytoplasm. IPs coming from PPI may regulate hundreds of different cell proteins (43).

2.2.4.1 Signal transduction proteins need membrane phospholipids

Certain receptor proteins of the OM are known to require membrane PL other than PI for full activity. The G-protein-coupled receptors (GPCR) are a large category that have been intensively studied; they account for some 3% of the proteins programmed by the human genome (44). They also have been productive drug targets for pharmacology researchers: an estimated 40–60% of drugs currently on the market are believed to act on GPCR (44,45). Among these are drugs for pain, heart failure, hypertension, asthma, peptic ulcer and prostate cancer (45).

The GPCRs are stimulated by chemicals (also called 'ligands') such as amino acids, peptides, proteins, lipid messengers (e.g. platelet-activating factor, eicosanoids and cannabinoids) and even entities as small as calcium ions, protons and light photons (46).

As their full name implies, these GPCR proteins must become coupled to G proteins in order to initiate ST. When a signalling ligand binds to the GPCR it triggers a shape change ('conformational change') in the GPCR that allows it to then directly link with a G protein (45).

The G proteins are so termed because they bind with molecules of GTP (guanosine triphosphate) or GDP (guanosine diphosphate), which then drives them to activate such membrane enzymes as adenylate cyclase, the phospholipase C family (PLC) and kinases that also can propagate signals (46,47). PLC are the main enzymes that physiologically degrade membrane PPI. The ST cascade GPCRs ➛ G proteins ➛ PLC ➛ PPI ➛ IP helps control such physiological processes as vision, smell, taste, brain neurotransmitter actions, mood, hormone actions and even body weight (11,45). For more details of this highly complicated array of processes, consult Figure 6 in Escriba and collaborators (11).

Membrane PLs are deeply involved in these ST molecular dynamics, and a good example for their involvement is the rhodopsin-transducin protein complex that is central to the ability of the eye's retina to detect and respond to light (47). Rhodopsin is a GPCR and the retina's major light-sensing protein, densely packed into the OM of the photoreceptor cells (46,47). As each rhodopsin molecule is activated by light it rapidly activates multiple transducin G protein molecules, thereby propagating and amplifying the light signal. The presence of PS in the membrane provides the best support for rhodopsin's physiological activation of transducing (48).

Membrane enzymes of the protein kinase C (PKC) family also are important in signal transduction. For full activation they need diacylglycerol (DAG), which can be produced by PPI breakdown. This suggests a likelihood of cross-activation between the PPI cascades and PKC cascades. Some PKCs likely require PS. Other research suggests that a PE molecule with tails 18:1/22:6 works particularly well to activate certain PKCs (11). Anti-cancer researchers are pursuing some of these findings.

PPI in the inner (cytoplasmic) half of the OM also contribute to important membrane functions other than ST (42). For example, they are involved in endocytosis, an inward membrane 'gulping action' that helps the cell acquire nutrients and recycle membrane molecules. PPI help anchor cytoplasmic skeletal molecules to the membrane (actin, for example) to help the cell maintain its shape. PPI anchoring of proteins on the cytoplasmic side of the OM is also believed important for phagocytosis, the process by which cells ingest and kill other cells. Phagocytosis is physiologically important for disposing of infectious agents or clearing debris from sites of tissue damage (1,8,42).

Signal transduction continues to fascinate researchers with interests as diverse as cancer, cell biology, immunology, neuroscience and vision. Drug development research involving ST is currently very intense, especially with respect to cancer, immunity, obesity and hypertension. Phospholipids of the PI and PS classes seem especially crucial for signal transduction's contributions to human physiology.

2.2.5 Phospholipids generate major cell messengers

Since the human body has many trillions of cells, efficient cell-to-cell communication likely confers survival advantage (6). The main tools that our cells use for communication are mostly small molecules, aptly termed 'cell-to-cell messenger substances' or simply 'cell messengers' (1,6). Currently experts agree that a molecule is a cell messenger if it has one or more identified receptors in the outer membranes of cells other than those cells that produce it (46). Many of the known cell messengers are derived from cell membrane PL (10,11,46).

In many cell types the OM has enzymes able to convert practically any part of a membrane PL molecule to make it water-compatible. Some also can convert membrane SM. This conversion produces a smaller molecule capable of rapidly migrating out of the membrane and into the water-phase environment that surrounds the cell. From there it can diffuse further to activate other cells (11,46). Some cell messengers are relatively unstable and therefore can only influence cells relatively close to their point of origin before they lose their potency. Others are sufficiently stable to reach the blood vessels and become circulated to other tissues.

A broad range of cell messengers can be derived from membrane PL or SM. Those that have been most researched are summarized below.

Eicosanoids (prostaglandins, thromboxanes and leukotrienes) are the most studied of all the cell messengers. They are generated from fatty acids, mostly of the omega-3 or omega-6 class. First the enzyme PLA2 (phospholipase A2) detaches the fatty acid from a PL molecule within the membrane (11). This free fatty acid is then available either to cyclo-oxygenase enzymes that route them into processing to prostaglandins (49), or lipo-oxygenase enzymes that process them to leukotrienes (6,11). Eicosanoids have a vast array of proven cell messenger effects, making a full description of them beyond the scope of this review. The reader is referred to recent comprehensive reviews by Lands (49), Korbecki (50) and Mashima (51).

Protectins, resolvins and other specialized pro-resolving mediators (SPM) are cell messengers derived from long-chain omega-3 fatty acids – DHA (docosahexaenoic acid), EPA (eicosapentaenoic acid) or DPA (docosapentaenoic acid) (52–54). The SPMs provide potent protection against tissue damage, and positively influence inflammation by promoting clearance of debris, dying cells and bacteria from damaged tissue sites. Protectins, resolvins, maresins and an ever-increasing array of known SPMs contribute to the prompt termination of the inflammation event once healing has begun (53,54).

Lysophospholipids (LPL) are PL molecules that have had one of their fatty acid tails removed, making them water-miscible and therefore able to mix into the water phase of the cell's outside environment. Several LPLs are well documented for their cell messenger actions in humans.

Lysophosphatidylcholine (LPC) can be generated from cell membranes or from circulating lipoproteins (LDL or HDL) by PLA2 or other enzymes, and is stable enough to circulate with the blood (55). LPC's messenger effects vary considerably, depending on its blood concentration and the type of fatty acid in its single tail (55,56).

The physiological concentration of LPC in human plasma ranges from a low of about 100–170 micromolar in healthy individuals, to 1,000 micromolar or higher in those with abnormally high LDL levels (56). At lower concentrations practically all the LPC is bound to proteins and functionally unavailable, but at higher concentrations some of the LPC is not bound and reacts with the delicate blood vessel lining ('endothelium'), often with pro-inflammatory consequences (55,56). This is another reason for healthcare practitioners to recommend strict management of LDL.

Lysophosphatidic acid (LPA) is a phospholipid that has lost one fatty acid tail as well as the unique portion of its head group (choline, ethanolamine, serine or inositol). LPA can get to very high concentrations in the blood, and may be an important messenger for the brain, which is enriched with LPA receptors coupled to G proteins (57,58). LPA is likely involved in the regulation of nerve cell birth, migration and subsequent differentiation into functioning brain tissue.

Lysophosphatidylserine (LPS) is PS with just one fatty acid tail. It is generated by a PLA1 enzyme (phospholipase A1) specific for PS (59). Unlike LPA, LPS does not reach high blood concentrations. The physiological roles of LPS seem to include helping to initiate inflammation at the site of an injury, and helping terminate it once the healing process has begun (59,60).

Lysophosphatidyinositol (LPI) is less well understood than LPC, LPA and LPS, but a receptor for it has been identified (44,61). Known as GPR55, this receptor has been found in the body's central nervous system, liver, bone, small intestine, adrenals, adipose tissue and pancreatic beta cells. LPI may help stimulate insulin release from the pancreas and may even have a broader role in regulating the body's energetics (44). LPI also may help regulate motor coordination via GPR55 receptors in the brain (61).

Platelet activating factor (PAF) is a PC molecule with one fatty acid tail replaced by an acetyl group at the C-2 (carbon-2) position (62). PAF has powerful effects on platelets, hence its name, and is produced by a variety of immune cells as well as by blood vessel linings (endothelia). Acting via its G-protein-coupled receptor, PAF can trigger clotting, regulate blood vessel functioning, and trigger apoptosis (cell suicide) (63). PAF also is important in both the initiation and termination of inflammation (64).

PAF is stable enough to circulate with the blood. The PAF signalling system exists even in unicellular organisms (protozoans) and likely developed as a protective mechanism for host defence (refer to Figure 1 in Yost et al. (63)). Recently PAF also was found to regulate synaptic activity in nerve cells cultured from the rat brain (hippocampus) (65).

Sphingosine-1-phosphate (S1P) is made in cells from ceramide, a sphingolipid that derives from membrane SM (66,67). Similar to PAF, S1P also can link inflammation with blood coagulation. S1P helps regulate cell proliferation, maturation, movement and migration during fetal development. S1P may be important for heart and brain development and for maturation of the circulatory system (66). S1P has at least five known receptors, all of the GPCR family (67). These make possible S1P's diverse involvement in embryonic development, cardiovascular physiology, tissue regeneration, inflammation and immunity (66,67).

Cell messengers are currently under intensive investigation, especially for their pharmacologic potential as 'drug targets' (46,64,67). From a physiological perspective, it is well established that cell messengers derived from membrane PL help coordinate the healthy functioning of the entire body.

2.2.6 Phospholipids – essential for electrical signal transmission

Electrical signal transmission is the specialized responsibility of nerve cells ('neurons') (1). The OM of a neuron generates an electrical impulse that rapidly self-propagates along the length of the cell. It quickly arrives at a junction with another neuron, where it can cross over and continue to propagate itself. This junction is called a synapse (6). Here the initiating ('presynaptic') neuron is specialized to release packets of neurotransmitters, and the responding cell ('postsynaptic') is specialized to respond to them. Electrical signal transfer across the synapse is a series of membrane events, and membrane PLs are centrally involved in these events.

Within the pre-synaptic neuron, located just under its OM are collections of tiny membrane microspheres ('vesicles'), loaded with neurotransmitter molecules (1). As the electrical signal arrives at the synapse, it stimulates these vesicles to rapidly fuse with the OM and release their transmitter molecules into the synaptic zone between the two cells. These bind with receptors on the OM of the postsynaptic cell, causing it to become electrically excited. The fusion process involves mainly two PLs: PS, and the inositol phospholipid PIP2 (phosphatidylinositol-4,5-bisphosphate) (68,69).

On the presynaptic side a protein called synaptotagmin-1 (Syt1) is responsible for 'zippering' the vesicle membrane onto the OM to ensure the two membranes fuse (68). Syt1 on the vesicle membrane binds to PIP2 on the OM to initiate the fusion process. Calcium flooding into the cell due to the electrical signal then further activates the Syt1-PIP2 molecular complex, after which Syt1 binds with PS in the outer membrane to complete the fusion process and release the neurotransmitters (69).

In the brain, synapses are the most basic functional units of information transfer because they link neurons into networks that manage the brain's functions (1,6). These networks must continually adapt to the body's ever-changing internal and external environments, and this is made possible by adaptability of the synapses, also called synaptic plasticity. Synaptic plasticity is highly responsive to the lysophospholipid LPA (lysophosphatidic acid) (57).

Rapid propagation of the electrical impulse along the length of a neuron requires a conducting sheath, analogous to the insulation of a wire. This sheath is made mainly of myelin, a lipid and protein material of which the main lipid component is SM (6). Previously myelin was thought to be simply a passive insulator, but now it is known to be highly dynamic and a source of messengers that help regulate neuron electrical activity. Some of these messengers are sphingolipids related to SM, such as ceramide and sphingosine-1-phosphate. Imbalances in their metabolism can result in loss of myelin from neurons and to cell death, linked to Alzheimer's, multiple sclerosis and other neurodegenerative diseases (13).

Researchers are continuing to unravel the complex involvements of synapses and myelin in nervous system physiology, but already it is clear that these crucial information transfer structures are exquisitely dependent on membrane PLs and cell messengers derived from them.

2.2.7 Phospholipids are building blocks for the circulating lipoproteins

The lipoprotein particles that circulate with the blood are essential for transporting nutrients to and from the tissues (1,14,70). Phospholipids are essential molecular building blocks for lipoproteins, due mainly to the same amphiphilic molecular character (charged at one end, uncharged at the other) that makes them building blocks for cell membranes.

The circulating lipoproteins fall into five main types: chylomicrons (CM), very low-density lipoproteins (VLDL), intermediate-density lipoproteins (IDL), low-density lipoproteins (LDL) and high-density lipoproteins (HDL) (1,14). All these particles are organized into a surface layer that contains PL molecules (mainly PC), cholesterol and proteins (termed apolipoproteins, or more simply apoproteins). The surface layer encloses a core that contains triglycerides (storage forms of fatty acids) and cholesterol esters (cholesterol combined with fatty acids). But the five types differ in their sizes, densities, shapes, detailed composition and physiological roles (14,70).

Most dietary PLs become assembled into lipoproteins very soon after they are absorbed. Intestinal epithelial cells (enterocytes) lining the duodenum of the small intestine absorb PL from the intestinal contents (71). Prior to being absorbed, most of the PL molecules have one or both of their tail groups removed by digestive enzymes (72). The enterocytes re-equip them with new fatty acid tails, then use them to make chylomicrons (72,73).

2.2.7.1 Chylomicrons, VLDL deliver nutrients to the tissues

Chylomicron (CM) particles are microscopic spherules substantially larger than the other lipoproteins. Their envelope is a single layer of PL molecules (mostly PC) interspersed with a smaller number of cholesterol molecules. The CM envelope also carries an apoprotein molecule, named apoprotein B48 or just apoB48 (14,73). Both the envelope and the core of the CM particle also accommodate fat-soluble nutrients, including vitamin A, carotenoids and vitamin E (73,74). Once mature, the CM particles are released by the enterocytes into the lymphatic circulation (74).

The CMs are carried with the lymph through the thoracic lymph duct and reach the bloodstream (venous circulation) (14). As they circulate with the blood through the muscles, adipose and other tissues, enzymes on the capillary walls (lipoprotein lipases) break down much of their triglyceride (TG) to release fatty acids which the tissues then consume, mostly for energy (14,73,74). Much of their fat-soluble nutrient content may also be removed (74). Less than five minutes after exiting the enterocyte the CM particle is reduced to a 'chylomicron remnant' which the liver then picks up and recycles into other lipoproteins (14,73).

Liver cells (hepatocytes) recycle some of the leftover TG and other materials from the CM remnants into VLDL (very low-density lipoprotein) particles. The envelope of VLDL is overwhelmingly PC with lesser amounts of SM>PE>PI>lyso-PC in that order (70). It also has an apoprotein, apoB100 (75). Its core is mostly TG. VLDLs also carry some fat-soluble nutrients, and (as happens with CM particles) as they pass through the tissues their TG and other contents are depleted by lipoprotein lipases (70).

As VLDLs are progressively depleted of their TG they become IDL (intermediate-density lipoprotein) particles. These are either picked up by the liver to be recycled, or continue to circulate and continue being depleted of their TG until they become classifiable as LDL (14,70). The LDL envelope resembles that of VLDL and IDL in having the same order of PC>SM>PE>PI> lyso-PC. The LDL core, however, has relatively little TG and is predominantly cholesterol esters. LDL particles can circulate for as long as three days and are the main vehicles for delivering cholesterol to the tissues (14,74).

The intestinal enterocytes and the liver hepatocytes need adequate dietary intakes of PC and other PL to produce CM, VLDL, IDL and LDL. In experiments with rats, limiting the dietary availability of PC to the enterocytes caused them to make less CM (71) and VLDL (75). This raises the issue of whether today's human populations have adequate dietary intakes of PL. Our modern food consumption has trended away from high-PL foods such as eggs, meat, organs and other animal foods (76). Also, the relatively common dietary deficiencies of nutrients such as fatty acids, vitamins A and E, folate and magnesium all can impair the body's capacity to utilize PL from the diet (77).

2.2.7.2 Lipoproteins clear excess cholesterol from the tissues

Phospholipids also are fundamental to HDL (high-density lipoprotein), the so-called 'good cholesterol' (14). Their relative abundance in HDL parallels their patterns in CM, VLDL, IDL and LDL: PC>SM>PE>PI>lyso-PC. HDLs are produced mostly by hepatocytes but also by enterocytes, and have properties that make them important for cardiovascular protection. Among these are antioxidant properties, inhibition of blood vessel inflammation, inhibition of thrombosis, promotion of blood vessel repair, and promotion of angiogenesis (formation of new blood vessels as needed by tissues) (78). But HDLs' most important physiological function may be 'reverse cholesterol transport' (RCT) (79,80). This involves HDL particles pulling cholesterol molecules from cells throughout the tissues and transporting them to the liver for subsequent disposal. The very high PC content of HDL particles seems especially crucial to this function.

The HDL particle begins its existence as immature HDL, a flattened disc-shaped structure very dense with PLs but also with multiple apoprotein molecules, including apoA1, apoA2 and others. It also can have relatively low amounts of cholesterol, triglycerides and fat-soluble vitamins (79). This particle works together with the enzyme LCAT (lecithin – cholesterol acyltransferase), which also circulates with the blood, to do RCT (78,80).

The RCT process begins as an HDL particle circulating through a tissue makes contact with a cell. It readily attaches to ('docks' with) the cell's outer membrane, then a nearby LCAT enzyme removes the fatty acid tails from PC molecules in the HDL and rapidly attaches them to cholesterol molecules within the cell OM. This converts the cell's OM cholesterol to cholesteryl esters. These ester molecules accumulate in the HDL particle and progressively balloon it from a discoid shape into a sphere (78,80). This spherical particle is then a mature HDL. It subsequently is removed from the blood after it arrives at the liver. The cholesterol ester molecules are then either excreted from the liver or recycled for other physiological needs.

Reverse cholesterol transport by HDL is a major factor in health maintenance and disease prevention. There is overwhelming evidence that the higher the HDL in the blood, the lower the risk for atherosclerotic cardiovascular disease (81,82). And most of the PL classes found in lecithins – PC, SM, PE, PI though not PS – are essential for HDL particle production and for their RCT and other beneficial properties (70,79,80).

2.2.8 Phosphatidylcholine supplies choline, essential and versatile nutrient

Choline is now classified as an essential dietary nutrient for most people, since removing it from the diet results in liver dysfunction and/or muscle damage for most men and for some women (83). Yet a high proportion of the citizens of industrialized nations, including more than 90% of Americans and perhaps a similarly high percentage of Canadians, southern Europeans, Chinese and Taiwanese, are consuming less dietary choline than the minimum recommended for good health (83,84).

PC, the phospholipid most abundant in lecithins, has choline in its head group (phosphatidyl*choline*) and is a rapid and efficiently utilized source of choline (86). Breast milk supplies not just PC but glycerophosphocholine, phosphorylcholine and other choline phospholipid derivatives, a further measure of the advantages of breastfeeding. The need to maintain adequate dietary intake of choline is evident from the numerous ways this nutrient contributes to our physiology, as detailed below.

2.2.8.1 Choline maintains nervous system function across the lifespan

Acetylcholine (ACh) is the body's most widely distributed nerve transmitter, essential for neurons not only in the brain but in virtually all the other organs. It is used for neuron-to-neuron synapses as well as for neuron – muscle synapses (neuromuscular junctions). Neurons that use choline are classified as cholinergic. The autonomic nervous system that regulates the organs other than the brain relies heavily on cholinergic neurons. The body's large populations of cholinergic nerve cells create high ongoing demand for choline as an ACh building block (87).

Cell populations other than neuronal also can be cholinergic. Macrophages and some other immune cells have ACh receptors able to respond to ACh released from other immune cells. Here ACh is acting more like a cell messenger to help manage inflammation and other aspects of immune response (88).

Aside from its neurotransmitter and cell messenger activity, ACh is also a tissue growth regulator for the brain's early development (89,90). Cholinergic nerve cells appear along with ACh receptors in the fetal human brain as early as the first trimester (89). This helps explain why dietary choline is important all the way through pregnancy. Studies indicate that pregnant women consuming choline in the lowest 25% range of dietary intake have at least twice the risk of delivering offspring with neural tube birth defects, compared to those in the highest 25% of choline intake (85,91). But brain development even after birth is linked to the mother's choline status: higher maternal plasma choline levels at week 16 were statistically correlated with higher cognitive function in the child at 18 months of age (92).

The body's reliance on ACh as a neurotransmitter, cell messenger and growth regulator may at times create demands for choline that exceed dietary choline intake. Experiments with cultured nerve cells indicate that when they are not obtaining sufficient choline from the growth medium, they can activate enzymes to detach choline from PC in their membranes, then use it to make ACh. This 'auto-cannibalization' of choline from cell membranes can generate at least 15% of the cell's total ACh pool (87). But since PC is essential to all our cell membranes, extended auto-cannibalization could threaten nerve cell performance. Including lecithin in the daily diet helps ensure against this possible scenario.

2.2.8.2 Choline supplies methyl groups for gene regulation, other metabolism

The phenomenon of epigenetics is currently one of the most actively researched in biology and medicine. As the term suggests, epigenetics is regulation of gene activity at a higher level than the DNA of which they are made. Epigenetics permeates all human life processes (6). The body uses small chemical groups called methyl (CH3) groups as a major epigenetic tool. Specialized enzymes can turn genes on or off by attaching methyl groups to their DNA or to proteins associated with the DNA. Choline can be an important source of methyl groups for epigenetics (15,93).

Under ideal conditions, the primary source of methyl for gene regulation is SAMe (S-Adenosylmethionine) (94). SAMe is a highly energized methyl donator made mostly in the liver by a reaction of the amino acid methionine with ATP. However the body's consumption of SAMe is high – on the order of 2–3 grams per day – so that the liver and other organs continually need methyl to regenerate SAMe once its methyl has been removed (95). The main pathway for regenerating SAMe requires methylfolate methyl-vitamin B12 as essential enzyme cofactors. When either of these vitamins is not sufficiently available, choline can be used as a backup (95). Each choline molecule can donate up to three methyl groups.

Epigenetics is just one of methyl's roles in cell-level physiology. Methyl is necessary for the body to make hormones, brain neurotransmitters, even phospholipids: much of the body's SAMe normally goes into converting PE to PC, which is the major PL needed for making cell membranes, lipoproteins, even the bile digestive fluid (94). Substantial dietary intake of lecithin would help supply the substantial amounts of PC needed for cell membrane maintenance, while also supplying choline for methyl homeostasis.

2.2.8.3 Choline helps regulate potentially toxic homocysteine

As SAMe is consumed, homocysteine (HCY) is generated (95). This is a substance with high toxic potential (96,97). Under optimal conditions HCY is rapidly removed once it is formed. It can be removed either via a metabolic pathway that uses it to make antioxidants, with vitamin B6 as an essential cofactor, or via a pathway that uses methyl groups to convert it to methionine, with the methyl coming from methylfolate and methyl-B12. If these two pathways fail to work optimally, HCY builds up in the blood and tissues, with potentially grave consequences.

Abnormally high blood HCY is linked to increased risk of vascular disease and stroke, depression, cognitive decline, Parkinson's disease, gastrointestinal conditions, bone loss, cancer development and diabetes (96,97). Adequate supplies of vitamins B6, B12 and folate are required for healthy HCY homeostasis, but when these are not sufficient choline can provide the crucial methyl groups needed to convert HCY to methionine (98,99). The clinical research with lecithins for this purpose is reviewed in Chapter 6.6.

2.2.8.4 Phosphatidylcholine (PC) – a better choline source than choline

Surprisingly, dietary choline in the form of free (unbound) choline or choline salts is a poor source of choline for metabolism. But PC is an excellent dietary source of choline (85,86).

Choline is not stable in water and therefore rapidly degrades in the intestinal contents, blood or body fluids. This helps explain why the blood has very little free choline and carries most of it in the form of PC or stable metabolites related to PC. The choline in our foods is rarely present as free choline, rather as PC or derivatives of PC (76).

Whereas free choline or the dietary supplements choline bitartrate or choline chloride do not produce a lasting increase in blood choline levels, PC is more than 90% absorbed and produces a sustained increase in blood choline (86). Choline taken by mouth can cause vomiting, other gastrointestinal adverse effects, sweating and lowering of blood pressure, none of which happens after oral consumption of lecithin or PC (86).

Also, free choline that is not absorbed after intake by mouth can reach the large intestine, where resident bacteria convert it into trimethylamine (TMA), a substance that can give the body an unpleasant fishy odour (85). PC is not substantially converted into TMA, and even large oral doses of PC do not create this unpleasant body odour (85).

But TMA is more than just cosmetically unpleasant. After being generated from free choline in the large intestine it can be absorbed into the bloodstream, reach the liver and there become converted to TMAO (trimethylamine oxide), which can have toxic effects (85). Taking choline as lecithin is safe even at very high daily intakes, up to about 2.6 grams/day of choline equivalent (more than 20 grams per day of PC) (99).

Pioneering choline researcher Steven Zeisel has recommended that individuals who do not consume substantial amounts of choline-rich foods (eggs, liver, meat, milk for example) should consider supplementing their diet with lecithin to provide PC (85). Lecithin is a safe and cost-effective dietary choline source, and makes an excellent base for shakes, smoothies or other convenient dietary dosage forms.

2.2.9 Phospholipids – essential to epithelial structure and function

Certain PLs, especially PC, have the important physiological function of coating a variety of the body's epithelia, which are delicate cell sheets that line open spaces in the organs. These PLs are called surface-active agents or surfactants, because their charged head groups can improve the spreading of water over the epithelial surface. But also, since they are amphiphilic, their uncharged tail groups can be employed to repel water from epithelia as needed. The contributions of PC and other PLs to epithelial structure and function are best understood for the lung's air sacs, the gastrointestinal (GI) tract and the joint surfaces.

2.2.9.1 Phospholipid surfactants preserve lung function

Surfactants are essential for the lung's gas exchange functions, which are crucial to our entire physiology. The lung's alveoli ('air sacs') are their main gas exchange units and cannot function without ample quantities of surfactant (1,16). The alveolar epithelium makes a surfactant lipoprotein complex with PLs as major ingredients.

Air sacs must expand considerably as they fill with air during each breath. Oxygen diffuses from the inhaled air into the blood vessels just under the epithelium, while simultaneously carbon dioxide diffuses from the blood vessels into the air space of the sac (1,16). But between breaths the sac's epithelium is extremely folded, with a very large surface area (1). This results in a very high surface tension that creates a high mechanical resistance against the lung becoming inflated at the next breath. Surfactant spread over the epithelial surface at the air – water interface lowers this surface tension and allows the lung to readily inflate.

The lung surfactant is a lipoprotein complex that contains 86% PL by weight (16,100). Other lipids (triglycerides, cholesterol, free fatty acids) make up another 4%, and the remaining 10% consists of several proteins. Of the PLs, PC is about 70% by weight, PG (phosphatidylglycerol) and PI (phosphatidylinositol) are about 9%, and PS, PE and SM take up another 6%. Most of the PC is the dipalmitoyl species (DPPC, 16:0/16:0 PC), which is extremely effective at lowering surface tension.

Besides its essentiality for gas exchange, the lung surfactant medium traps foreign bodies and dead cells or other cell debris for subsequent clearance from the lung tissue. It also assists in immune defence. For details, refer to Figure 2 in Han and Mallampalli (100).

PG is far more concentrated in the lung than in other tissues, and is important for regulating immunity and opposing viral infection of the lungs (100). The PG in lung surfactant has potent antiviral activity against influenza A virus (101) and respiratory syncytial virus (RSV) (101,102). It also helps fight inflammation triggered by *Mycoplasma pneumoniae* infection (103). The PI in the surfactant contributes to the antiviral host defence benefits of PG (100,104).

Ongoing production of lung surfactant is essential to respiratory health, and surfactant abnormalities contribute to a diverse array of respiratory illnesses (16,100). Inherited or acquired deficiencies of surfactant production are linked to neonatal respiratory distress syndrome (RDS), cystic fibrosis, lung infections, asthma and inflammatory illness, COPD (chronic obstructive lung disease) and sepsis. Tobacco smoke and other lung toxins all can deplete surfactant and threaten survival. The central role played by PLs in lung surfactant function is one dramatic example of their absolute importance to human life.

2.2.9.2 Phospholipids reinforce the digestive epithelial barrier

At epithelial surfaces elsewhere than the lung, PLs are used for an opposite purpose than as surfactant. Their hydrophobic tail groups make them suited to serve a purpose functionally opposite to a surfactant role: rather than encouraging mixing with water, here they are used to repel water. Researchers believe that PL molecules are oriented in the intestinal mucus layer with their hydrophobic tail groups pointing away from the mucus and towards the intestinal lumen (17,18). This effectively forms a hydrophobic barrier that restricts water and strong stomach acid from reaching the epithelium. In the stomach this protects the epithelium against being 'auto-digested' by the strong acid it has created (1).

Moving past the stomach, most if not all the epithelium lining of the (GI) tract has a mucus layer (18). Where stomach acid is not a problem it serves as a lubricant, lowering friction to ease movement of the intestinal contents. It also serves to trap microbes or other foreign agents, which can later be disposed of as the epithelial cells shed old mucus and replace it with new (1). PC is integral to the structure and functioning of this GI mucus layer.

The GI mucus consists primarily of a thick layer of glycoproteins called mucins, with embedded protective biomolecules such as secretory IgA, lactoferrin, trefoil peptides and PL (17). Most of the mucus PLs are PC molecules. They are most likely positioned at the outermost edge of the mucus, packed closely side by side in a single molecular layer with their positively charged head groups embedded in the mucins and their uncharged hydrophobic tails extending into the intestinal lumen (Fig. 4). This sheet-like distribution would create a continuous hydrophobic barrier between the semi-solid mucus zone and the liquid-phase intestinal contents.

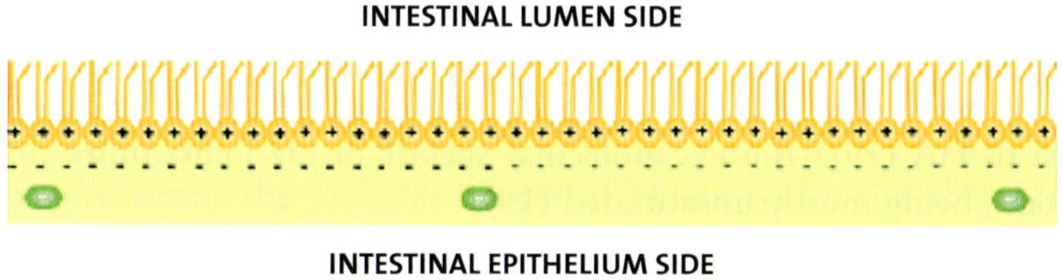

Figure 4.

Close packing of PC molecules 'standing on their heads' generates a hydrophobic barrier for the GI mucus lining. The yellow zone represents the outermost layer of mucin, carrying negative charges, while the green bodies represent accumulations of PC ('PC lamellae'). From Schneider and collaborators (18).

The intestinal epithelium benefits from this hydrophobic mucus barrier, especially because of its vulnerability to penetration by water-borne viruses or bacteria, or water-soluble chemical toxins that are brought in with food and drink. In animal experiments, removal of the PL layer allows such agents to breach the epithelium and (either separately or together) cause ulceration or other GI disease (105).

Along the GI tract, the stomach and colon epithelia are most vulnerable to attack by toxins, microbes or other potentially harmful agents (17). The troublesome *Helicobacter pylori* can use phospholipase enzymes (PLases) to degrade the mucus PL barrier and then take up residence in the mucus layer (106).

Non-steroidal anti-inflammatory drugs (NSAIDs) are used extensively in medicine. But some 20–40% of chronic users experience GI injury, mainly in the stomach (107). In rat experiments, administration of PL by mouth blocked the injurious effects of aspirin and other NSAIDs on the epithelium, and even promoted the healing of ulcers (105). Exciting progress with the use of PC to protect the GI tract of humans is discussed in Chapter 6.6.

2.2.9.3 Phospholipids help keep joint friction manageable

In the weight-bearing joints (hips and knees), PC and other PLs are used together with hyaluronic acid and other biomolecules to make a hydrophobic 'lubricant' that allows the apposing joint surfaces to slide freely over each other. This biological lubricant coats the epithelium of the cartilage that caps the bony surfaces. It has a thick, gel-like consistency that allows the joint surfaces to slide past each other with extremely low levels of friction, even though the fluids that fill the joint are maintained at very high pressures (108).

The lubricant material contains water, proteins, the linear polysaccharide lubricin, chondroitin sulphate, hyaluronic acid (HA) and phospholipids (mostly PC, with lesser amounts of PE, SM, PG and PS) (108). The PC is mostly bonded to HA (19). The PC molecular species in the joints differ from lung surfactant, being mostly unsaturated (19).

Evidence that PC is required for this joint lubricant to work comes from experimental studies. The HA at the cartilage boundary has a high intrinsic frictional coefficient, but when bound with PC its friction becomes very low, which is the normal state in the healthy joint (108). Depletion of this lubricating gel, or more specifically depletion of its PC content, can increase friction within the joint and usher in degenerative joint disease.

Clinically, dietary PC supplementation has shown promise for damaged joints (17).

2.2.10 Glycerophosphocholine (GPC) – unique cell volume stabilizer

Most of the cell interior is water, and water molecules are highly diffusible because they are so small. The cell's outer membrane is permeable to water but not to many larger molecules. When there are differing concentrations of such larger molecules between the cell interior and the outside, there will also be unequal concentrations of water between the interior and the outside. The water molecules then will have a tendency to move across the membrane to equalize their concentration, and this is termed osmotic pressure. Cells must homeostatically regulate osmotic pressure to avoid destruction (1,109), and one of their most useful tools for this purpose is the phospholipid GPC (glycerophosphocholine, alpha-GPC, choline alphoscerate). GPC is a major osmotic regulator or 'osmolyte' in humans (109,113).

Osmosis can be defined as the movement of water from a zone of higher water concentration to one of lower water concentration, across some semipermeable barrier (1). For cells the outer membrane is the semipermeable barrier. Osmotic pressure challenges cell survival because uncontrolled loss of water from the cell interior can lead to shrinkage, or uncontrolled water entry can lead to swelling, either of which can be lethal to the cell (1,21,109). But our cells also face a more subtle threat from osmotic pressure: metabolic damage from toxin build-up.

Our cells can be metabolically efficient because their proteins, DNA and other biomolecules are optimized to function within a relatively narrow range of chemical conditions (tolerable ranges of acidity – alkalinity, ionic charge distributions, reduction/oxidation balance, for example) (1,6). In particular, when water is drawn out of the cell its contents become more concentrated, and especially ions and metabolites become more concentrated. Eventually these rising concentrations exceed the tolerable ranges of biomolecules and they become damaged or irreversibly destroyed (1,21,109,111).

GPC is particularly useful to prevent this scenario because it is a 'compatible osmolyte', meaning it can get to very high levels in cell cytoplasm without having toxic effects on nearby DNA, proteins or other biomolecules (110,111). It is a zwitterion, having both a plus and a minus charge on the same molecule, and this gives it useful buffering capacity against ions such as sodium and chloride. GPC is a highly effective buffer against urea build-up, and is unique among human osmolytes for its capacity to protect proteins against damage by urea (110–112).

Urea is a metabolic waste product that is concentrated by the kidneys and then excreted in the urine. The kidney's medullary cells are especially subject to extreme osmotic stress from urea as well as sodium and chloride ions. However, they have the capacity to accumulate GPC to very high concentrations – over 100 millimoles in the rat kidney (110). These cells carry enzymes both to make GPC from PC, and to degrade GPC when necessary (112). By simultaneously regulating the activities of both sets of enzymes, the kidney cells fine-tune their internal GPC levels in lock-step with fluctuating concentrations of urea, sodium, chloride and other waste products.

All human cells are subject to osmotic challenges. They can make GPC using enzymes that remove PC's two tail groups (111,112). Since the GPC molecule has no fatty acid tails it is freely water-soluble and readily becomes distributed in the cytoplasm. GPC also contributes to osmotic regulation in brain cells (113), and this property – along with its being a ready source of choline (see Section

2.2.8) – helps account for its benefits to human brain function when used as a dietary supplement (114).

2.2.11 Phosphatidylcholine (PC) – vital constituent of bile

Bile is a digestive fluid produced by the liver (1). Bile has a range of functions in digestive physiology (1,22,115,116):

- Emulsifies food fats and other lipids into water. This emulsification step is essential for the body to absorb lipids.
- Transfers cholesterol out of the liver. This is vital for cholesterol homeostasis, the processes that manage cholesterol in our cells, tissues and organs.
- Excretes bilirubin, a potentially toxic haemoglobin breakdown product.
- Eliminates excess amounts of copper, a mineral essential for human metabolism but potentially toxic if it accumulates in the tissues.
- Helps eliminate normal metabolites that could become toxic, e.g. steroid hormones.
- Helps clear xenobiotics (synthetic and other foreign substances) from the body. Most if not all xenobiotics are potentially toxic.

Bile is produced mostly by epithelial cells of the liver. It is first 'assembled' in the bile canaliculi, microchannels that course between the liver cells (1,22). The canaliculi are organized into a network that moves the bile fluid to the gall bladder, where it is stored and concentrated for later release during a meal.

Bile fluid is about 95% water and 5% non-water (22). Of its non-water constituents approximately 67% are bile acids, 22% phospholipids, 4.5% proteins, 4% cholesterol and 0.3% bilirubin. The bile acids are generated from the breakdown of cholesterol by liver cells, and have powerful detergent properties. They work in concert with the phospholipids to give bile its considerable emulsifying character.

The bile PLs are overwhelmingly PC, with small amounts of SM present (22). The absolute amount of PC in human bile at any time is about 5% of the liver's total PC. PC is crucial to the functionality of bile – its emulsifying action works together with the detergent action of the bile salts to increase cholesterol's solubility by more than 1 million times and thereby fully disperse it into the bile fluid (22).

The bile acids are pumped into the bile canaliculi by proteins located in the liver cell outer membranes, called bile salt export pumps (22,117). This is a mass molecular transfer that also osmotically 'draws' water into the canaliculus to make the bile fluid. The liver cell outer membranes are also equipped with trans-membrane proteins that 'flip' PC molecules, from the membrane's inner half (facing the cytoplasm) to its outer half that faces the canalicular contents. As PC molecules concentrate in this outer membrane half, the strong detergent actions of bile acids already in the canaliculus extract them from the membrane – along with cholesterol and SM – into the canalicular contents (22).

As they become dispersed into the canaliculus, the bile salts, PC, cholesterol and SM come together and form aggregates. Depending on their size and complexity these are called micelles, mixed micelles (larger) or vesicles (the largest). These molecular aggregates are the bases for the physiology of bile fluid.

With each meal, the gall bladder contracts and empties hundreds of millilitres of bile fluid into the duodenal segment of the GI tract (22,117). As the bile aggregates are diluted into the duodenal contents, they disperse the food lipids (triglycerides, phospholipids, cholesterol) into smaller aggregates within the intestinal contents. Within these microscopic aggregates each PC molecule has its charged head group extending out into the water phase, while its uncharged FA tails likely come together with the tails of other PC molecules to create a nanoscale non-aqueous phase. This phase attracts cholesterol, fat-soluble vitamins and other fat-soluble nutrients (1). It also may accommodate uncharged xenobiotics (foreign molecules), endobiotics (hormones or other molecules produced in the body and no longer needed) or miscellaneous other uncharged molecules.

The PC – bile acid complexes derived from bile fluid also can absorb breakdown products of fats digested by lipase enzymes. These products include fatty acids, monoglycerides and other metabolites of triglycerides. Also present will be fat-soluble vitamins such as A, D, E and K, along with other fat-soluble nutrients. Subsequently these PC – bile acid – nutrient 'supercomplexes' become absorbed into the enterocytes that line the intestinal epithelium (1). Once inside the enterocyte, the complexes disintegrate and the nutrients are routed along various transport pathways, such as into chylomicrons (71–73). Much of the PC and bile salts end up back in the liver and can be reused in bile fluid.

PC is essential to the molecular organization and functionality of bile, and this makes it a major contributor to the physiology of digestion, absorption and detoxication (1,22,117).

2.2.12 Phosphatidylserine (PS) – essential for eliminating unwanted cells

The body's routine maintenance includes disposing of cells that are dead, dying or otherwise not wanted (119). Unwanted cells can include those infected with pathogenic bacteria or viruses, malignant cells, or cells damaged by toxins. The body's mechanisms for cell disposal have one consistent theme: exposure of PS on the outer surface of a cell, which acts as a 'flag' that signals immune cells to promptly and efficiently eliminate it (23,119).

Healthy cells do not have PS on their outer surface. PS is normally located in the inner half of the OM that faces the cytoplasm, not the outer half that faces the outside environment. Considering its molecular properties, PS would tend to be dispersed within both halves of the OM. In healthy cells this does not happen, but in unhealthy cells it does (6).

The factor that normally prevents PS being distributed on the outer surface is PS transport enzymes located in the outer membrane. These are formally known as P4-ATPases and informally as 'flippases' (34,35,118,119). As the 'ATPases' part of their formal name implies, these enzymes require ATP to do their job. In healthy cells they obtain sufficient energy from ATP to routinely relocate or 'flip' PS molecules from the outer half of the membrane to the inner half. But for thermodynamic reasons PS molecules naturally try to 'flop' back to the outer half. When a cell becomes unhealthy its ATP production falls and the flippases fail to keep up with PS that has flopped into the outer half of the membrane. PS molecules then come to protrude into the surroundings, where immune cells sense them as a signal that the cell should be destroyed (23).

The immune cells that carry out this task are usually phagocytes, so called because they dispose of unwanted cells by phagocytosis, first surrounding the cell, then internalizing it, and eventually digesting it – recall the amoeba of introductory biology (6). The body's 'professional' phagocytes (macrophages, dendritic cells, neutrophils) are equipped on their outer surfaces with receptor proteins capable of recognizing PS (23,119). Phagocyte PS receptors bind with PS molecules on the surface of a nearby unwanted cell, then activate the phagocyte to destroy the cell.

Some dying cells even hasten their own demise via cell suicide ('apoptosis') (6,11). They can use other ATPase enzymes called ABC transporters or 'floppases' to actively flop PS from the outer membrane's inner half to its outer half (34,35). Some apoptotic cells use related 'scramblase' enzymes that actively randomize PS in both halves of the OM and thereby further increase its surface density (118,119).

By actively displaying PS on their outer surface, apoptotic cells send an 'eat me' message to phagocytes (23,119). This action helps ensure the body's greater good, because swift removal of doomed cells helps the body avoid infection, abnormal immune activation (autoimmunity), unnecessary inflammation and even the spread of malignant cells. PS is therefore indispensable to the physiology of cell turnover (119).

2.2.13 Phosphatidylserine (PS) and phosphatidylethanolamine (PE) – fundamental to blood coagulation and clot formation

Often when the body is injured, bleeding occurs. Survival requires that the bleeding be stopped as quickly as possible, and the processes of coagulation and clotting make this happen (1). Blood coagulation and clotting (CC) are tightly coupled, and their fundamental mechanism resembles that for the elimination of dead or dying cells just described, since they also feature externalization of PS on the cell surface.

Blood coagulation is typically initiated when events of tissue injury activate platelets at the injury site (1,6). Platelets are cell fragments that circulate with the blood – smaller than other cells, lacking a nucleus but enclosed by a typical bilayer outer membrane (6). As a platelet sustains damage due to some form of tissue injury, PS molecules flop from the inner half of its OM to the outer half. Phosphatidylethanolamine (PE), also normally located in the inner half, also flops to the outer half (24,120).

As they accumulate in the outer half of the platelet's OM, some of these PS and PE molecules come to protrude from the platelet surface into the outer environment. These protruding molecules then can make direct contact with protein 'coagulation factors' that circulate with the blood. These factors are actually enzymes, known as factors VIIa, IX and X (1). Factor VIIa interacts with PS exposed on the platelet surface, then (in the presence of calcium ions) this factor VIIa – PS complex interacts with the other factors to drive the coagulation cascades that ultimately result in clot formation (120).

Though PS has long been thought to be important for the cascade of coagulation and clotting, experimentally the cascade was found to require a higher concentration of PS than is likely to occur under physiological conditions. This puzzle was solved when it was discovered that the presence of PE dramatically lowers this PS requirement (24,120). Exposure of PE on the surface of the damaged platelet allows for a relatively small amount of PS to successfully trigger clotting. As much as 4% of the platelet's total content of PS+PE can appear on its surface during this process (24).

The central role of PS in coagulation and clotting is clinically striking in Scott syndrome. This is a human genetic defect that causes platelets to externalize less PS when stimulated, and markedly diminishes the individual's capacity for coagulation and clotting (121). Scott syndrome emphasizes the physiological importance of PS working in synergy with PE, the coagulation factors and calcium to drive the survival cascades of coagulation and clot formation.

2.2.14 Concluding remarks

The phospholipids found in lecithins permeate every aspect of human physiology. Their amphipathic molecular structure combined with their other unique properties make them fundamental constituents of membranes, the cell components that power life as we know it. Phospholipids in our cell membranes mediate stimulation, activation, eating, drinking, secretion, excretion, detoxication and even the genetic programming of our cells. Phospholipids' molecular organization also destined them to be major organizing components of the lipoprotein particles.

On the more integrated level of tissue organization, individual PLs make important contributions to electrical signal transmission, for efficient disposal of unwanted cells, to guard against toxin build-up, and to protect against haemorrhage following an injury. At the organ level, the unique properties of PL help the lungs, joints, digestive system, kidneys and liver do their important work. The pivotal importance of PLs for the emulsifying actions of bile fluid foresaw their importance as commercial emulsifiers.

Wherever there is physiology there can be pathophysiology – abnormal function and disease. Drawing on the documented contributions of PL to physiology, generations of researchers have developed a considerable range of applications of lecithins as dietary supplements for enhancing human health, as reviewed in Chapter 6.6.

References

1. Hall, J.E., 2016. Guyton and Hall Textbook of Medical Physiology. 13th Edition, International. Philadelphia, PA, Elsevier.
2. Schneider, M., Schmitt, H., 2013. Nutritional aspects of polar lipids from milk phospho- and glycolipids. Agro FOOD Industry Hi Tech 24, 50–52.
3. Schneider, M., 2006. Phospholipids – what's new? Lipid Tech. 18, 250–255.
4. Kidd, P.M., 2019. PhosphatidylSerine (PS). In Pizzorno, J., Murray, M. (Ed.), Textbook of Natural Medicine, Fifth Edition. New York, Elsevier.
5. Hellhammer, J., Vogt, D., Franz, N., Freitas, U., Rutenbergothers, D., 2014. A soy-based phosphatidylserine/phosphatidic acid complex (PAS) normalizes the stress reactivity of hypothalamus-pituitary-adrenal-axis in chronically stressed male subjects: a randomized, placebo-controlled study. Lipids Health Dis. 13, 121.
6. Alberts, B., Johnson, A., Lewis, J., Morgan, D., 2015. Molecular Biology of the Cell. 6th Edition. New York, Garland Science.
7. Kuellenberg, D., Taylor, L.A., Schneider, M., Massing, U., 2012. Health effects of dietary phospholipids. Lipids Health Dis. 11(3).
8. Nicolson, G.L., 2014. The fluid-mosaic model of membrane structure: still relevant to understanding the structure, function and dynamics of biological membranes after more than 40 years. Biochim. Biophys. Acta. 1838, 1451–1466.
9. Lu, W., Claypool, S.M., 2015. Disorders of phospholipid metabolism: an emerging class of mitochondrial disease due to defects in nuclear genes. Frontiers Genetics 6, Article 3.
10. Van Meer, G., Voelker, D.R., Feigenson, G.W., 2008. Membrane lipids: where they are and how they behave. Nat. Rev. Mol. Cell Biol. 9, 112–124.
11. Escriba, P.V., Gonzalez-Ros, J.M., Goni, F.M., Kinnunen, P.K.J., Vigh, L., Sanchez-Magraner, L., Fernandez, A.M., Busquets, X., Horvath, I., Barcelo-Coblijn, G., 2008. Membranes: a meeting point for lipids, proteins and therapies. J. Cell Mol. Med. 12, 829–875.
12. Liu, S., Alexander, R.K., Lee, C.H., 2014. Lipid metabolites as metabolic messengers in inter-organ communication. Trends Endocrinol. Metab. 25(7), 356–363.
13. Czubowicz, K., Jesko, H., Wencel, P., Lukiw, W.J., Strosznajder, R.P., 2019. The role of ceramide and sphingosine-1-phosphate in Alzheimer's disease and other neurodegenerative disorders. Mol. Neurobiol. Retrieved from doi.org/10.1007/s12035-018-1448-3.
14. Daniels, T.F., Killinger, K.M., Michal, J.J., Wright Jr., R.W., Jiang, Z., 2009. Lipoproteins, cholesterol homeostasis and cardiac health. Int. J. Biol. Sci. 5(5), 474–488.
15. Zeisel, S.H., 2012. Dietary choline deficiency causes DNA strand breaks and alters epigenetic marks on DNA and histones. Mutation Res. 733, 34–38.
16. Agassandian, M., Mallampalli, R.K., 2013. Surfactant phospholipid metabolism. Biochim. Biophys. Acta. 1831(3), 612–625.
17. Gibson, P.R., Muir, J.G., 2005. Reinforcing the mucus: a new therapeutic approach for ulcerative colitis? Gut 54, 900–903.
18. Schneider, H., Braun, A., Fuellekrug, J., Stremmel, W., Ehehalt, R., 2010. Lipid based therapy for ulcerative colitis – modulation of intestinal mucus membrane phospholipids as a tool to influence inflammation. Int. J. Mol. Sci. 11, 4149–4164.
19. Kosinska, M.K., Ludwig, T.E., Liebisch, G., et al., 2015. Articular joint lubricants during osteoarthritis and rheumatoid arthritis display altered levels and molecular species. PLoS ONE 10(5), e0125192 (18).
20. Vecchio, P., Thomas, R., Hills, B.A., 1999. Surfactant treatment for osteoarthritis (letter to the editor). Rheumatology (Oxford) 38(10), 1020-1.
21. Hoffmann, E.K., Lambert, I.H., Pedersen, S.F., 2009. Physiology of cell volume regulation in vertebrates. Physiol. Rev. 89, 193–277.
22. Reshetnyak, V.I., 2013. Physiological and molecular biochemical mechanisms of bile formation. World J. Gastroenterol. 19(42), 7341–7360.
23. Yang, H., Chen, Y.-Z., Zhang, Y., et al., 2015. A lysine-rich motif in the phosphatidylserine receptor PSR-1 mediates recognition and removal of apoptotic cells. Nature Communications 6, 5717 (29).
24. Clark, S.R., Thomas, C.P., Hammond, V.J., et al., 2013. Characterization of platelet aminophospholipid externalization reveals fatty acid as molecular determinants that regulate coagulation. PNAS 110(15), 5875–5880.

25. Hulbert, A.J., 2007. Membrane fatty acids as pacemakers of animal metabolism. Lipids 42, 811–819.
26. Abbott, S.K., Else, P.L., Atkins, T.A., Hulbert, A.J., 2012. Fatty acid composition of membrane bilayers: importance of diet polyunsaturated fat balance. Biochim. Biophys. Acta. 1818, 1309–1317.
27. Contreras, F.X., Ernst, A.M., Wieland, F., Brugger, B., 2011. Specificity of intramembrane protein-lipid interactions. Cold Spring Harbor. Perspect. Biol. 3, a004705 (18).
28. Landreh, M., Robinson, C.V., 2015. A new window into the molecular physiology of membrane proteins. J. Physiol. 593(2), 355–362.
29. Radoicic, J., Lu, G.J., Opella, S.J., 2014. NMR structures of membrane proteins in phospholipid bilayers. Q. Rev. Biophys. 47, 249–283.
30. Cornelius, F., Habeck, M., Kanai, R., et al., 2015. General and specific lipid-protein interactions in Na,K-ATPase. Quelle: https://www.sciencedirect.com/science/article/pii/S0005273615000917] Biochim. Biophys. Acta. 1848(9), 1729–1743.
31. Sturgeon, R.M., Baenziger, J.E., 2010. Cations mediate interactions between the nicotinic acetylcholine receptor and anionic lipids. Biophys. J. 98, 989–998.
32. Shinzawa-Itoh, K., Aoyama, H., Muramoto, K., et al., 2007. Structures and physiological roles of 13 integral lipids of bovine heart cytochrome c oxidase. EMBO J. 26, 1713–1725.
33. Stefkova, J., Polende, R., Hubacek, J.A., 2004. ATP-binding cassette (ABC) transporters in human metabolism and diseases. Physiol. Res. 53, 235–243.
34. Daleke, D.L., 2003. Regulation of transbilayer plasma membrane phospholipid asymmetry. J. Lipid. Res. 2003, 233–242.
35. Hankins, H.M., Baldridge, R.D., Xu, P., Graham, T.R., 2015. Role of flippases, scramblases, and transfer proteins in phosphatidylserine subcellular distribution. Traffic 16(1), 35–47.
36. Krebs, J.J.R., Hauser, H., Carafoli, E., 1979. Asymmetric distribution of phospholipids in the inner membrane of beef heart mitochondria. J. Biol. Chem. 254(12), 5308–5316.
37. Brand, M.D., 2014. The role of mitochondria in longevity and lifespan. Longevity & Healthspan 3, 7 (3).
38. Schenkel, L.C., Bakovic, M., 2014. Formation and regulation of mitochondrial membranes. Int. J. Cell Biol. Vol. 2014, Article ID 709828 (13).
39. Paradies, G., Paradies, V., De Benedictis, V., et al., 2014. Functional role of cardiolipin in mitochondrial bioenergetics. Biochim. Biophys. Acta. 1837(4), 408–417.
40. Reynolds, S., 2015. Successful management of Barth syndrome. J. Multidisc. Healthcare 8, 345–358.
41. Swerdlow, R.H., 2014. Bioenergetic medicine. Brit. J. Pharmacol. 171, 1854–1869.
42. Balla, T., 2013. Phosphoinositides: tiny lipids with giant impact on cell regulation. Physiol. Rev. 93, 1019–1137.
43. Hammond, G.V., Balla, T., 2015. Polyphosphoinositide binding domains: key to inositol lipid biology. Biochim. Biophys. Acta. 1851(6), 746–758.
44. Shore, D.M., Reggio, P.H., 2015. The therapeutic potential of orphan GPCRs, GPR35 and GPR55. Frontiers Pharmacol. 6, Article 69 (22).
45. Chung, K.Y., 2013. Structural aspects of GPCR-G protein coupling. Toxicol. Res. 29, 149–155.
46. Im, D.S., 2013. Intercellular lipid mediators and GPCR drug discovery. Biomol. Ther. 21, 411–422.
47. Cattaneo, F., Guerra, G., Parisi, M., et al., 2014. Cell-surface receptors transactivation mediated by G protein-coupled receptors. Int. J. Mol. Sci. 15, 19700–19728.
48. Jastrzebska, B., Goc, A., Golczak, M., Palczewski, K., 2009. Phospholipids are needed for proper formation, stability and function of the photoactivated rhodopsin-transducin complex. J. Biochem. 48(23), 5159–5170.
49. Lands, B., 2014. Historical perspective on the impact of n-3 and n-6 nutrients on health. Prog. Lipid Res. 55, 17–29.
50. Korbecki, J., Baranowska-Bosiacka, I., Gutowska, I., Chlubek, D., 2014. Cyclooxygenase pathways. Acta Biochimica Polonica 61, 639–649.
51. Mashima, R., Okuyama, T., 2015. The role of lipoxygenases in pathophysiology; new insights and future perspectives. Redox Biology 6, 297–310.
52. Isobe, Y., Arita, M., 2014. Identification of novel omega-3 fatty acid-derived bioactive metabolites based on a targeted lipidomics approach. J. Clin. Biochem. Nutr. 55, 79–84.
53. Dalli, J., Colas, R.A., Serhan, C.N., 2013. Novel n-3 immunoresolvents: structures and actions. Scientific Reports 3, 1940 (13).
54. Orr, S.K., Palumbo, S., Bosetti, F., et al., 2013. Unesterified docosahexaenoic acid is protective in neuroinflammation. J. Neurochem. 127, 378–393.

55. Kozina, A., Opresnik, S., Wong, M.S.K., et al., 2014. Oleoyl-lysophosphatidylcholine limits endothelial nitric oxide bioavailability by induction of reactive oxygen species. PLoS ONE 9(11), e113443 (20).
56. Rao, S.P., Riederer, M., Lechleitner, M., et al., 2013. Acyl chain-dependent effect of lysophosphatidylcholine on endothelium-dependent vasorelaxation. PLoS ONE 8(5), e65155 (10).
57. Garcia-Morales, V., Montero, F., Gonzalez-Forero, D., et al., 2015. Membrane-derived phospholipids control synaptic neurotransmission and plasticity. PLoS Biol 13(5), e1002153 (30).
58. Choi, J.W., Chun, J., 2013. Lysophospholipids and their receptors in the central nervous system. Biochim. Biophys. Acta 1831(1), 20–32.
59. Frasch, S.C., Bratton, D.L., 2012. Emerging roles for lysophosphatidylserine in resolution of inflammation. Progr. Lipid Res. 51(3), 199–207.
60. Frasch, S.C., Fernandez-Boyanapalli, R.F., Berry, K.A., et al., 2013. Neutrophils regulate tissue neutrophilia in inflammation via the oxidant-modified lipid lysophosphatidylserine. J. Biol. Chem. 288(7), 4583–4593.
61. Wu, C.S., Chen, H., Sun, H., et al., 2013. GPR55, a G-protein coupled receptor for lysophosphatidylinositol, plays a role in motor coordination. PLoS ONE 8(4), e60314 (14).
62. McIntyre, T.M., 2012. Bioactive oxidatively truncated phospholipids in inflammation and apoptosis; formation, targets, and inactivation. Biochim. Biophys. Acta 1818, 2456–2464.
63. Yost, C.C., Weyrich, A.S., Zimmerman, G.A., 2010. The platelet activating factor (PAF) signaling cascade in systemic inflammatory responses. Biochimie J. 92(6), 692–697.
64. Yag, D., 2014. Macrophage derived platelet activating factor implicated in the resolution phase of gouty inflammation. Int. J. Inflamm. Vol. 2014, Article ID 5264696 (6).
65. Hammond, J.W., Lu, S.M., Gelbard, H.A., 2016. Platelet activating factor enhances synaptic vesicle exocytosis via PKC, elevated intracellular calcium, and modulation of Synapsin 1 dynamics and phosphorylation. Frontiers Cell. Neurosci. 9, Article 505 (13).
66. Furst, H., Saba, J.D., 2010. An update on sphingosine-1-phosphate and other sphingolipid metabolites. Nat. Chem. Biol. 6(7), 489–497.
67. Mahajan-Thakur, S., Bohm, A., Jedlitschky, G., et al., 2015. Sphingosine-1-phosphate and its receptors: a mutual link between blood coagulation and inflammation. Mediators Inflamm. Vol. 2015, Article ID 831059 (11).
68. Lai, Y., Lou, X., Diao, J., Shin, Y.K., 2015. Molecular origins of synaptotagmin 1 activities on vesicle docking and fusion pore opening. Scientific Reports 5, 9267 (7).
69. Lai, Y., Shin, Y.K., 2012. The importance of an asymmetric distribution of acidic lipids for synaptotagmin 1 function as a Ca2+ sensor. Biochem. J. 443, 223–229.
70. Dashti, M., Kulik, W., Hoek, F., et al., 2011. A phospholipidomic analysis of all defined human plasma lipoproteins. Scientific Reports 1, 139 (11).
71. Kuksis, A., Shaikh, N.A., Hoffman, A.G.D., 1979. Lipid absorption and metabolism. Environ. Health Perspec. 33, 45–55.
72. Parthasarathy, S., Subbaiah, P.V., Ganguly, J., 1974. The mechanism of intestinal absorption of phosphatidylcholine in rats. Biochem. J. 140, 503–508.
73. Iqbal, J., Hussain, M.M., 2009. Intestinal lipid absorption. Am J. Physiol. Endocrinol. Metab. 296, E1183–E1194.
74. Harrison, E.H., 2012. Mechanisms involved in the intestinal absorption of dietary vitamin A and provitamin A carotenoids. Biochim. Biophys. Acta 1821(1), 70–77.
75. Yao, Z., Vance, D.E., 1989. Head group specificity in the requirement of phosphatidylcholine biosynthesis for very low density lipoprotein from cultured hepatocytes. J. Biol. Chem. 264(19), 11373–11380.
76. Zeisel, S.H., Mar, M.H., Howe, J.C., Holden, J.M., 2003. Concentrations of choline-containing compounds and betaine in common foods. J. Nutr. 133, 1302–1307.
77. Gimenez, M.S., Oliveros, L.B., Gomez, N.N., 2011. Nutritional deficiencies and phospholipid metabolism. Int. J. Mol. Sci. 12, 2408–2433.
78. Rye, K.A., Barter, P.J., 2014. Cardioprotective functions of HDLs. J. Lipid Res. 55, 168–179.
79. Martinez-Beamonte, R., Lou-Bonafonte, J.M., Martinez-Gracia, M.V., Osada, J., 2013. Sphingomyelin in high-density lipoproteins: structural role and biological function. Int. J. Mol. Sci. 14, 7716–7741.

80. Fisher, E.A., Feig, J.E., Hewing, B., et al., 2012. HDL function, dysfunction, and reverse cholesterol transport. Arterioscler. Thromb. Vasc. Biol. 32, 2813–2820.
81. Mora, S., Glynn, R.J., Ridker, P.M., 2013. High-density lipoprotein cholesterol, size, particle number, and residual vascular risk after potent statin therapy. Circulation 128, 1189–1197.
82. Feig, J.E., Hewing, B., Smith, J.D., et al., 2014. High-density lipoprotein and atherosclerosis regression. Circ. Res. 114, 205–213.
83. Wiedeman, A.M., Barr, S.I., Green, T.J., et al., 2018. Dietary choline intake: current state of knowledge across the life cycle. Nutrients 10, 1513 (24).
84. Blumberg, J.B., Frei, B., Fulgoni, V.L., et al., 2018. Contribution of dietary supplements to nutritional adequacy by socioeconomic subgroups in adults of the United States. Nutrients 10, 4 (11).
85. Zeisel, S.H., 2013. Nutrition in pregnancy: the argument for including a source of choline. Int. J. Women's Health 5, 193–199.
86. Wurtman, R.J., Hirsch, M.J., Growdon, J.H., 1977. Lecithin consumption raises serum free-choline levels. Lancet 310(8028), 68–69.
87. Lee, H.C., Fellenz-Maloney, M.P., Liscovitch, M., Blusztajn, J.K., 1993. Phospholipase D-catalyzed hydrolysis of phosphatidylcholine provides the choline precursor for acetylcholine synthesis in a human neuronal cell line. Proc. Natl. Acad. Sci. USA 90, 10086–10090.
88. D'Elia, R.V., Harrison, K., Oyston, P.C., et al., 2013. Targeting the "cytokine storm" for therapeutic benefit. Clinical Vaccine Immunol. 20(3), 319–327.
89. Dwyer, J.B., McQuown, S.C., Leslie, F.M., 2009. The dynamic effects of nicotine on the developing brain. Pharmacol. Ther. 122(2), 125–139.
90. Picciotto, M.R., Higley, M.J., Mineur, Y.S., 2012. Acetylcholine as a neuromodulator: cholinergic signaling shapes nervous system function and behaviour. Neuron 76(1), 116–129.
91. Shaw, G.M., Carmichael, S.L., Yang, W., et al., 2004. Periconceptional dietary intake of choline and betaine and neural tube defects in offspring. Am J. Epidemiol. 160(2), 102–109.
92. Wu, B.T.F., Dyer, R.A., King, D.J., et al., 2012. Early second trimester maternal plasma choline and betaine are related to measures of early cognitive development in term infants. PLoS ONE 7(8), e43448 (8).
93. Zeisel, S.H., 2012. Diet-gene interactions underlie metabolic individuality and influence brain development: implications for clinical practice derived from studies on choline metabolism. Ann. Nutr. Metab. 60(Suppl. 3), 19–25.
94. Stead, L.M., Brosnan, J.T., Brosnan, M.E., et al., 2006. Is it time to reevaluate methyl balance in humans? Am. J. Clin. Nutr. 83, 5–10.
95. Mato, J.M., Martinez-Chantar, M.L., Lu, S.C., 2013. S-adenosylmethionine metabolism and liver disease. Ann. Hepatol. 12(2), 183–189.
96. McKully, K.S., 2007. Homocysteine, vitamins, and vascular disease prevention. Am. J. Clin. Nutr. 86(Suppl.), 1563S–1568S.
97. Schalinske, K.L., Smazal, A.L., 2012. Homocysteine imbalance: a pathological metabolic marker. Adv. Nutr. 3, 755–762.
98. Cederbaum, A.I., 2010. Hepatoprotective effects of s-adenosyl-l-methionine against alcohol- and cytochrome P450 2E1-induced liver injury. World J. Gastroenterol. 16, 1366–1376.
99. Olthof, M.R., van Vliet, T., Verhoef, P., et al., 2005. Effect of homocysteine-lowering nutrients on blood lipids: results from four randomised, placebo-controlled studies in healthy humans. PLoS Medicine 2(5), e135 (11).
100. Han, S., Mallampalli, R.K., 2015. The role of surfactant in lung disease and host defense against pulmonary infections. Ann. Am. Thoracic. Soc. 12(5), 765–774.
101. Numata, M., Kandasamy, P., Voelker, D.R., 2012. Anionic pulmonary surfactant lipid regulation of innate immunity. Expert Rev. Respir. Med. 6(3), 243–246.
102. Numata, M., Nagashima, Y., Moore, M.l., et al., 2013. Phosphatidylglycerol provides short-term prophylaxis against respiratory syncytial virus infection. J. Lipid Res. 54, 2133–2143.
103. Kandasamy, P., Zarini, S., Chan, E.D., et al., 2011. Pulmonary surfactant phosphatidylglycerol inhibits *Mycoplasma pneumoniae*-stimulated eicosanoid production from human and mouse macrophages. J. Biol. Chem. 286(10), 7841–7853.
104. Numata, M., Kandasamy, P., Nagashima, Y., et al., 2015. Phosphatidylinositol inhibits respiratory syncytial virus infection. J. Lipid Res. 56(3), 578–587.

105. Tovey, I., 2015. Role of dietary phospholipids and phytosterols in protection against peptic ulceration as shown by experiments on rats. World J. Gastroenterol .21(5), 1377–1384.
106. Radosz-Komoniewska, H., Bek, T., Jozwiak, J., Martirosian, G., 2005. Pathogenicity of *Helicobacter pylori* infection. Clin. Microbiol. Infect. 11, 602–610.
107. Lanza, F.L., Marathi, U.K., Anand, B.S., Lichtenberger, L.M., 2008. Clinical trial: comparison of ibuprofen-phosphatidylcholine and ibuprofen on the gastrointestinal safety and analgesic efficacy in osteoarthritic patients. Aliment. Pharmacol. Ther. 28, 431–442.
108. Seror, J., Zhu, L., Goldberg, R., et al., 2015. Supramolecular synergy in the boundary lubrication of synovial joints. Nature Communications 6, 6497 (7).
109. Burg, M.B., Ferraris, J.D., 2008. Intracellular organic osmolytes: function and regulation. J. Biol. Chem. 283(12), 7309–7313.
110. Wirthensohn, G., Beck, F.-X., Guder, W.G., 1987. Role and regulation of glycerophosphocholine in rat renal papilla. Pfluegers. Arch. 409, 411–415.
111. Gallazzini, M., Burg, M.B., 2009. What's new about osmotic regulation of glycerophosphocholine. Physiology 24, 245–249.
112. Topanurak, S., Ferraris, J.D., Li, J., et al., 2013. High NaCl- and urea-induced posttranslational modifications that increase glycerophosphocholine by inhibiting GDPD5 phosphodiesterase. PNAS 110(18), 7482–7487.
113. Videen, J.S., Michaelis, T., Pinto, P., Ross, B.D., 1995. Human cerebral osmolytes during chronic hyponatremia. A proton magnetic resonance spectroscopic study. J. Clin. Invest. 95, 788–793.
114. Carotenuto, A., Rea, R., Traini, E., et al., 2017. The effect of the association between donezepil and choline alphoscerate on behavioral disturbances in Alzheimer's disease: interim results of the ASCOMALVA trial. J. Alz. Dis. 56(2), 805–815.
115. Wu, F., Wang, J., Pu, C., et al., 2015. Wilson's disease: a comprehensive review of the molecular mechanisms. Int J Mol Sci 16, 6419–6431.
116. Raunio, H., Kuusisto, M., Juvonen, R.O., Pentikainen, O.T., 2015. Modeling of interactions between xenobiotics and cytochrome P450 (CYP) enzymes. Frontiers Pharmacol. 6, Article 123 (14).
117. Esteller, A., 2008. Physiology of bile secretion. World J. Gastroenterol. 14(37), 5641–5649.
118. Coleman, J.A., Quazi, F., Molday, R.S., 2013. Mammalian P4-ATPases and ABC transporters and their role in phospholipid transport. Biochim. Biophys. Acta 1831(3), 555–574.
119. Hochreiter-Hufford, A., Ravichandran, K.S., 2013. Clearing the dead: apoptotic cell sensing, recognition, engulfment, and digestion. Cold Spring Harb. Perspect. Biol. 5, a008748 (19).
120. Tavoosi, N., Davis-Harrison, R.L., Pogorelov, T.V., et al., 2011. Molecular determinants of phospholipid synergy in blood clotting. J. Biol. Chem. 286, 23247–23253.
121. Kojima, H., Newton-Nash, D., Weiss, H.J., et al., 1994. Production and characterization of transformed B-lymphocytes expressing the membrane defect of Scott syndrome. J. Clin. Invest. 94, 2237–2244.

3

Origin and Production of Lecithin

3.1 Production of vegetable lecithins. An overview

Willem van Nieuwenhuyzen

3.1.1 Introduction

Phospholipids (PLs) are important biochemical intermediates in the growth and functioning of cells in plants and animals. All living cells contain PLs. Phospholipids are important constituents of the vegetable lecithins derived commercially from oil-bearing seeds such as soybeans, sunflower kernels and rapeseed (canola). Vegetable lecithins contain predominantly the PL classes phosphatidylcholine (PC), phosphatidylethanolamine (PE), phosphatidylinositol (PI), phosphatidic acid (PA), minor amounts of lysophosphatidylcholine (LPC), occasionally phosphatidylserine (PS), and other glycerol phospholipids of complex fatty acid composition. The molecular structure of the main phospholipids is given in Fig. 1. Animal lecithins are sourced from egg, dairy, milk and marine species including fish and krill (1).

Phospholipids with their surface-active properties are used as emulsifiers in a vast range of food, pharmaceutical, cosmetic, feed and technical applications. PLs also have nutritional functions as sources of organic bound choline, inositol and fatty acids. In biochemistry and medicine, the name lecithin is exclusively given to the 3-sn phosphatidylcholine. In this chapter lecithin is defined as a mixture of phospholipids with adherent glycolipids and triglyceride oil.

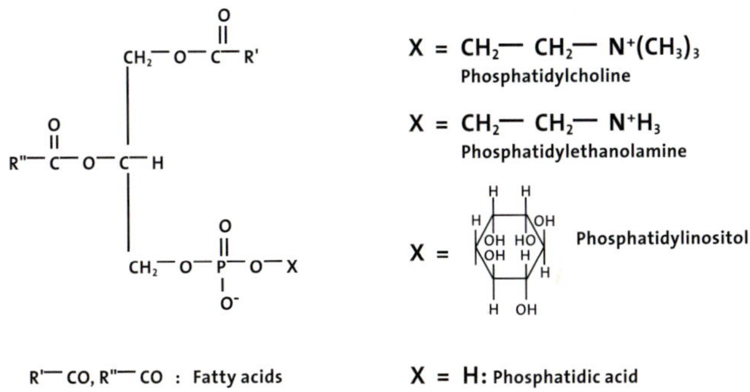

Figure 1.
Molecular structures of the main phospholipids (3).

3.1.2 Occurrence

The volumes of oilseeds grow progressively because of increasing agricultural areas with seed varieties having higher yields per acre/hectare. Vegetable oil from the seeds needs to be refined, in order to make it suitable and pleasant for human consumption. The removal of the phosphatides as lecithin gum, which is the first refining step of the extracted crude oil, is beneficial for economic oil refining. Quantity-wise, vegetable lecithin gums are available in abundance worldwide. In the oil mill the surplus gums are often sprayed on the meal, before the toaster. This addition reduces the protein content in soybean meal (trade specification minimum 44% protein) and high-protein (HP) soybean meal (trade specification minimum 48% protein), which when too high may cause protein quality problems.

For the dedicated uses required by many products and industries, only a part of the available lecithin gum is needed and produced into dried commercial lecithin to comply with high-quality technological properties. Quality-wise, some vegetable lecithin sources are scarce due to special sourcing requirements and the need for identity preservation (2).

Soya or soy lecithin has been the primary source of vegetable lecithin for decades, making it a leader of the food emulsifiers group. Large crops of soybeans grow in regions of North and South America, China and India. Crushing plants are spread all over the world. Quite a number of crushing and refinery plants have production units for processing gums into dried crude lecithin. Basic lecithin with 0.5% yield is a co-product of the soy crushing industry.

European retailer and consumer requirements for non-GM food have induced the search for Identity Preserved (IP) non-GM lecithins since 1996. Currently countries on other continents also require increasing quantities of the IP non-GM lecithins with highest technological functional properties in food and feed. This (arbitrary) requirement has changed lecithin sourcing worldwide. Traditional non-GM soybean crops are becoming scarcer, which presents market opportunities for high-quality IP soy, rapeseed and sunflower lecithins.

Rapeseed lecithin, also called canola lecithin, is derived from the rapeseed plant. In the 1970s new rapeseed varieties with low erucic fatty acid (C22:0) composition were bred, and were marketed in Canada as Canadian Low Erucic Acid Rapeseed (CLEAR). Later on, rapeseed plant varieties with low glucosinolate content followed, as 'double zero' rapeseed with positive nutritive performance in food (oil) and feed (meal). These products are traded as 'canola' in North America, while elsewhere the name 'rapeseed' is mostly used. Today rapeseed crops grow in Canada, Europe, China and India in areas with good soil and favourable weather conditions, and from this canola lecithin is produced, depending on market demand.

Interestingly, lecithin from the traditional high erucic rapeseed varieties was already being processed in Germany and Eastern European countries from the 1940s, for use in technical industries and sometimes also in feed and food. The low lecithin content of less than 1% needed in feed and food recipes made the product attractive as an economically priced alternative to soy lecithin.

Sunflower lecithin is processed from the black oil-bearing sunflower seed varieties. Whereas high wax content in these lecithin gums caused viscous handling problems in the past, improved seed treatment and degumming technology now allows the processing of good-quality sunflower lecithin with low wax content. Sunflower oil crushing is often located in the harvesting regions, with sunflower lecithin produced in increasing quantities in Argentina, Central Europe (Hungary) and Eastern Europe (Ukraine, Russia). Now sunflower lecithin with its tailor-made qualities is established in the European food market.

Table 1.

Lecithins produced from vegetable sources.

Lecithin source	Market availability	Potential	Limitations
Soybeans	Large over decades	Soy (GMO quality) is largest oil seed crop. Gums available in surplus	IP non-GM lecithin quality is scarce for food use in some regions
Sunflower seeds	Growing availability	High; some state-of-the-art plants	High food use interest
Canola/rapeseed	Limited commercial quantities	Sufficient raw material; process optimization possible	Greenish colour; critical sensory, flavour characteristics

3.1.3 Processing technologies for vegetable lecithin

Lecithin is recovered from the crude oil (Fig. 2) and processed in four subsequent unit operations (Fig. 3), which are extensively described in a review publication (3).

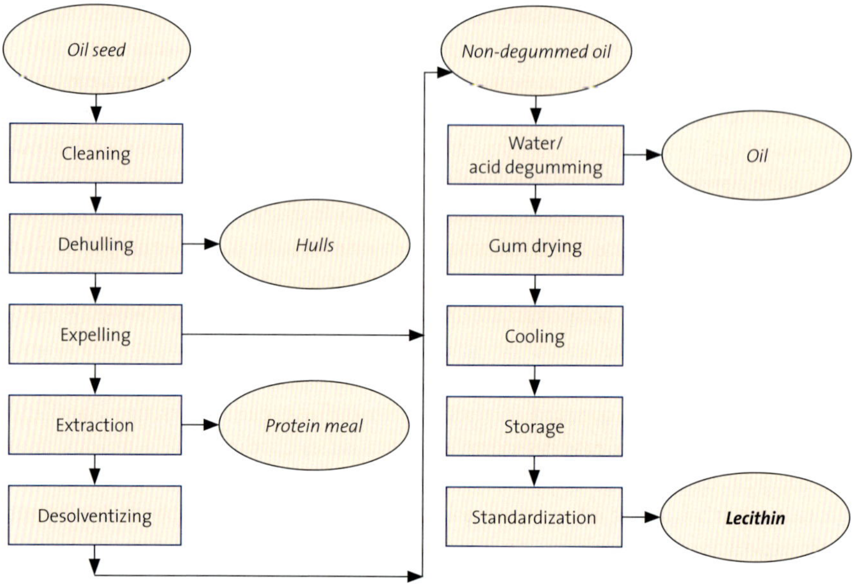

Figure 2.
Vegetable lecithin recovery from plant seeds.

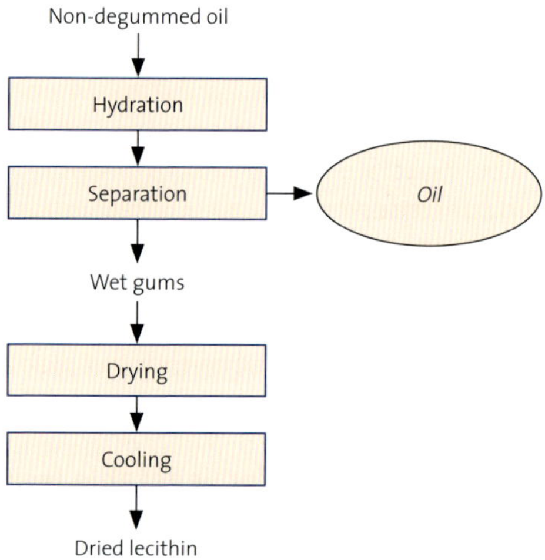

Figure 3.
Block diagram of lecithin processing.

Hydration of phospholipids

In the degumming process the polar phospholipids hydrate within one hour to form a gum with a higher specific density than the oil. Water (1–3%) is mixed thoroughly with the oil at 50–80°C. The easily hydratable phospholipids are PC, PI and LPC, while PE and PA have low hydrating properties and are therefore marked as non-hydratable phospholipids (NHP). Dry bean or seed storage and efficient front-end dehulling of beans or seeds prior to grinding/flaking are important for keeping the phospholipase D (PLD) enzyme activity low in the seed. This prevents the undesired hydrolysis of PC (hydratable) into PA (non-hydratable) and promotes smooth degumming of the oil. Water degumming gives a crude oil with 50–100 ppm residual phosphorus. In practice, a mix of the various phospholipids in the gum is obtained. The remaining NHP (PE, PA) in the water-degummed oil can be made hydratable in a second step with the citric acid through so-called super-degumming (4), or enzymatic degumming with the use of the phospholipases PLA_1 and PLA_2 (5). The principles of degumming for the removal of phospholipids from the oil are still a topic of research and economic-technological considerations. These optimized degumming processes result in oils with residual 5–10 ppm P, to be used in economically efficient physical refining. Mostly the hydrolysed lecithin gums from the second step are not utilized for lecithin production, for economic and technical reasons.

Separation of lecithin gums

After the hydration period the gums are separated from the oil in continuously operating centrifuges at 50–70°C. The adjustment of optimal back-pressure conditions in the centrifuge gives efficient degumming yields with a crude soy oil with maximal 0.3% NHP (maximal 100 ppm P) and a lecithin gum with a low oil content (30–39% oil on dry basis). Gums often contain maximum 50% water, minimum 33% acetone-insoluble matter and maximum 17% oil. Leading manufacturers of dedicated centrifuges are Alfa Laval and GEA Westfalia, while engineering companies may also mention other equipment producers. Gums from acid and enzymatic degumming processes have a lower viscosity (more liquid) than gums from water degumming. Reasons are the modified phospholipid structures and changed gum composition.

Drying

Lecithin is dried to a content of <1% moisture, achieving long-time shelf life and convenient liquidity. Batch, semi-batch and continuous drying film evaporators are used.

Commercial lecithin production was invented by Hermann Bollmann, Germany, in 1921, who developed a rotating cylindrical ball with heating coil as reactor to evaporate the 40–50% water from the gums at 70–80°C and 25–50 mbar vacuum in 6–10 hours, to a residual moisture content of 0.5–2% (6). Bollmann dryers were used for gum drying until the 1970s.

Today the DE SMET type batch dryer is made out of a horizontal cylindrical jacketed body with a dome. Hot water circulates in the jacket. A heavy-duty, rotating coil mounted on a shaft provides the necessary heat, while at the same time it agitates the lecithin gums. The dryer operates in a 6- to 8-hour cycle with 4–6 hours' drying under low vacuum and 1 hour final drying under high vacuum. In Latin America and other continents this type of batch dryer is still frequently installed, as it requires low investment costs.

Often long batch drying times lead to severe darkening due to Maillard reaction of the adherent sugars and Amadori reaction between sugars and PE. In the 1960s European oil mills switched from batch Bollmann dryers to continuous rotating horizontal thin-layer evaporators. Thin-layer evaporators have the advantage of high performance capacity per unit drying surface of for example 70–90 kg gums/sqm/h and a short drying time of 1–2 minutes, which is adequate for achieving a good colour quality (7). Experience showed that a gap of several millimetres between rotor and stator was beneficial for achieving a stable thickness of the film. The viscosity of the gums during drying increases rapidly, with highest values at 5–15% residual water (Fig. 4). This is due to a change in the structure of the gums during the transition from O/W to W/O organization. Further drying to below 1% moisture reduces viscosity values and gives a stable microbial and shelf life quality. Typical process conditions are: heating temperature 90–110°C, rotor–stator gap of 1–2 millimetres, residence time 1–2 minutes and vacuum 30–150 mbar.

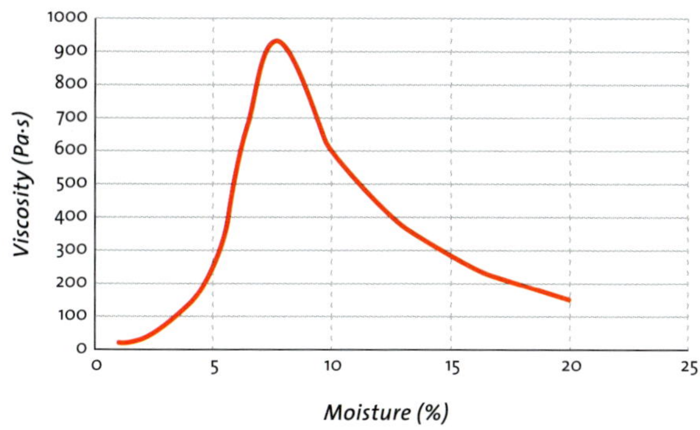

Figure 4.
Rheological behaviour of lecithin gums during drying (7).

North American oil mills often use vertical thin-film evaporators, which operate very well for the production of light-coloured lecithins. Recently an Argentinean equipment manufacturer offered a falling film evaporator in which the rotor with flexible scrapers agitates and transports the material. Unfortunately, comparative data on the lecithin colour quality between the various types of driers are not published.

Lecithin has a naturally brown colour. This colour can be attributed to its component carotenoids, melanoids and porphyrins in addition to the Maillard and Amadori reactions. Slight bleaching of preferably the gums prior to drying with a 35% H_20_2 solution is possible, but the product should comply with the EU E322 additive regulation for food uses with the legal specification POV max. 10 mEq/kg. Food processors may even desire a lower peroxide value for achieving optimal shelf life properties of the processed foods. Thus the bleaching effect on colour is limited, but it has a positive impact on the microbiological count.

Cooling

Rapid cooling of the lecithin to below 50°C is effective to prevent post-darkening by Maillard reactions. Therefore, it is advised to use a heat exchanger just after the drier. Otherwise, the results of careful drying will be counteracted by post-darkening during a slow cooling regimen from 90–100°C to ambient temperature. For bulk storage a temperature of around 40°C under dry conditions is recommended. It is advised to use adequate tanks with stirring facilities for keeping stored lecithin homogeneous. At 20–30°C lecithin can easily be stored for over a year without significant changes in product quality and functional properties.

3.1.4 Differences in soy, rapeseed and sunflower lecithin processing and composition

Soy lecithin
Most vegetable lecithin is derived as gums from soybean oil. Extracted soybean oil contains 800–1,200 ppm phosphorus, equivalent to 2–3% commercial lecithin with an acetone-insoluble content of 60%.

A number of oil mills have improved the filtration of the extracted oil-hexane miscella in continuous operating filtration units, with filtration/bleaching earth for optimizing the production of lecithins with low residual dirt (earth, seed fines) expressed as hexane (HI) or toluene insoluble (TI), complying with the legal specification of 0.3%. Well-filtered lecithins with a transparent glossy appearance have HI far below 0.1% and a low iron content of Fe <50 ppm. Intensive filtration of the non-degummed oil also supports the cleanness of the centrifuges in the degumming step. Highly clarified lecithin products can be obtained only by filtering the crude oil by these plate and frame filters, whereby the dry oil is heated to 80°C and 0.1% filter aid is added. It is not possible to effectively filter the high viscous turbid lecithin on plant scale for the removal of residual proteins to yield a visually clear lecithin.

Rapeseed lecithin
Lecithin recovery from rapeseed starts with seed preparation, often followed by expelling (pressing) to squeeze out the seed oil content from 42% down to 15–20% in the press cake; afterwards the grinded cake is usually hexane-extracted for removing the oil to below 0.5% (Fig. 1). The press oil and extraction oil streams are combined for degumming, in which process step the lecithin gums are separated in continuous decanters. As with soy lecithins, well-filtered lecithins with a transparent glossy appearance have HI far below 0.1% and a low iron content of Fe <50 ppm. Also, as with soy lecithins it is not possible to effectively filter high-viscosity turbid lecithin at large plant scale operation for achieving a visually clear rapeseed lecithin.

Rapeseed contains chlorophyll components, as a result of which the rapeseed lecithin colour may have a greenish-brown tone. The natural tocopherol content is about 400 ppm, consisting of 100 ppm α-tocopherols and 200 – 300 ppm γ-tocopherols. Rapeseed lecithin yields are about 0.2 – 0.3% on seed basis, compared to 0.5% for soy lecithin from soy on bean basis.

Rapeseed lecithin can be standardized by adjusting the acetone-insoluble content to AI = 62 – 64% by oil and fatty acid addition, giving a liquid consistency at room temperature.

Research in the 1960s and 1970s with rapeseed lecithin gums established good potential for modification of rapeseed lecithin by hydrolysis and alcohol fractionation, as a potential functional emulsifier in food (8). Full acetylation of the ethanol-insoluble fraction gives products with improved emulsifying properties (9).

Sunflower lecithin

Sunflower lecithin is produced in a similar way as rapeseed lecithin, usually combining the pressed oil and extracted oil streams for the degumming step. In addition to the monograph of Pardun from 1988 (8), production and modification processes are also reported by Hollo (10). High-quality sunflower lecithins are now available on the market. Some decades ago the waxes from the sunflower hulls were often concentrated in the lecithin, contributing to a paste-like consistency with difficult dosing properties. Nowadays the separation of the hulls from the sunflower seeds (dehulling) prior to extraction is more efficient, giving also a sunflower lecithin with low wax content and convenient viscosity properties. Sunflower lecithin can be modified by enzymatic hydrolysis, making it suitable as a functional emulsifier (11,12).

Table 2 gives the typical total composition of soy lecithin in comparison to sunflower and rapeseed lecithin. The PL composition of the three varieties does not differ that much, which is also due to the quite similar oil extraction processes. The fatty acid pattern largely follows the fatty acid composition of the triglycerides, as shown in Table 3. This fatty acid division is measured in the liquid lecithin, consisting of phospholipids and triglycerides.

Table 2.
Average phospholipid compositions of vegetable lecithins (2,13; modified).

Phospholipids	Rapeseed lecithin %	Soy lecithin %	Sunflower lecithin %
Phosphatidylcholine	17	15	16
Phosphatidylethanolamine	9	11	8
Phosphatidylinositol	10	10	14
Phosphatidic acid	4	4	3
Other phospholipids	6	7	6
Total: All phospholipids	**46**	**47**	**47**

Table 3.
Average fatty acid compositions of vegetable lecithins (2,13).

Fatty acid	Rapeseed lecithin %	Soy lecithin %	Sunflower lecithin %
Palmitic acid (C16:0)	7	16	11
Stearic acid (C18:0)	1	4	4
Oleic acid (C18:1)	56	17	18
Linoleic acid (C18:2)	25	55	63
Linolenic acid (C18:3)	6	7	0
Others	5	1	4

3.1.5 Future trends

Lecithin is a co-product of the oil seeds milling industry, a sector which has undergone a strong concentration of ownership of oil mills in the last decades. The availability of large volumes of lecithin gums from seed oil degumming and refining challenges the further economic valorization of lecithins as sustainable surface-active agents for their many applications in the food, feed, cosmetic, pharmaceutical and technical industries.

European sourcing requirements for Identity Preserved (IP) non-GM soy lecithin for the European food market has drastically changed the lecithin world market supply system since 1996. Traditional non-GM soybean availability will become scarcer, which presents a market opportunity for high-quality IP rapeseed and sunflower lecithins.

Phosphatidylcholine fractions with specific essential Ω-3/6/7/9 fatty acid profiles, sourced from milk, fish, krill, fungi and algae, are in the R&D pipeline or have reached the early market stage for use as nutritional ingredients and health supplements. Extensive clinical studies are required for proving and substantiating of health claims, and the resulting dossiers have to be approved by the regulatory safety authorities before these dedicated lecithins can be successfully positioned in the market.

References

1. Schneider, M., 2008. Major sources, composition and processing. In Gunstone, F.D. (Ed.), Phospholipid Technology and Applications. The Oily Press, Bridgwater, 21-39.
2. Van Nieuwenhuyzen, W., 2014. The changing world of lecithins. INFORM 25(4), 254-259.
3. Van Nieuwenhuyzen, W., Tomás, M.C., 2008. Update of vegetable lecithin and phospholipid technologies. Eur. J. Lipid Sci. Tech. 5, 472-486.
4. Ringers, H.J., Seegers, J.C., 1977. Degumming process for triglyceride oils. United States Patent 4,049,686.
5. Buchold, H., Boensch, R., Schroeppel, J., 1994. Process for enzymatic degumming of vegetable oil. European Patent 0654527.
6. Bollmann, H., 1921. German Patent 382912.
7. Van Nieuwenhuyzen, W., 1976. Lecithin production and properties. JAOCS 53, 425-427.
8. Pardun, H., 1988. Die Pflanzenlecithine. Verlag für chemische Industrie H. Ziolkowsky, Augsburg, 271-272, 247, 286, 296.
9. Sosada, M., Pasker, B., Gabzdyl, R., 2003. Optimization by full factorial design of the emulsifying properties of ethanol insoluble fraction from rapeseed lecithin. Eur. J. Lipid Sci. Tech. 105, 672-676.
10. Hollo, J., Peredi, J., Ruzics, A., Jeranek, M., Erdelyi, A., 1993. Sunflower lecithin and possibilities for utilisation. JAOCS 70, 997-1001.
11. 't Hooft, C. , Van den Kommer, M., Seegers, J.C., 2005. Food composition suitable for shallow frying comprising sunflower lecithin. European Patent EP1607003.
12. Cazebas, D.M., Madoery, R., Diehl, B.W.K., Tomás, M.C., 2011. Application of enzymatic hydrolysis on sunflower lecithin using a pancreatic PLA2. JAOCS 88, 443-446.
13. Van Nieuwenhuyzen, W., 2015. Production and Utilization of Natural Phospholipids. In Ahmad, M., Xu, X., Polar Lipids, Biology, Chemistry, and Technology. AOCS Press, Urbana, Illinois.

3.2 Animal lecithins/phospholipids from egg and milk

Michael Schneider

Traditionally, commercially available natural phospholipids were – with the exception of egg and milk phospholipids – almost exclusively made from vegetable lecithins, by products of large-scale vegetable oil refining. Lecithins are complex mixtures of phospholipids, optionally containing glyco- and neutral lipids, obtained by physical procedures from vegetable and animal foodstuffs (E322).

Different from those relatively low-cost vegetable by-products, phospholipids of animal origin have to be selectively extracted by source-specific solvent processes from animal biomasses, for example from egg yolk, buttermilk, fish roe or krill. This has of course serious consequences for the cost of those products.

There is a clear discrepancy in volume and market recognition between animal and vegetable phospholipids. To a certain extent this can be explained by the price differences. The plant-derived lecithin and phospholipid products found applications as affordable and effective food additives, for technological as well as cost reasons. Today egg and milk phospholipids play important roles mainly because of their specific physiological properties and for certain selected pharmaceutical uses. This chapter reviews their compositions and clinical usefulness.

3.2.1 Egg lecithins/phospholipids

3.2.1.1 Raw materials and processing

Lecithin was discovered by Gobley in 1846/47 when he identified a phosphorus-containing lipid in egg yolk. Following his application of the Greek word *lekithos* for egg yolk to his preparation, the term *lecithin* became popular.

Nowadays egg lecithins with their complex phospholipid make-up are mainly produced from dried egg yolk by solvent (acetone or ethanol) extraction, and sometimes even by column chromatography. The many attempts to use liquid egg yolk as raw material, thus saving drying costs, have failed so far. One problem is the fact that phospholipids are firmly associated with the protein fractions of egg yolk, requiring a polar solvent such as ethanol to break up this association and liberate the phospholipids.

The schematic diagram of egg lecithin processing in Fig. 1 can stand as an example for animal lecithin processing in general, though several less important alternatives have also found industrial application.

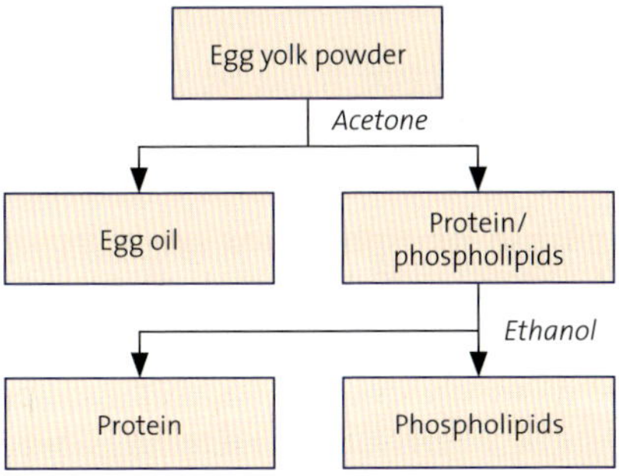

Figure 1.
Egg lecithin processing.

Alternatively egg yolk powder can be extracted directly with ethanol, leading to an extract containing phospholipids and egg oil in a ratio of approximately 30 to 70.

3.2.1.2 Composition

The phospholipid as well as the fatty acid composition of different commercial lecithins is very much source-specific, as shown in Tables 1 and 2. Important to note is the fact that egg phospholipids – whether from chicken or fish eggs – are particularly rich in phosphatidylcholine (PC).

Table 1.
Polar lipid compositions of different lecithins (on oil-free basis).

	Soy wt %	Egg wt %	Milk wt %	Fish (salmon roe) wt %
PC	23	73	27	82
PE	22	18	22	4
PI	14	2	8	3
PA	7	–	–	–
PS	Traces	–	12	1
SPM	–	3	27	2
Glycolipids	12	–	7	–

PC = Phosphatidylcholine, PE = Phosphatidylethanolamine, PI = Phosphatidyilnositol, PA = Phosphatidic acid, PS = Phosphatidylserine, SPM = Sphingomyelin

Table 2.
Fatty acid compositions of different lecithins.

	Soy wt %	Egg wt %	Milk wt %	Fish (salmon roe) wt %
Saturated	22	41	55	17
Monounsaturated	12	35	35	21
Polyunsaturated	66	24	10	62

Even though soy and fish phospholipids are quite rich in polyunsaturated fatty acids, the difference is that fish phospholipids mainly contain omega-3 fatty acids whereas soy contains predominantly omega-6 fatty acids.

3.2.1.3 Nutritional applications of egg phospholipids

The more or less exclusive nutritional application of egg phospholipids is infant feeding formulas. The commercial product of choice is a compound almost identical to the natural lipid composition of egg yolk with approximately 70% triglycerides (egg oil) and 30% phospholipids, obtained by ethanol extraction of egg yolk powder.

The main reason to incorporate egg lipids into infant formula products is their content of two long-chain polyunsaturated fatty acids (LC-PUFA), namely the omega-6 arachidonic acid (ARA, C 20:4 ω-6) and the omega-3 docosahexaenoic acid (DHA, C 22:6 ω-3), as shown in Fig. 2. Both of these are exclusively bound to the phospholipid backbone and are not substantially present in the triglyceride fraction.

In human breast milk these fatty acids occur naturally and have fundamental importance for the cells of the developing baby, as well as being clinically proven nutrients for the proper development of the infant's eyes, brain and immune system.

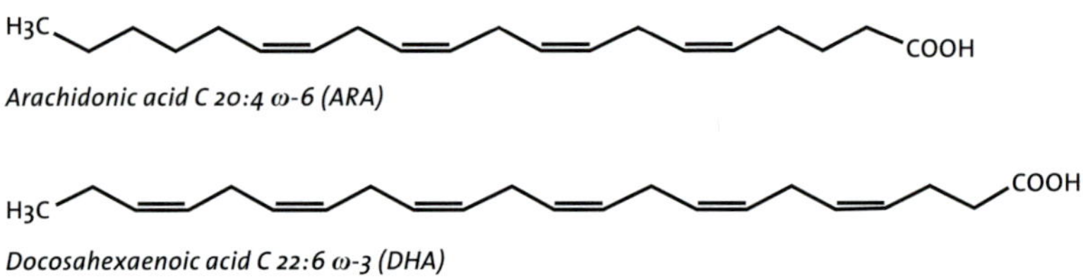

Figure 2.
Molecular layouts of ARA and DHA.

Though in principle both ARA and DHA can be self-synthesized by the human body, starting from the essential fatty acids linoleic acid (C 18:2 ω-6) and α-linolenic acid (C 18:3 ω-3), the complex enzyme system required is not yet fully developed in newborn babies (1). The following schematic (Fig. 3) shows the sequence of this enzymatic desaturation and chain elongation process:

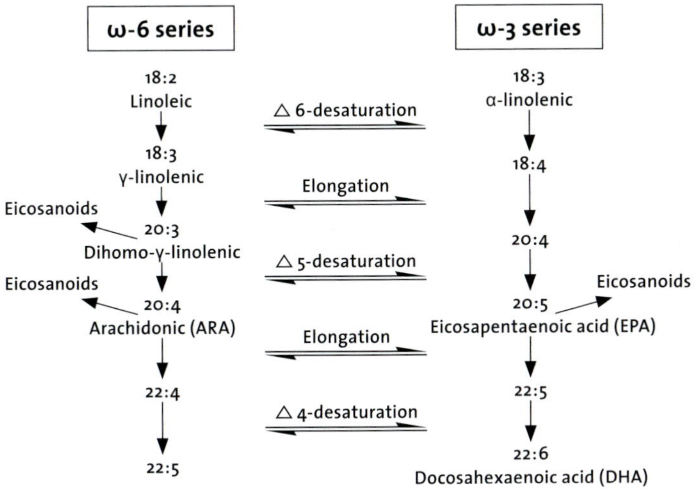

Figure 3.
Fatty acid elongation and desaturation (2).

ARA is a precursor of highly bioactive metabolites, so-called eicosanoids, which function as hormone-like lipid mediators regulating a large number of cell and metabolic functions such as inflammation and blood pressure.

DHA is an important building block for cell membranes in the brain tissues, the retina of the eye, and other tissues, and is a major contributor to optimal fluidity and overall functionality of all human cell membranes (3). It is the most abundant fatty acid in the optic nerve, and is centrally involved in the transmission of signals by the photoreceptors of the human eye. Looking at the high number of clinical studies investigating the effect of DHA addition into infant formula products, it is more than obvious that the development of visual acuity is significantly better with DHA supplementation, as compared to a non-DHA supplemented product (4-6). DHA supplementation increases the visual maturation especially in preterm infants; it helps photoreceptor differentiation and the activation of the visual pigment rhodopsin (7). DHA supplementation is also of great importance during pregnancy, because the fast-developing brain and nervous system of the fetus has very substantial requirements for DHA in order to develop cell membranes and link them into synaptic connections (8).

Usually breastfeeding supplies the baby with sufficient amounts of ARA and DHA by way of the mother. Artificial infant formula however, based on cow's milk components and vegetable oils, contains no or only hardly measurable contents of ARA and DHA.

Consequently the baby's requirements are unlikely to be satisfied. Even though such products contain a reasonable amount of linoleic and linolenic acid, which theoretically could serve as precursors, their conversion to ARA and DHA is minimal since – as already mentioned – the baby's enzymatic system doesn't function properly (1). Because of these reasons ARA and DHA are considered "essential" nutrients for preterm and even full-term babies.

Human milk contains only small amounts of ARA and DHA, but these are significant. ARA occurs at approximately 0.15 grams per 100 grams and DHA at approximately 0.09 grams per 100 grams of infant formula solids. Especially with regard to ARA and DHA it is important for infant formula products to get as close as possible to the amounts found in mothers' milk.

Egg lipid preparations are attractive for addition to infant formulas, as they are well suited to facilitate fortification with both ARA and DHA in the required amounts and ratios (Table 3).

Table 3.

Arachidonic and docosahexaenoic acid content of typical egg lipid preparations (30% phospholipids).

Fatty acid	% of total fatty acids	g/100 g of egg lipids
Arachidonic acid (C 20:4, ω-6)	min. 1.80	1.58
Docosahexaenoic acid (C 22:6, ω-3)	min. 0.90	0.79

As sources of nutritional phospholipids and fatty acids, egg lipids are closely comparable to lipids from mothers' milk. Thanks to their phospholipid composition, their content of ARA and DHA, their enrichment with choline (from phosphatidylcholine) and their natural cholesterol content, the egg lipids very closely mimic the breast milk lipid profile.

The similarity in the molecular positioning of ARA, DHA and other LC-PUFAs between the breast milk and egg lipids is consistent with their highly efficient uptake and metabolism in the infant – functionally distinct from the LC-PUFA-containing triglycerides coming from microalgae sources via fermentation (Fig. 4). Animal studies indicate that the brain uptake of both ARA and DHA is approximately twofold higher when these are supplied in phospholipid form, compared to triglyceride form (9).

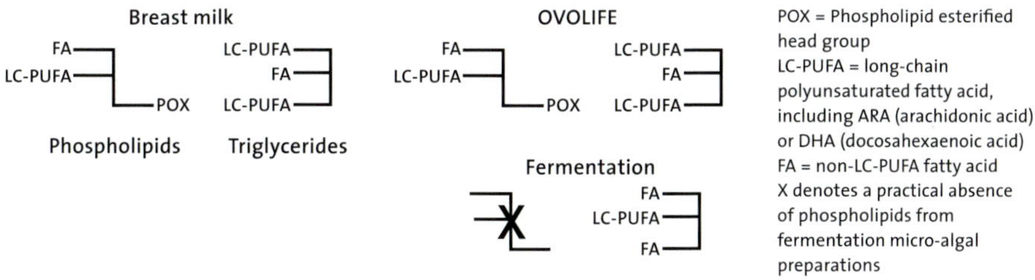

Figure 4.

Fatty acid molecular positioning in human breast milk lipids, similarity to a typical egg lipid ('OVOLIFE'), and difference from a fermentation (microalgae) lipid preparation.

Some randomized human trials have examined the effects of postnatal dietary LC-PUFAs on specific measures, such as visual attention, problem-solving and language skills (10,11). The results from a number of trials with preterm and term infants are still not conclusive, but overall there is substantial evidence that supplementation with ARA and DHA can improve the behavioural development of infants. It is often speculated that better visual acuity might promote faster and better development of higher brain functions, like problem-solving capability.

Egg lipids are easy to use in the process of manufacturing infant formula. They are amber-coloured, viscous fluids which easily mix with the other lipids of the recipe. Their phospholipid content is 30% and their cholesterol ranges between 3.0 and 3.5%. Their choline content (in the form of PC, which is highly bioavailable) is 3.5 grams per 100 grams.

Besides their importance to normal physiology, egg phospholipids serve as surfactant emulsifiers that augment the phospholipid surfactants of the healthy gastrointestinal tract. Clinical studies indicate they protect the baby's stomach and intestinal mucosa against diarrhea and the life-threatening condition of necrotic enterocolitis, both often observed in preterm babies (12,13).

3.2.1.4 Pharmaceutical applications of egg phospholipids

Oil-free egg phospholipid preparations are traditionally used as emulsifiers in intravenous fat emulsions (10 or 20% oil) for parenteral nutrition – the by far biggest share of phospholipid pharmaceutical use. Two types of egg lecithins are commercially available for this purpose – one with phospholipid composition similar to that naturally occurring in egg yolk (73% phosphatidylcholine), the other slightly concentrated via fractionation, with an 80% PC content.

These emulsions serve mainly as sources of energy with a balanced fatty acid profile for patients who are dependent on intravenous nutrition. More modern parenteral emulsions can contain a more optimized fatty acid profile due to enrichment with omega-3 fatty acids from fish oils.

Second-generation emulsions containing egg phospholipids are on the market for use as carrier vehicles for lipophilic drugs, for example the anaesthetic propofol dissolved inside the dispersed lipid droplets.

3.2.1.5 Summary

Egg lecithin phospholipids are close to ideal lipid preparations for the nutritional enrichment of infant formula products. Not only are they a functionally attractive combination of ARA and HA LC-PUFAs, choline and cholesterol, but also these are in natural rather than chemically modified forms and have superior stability and bioavailability. Egg lecithin phospholipids are safe ingredients with a long history of clinical application in human and infant nutrition.

3.2.2 Milk phospholipids

Milk phospholipids are a new class of natural phospholipids of animal origin, now commercially available. They offer a broad spectrum of both technological and physiological properties which are unique to this kind of polar lipid extract and which are totally different in composition and application profile from all other phospholipid products on the market.

3.2.2.1 What are milk phospholipids made from?

As raw cow's milk contains only 0.03% phospholipids, milk fractions like for example butter serum, buttermilk or cream are more appropriate sources for commercial production.

These materials are significantly higher in their content of the so-called milk fat globule membrane (MFGM) – the part of milk specifically carrying the phospholipids.

The MFGM membrane surrounding and stabilizing the dispersed milk fat droplets has an extremely complex organization (Fig. 5). It consists of three layers containing a complex composition of proteins, peptides, enzymes, nucleotides, cholesterol, glycolipids (carbohydrate-containing polar lipids) and also phospholipids (14).

Because of the low phospholipid content of more or less all the milk raw materials, including the milk fractions mentioned, traditional solvent extraction processes as applied for other phospholipid sources would have been too complicated and especially too expensive to allow commercial uses. Therefore new technologies had to be developed to obtain products at reasonable purity and cost.

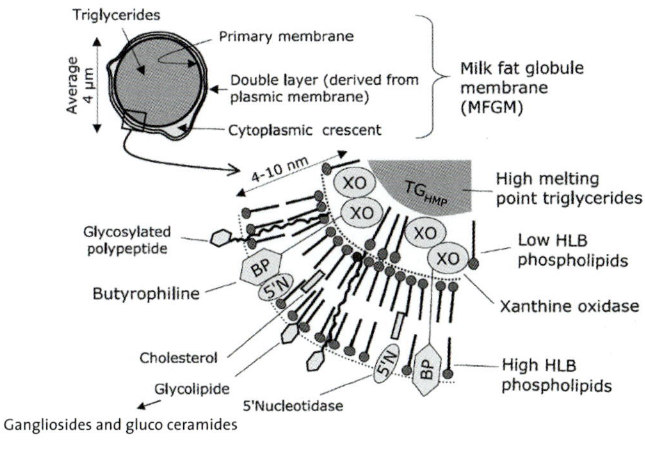

Figure 5.
Milk fat globule membrane (MFGM).

The process of choice is a multistep approach to concentrate the polar lipids (phospho- and glycolipids) from the milk fat globule membrane – in the first step using standard milk processing technology like ultra- and diafiltration (15) leading to polar lipid concentrates of 20 or even some more per cent. Several process alternatives are described especially in the patent literature (16–18). In the case of highly concentrated phospholipids (more than 50% content), solvent processing is unavoidable.

A serious complication for the manufacture of milk phospholipids is the low phospholipid concentration in milk, even in milk fractions. Substantial volumes of raw material are needed to produce commercial quantities of phospholipid concentrates (for a summary of the process, see Fig. 6). This is a strong barrier for small or medium-sized dairy companies with too small volumes of suitable raw materials.

A typical milk phospholipid concentrate made with solvent-free membrane processing is a fully natural polar lipid concentrate of milk fat globule membrane (MFGM) components.

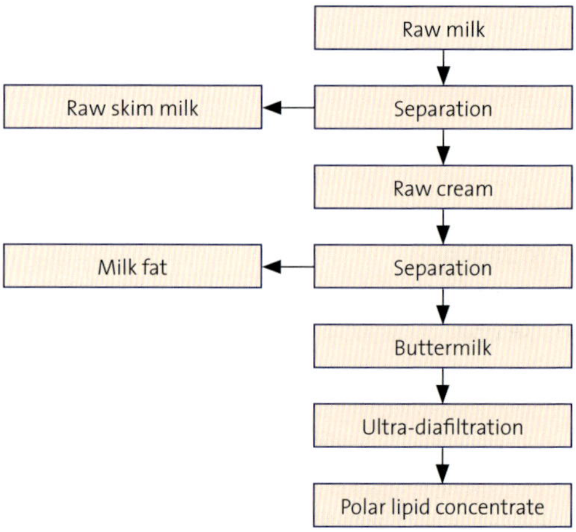

Figure 6.
Milk phospholipid processing.

As already shown before (see Table 1), milk polar lipid composition is significantly different from all other commercial lecithin and phospholipid products, both in phospholipid and polar lipid pattern but also fatty acid profile. Both differences make these products very attractive for a variety of new and innovative applications.

Table 4.

Composition of a commercial 20% milk polar lipid concentrate.

Phospholipids	18	Gangliosides	0.6
PC (Phosphatidylcholine)	5	Sat. fatty acids	55
PE (Phosphatidylethanolamine)	5	Mono-unsat. fatty acids	35
PI (Phosphatidylinositol)	1	Poly-unsat. fatty acids	10
PS (Phosphatidylserine)	2	Protein	50
SPM (Sphingomyelin)	5	Lactose	10
Ceramides	2	Triglycerides	12
		Minerals	8

3.2.2.2 Technological applications of milk phospholipids

Because of the relatively high degree of saturated (approximately 55%) or mono-nounsaturated fatty acids (approximately 35%), milk phospholipids are – importantly for food and cosmetic applications – quite stable against oxidation and very stable against hydrolytic breakdown in aqueous environments. In consequence the contribution to the taste and flavour profile of a finished product is not negatively affected by oxidation or liberated free fatty acids (as with soya phospholipids for example).

An interesting application of low-concentrated milk phospholipids (e.g. 20%) is the use as efficient emulsifiers in a variety of food applications like ice cream, coffee creamers and other dairy-based compositions where they can substitute synthetic emulsifiers, avoid stabilizers and would allow green labelling (19). The taste and smell of these concentrates are absolutely comparable to milk powder.

As the products contain a significant portion of protein they can also substitute whole milk powder in milk chocolate, improving its nutritional profile. As will be shown in the physiological properties Chapter 2.2, one can think about a much wider range of functional foods making not only use of the technological properties but especially of the multitude of attractive nutritional benefits (20).

Milk phospholipids/polar lipids are versatile ingredients for functional cosmetics. They are not only excellent emulsifiers, creating a very good and soft skin feel. The ceramides (see Fig. 7) as part of milk polar lipid composition contribute to the intercellular lipid barrier of the horny layer, thus preventing trans-epidermal water loss (21). Ceramides are also known to stop the penetration of foreign agents into the skin, avoiding skin irritation.

Ceramides from milk are currently being clinically investigated for possible activity against the symptoms of atopic dermatitis.

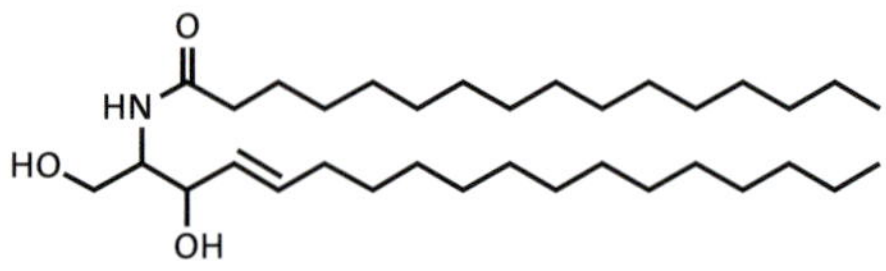

C16-Ceramide

Figure 7.
A Ceramide.

In milk polar lipid concentrates the ceramides are bound to low molecular carbohydrate structures (glucosyl- and lactosylceramides).

Higher concentrated, especially isolated milk phospholipid products can be used to prepare efficient liposomal systems with good entrapment stability (22).

Liposomes are tiny, artificial cell-like structures, surrounded by either one (unilamellar) or many (multilamellar) phospholipid double layers (Fig. 8). Inside the vesicular structure, or between the different phospholipid membranes, water compartments are entrapped, carrying and protecting water-soluble actives. But also lipophilic payloads can be entrapped and protected from the environment inside the lipophilic fatty acid domains of the phospholipid double layers (23).

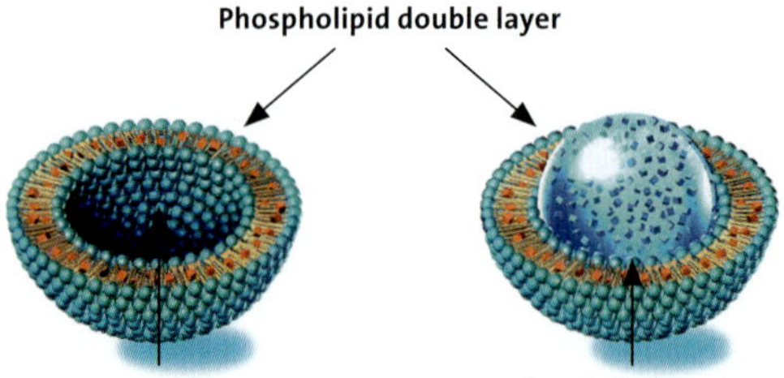

Figure 8.
Liposome structure.

Liposomes are versatile delivery systems, originally for pharmaceutical actives, but also widely used in cosmetics.

Milk phospholipids have clear advantages over all other phospholipids so far used for liposome production because the content of sphingomyelin as well as the high amount of saturated fatty acids create a quite rigid, less permeable membrane structure. The benefits for liposome preparations are the stability against oxidation and a moderate so-called phase transition temperature of approximately 28°C. The phase transition temperature is the temperature at which the membrane undergoes a transition from an organized fatty acid domain (a kind of crystalline structure) to an unorganized one (a kind of liquid structure). At ambient temperatures milk phospholipid liposome membranes are therefore crystalline with excellent entrapment characteristics and stability – at higher temperatures (e.g. at skin temperature) they tend to release their payload. Liposomes are a simple approach to protect sensitive ingredients and to release them under targeted conditions.

3.2.2.3 Nutritional properties of milk phospholipids

Specific phospholipids

Milk phospholipids are particularly rich in three nutritionally important phospholipids, namely phosphatidylcholine (PC), sphingomyelin (SPM) and phosphatidylserine (PS).

PC and PS are so-called glycerophospholipids, different from SPM, a sphingophospholipid, where the backbone is not glycerol but the long-chain amino alcohol sphingosine (Fig. 9).

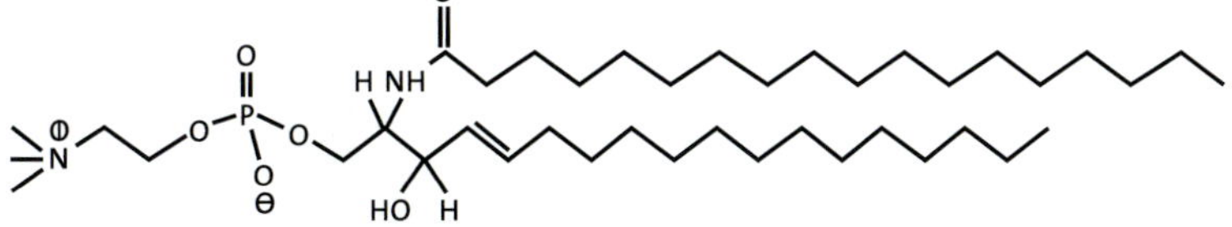

Figure 9.
Structure of sphingomyelin (SPM).

Phosphatidylcholine (PC) serves as a highly bioavailable source of the vitamin-like nutrient choline, being an important part of all infant formula recipes (24) for example. Milk PC also has a well demonstrated effect to protect the mucosa of stomach and intestine against aggressive substances with toxic potential like non-steroidal anti-inflammatory drugs (NSAIDs) like aspirin or indomethacin (25,26), to treat inflammatory bowel disease, and to prevent or even cure chronic ulcerative colitis (27,28). The importance of phospholipids for digestive health is well documented but it is not clear if this is simply a physicochemical function of mucosa protection or if there are other underlying biochemical mechanisms.

Some literature reports about PC's cholesterol reducing function, obviously especially when it is saturated (29). That means milk PC with its high degree of saturation should be more effective than the unsaturated PC from soy.

Phosphatidylserine (PS) has an outstanding record of more than 60 clinical studies as nutrient of choice to improve cognitive functions and to alleviate stress symptoms (30). The first PS clinical investigations were done using an extract from bovine brain cortex (31). But with the outbreak of BSE, the so-called 'mad cow' disease, bovine PS could no longer be certified completely safe for use. An alternative was developed from soybean lecithin using enzymatic head group exchange (32). This soy PS has since proven safe and effective in numerous clinical trials. For a further, in-depth analysis of PS as a dietary supplement, refer to Chapter 6.6.

Many clinical studies have shown that there is no difference in clinical performance between brain- and soy-derived PS, even though there are substantial differences in their fatty acid profiles. That means fatty acids likely don't play a major role in the physiological activity of PS, which could now also be confirmed with studies using milk-derived PS (33,34).

Phosphatidylserine currently is the best nutrient of choice for attempts to revitalize memory, learning, concentration and even vocabulary skills – all these functions decline with age. PS is by far best-studied phospholipid for brain performance (35).

As people age they lose some sharpness in memory and the other higher-level cognitive functions; this has been called age-associated memory impairment (AAMI) and related cognitive decline (ARCD). AAMI and ARCD likely affect the large majority of the over-50 population, gradually developing from early adulthood. Individuals suffering from this condition might experience problems at work and in their personal lives as their ability to perform demanding everyday memory tasks is negatively affected.

A clinical trial by Crook and colleagues on AAMI individuals addressed these issues and used name/face acquisition and remembering phone numbers as their primary study parameters. After just three weeks of PS supplementation to subjects with memory problems, after comparing the PS group with the statistically observed decline in a control population of the same age range, they concluded that PS supplementation had 'rolled back the clock' of their cognitive age by as much as 12 years (Fig. 10) (36).

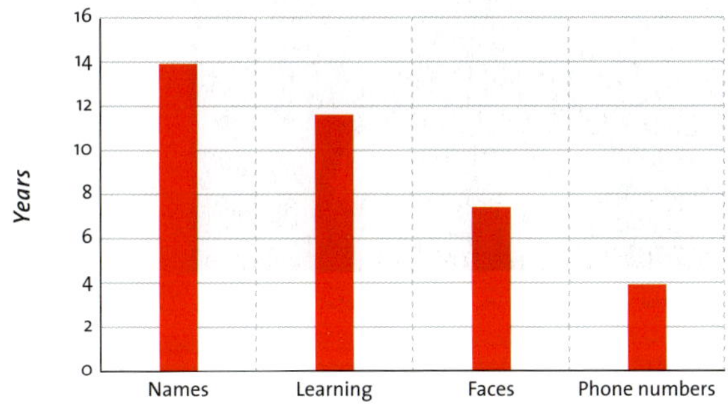

Figure 10.
Improvement of memory skills after PS treatment. From Crook et al. (36).

But PS also benefits other brain activities like coping with stress and fighting depression. In young and healthy men it lowered the production of stress hormones linked to strenuous exercise, and it also eased stress-related mood symptoms in the elderly. Quite recent studies with soy PS supplementation showed stress-dampening effects and improvements in concentration parameters in younger and healthy study groups. It is interesting to note that the effects have been more pronounced in patients with introverted personal behaviour (37).

Still today there are no convincing published data about the metabolism, absorption and body distribution of PS in humans. However we have a number of indirect indicators of how PS is working.

Phosphatidylserine helps the brain to process energy. Our brains require a lot of energy to carry out their functions properly. Even though the human brain weighs only about 1.35 kg (about 3 lbs) it uses about 20% of all the body's energy at rest and perhaps as much as 60% when we are doing hard mental work. As researchers can take advantage of sophisticated imaging instruments like positron emission tomography (PET) to measure brain energy generation, it could be shown that after oral intake of PS energy turnover in human brains was drastically increased.

Under normal physiological conditions the human brain gets its energy predominantly from glucose metabolism. In one study with Alzheimer's patients, PET imaging impressively showed how PS could increase brain glucose metabolism over a period of weeks (Fig. 11) (38).

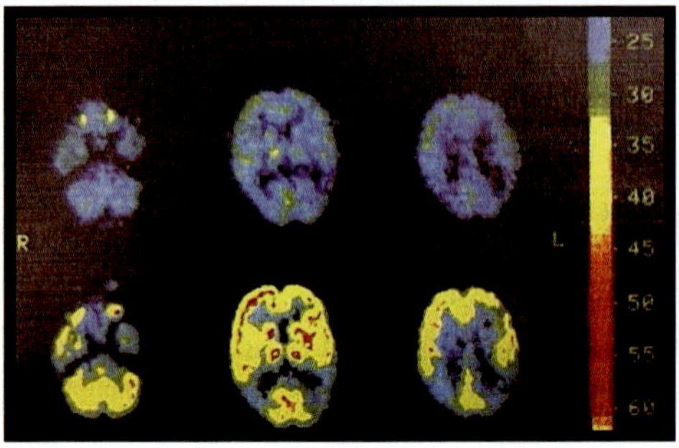

Before PS (low activity)

After 500 mg PS (high activity)

(Blue colour = low, yellow = medium, red = high metabolic activity)

Figure 11.

Stimulation of glucose metabolism in an Alzheimer's brain, after 500 mg PS for three weeks. From Klinkhammer et al. (38).

In animal brains, the increase of glucose metabolism levels after PS administration may be as much as threefold. Another effect of PS, as observed in animals, is that it stimulates acetylcholine synthesis and triggers the release of neurotransmitters from the vesicles at the synaptic gap (Fig. 12). Synapses tend to be enriched in PS, which is essential for their functioning (for details, refer to Chapter 2.2).

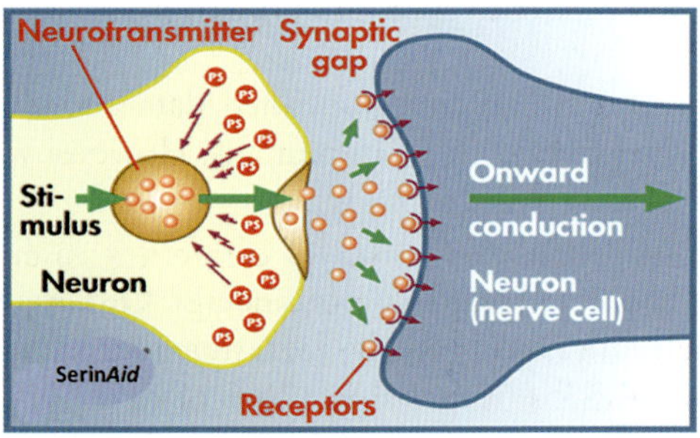

Figure 12.

Overview of neurotransmitter release and transport.

This effect on acetylcholine may explain why physical performance (39) was significantly improved in cycling tests (40). Also in golfers concentration and improved play performance was observed (41). Daily dosages of PS in most clinical trials ranged from 100 to 300 mg per day, and up to 800 mg per day when physical performance and stress were studied.

Sphingomyelin (SPM) is known as a precursor of a number of highly biologically active metabolites such as ceramides (fatty acid amides of sphingosine), sphingosine and sphingosine phosphates. These metabolites function as a kind of second messenger and – amongst a long list of other biochemical effects – influence the behaviour of growth factors, induce cell apoptosis (42) and contribute to an anti-colon cancer effect, as has been shown with experimental animals (43,44).

Another aspect confirmed in human clinical trials is SPM's cholesterol-lowering activity by inhibiting intestinal absorption of food-based cholesterol. Sphingomyelin – similar to some findings with saturated PC – can reduce LDL cholesterol and elevate HDL cholesterol.

Sphingomyelin also has similar effects as phosphatidylcholine to physically protect stomach and gut mucosa. All these data suggest that sphingolipids represent a 'functional' constituent of food.

3.2.2.4 Nutritional properties of ceramides and gangliosides

The growing interest in milk polar lipid concentrates is not only because of their unique phospholipid composition but specifically for infant nutrition purposes, because of their content of phosphorus-free polar lipids, namely ceramides, ceramide derivatives and especially gangliosides.

Ceramides and cerebrosides

Cerebrosides are carbohydrate derivatives of ceramides, another sphingolipid class of molecules. Cerebrosides are the major glycosphingolipids of human milk. They are found in all dairy MFGMs, where they are of great physiological importance. One of the most studied roles of cerebrosides relates to their function as pro-apoptotic molecules. Apoptosis, a form of programmed cell death, is essential for the maintenance of normal tissue homeostasis and is an important physiological response to many forms of cellular stress (45).

Galactosylceramide was originally described as part of the central nervous system. It is expressed in glial cells and it is the major lipid component of myelin, the protective sheath of axons that insulate the axon from electrical activity. This insulation is necessary to increase the rate of transmission of neuronal signals (46).

artificial infant milk formulas to human milk composition. Currently lacking from such formulas, and though relatively minor by quantity, these nutrients are extremely important to get artificial formulas closer to the 'gold standard' of human milk.

Phospholipids contribute to digestive health, function as a bioavailable choline source, alleviate stress, improve cognition and may even fight colon cancer.

Ceramides suppress cancer cells and improve the myelination of the central nervous system.

Gangliosides protect the baby's digestive tract in many ways, but especially by reducing the burden of pathogenic microorganisms and by binding enterotoxins, thus contributing to improved intestinal immunity. Important also to note is that gangliosides promote the healthy development of the baby's brain.

References

1. Friedman, Z., 1979. Polyunsaturated fatty acids in the low-birth-weight infant. Semin. Perinatol. 3, 341–361.
2. Sprecher, H., 2000. Metabolism of highly unsaturated n-3 and n-6 fatty acids. Biochim. Biophys. Acta 1486, 219-231.
3. Salem, N., Litman, B., Kim, H.Y, et. al., 2001. Mechanisms of action of docosahexaenoic acid in the nervous system. Lipids 36 (9), 945–959.
4. Uauy, R.D., Birch, D.G., Birch, E.E., et.al., 1990. Effect of dietary omega-3 fatty acids on retinal function of very-low-birth-weight neonates. Pediatr. Res. 28 (5), 485-492.
5. Birch, D.G., Birch, E.E., Hoffman, D.R., 1992. Retinal development in very-low-birth-weight infants fed diets differing in omega-3 fatty acids. Vis. Sci. 33, 2365-2376.
6. SanGiovanni, J.P., Berkey, C.S, Dwyer, J.T., Colditz, G. A., 2000. Dietary essential fatty acids, long-chain polyunsaturated fatty acids, and visual resolution acuity in healthy fullterm infants: A systematic review. Early Hum. Dev. 57 (3), 165-188.
7. Hoffman, D.R., Theuer, R.C., Castañeda, Y.S., Wheaton, D.H., et.al., 2004. Maturation of visual acuity is accelerated in breast-fed term infants fed baby food containing DHA-enriched egg yolk. J. Nutr. 134 (9), 2307-2313.
8. Koletzko, B., Cetin, I., Brenna, J. T., 2007. Dietary fat intakes for pregnant and lactating women. Br. J. Nutr. 98 (5), 873 – 877.
9. Wijendran, V., Huang, M.C., Diau, G.Y., Boehm, G., et.al., 2002. Efficacy of dietary arachidonic acid provided as triglyceride or phospholipid as substrates for brain arachidonate accretion in baboon neonates. Pediatr. Res. 51, 265–272.
10. Makrides, M., Neumann, M. A., Simmer, K., Gibson, R. A., 2000. A critical appraisal of the role of dietary long-chain polyunsaturated fatty acids on neural indices of term infants: A randomized, controlled trial. Pediatrics 105 (1) 32-38.
11. Fewtrell, M.S., Morley, R., Abbott, R.A., et.al., 2002. Double-blind, randomized trial of long-chain polyunsaturated fatty acid supplementation in formula fed to preterm infants. Pediatrics 110, 73-82
12. Carlson, S.E., Montalto, M.B., Ponder, D.L., Werkman, S.H., et.al., 1998. Lower incidence of necrotizing enterocolitis in infants fed a preterm formula with egg phospholipids. Pediatr. Res. 44, 491-498.
13. Carlson, S.E., Werkman, S.H., Tolley, E.A., 1996. Effect of long-chain n–3 fatty acid supplementation on visual acuity and growth of preterm infants with and without bronchopulmonary sysplasia. Am. J. Clin. Nutr. 63, 687–697.

14. Dewettinck, K., Rombaut, R., Tienpont, N., et al., 2008. Nutritional and technological aspects of milk fat globule membranes. Int. Dairy J. 18, 436-457.
15. Sachdeva, B., Buchheim, W., 1997. Recovery of phospholipids from buttermilk using membrane processing. Kieler Milchwirtschaftliche Fortschrittsberichte 49, 47-68.
16. WO 2002/034062, Boone, M. Method for Obtaining Products Enriched in Phospho- and Sphingolipids.
17. WO 2006/128465, Arla. PS enriched milk fractions for the formulation of functional foods.
18. WO 2006/041316, Fonterra. Beta-serum dairy products, neutral lipid depleted and/or polar lipid enriched dairy products, and processes for their production.
19. Schneider, M., 2001. Phospholipids for functional foods. Eur. J. Lipid Sci. 103, 98-101.
20. Schneider, M., Schmitt, H., 2013. Nutritional aspects of polar lipids from milk phospho- and glycolipids. Agro Food Ind. Hi Tech. 24, 50-53.
21. Coderch, L., Lopez, O., de la Maza, A., et al., 2003. Ceramides and skin function. Am. J. Clin. Dermatol. 4, 107-129.
22. Thompson, A.K., Singh, H., 2006. Preparation of liposomes from milk fat globule membrane phospholipids using microfluidizer. J. Dairy Sci., 410-419.
23. New RRC, 1990. Liposomes – A Practical Approach, IRL Press, Oxford University Press.
24. Wurtman, R.J., Growdon, J.H., Hirsch, M.J., 1977. Lecithin consumption raises serum-free choline levels. Lancet 2, 68-69.
25. Dial, E.J., Lichtenberger, L.M., 1984. A role for milk phospholipids in protection against gastric acid. Studies in adult and suckling rats. Gastroenterol., 379-385.
26. Lichtenberger, L.M., Ulloa, C., Vanous, A.L., et al., 1996. J. Pharmacol. Exp. Ther., 1221-1227.
27. Stremmel, W., Merle, U., Zahn, A., et al., 2005. Retarded release phosphatidylcholine benefits patients with chronic ulcerative colitis. Gut, 966-971.
28. Stremmel, W., Ehehalt, R., Autschbach, F., et al., 2007. Phosphatidylcholine for steroid-refractory chronic ulcerative colitis: A randomized trial. Ann. Intern. Med., 603-610.
29. Jiang, Y., Noh, S.K., Koo, S.I., 2001. Egg phosphatidylcholine decreases lymphatic absorption of cholesterol in rats. J. Nutr. 131, 2358-2363.
30. Louis-Sylvestre, J., 1999. Phosphatidylserine et troubles mnesiques du sujet ages. Cah. Nutr. Diet. 34, 349-357.
31. Toffano, G., 1987. The therapeutic value of phosphatidylserine effect, in Lecithin: Technological, Biological and Therapeutic Aspects, Hanin, I. (Ed.). Ansell GB, 137-146.
32. DE 19917249C2, Meyer, L. Verfahren zur Herstellung von Phosphatidylserine-Produkten.
33. Hellhammer, J., Waladkhani, A.R., Hero, T., et al., 2010. Effects of milk phospholipid on memory and psychological stress response. Brit. Food J. 112, 1124-1137.
34. Schubert, M., Contreras, C., Franz, N., et al., 2011. Milk-based phospholipids increase morning cortisol availability and improve memory in chronically stressed men. Nutr. Res., 413-420.
35. Kidd, P., 2002. Func. Foods Nutraceut. 12, 30-40.
36. Crook, T., Petrie, W., Wells, C., et al., 1992. Effect of phosphatidylserine in Alzheimer's disease. Psychopharmacol. 28, 61-66.
37. Benton, D., Donohoe, R.T., Sillance, B., et al., 2001. The influence of phosphatidylserine supplementation on mood and heart rate when faced with an acute stressor. Nutr. Neurosci. 4, 169-178.
38. Klinkhammer, P., Szelies, B., Heiss, W.D., et al., 1990. Effect of phosphatidylserine on cerebral glucose metabolism in Alzheimer's disease. Dementia 1, 197-201.
39. Fahey, T.D., Pearl, M.S., 1998. The hormonal and perceptive effects of phosphatidylserine administration during two weeks of resistive exercise-induced overtraining. Biol. Sport 15, 135-144.
40. Kingsley, M.I., Miller, M., Kilduff, L.P., et al., 2006. Effects of phosphatidylserine on exercise capacity during cycling in active males. Sports and Exercise 36, 657-669.
41. Jäger, R., Purpura, M., Geiss, K.R., et al., 2007. The effect of phosphatidylserine on golf performance. J. Int. Soc. Sports Nutr. 4, 15-40.
42. Schmelz, E.M., 2001. Dietary sphingomyelin and other sphingolipids in health and disease. Brit. Nutr. Found. Nutr. Bull. 25, 135-139.
43. Schmelz, E.M., Dillehay, D.L., Webb, S.K., et al., 1996. Sphingomyelin consumption suppresses aberrant colonic crypt foci and increases the proportion of adenomas versus adenocarcinomas in CF1 mice treated with 1,2-dimethylhydrazine: implications for dietary sphingolipids and colon carcinogenesis. Cancer Res. 56, 4936-4941.
44. Berra, B., Colombo, I., Sottocornola, E., et al., 2002. Dietary sphingolipids in colorectal cancer prevention. Eur. J. Cancer Prev. 11, 193-197.

45. Hannun, Y.A., Linardic, C.M., 1993. Sphingolipid breakdown products: antiproliferative and tumor-suppressor lipids. Biochim. Biophys. Acta 21, 223-236.
46. Martinez, M., 1982. Myelin lipids in developing cerebrum, cerebellum and brain stem of normal and undernourished children. J. Neurochem. 39, 1684-1692.
47. Karlsson, K.A., 1989. Animal glycosphingolipids as membrane attachment sites for bacteria. Annu. Rev. Biochem. 58, 309-350.
48. Minagawa, M., Liai, T., Sato, Y., et al., 2005. Repeated administration of alpha-galactosyl ceramide attenuates natural killer T cells. Gan To Kagaku Ryoho 32, 1553-1555.
49. Wu, X.F., Wu, X.X., Guo, W.J., et al., 2012. Cerebroside D, a glucoceramide compound, improves experimental colitis in mice with multiple targets against activated T lymphocites. Toxicol. Appl. Pharmacol. 263, 296-302.
50. Oku, H., Wongtantintharn, S., Iwasaki, H., 2007. Tumor specific cytotoxicity of glucosylceramide. Cancer Chemother. Pharmacol. 60, 767-775.
51. Nakano, T., Sugawara, M., Kawakami, H., 2001. Sialic acid in human milk: composition and functions. Acta Pediatr. Taiwan. 42, 11-17.
52. Laegreid, A., Fuglesang, J., 1986. Human and bovine milk: comparison of ganglioside composition and enterotoxin-inhibitory activity. Pediatr. Res. 20, 416-421.
53. Miklavic, J.J., Schnabl, K.L., 2012. Dietary ganglioside reduces proinflammatory signaling in the intestine. J. Nutr. Metab. 2012, 280-286.
54. Rueda, R., Sabatel, J.L., Maldonado, J., 1998. Addition of gangliosides to an adapted milk formula modifies levels of fecal Escherichia coli in preterm newborn infants. J. Pediatr. 133, 90-94.
55. Laegreid, A., Otnaess, A.B.K., 1987. Trace amounts of ganglioside GM1 in human milk inhibit enterotoxins from Vibrio cholera and Escherichia coli. Life Sci. 40, 55-201262.
56. Rueda, R., Maldonado, J., Narbona, E., et al., 1998. Milk mucosal immunity and the microbiome: Impact on the neonate. Early Hum. Dev. 53, 135-147.
57. Vyas, A.A., Schnarr, R.L., 2001. Brain gangliosides: functional ligands for myelin stability and control of nerve regeneration. Biochemie 83, 677-682.
58. Colombaioni, L., Garcia-Gil, M., 2004. Sphingolipid metabolites in neural signaling and function. Brain Res. Rev. 46, 328-355.
59. McJarrow, P., Schnell, N., Jumpsen, J., et al., 2009. Influence of dietary gangliosides on neonatal brain development. Nutr. Rev. 67, 451-463.
60. Wang, B., 2012. Molecular mechanism underlying sialic acid as an essential nutrient for brain development and cognition. Adv. Nutr. 1, 465-472.
61. Koletzko, B., Aggett, P.J., Bindels, J.G., et al., 1998. Growth, development and differentiation: a functional food science approach. Br. J. Nutr., 35-45.

3.3 Phospholipids of milk

Gerd Konrad

3.3.1 The milk fat globule membrane

Milk lipids exist almost exclusively in the form of fat globules with a diameter of 0.1 – 15 µm, enveloped in an extremely thin milk fat globule membrane (MFGM) of 8 – 10 nm. The latter acts as an emulsifier, prevents flocculation and coalescence of the milk fat and protects it against enzymatic lipolysis. The MFGM undergoes considerable changes as a result of technical treatment such as heating, refrigeration, evaporation or homogenization. Thermal treatment of buttermilk has enormous effects on both the protein composition of the MFGM and the polar lipids. Through thiol – disulfide interchange reactions and disulfide bond formation, mainly $\beta 1$-lactoglobulin and to a much lesser extent α-lactalbumin are bound to the MFGM (1,2,3). Moreover, an approximately 20% loss of polar lipids occurs when milk is heated to 80°C for 3 minutes (1). But opinions differ on the thermostability of the phospholipids. Whereas Morin et al. (3) found less phospholipids in pasteurized buttermilk than in the unheated product, Gassi et al. (4) reported that the proportion of phospholipids in buttermilk powder rose as the heat stress increased. Homogenization breaks down milk fat globules, so that completely new membranes are formed in which the natural membrane proteins are greatly depleted, leading to a much higher proportion of caseins. Whey proteins are hardly incorporated into this new MFGM milk fat globule membrane (1,5).

Characteristically, the MFGM consists of a double layer of polar lipid membranes – an inner, inhomogeneous monomolecular layer made up of polar lipids and proteins surrounding the fat droplet and an outer membrane that is also made up of proteins and polar lipids. Together, these form a double membrane (6,7).

The MFGM double membrane consists of approximately equal parts of proteins and phospholipids (8), a few neutral lipids including cholesterol, and water (7). However, the protein content may vary considerably, between 25 and 60%, depending on the source of the milk, the isolation method used and the technical history of the samples (6). The most important MFGM proteins are listed in Table 1. The main protein of the coat fraction is butyrophilin, with a share of up to 40% (6,9). It appears to be firmly associated with the MFGM, since it cannot be extracted with detergents. Xanthine oxidase, on the other hand, which on average accounts for about 20% of the MFGM proteins, can be removed by washing with highly concentrated salt solutions or with detergents. It is therefore assumed to be a peripheral membrane protein (6) which may be bound to butyrophilin and serve as a spacer between the inner and outer membranes. Mucin 1 is very easily removed from the MFGM by cooling and stirring the milk, and Mather (9) therefore assumes that it is associated with the MFGM on the outside. PAS 6/7 (periodic acid/Schiff) can also be washed out of the MFGM easily, and it is therefore probable that they are only loosely associated with the MFGM. Adipophilin has a strong affinity with triglycerides and is to be found in the inner monolayer. Besides mucin 1 and butyrophilin, the membrane proteins CD36, PAS III and PAS 6/7 are situated in the outer double membrane (7). There are also many other membrane proteins whose nomenclature is not standardized (9), and at least 50 polypeptides with molar masses from 10 – 300 thousand daltons (kilodaltons, kDa) (10).

Table 1.

The most important protein components of the milk fat globule membrane (6).

Protein	Molecular mass (Daltons)
Mucin 1 (MUC 1)	160,000–200,000
Xanthine oxidase	150,000
PAS III	95,000–100,000
CD36 or PAS IV	76,000–78,000
Butyrophilin	67,000
Adipophilin (ADPH)	52,000
PAS 6/7	48,000–54,000
Fatty-acid binding protein (FABP)	13,000
BRCA 1	210,000

The polar lipids of the MFGM are the sum of the phospholipids and the sphingolipids. These are amphiphilic, i.e. they have a lipophilic hydrocarbon group and a polar, hydrophilic head group. In polar solvents like water they form micelles or membranes, which is very important to consider in connection with their technical isolation. Sphingolipids only occur in animal fats and are said to have health-promoting properties. Of these, sphingomyelin is the second most common phospholipid in the gut and is closely associated with cholesterol metabolism. It probably has an anticarcinogenic effect in the bowel (10 – 12). Components of sphingomyelin (ceramide, sphingosine 1-phosphate) play an important role in signal transduction (11).

Along with the glycosylated membrane proteins of the MFGM, the phospholipids are thought to have bioactive properties. For example, they bind pathogenic microbes, which are then excreted with the faeces. In vitro and dose-dependently, polar lipids inhibit the infective activity of the rotavirus (13). By various mechanisms, phospholipids may be able to alleviate inflammatory skin diseases, since they are said to have an anti-inflammatory effect through influencing glutathione metabolism in the body (14). The firm Arla has applied for a patent to protect a concentrate containing a high proportion of phospholipids as a basis for nutraceuticals (15). There are further patents relating to the use of milk phospholipids in baby food (16) or as food supplements (17).

Chemically, phospholipids are esters of glycerol with two fatty acids and an amino alcohol esterified with a phosphate group. According to Rombaut et al. (18) and Burling et al. (15), the most important phospholipids of milk are phosphatidylcholine (PC, with a share of 5.4% of the MFGM), phosphatidylethanolamine (PE, averaging 4.3%), phosphatidylserine (PS, with 2.3%) and phosphatidylinositol (PI, with 1.3%). The most significant sphingolipids are sphingomyelin, averaging 5.1% of the MFGM, and glucosylceramide and lactosylceramide, which together make up 1.3% of the MFGM (15,19). The sphingomyelin content increases over the lactation period and is highest in summer (20).

When cream is made into butter, nearly half of the polar lipids present in the MFGM pass into the buttermilk, which is why the latter is most often used as a source of milk phospholipids. Assessments of the phospholipid content of buttermilk differ. Whereas Elling et al. (21) found a level of phospholipids seven times higher than in cow's milk, namely 0.89 g/kg, Christie et al. (22) reported 0.72 g/kg in buttermilk and Sachdeva and Buchheim reported (23) only 0.25 g/kg. The phospholipid content of buttermilk depends greatly on the fat content of the butter-making cream. The figures for the proportion of phospholipids in the fat of the milk vary between 0.42% (24) and 0.8% (25). The most thorough analysis of the milk phospholipids to date was made by Rombaut et al. (18,19). In butter serum they found an average of 1.60 g/kg, in buttermilk 0.91 g/kg, in cream (39% fat) 1.39 g/kg and in whey 0.18 g/kg. At 20.5% the share of sphingomyelin in the milk phospholipids is much lower in buttermilk than in butter serum (28.8%).

3.3.2 Identification methods

The simplest way to identify phospholipids in milk is through the phosphate content (26). The phospholipids are extracted from the sample with a 2:1 mixture of chloroform and methanol, after which the extract is dried and ashed. The ash is hydrolysed with sulphuric acid and the phosphorus converted with molybdate and ascorbic acid into molybdenum blue, which is determined photometrically. The phospholipid content can be calculated approximately from the phosphorus content by multiplying by the factor 25.44 (27). With low-fat products especially the distribution of the phospholipids over the hydrophilic and lipophilic phases may present a problem.

Rombaut et al. (18,28) developed a high-performance liquid chromatography (HPLC) method with evaporative light scattering detection for analysing the individual phospholipids and sphingolipids in MFGM, buttermilk and butter. The evaporative light-scattering detector (ELSD) is desirable by being compatible with a number of solvents used in preparing the samples. ELSD also has further advantages: gradient elution is possible, and in contrast to the UV detector the signal seems to be independent of the degree of saturation and the chain length of the lipid acyl groups.

The separation results achieved with high-performance thin-layer chromatography (HPTLC) are less satisfactory (29). In this method the polar lipids are separated by means of chloroform, methanol and water (65:25:4) and the non-polar lipids with petroleum ether, ethyl ether and acetic acid (85:15:2). The lipids are then dissolved in 10 mg/ml chloroform and methanol (2:1) and applied to a ready-made silica gel coated glass plate. At least semi-quantitative determination of the individual lipids is possible with the use of standards. This method is very flexible, fairly simple and inexpensive. However, the most precise determination of the individual phospholipids is achieved by quantitative ^{31}P NMR spectroscopy according to method SAA043 (30). This method does not require complicated preparation of the samples because non-phospholipid components are not detected. Disadvantages are the high cost of the equipment and the need for highly specialized personnel. For these reasons there are very few specialist laboratories able to carry out such analyses. In Germany, an established company is Spectral Service GmbH in Cologne, which offers its services worldwide.

All these methods are very time-consuming and their suitability for efficient process development is therefore limited. A quick alternative method for processes based on the ultrafiltration of buttermilk has been developed by Konrad et al. (2012, unpublished). This method will be described in Section 3.3.3.2.2.

3.3.3 Extraction of phospholipids from buttermilk

3.3.3.1 Overview of the literature

Although several overviews exist of the isolation of MFGM and its significance for health, they do not contain very detailed discussions of the different technologies available (6,7,31).

The classic methods involved extraction and centrifugation and were based mainly on the use of buttermilk. Because of the solvents used, phospholipids acquired by extraction are not very suitable for food applications. These methods were therefore developed for isolation and analysis on a laboratory scale. Centrifugation methods, on the other hand, yield phospholipids with food quality. A typical method was developed by Kanno and Kim (32). As a pretreatment they precipitated some proteins from the buttermilk isoelectrically at pH 4.8 or salted them out with ammonium sulphate, and after centrifugation acquired a protein-rich precipitate and a supernatant enriched with lipids which was subsequently adjusted to a pH of 6.8. Thermo-calcium precipitation of the lipoproteins according to Lehmann (33) with subsequent solvent extraction of the phospholipids, or enzymatic treatment, proved unsuitable because of the presence of casein micelles in buttermilk. Apart from its higher phospholipid content, centrifuged buttermilk has roughly the same composition as skimmed milk. Rombaut and Dewettinck (34) treated whey at 55°C, pH 7.7, with 0.205 g calcium per litre of whey and then centrifuged the polar lipids out as compact, solid pellets. Under these conditions they succeeded in extracting 91% of the phospholipids with a phospholipid concentration of 8.34% in the dry matter.

A method of obtaining the entire MFGM with a high yield was patented by Gorewit (35). In this method, pasteurized, non-homogenized milk is whipped until a phase inversion occurs. The fat phase is washed again with water at 37°C for 24 h, then microfiltered, and the two phases are dried together. Many modern methods are based on the use of microfiltration, in which the MFGM is separated into a protein-rich phase and a phase enriched with phospholipids. A problem in membrane filtration is the fairly similar particle sizes of the MFGM components – the lipids and the casein micelles (6,36). Without pretreatment of the buttermilk, the extraction of the phospholipids by microfiltration therefore poses a challenge. The greatest difficulties are caused by the casein present in the buttermilk, which very quickly blocks the filtration membranes. Sachdeva and Buchheim (23) solved this filtration problem by precipitating the casein with rennet or acid before membrane filtration. On the other hand, this caused losses of polar lipids in the range of at least 20%. Morin et al. (37) therefore used whey buttermilk, which results when whey butter is made from whey cream; unlike normal buttermilk it is practically free from casein. But at about 25 l/hm^2 and a filtration temperature of 8 – 10°C, the flux achieved was not much greater than with microfiltration of the usual buttermilk. Moreover, the concentration of the phospholipids was lower when whey buttermilk was used than with normal buttermilk.

Satisfactory separation of the proteins from the phospholipids cannot be achieved by microfiltration alone (38). Complexing agents like Na (sodium) citrate can be used to dissolve caseins so that they pass through the microfiltration membranes. There have been conflicting reports on the use of Na citrate. Corredig et al. (39) were able to successfully separate casein from the phospholipids by adding Na citrate to the dry matter of the buttermilk at a ratio of 1:4. Rombaut et al. (19) achieved better transmission of the caseins in the microfiltration of butter serum in the presence of 80 mM citrate, but at the cost of a considerable loss of phospholipids. Nor were convincing results achieved with dissociating agents such as urea or EDTA, or by ethanol precipitation (19,38). Morin et al. (38) succeeded in separating the phospholipids concentrated by microfiltration into an alcohol-insoluble fraction (phosphatidylethanolamine and cerebrosides) and an alcohol-soluble fraction; the two had different functional properties.

A further possibility of increasing the phospholipid concentration by microfiltration of buttermilk is through enzymatic protein hydrolysis. Roesch and Corredig (40) hydrolysed buttermilk concentrates produced by microfiltration with trypsin and chymotrypsin, and acquired a fraction enriched with phospholipids after repeated microfiltration and diafiltration.

Morin et al. (38) investigated the most important factors influencing the microfiltration of reconstituted buttermilk using uniform transmembrane pressure (UTP mode). It emerged that the process temperature had no effect on protein transmission, but an increase in the temperature from 7°C to 25°C or 50°C reduced the transmission of both lipids and phospholipids. Maximum lipid concentration was found at 25°C. In further trials, the study group led by Jiménez-Flores showed that the pasteurization of cream results in a higher yield of polar lipids (3). By washing cream before making it into butter, Morin et al. (37,41) reduced the casein and protein content of the buttermilk by as much as 74% and thus doubled the flux during microfiltration; however, only a low phospholipid concentration of about 4% of the dry matter was achieved even after intensive diafiltration. Moreover, the phospholipid losses during microfiltration were >10%.

Astaire et al. (29) coupled microfiltration with supercritical CO_2 extraction to increase the polar phospholipid content, as an alternative to conventional lipid fractionation with potentially toxic solvents. Supercritical CO_2 extraction removes just the non-polar lipids from the microfiltration retentate, which greatly increases its polar lipid content (42). The proteins, on the other hand, remain present and unchanged. The denatured whey proteins and caseins can be eliminated additionally by precipitation at pH 4.6. The firm Fonterra (New Zealand) has patented the concentration of phospholipids by 'near-supercritical' CO_2 extraction (43) and sells the concentrates commercially under the trade name Phospholac 600, with a lipid content of 83% (44).

The study group under Jiménez-Flores has also experimented with supercritical CO_2 extraction. De Rezende Costa et al. (45) describe the ultrafiltration of whey buttermilk coupled with supercritical CO_2 extraction. Without supercritical CO_2 extraction they achieved a maximum phospholipid concentration of 7.17% of the dry matter after ultrafiltration, up to a volume reduction ratio of 1:10 at 25°C, with a 10 kDa membrane and after fivefold diafiltration. With supercritical CO_2 extraction they were able to increase the phospholipid concentration to as much as 11.95% of the dry matter. The whey cream had a comparatively high fat content of 38.5%, which guarantees a high phospholipid concentration in the whey buttermilk. However, a method of coupling microfiltration with supercritical CO_2 extraction for acquiring phospholipids from buttermilk, already published by the same study group, produced much poorer results (46,47,48).

Researchers at the Ghent University (Belgium) carried out intensive investigations into the isolation of milk phospholipids. Butter serum and sour buttermilk quark whey were tested as raw materials in addition to traditional buttermilk. Compared with whey, the latter has a high phospholipid content (1.04 g/kg) and contains practically no casein (34,49). Both the use of ultra- and microfiltration for concentrating the phospholipids, and thermo-calcium precipitation of the lipids were investigated (49). It was found that with microfiltration of sour buttermilk quark the retention of the polar lipids was extremely low, but that it could be improved greatly by thermo-calcium precipitation of the lipids. When the pH was raised from 4.6 to 7.5 and the temperature to 50°C, the retention of the polar lipids by microfiltration increased from 64 to 98% and membrane fouling (soiling of the membranes during filtration) was simultaneously minimized. As Sachdeva and Buchheim (23) had already reported in connection with the ultra- and microfiltration of decaseinated buttermilk, in these Ghent tests ultrafiltration also retained nearly all the polar lipids. After three-stage diafiltration, a polar lipid concentration of 6.79% of the dry matter was achieved. The influence of the pH of the buttermilk on ultrafiltration performance had already been investigated by Ramachandra Rao et al. (50). In each case the flux was deposit-controlled by the top fouling layer, and the optimum pH was 6.0.

3.3.3.2 Ultrafiltration of whey buttermilk

3.3.3.2.1 Description of whey buttermilk

Konrad et al. (36) analysed the particle sizes of the MFGM components in whey buttermilk. To do this they washed the whey cream thoroughly with a 0.02 M KCl solution in order to remove all the whey proteins. The fat content was then further increased by centrifugation, and the washed whey cream made into butter. The resulting whey buttermilk showed a bimodal particle size distribution (Fig. 1). The components of the MFGM had a Sauter diameter of approx. 1.2 µm, whereas the fat globules were about 8 µm in diameter. In order to measure the particle size of the purified phospholipids, the MFGM proteins were subjected to a peptic hydrolysis ('peptized'). As protein hydrolysis progressed, the remaining phospholipids took on a particle size of about 0.8 µm. This indicated that the latter always form micelles in aqueous solution and do not occur in a monomolecular form. The phospholipid losses discussed in connection with the microfiltration of buttermilk confirm this result, since predictably most of the phospholipids are retained by the microfiltration membranes. So the working hypothesis deduced for successful development of the process was that ultrafiltration is the method of choice for concentrating the phospholipids of the milk, and that as many macromolecular components (fat, proteins) as possible must be removed from the buttermilk before ultrafiltration.

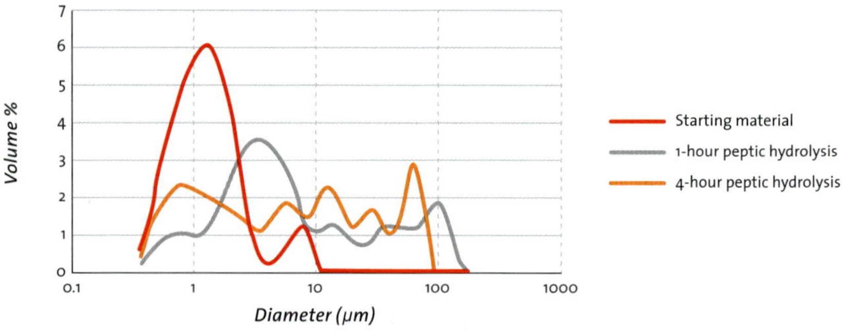

Figure 1.

Change in the particle size distribution of whey buttermilk (derived from washed whey cream) as a result of peptic hydrolysis.

Table 2 lists the typical chemical-analytical values of the raw materials used by Konrad et al. (51) for obtaining milk phospholipids. It is noticeable that apart from its higher phospholipid content, whey buttermilk has a similar composition to that of whey, only diluted by about 10%. Through centrifugation, denatured whey protein is separated off as well as the fat. The values also show that butter serum is the ideal raw material for concentrating the phospholipids of milk, since the phospholipids already account for about 9% of the dry matter.

The higher the initial phospholipid content of the raw material, the higher the purities that can be expected in the retentate after concentration using ultrafiltration, as the ultrafiltration membranes hold back practically all the phospholipids. Unfortunately, very few dairies can offer butter serum, so it was necessary to use buttermilk produced from whey cream for developing the process. A further limitation resulted from the fact that most cheese factories only produce whey cream with a low fat content of about 22% in order to save energy. Usually, it is used as a rework in cheese production. Whey cream with a high fat content is only made in the few factories that produce whey butter.

Table 2.

Chemical-analytical values of the raw materials used.

Whey constituent	Mass unit	Whey buttermilk[1]	Whey buttermilk[1], centrifuged	Butter serum
Dry matter	g/100 g	5.9	5.7	1.9–2.1
Lactose	g/100 g	3.2	3.2	0.1–0.3
Total N	mg/100 g	132	110	65
Protein[2]	g/100 g	0.84	0.70	0.40
Fat	g/100 g	0.7–0.8	0.1–0.3	0.1–0.5
Ash	g/100 g	0.5–0.6	0.5–0.6	0.1–0.2
Phosphorus	mg/kg	750–900	580–870	3,000
Phospholipids	g/kg	0.30	0.30	1.4–1.8
Phospholipids	% of dry mass	0.50	0.52	7.3–9.5

[1] Obtained from whey cream with 22.0% fat
[2] Total N × 6.38

As Konrad et al. (36) further found, the initial concentration of the phospholipids in whey buttermilk is direct-linearly dependent on the fat content of the used whey cream (Fig. 2). The exact content was confirmed by means of repeated ^{31}P NMR analyses. In regular cream for butter-making the phospholipid content is about 0.4% of lipids (28). Using whey cream a constant part of about 35% of the phospholipids (PL) passes into the whey buttermilk after churning, thus the expected phospholipid level in whey buttermilk can be calculated by multiplying the fat content of the initial whey cream with the factor 0.00137, as in Equation 1, below (36):

$$\mathbf{PL\ (\%) = CF\ (\%) \times 0.00137} \quad \mathbf{(1)}$$

PL stands for phospholipids, CF for the percentage of cream fat.

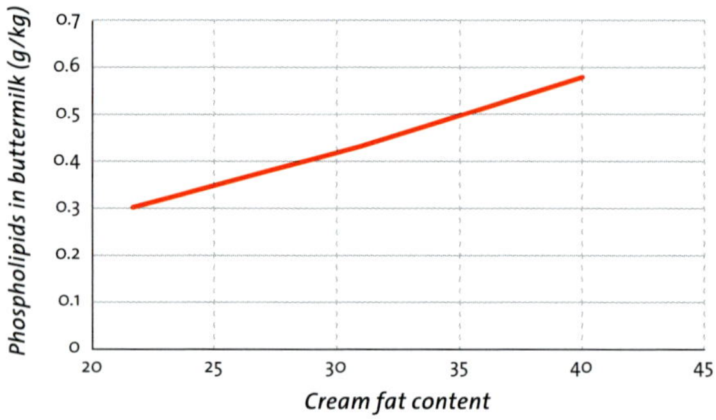

Figure 2.

Calibration curve for the transition of phospholipids from whey cream into whey buttermilk.

3.3.3.2.2 Rapid method for determining the phospholipid concentration

On the assumption that practically no macromolecular constituents – including phospholipids – can be lost during ultrafiltration, Konrad et al. (51) developed a rapid method to estimate the current phospholipid concentration during concentration steps. If the objective is to assess the phospholipid concentration reached in the dry matter during ultrafiltration or diafiltration, the known phospholipid concentration of the starting substrate must be expressed in relation to the current mass of the dry matter in the ultrafiltration/diafiltration retentate, as in Equation 2, below:

$$\mathbf{PL\ (\%\ DM) = \frac{m_o \times (CF \times 0.00137)}{m_R \times DM} \times 100} \quad \mathbf{(2)}$$

m_o being the initial mass of the feed, m_R the mass of the current retentate, DM the dry matter of the retentate (in %) and CF the cream fat percentage.

This method can be applied to all starting substrates if the initial phospholipid concentration is known. The values calculated coincide well with the results of the complex method of determining the phospholipids, and the deviations were smaller than in the molybdate method. Whereas the effects of sample preparation or technical variations on phospholipid purity could only be tentatively estimated using the molybdate method, applying this new method made it possible to deduce statistically significant differences and to develop the process described below.

3.3.3.2.3 Pretreatment

Untreated whey buttermilk very quickly clogs the ultrafiltration membranes. As Fig. 3 shows, the prerequisite for a good permeation flux is the removal of fat and insoluble whey proteins as well as cheese fines by centrifugation. The result is a considerable increase in the flux during ultrafiltration, making it no problem to concentrate the whey buttermilk and achieve a volume reduction of 1:20.

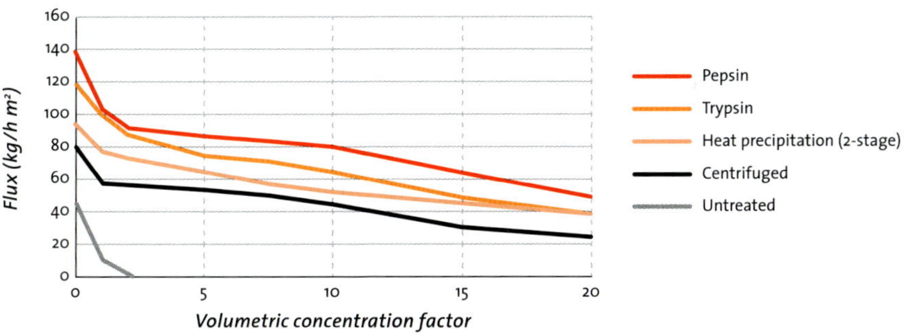

Figure 3.
Ultrafiltration of whey buttermilk as a function of pretreatment.

A further deproteinization of whey buttermilk before the ultrafiltration step increases the flux and the final phospholipid concentration still further. Untreated whey buttermilk contains the MFGM proteins and about twenty times the amount of whey proteins which would be concentrated together with the phospholipids during ultrafiltration. Deproteinization can be carried out by enzymatic and/or thermal means. The digestive enzymes pepsin and trypsin lend themselves to this, but it must be taken into account that neither pepsin nor trypsin is able to break down all the individual native whey proteins. Whereas pepsin is unable to digest β-lactoglobulin in its native state (52), trypsin is very slow to attack native α-lactalbumin (53). Furthermore, they are not able to hydrolyse all the MFGM proteins. Pepsin is very slow in hydrolysing butyrophilin (54), and in general the glycosylated MFGM proteins, especially mucin, are hard to digest with pepsin (55). On the other hand, butyrophilin and adipophilin are digested well by trypsin, though lactadherin and xanthine oxidase are not well digested (56). The phospholipid concentration can be increased by about 2.5% in the dry matter by protein hydrolysis, whichever enzyme is used (51). Since pepsin is used at a pH <2, it should be favoured for microbiological reasons. A positive additional effect of enzymatic protein hydrolysis is that it doubles the filtration flux.

Besides enzymatic treatment, thermal protein precipitation can be used to increase the attainable phospholipid concentration. In this case a two-stage method offers the greatest advantages: in the first stage the whey protein is unfolded thermally at a pH of <4.0 and temperatures >80°C, and precipitation does not take place until the pH has increased to between 4.5 and 5.5. At the low pH values, no aggregation takes place because all the whey proteins carry a uniform positive charge below their isoelectric point and repel each other. Moreover, under the acidic conditions, no disulphide exchange takes place. A great additional benefit is that the whey proteins precipitate in the form of microparticles without the need for stirring or shearing (57). These microparticulated proteins have a mouth feel similar to that of fat, since the average particle size of 1 – 10 µm is practically identical to that of the fat globules. The heating conditions during primary unfolding of the molecules mainly influence the particle size, whereas the precipitation conditions have more effect on the yield. Phospholipid losses do not occur until after extensive heat treatment of >30 min at 90°C.

The thermal separation of proteins out of whey buttermilk is slightly more effective than enzymatic separation because nearly all the whey proteins are precipitated. Therefore, it was possible to increase the phospholipid concentration after ultra- and diafiltration from 4.2% to 11.2% of the dry matter. Additional peptic treatment increases the purity by a further 1 – 2% of the dry matter. These findings are related to whey buttermilk that had been obtained from whey cream with 22% fat and therefore contained only 0.30 g/kg phospholipids. If whey cream with a higher fat content is used it is possible to produce concentrates with a phospholipid content as high as 20% of the dry matter in this manner.

3.3.3.2.4 Ultrafiltration

Further concentration of the pretreated whey buttermilk is carried out by ultra- and diafiltration. Both the flux and the attainable phospholipid concentration of the retentates will depend on the pH of the whey buttermilk, the filtration temperature, the transmembrane pressure and the ultrafiltration membrane used.

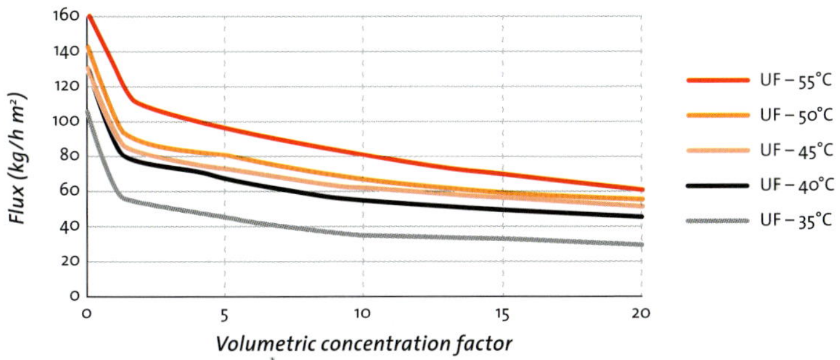

Figure 4.

Filtration flux of deproteinized whey buttermilk as a function of the ultrafiltration temperature (UF). Conditions: 300 kDa membrane; 2.0 bar transmembrane pressure; pH 2.0.

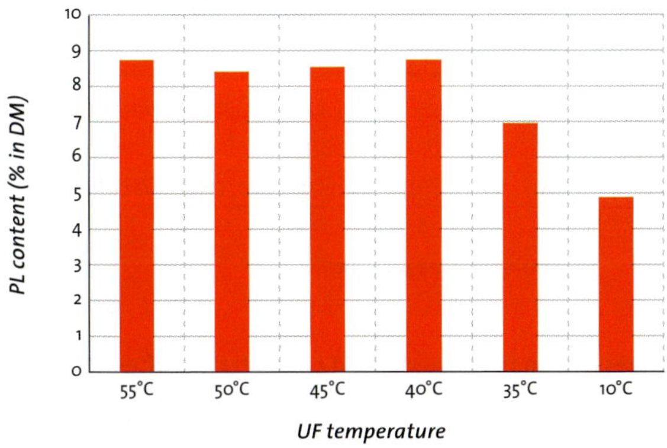

Figure 5.

Influence of the temperature during ultrafiltration on the final phospholipid concentration.

Ultrafiltration of untreated, centrifuged whey buttermilk in the pH range 4–5 is very difficult. After the proteins have been separated off, it can be filtered even in the isoelectric range of the whey proteins. Concerning the final phospholipid concentrations that can be achieved, there are two optimum pH ranges: pH 2.0–3.0 and pH 4.8–5.5. In these ranges the associated low-molecular-weight substances are most easily separated from the phospholipids. Ultrafiltration is best conducted in the acidic pH range after pepsin hydrolysis, whereas it is best to work at pH 5 after thermal protein precipitation.

As the filtration temperature rises the flux also increases, in principle, since the viscosity of the feed decreases (Fig. 4). But there is an upper limit to the temperature that can be used because of specific properties of the membrane and the risk of damage to the product.

The final phospholipid concentration that can be achieved after ultra- and diafiltration also depends on the filtration temperature. As Fig. 5 shows, filtration temperatures below 40°C result in a linear decline of the phospholipid concentration in the ultrafiltration retentates, the reason being that low-molecular-weight substances like lactose, peptides or minerals do not pass into the permeate completely because of the high viscosity of the feed. Upwards of about 40°C most of those substances can be removed, and a further rise in temperature increases the phospholipid concentration slightly.

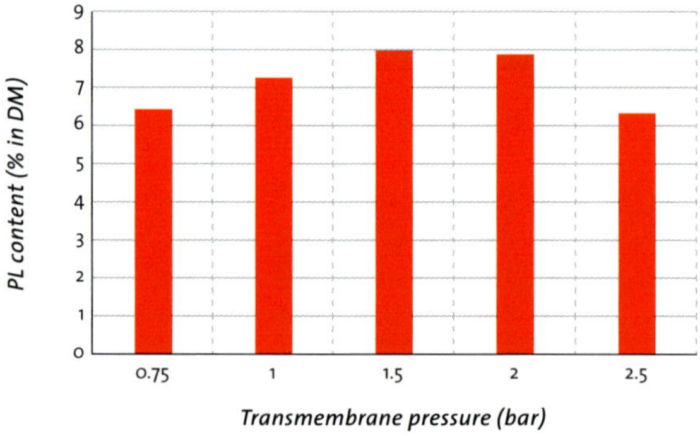

Figure 6.

Influence of transmembrane pressure on the phospholipid concentration after ultrafiltration of heat-precipitated whey buttermilk. Conditions: 300 kDa membrane, 2.0 bar, 50°C, pH 5.5.

Transmembrane pressure is one of the most important parameters in membrane filtration. The transmembrane pressure correlates with the optimum ultrafiltration flux and the highest degree of phospholipid purity that can be achieved; the optimal range for the process is 1.5–2.0 bar (Fig. 6). If the pressure is too low it is inadequate for efficient separation of the low-molecular substances because mass transfer is mainly contributed by diffusion (58). If it is too high, above the critical flux at about 2.0 bar, compact fouling layers form in the course of ultrafiltration, and these lower the flux and increase the total retention of the constituents of the whey buttermilk.

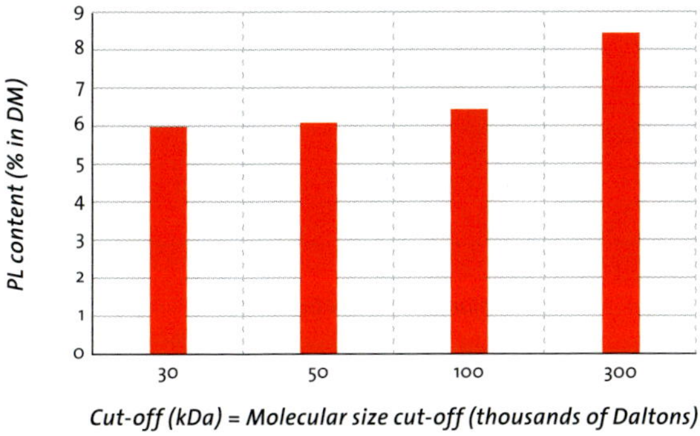

Figure 7.

Influence of the molecular cut-off size on the attainable purity of the phospholipids after ultrafiltration of deproteinized whey buttermilk (36).

Besides the variables discussed above, the molecular mass cut-off of the membranes also affects the flux and the attainable purity of the phospholipids. As can be seen in Fig. 7, the attainable purity enhances with increasing cut-off of the membranes. This can be explained by the more efficient transmission of the associated substances into the permeate. As expected, the permeation flux increases also during ultrafiltration with increasing cut-off. The upper limit of ultrafiltration is reached with 1,000 kDa membranes, since microfiltration cannot be used as a concentration method because considerable losses of phospholipids would occur as the pore size of the microfiltration membranes and the particle size of phospholipids are very similar.

The use of different ultrafiltration systems such as membrane cassettes, spiral wound modules or ceramic membranes has no significant effect on the attainable purity of the phospholipids if the molecular mass cut-off of membranes is the same. Even the flux is similar if all the general technical conditions discussed are complied with.

It is no problem to concentrate pretreated whey buttermilk to a volume reduction of 1:20; in this process the dry matter and the phospholipid content increase linearly in relation to the volume reduction (Fig. 8). At the approximately 16% dry matter thus achieved, the viscosity of the retentate is still low enough to prevent clogging of the ultrafiltration membranes.

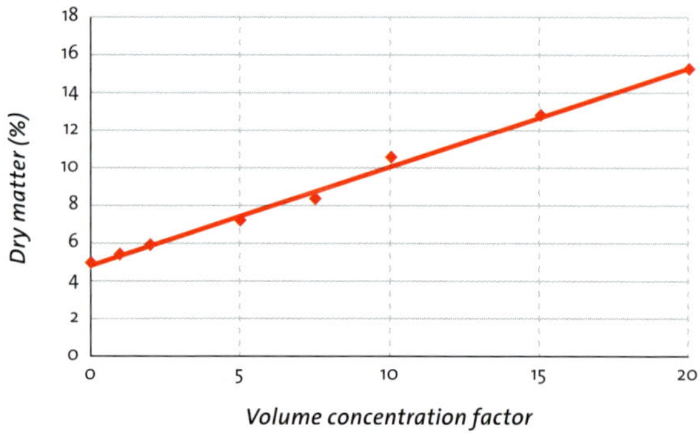

Figure 8.
Increase in the dry matter of whey buttermilk in the course of ultrafiltration with 300 kDa membranes.

Diafiltration

After ultrafiltration, the retentate still contains considerable amounts of low-molecular-weight substances such as lactose, peptides and milk salts that can be removed by diafiltration. Diafiltration is a technique using ultrafiltration membranes to remove those species from macromolecules by 'washing'. Diafiltration can be carried out in either the continuous or the batch mode. The continuous mode has the advantage that less washing water is needed, since it works with a high volume reduction factor. Because of the extremely high concentration gradient at the membranes, the small molecules permeate more efficiently than in the diluted medium. The mode has no influence on the yield or purity. However, the continuous mode does have the disadvantage of greater technical complexity, since the permeate constantly has to be compensated with an equal amount of fresh water.

In the case of batch diafiltration, the feed is first diluted and then concentrated back to the starting volume. Each additional diafiltration volume (DF) reduces the smaller molecules further. Tests were carried out to determine how many diafiltration volumes were necessary in order to achieve maximum phospholipid purity. This is achieved when the mass of the dry matter in the retentate no longer changes, namely after four diafiltration steps (Fig. 9). Finally, the phospholipid concentrate is further concentrated as far as possible by ultrafiltration before being dried.

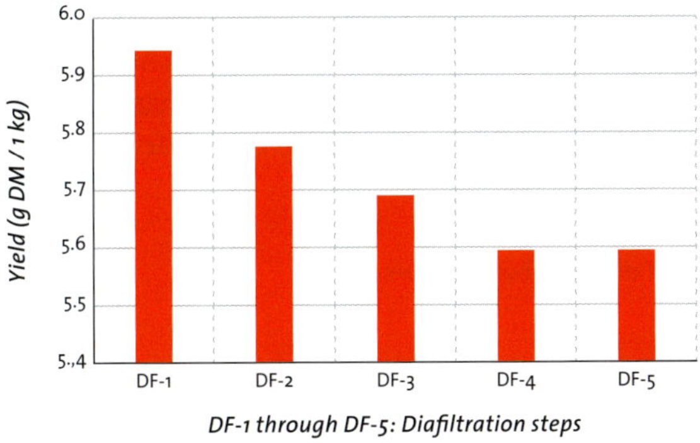

Figure 9.
Decrease in the mass of the dry matter resulting from batch diafiltration (in relation to 1 kg whey buttermilk).

3.3.3.3 Butter serum as a raw material

As discussed above, only about 35% of the phospholipids passes into the whey buttermilk after churning and thus most of them remain in the butter. Therefore, it is a great advantage to concentrate phospholipids out of butter serum, since it naturally retains high phospholipid concentrations in its dry matter. Bezelgues et al. (59) washed a cream with a fat content of 80%, concentrated it back to the original fat content and destabilized the cream matured at 4°C in order to obtain an MFGM fraction with a phospholipid concentration of 25%. The sphingomyelin content in this concentrate was 26.8%, as much as in the serum phospholipids listed in Table 4.

Catchpole et al. (60) achieved higher purities either by extracting so-called β-serum with dimethyl ether or by treating it with supercritical CO_2 extraction. In this way they achieved high phospholipid concentrations of 31.0% and 65.7% of the dry matter, respectively. The β-serum used arises in the production of anhydrous milk fat by the modern separator method. Table 3 compares the composition of classic butter serum with that of β-serum from the point of view of obtaining phospholipids. Classic butter serum is produced by thermal treatment of butter. It contains little residual protein because the thermolabile proteins are precipitated by heating.

According to Table 3, at 1.8 g/kg or 6.5 g/kg the initial phospholipid concentration of butter serum is much higher than that of whey buttermilk (0.3 – 0.5 g/kg). The values stated correspond to the results published by Rombaut et al. (18). If the phospholipid concentrations are viewed in relation to the dry matter, the difference becomes even more obvious. Whereas butter serum has a phospholipid content of about 9–10% of the dry matter, the value for whey buttermilk is much lower, at about 0.5–0.9% of the dry matter.

Table 3.

Average compositions of different butter serums.

Constituent	Serum, classic	β-serum
Dry matter, %	2.0	6.4
Protein, %	0.54	1.90
Fat, %	1.1	1.3
Phosphorus, mg/kg	120	830
Phospholipids, g/kg	1.8	6.5
Phospholipids, % of dry matter	9.0	10.1

Before classic butter serum is ultrafiltered, the residual fat should be centrifuged off; this has the simultaneous effect of greatly increasing the phospholipid concentration in the dry matter. The flux in the quasi-stationary range of ultrafiltration is then about 60 kg/hm^2, which is comparable to that assessed during the ultrafiltration of pretreated whey buttermilk (see Figs. 3 and 4). Since classic butter serum is low in protein (because of the heat treatment), additional pepsin hydrolysis increases the flux of butter serum only marginally and thus only slightly increases the final phospholipid concentration. Using membranes with a molecular weight cut-off of 300 kDa, concentrates with a phospholipid content of 20–32% of the dry matter can be obtained from this butter serum by ultrafiltration. Optimum conditions for the ultrafiltration of butter serum are temperatures >45°C, a neutral pH and a transmembrane pressure of 2.0 bar.

For classic butter serum, the phospholipid yield is between 0.5 and 0.6 g per kilogram of butter. A further result to be emphasized is that the sensory and storage properties of the anhydrous milk fat derived from whey butter do not differ from those of commercially available clarified butter. Therefore, whey butter is not a waste product, and can be marketed profitably.

Use of β-serum

As the chemical-analytical data in Table 3 show, β-serum is the ideal substrate for the production of milk phospholipids. β-serum also has to be defatted by centrifugation before ultrafiltration. Since the initial level of dry matter is much higher than with classic butter serum, ultrafiltration of β-serum is more difficult.

Fig. 10 shows that the critical flux in the ultrafiltration of β-serum is reached at a volume reduction of about 1:7. That is, from this deviation point onwards the filtration is no longer membrane-controlled; the active filter medium is now the top fouling layer and so it is deposit-controlled. At this stage, at the latest, ultrafiltration should stop, otherwise the membranes will be blocked and clogged filters are very difficult to clean. After diafiltration with 5 diavolumes of water, concentrates with phospholipid concentrations >50% of the dry matter can be attained (Table 4). Previously, such purities had only been achieved by membrane filtration in conjunction with other purification methods. In order to determine the maximum level of purity that can be achieved by this method, the protein present in the β-serum was additionally hydrolysed with trypsin after centrifugation (see "Tryptically" plot in Fig. 10). Unlike classic butter serum, β-serum still contains large amounts of residual protein. Purity can be increased by about 10% to finally reach 60% of the dry matter, by tryptic hydrolysis (Konrad et al., 2012a, unpublished).

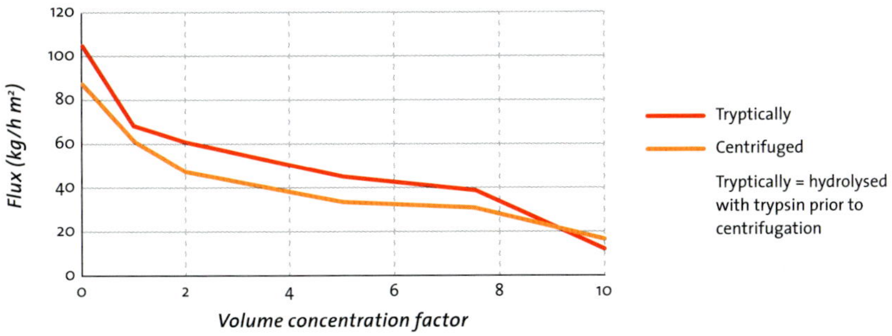

Figure 10.
Ultrafiltration of ß-serum at pH 7.5 and 45°C with 300 kDa membranes (transmembrane pressure 2.0 bar).

In order to improve the filtration properties of butter serum, Rombaut et al. (19) added 80 mM Na citrate and thus increased the flux during microfiltration. Na citrate dissociates the low concentration of casein micelles present, thus facilitating casein transmission, but this is at the cost of the phospholipid yield. As already discussed, some of the phospholipids also pass through the microfiltration membranes because of their particle size, which meant that the yield was only 64%.

3.3.3.4 ^{31}P NMR analyses

The ^{31}P NMR analyses of typical samples produced by the ultrafiltration methods described above are shown in Table 4. The ^{31}P NMR method also permits conclusions as to the quantitative distribution of the individual phospholipids.

It is noticeable that phospholipids derived from butter serum have a significantly higher sphingomyelin (SM) content than those derived from whey buttermilk. This result coincides with findings from the literature (19,28,59,61) and suggests that sphingomyelin is the most intensively bound individual phospholipid to butter fat and therefore has the best emulsifying properties of all the phospholipids present. During butter-making it passes into the buttermilk only to a limited extent.

Table 4.

Results of the ^{31}P NMR analysis of phospholipid concentrates obtained from whey buttermilk and butter serum.

Phospholipid Profile	Whey buttermilk g/kg	Percentage of total phospholipids %	Serum, class. g/kg	Percentage of total phospholipids %	β-serum g/kg	Percentage of total phospholipids %
PC	2.99	25.6	7.46	32.1	14.74	27.4
2-LPC	0.20	1.7	0.21	0.9	0.59	1.1
PI	0.88	7.5	1.78	7.7	4.04	7.5
PS-Na	1.14	9.8	2.34	10.0	5.8	10.8
SPH	**2.45**	**20.9**	**5.80**	**24.9**	**13.49**	**25.1**
PE	3.72	31.8	5.48	23.5	14.05	26.2
LPE	0.31	2.7	0.20	0.9	0.99	1.9
Total	**11.69**	100.0	**23.27**	100.0	**53.7**	100.0

* PC = phosphatidylcholine, 2-LPC = 2-lysophosphatidylcholine; PI = phosphatidylinositol; PS-Na = phosphatidylserine-sodium; SPH = sphingolipids; PE = phosphatidylethanolamine, LPE = lysophosphatidylethanolamines

3.3.3.5 Functional properties

In the food industry, phospholipids (mainly of vegetable origin) are used for their emulsifying and structure-enhancing functionalities. Both the phospholipids and the proteins of the MFGM can have functional effects on foods. Wong and Kitts (62) investigated the functional properties of buttermilk powder and found that both its water-binding capacity and its foaming and emulsifying properties were inferior to those of dried egg white. On the other hand, the emulsifying effects of buttermilk are superior to those of skimmed milk or whey powder, probably because of its higher phospholipid content (63). In studies comparing the emulsifying and foaming properties of different buttermilks with whey protein concentrates and skimmed milk powder, Sodini et al. (63) found that whey buttermilk had the best emulsifying properties but also generated the lowest foam overrun and the poorest foam stability. At 1.87% of the dry matter, the phospholipid content of the whey buttermilk was somewhat higher than in the other samples. On the other hand, an MFGM fraction concentrated from buttermilk by microfiltration has less satisfactory emulsifying properties than a protein isolate consisting of the milk proteins and MFGM proteins (64) but better properties than skimmed milk powder or Na caseinate (65). But emulsions with low MFGM concentrations have relatively good foam stability (66). A thermally induced partial denaturation of the whey proteins in buttermilk may also help to increase the stability of buttermilk emulsions (67).

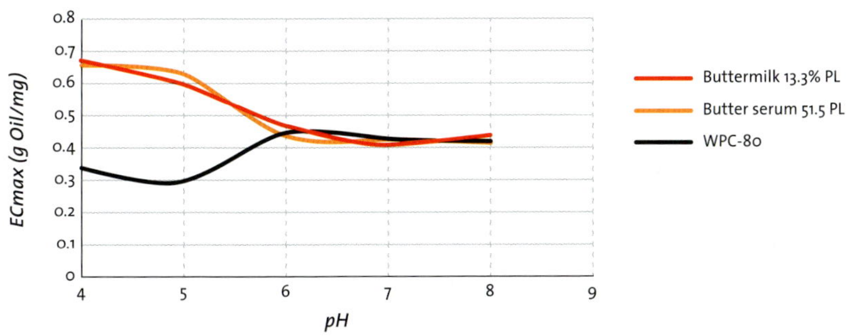

Figure 11.

Maximum emulsifying capacity (ECmax) of phospholipid concentrates in comparison with a whey protein concentrate (WPC-80) (68).

The results of the tests for the emulsifying properties of phospholipid concentrates with greatly differing degrees of purity in the model system are shown in Fig. 11 (68). The maximum emulsifying capacity (ECmax) was determined in 0.1% protein or phospholipid solutions by emulsifying oil continuously under standardized conditions until phase reversal occurred (69). At the inversion point of the emulsion the electrical resistance suddenly rises. The substances used were a phospholipid concentrate derived from whey buttermilk, with a phospholipid concentration of 13.3% of the dry matter, and a preparation obtained from β-serum with a phospholipid concentration of 51.5% of the dry matter; they were compared with a commercially available whey protein concentrate containing about 80% protein (WPC-80). In spite of the great differences in purity, the maximum emulsifying capacity of the phospholipid concentrates derived from whey buttermilk and butter serum did not differ significantly (Fig. 11). In the pH range 6–8, WPC-80 was able to emulsify much the same amount of oil, whereas under slightly acidic conditions the whey protein was only able to emulsify about half as much oil as the phospholipid concentrates.

Moreover, phospholipids result in much more stable emulsions than milk proteins. Most food emulsions are in the slightly acidic pH range, and the stability of phospholipid emulsions is unsurpassed in precisely the pH 4–6 range (68).

As well as being good emulsifiers, phospholipids are generally recognized as having an antioxidative effect. They increase the resistance of the milk fat to oxidation during storage (70), and in the tests carried out by Hamzawi (71) they also strengthened the protective effect of α-tocopherol.

By adding MFGM to reconstituted or evaporated milk it is possible to considerably improve the stability of the products, and minimize fouling of the heating plant (72). In evaporated milk emulsions the MFGM fragments of buttermilk powder and β-serum increase their heat stability in direct proportion to their concentration (73).

Using a microfluidizer, from milk phospholipids Thompson and Singh (44) produced liposomes that can be used for microencapsulating bioactive substances. It emerged that such milk phospholipid liposomes are more suitable for microencapsulation than those produced from soy phospholipids (74). It is also possible to produce nano-liposomes as small as 100 nm in size from milk phospholipids (75). Their efficacy in enveloping ascorbic acid increases as a function of the phospholipid concentration. Nano-emulsions produced with phospholipids as an emulsifier are highly stable and show no signs of Ostwald ripening even after storage for 400 h (76).

Further applications for milk phospholipids could lie in improving the foaming properties of recombined whipped cream (77) or as a means of controlling the crystallization of anhydrous milk fat (78) or butter (79). Adding phospholipids delays crystallization of the milk fat. Milk phospholipids also enhance the structure of yoghurt (80) by increasing the firmness of the gel and the water-binding capacity; this was also observed in recombined rennet gels (81).

The possible applications of milk phospholipids have by no means all been examined. They meet with the greatest interest when additives in dairy products can be replaced by phospholipids, which are now available industrially, in order to avoid the need for a label declaration.

References

1. Lee, S.J., Sherbon, J.W., 2002. Chemical changes in bovine milk fat globule membrane caused by heat treatment and homogenization of whole milk. Journal of Dairy Research 69, 555-567.
2. Ye, A., Singh, H., Taylor, M.W., Anema, S., 2004. Interactions of whey proteins with milk fat globule membrane proteins during heat treatment of whole milk. Lait 84. 269-283.
3. Morin, P., Jiménez-Flores, R., Pouliot, Y., 2007a. Effect of processing on the composition and microstructure of buttermilk and its milk fat globule membranes. International Dairy Journal 15, 1179-1187.
4. Gassi, J.-Y., Famelart, H.-M., Lopez, C., 2008. Heat treatment of cream affects the physicochemical properties of sweet buttermilk. Dairy Science Technology 88, 369-385.
5. García-Risco, M.R., Ramos, M., López-Fandiño, M., 2002. Modifications in milk proteins induced by heat-treatment and homogenization and their influence on susceptibility to proteolysis. International Dairy Journal 12, 679-688.
6. Singh, H., 2006. The milk fat globule membrane – a biophysical system for food applications. Current Opinion in Colloidal & Interface Science 11, 154-163.
7. Dewettinck, K., Rombaut, R., Thienpont, N., Le, T.T., Messens, K., Van Camp, J., 2008. Nutritional and technological aspects of milk fat globule membrane material. International Dairy Journal 18, 436-457.
8. Kanno, C., 1990. Secretory membranes of the lactating mammary gland. Protoplasma 159, 184-208.
9. Mather, I.H., 2000. A review and proposed nomenclature for major proteins of the milk fat globule membrane. Journal of Dairy Science 83, 203-247.
10. Spitsberg, V.L., 2005. Bovine milk fat globule membrane as a potential nutraceutical. Journal of Dairy Science 88, 2289-2294.
11. Vesper, H., Schmelz, E.-M., Nikolova-Karakashian, N., Dillehay, D.L., Lynch, D.V., Merrill, A.H., 1999. Sphingolipids in food and the emerging importance of sphingolipids to nutrition. Journal of Nutrition 129, 1239-1250.
12. Parodi, P.W., 2004. Milk fat in human nutrition. Australian Journal of Dairy Technology 59, 3-59.
13. Fuller, K.L., Kuhlenschmidt, T.B., Kuhlenschmidt, M.S., Jiménez-Flores, R., Donovan, S.M., 2013. Milk fat globule membrane isolated from buttermilk or whey cream and their lipid components inhibit infectivity of rotavirus in vitro. Journal of Dairy Science 96, 3488-3497.
14. Innis, S.M., Davidson, A.G., Melynk, S., James, S.J., 2007. Choline-related supplements improve abnormal plasma methionine-homocysteine metabolites and glutathione status in children with cystic fibrosis. American Journal of Clinical Nutrition 85, 702-708.
15. Burling, H., Andersson, I., Schneider, M., 2006. Phosphatidylserine enriched milk fractions for the formulation of functional foods. US-PS 2006/0280779 A1 (14.12.2006).
16. Rutenberg, D., 2004 Infant formula supplemented with phospholipids, US Patent Application 20040022922.
17. Purpura, M., Jäger, R., Rabeler, R., Hoppe, H.-U., 2005. Physiologically active composition based on phosphatidylserine. WO 2005/023271 A1 (17.3.05).
18. Rombaut, R., Camp, J.V., Dewettinck, K., 2005. Analysis of phosphor- and sphingolipids in dairy products by a new HPLC method. Journal of Dairy Science 88, 482-488.
19. Rombaut, R., Dejonkheere, V., Dewettinck, K., 2006. Microfiltration of butter serum upon casein micelle destabilization. Journal of Dairy Science 89, 1915-1925.
20. Graves, E.L.F., Beaulieu, A.D., Drackley, J.K., 2007. Factors affecting the concentration of sphingomyelin in bovine milk. Journal of Dairy Science 90, 706-715.
21. Elling, J.L., Duncan, S.E., Keenan, T.W., Eigel, W.N., Bolingh, J., 1996. Composition and microscopy of reformulated creams from reduced-cholesterol butter oil. Journal of Food Science 61, 48-53.
22. Christie, W.W., Noble, R.C., Davies, G.,1987. Phospholipids in milk and dairy products. Journal of Society of Dairy Technology 40, 10-12.
23. Sachdeva, S., Buchheim, W., 1997. Recovery of phospholipids from buttermilk using membrane processing. Kieler Milchwirtschaftliche Forschungsberichte 49, 47-68.

24. Sánchez-Juanez, F., Alonso, J.M., Zancada, L., Hueso, P., 2009. Distribution and fatty acid content of phospholipids from bovine milk and bovine milk fat globule membranes. International Dairy Journal 19, 273-278.
25. Walstra, P., Wouters, J.T.M., Guerts, T.J., 2006. Dairy Science and Technology. Second Edition, Boca Raton, FL., USA, CRC Press.
26. IDF (International Dairy Federation), 1990. Determination of Total Phosphorus Content. Spectrometric Method. IDF Standard 21B. Brussels, Belgium: International Dairy Federation.
27. McDowall, A.K.R., 1958. Phospholipids in New Zealand dairy products. Journal of Dairy Research 25, 192-202.
28. Rombaut, R., Dewettinck, K., Van Camp, J., 2007a. Phospho- and sphingo-lipid content of selected dairy products as determined by HPLC coupled to an evaporative light scattering detector (HPLC-ELSD). Journal of Food Composition and Analysis 20, 308-312.
29. Astaire, J.C., Ward, R., German, J.B., Jiménez-Flores, R., 2003. Concentration of polar lipids from buttermilk by microfiltration and supercritical fluid extraction. Journal of Dairy Science 86, 2297-2307.
30. Diehl, B., 2002. ^{31}P NMR in study of phosphorous-containing lipids. Lipid Technology 14, 62-65.
31. Jiménez-Flores, R., Guillaume, B., 2008. The milk fat globule membrane as an ingredient: why, how, when? Dairy Science Technology 88, 5-18.
32. Kanno, C., Kim, D.H., 1990. A simple procedure for the preparation of bovine milk fat globule membrane and a comparison of its composition, enzymatic activities, and electrophoretic properties with those prepared by other methods. Agricultural and Biological Chemistry 54, 2845-2854.
33. Lehmann, H., 1989. Verfahren zur Dephospholipidation von Molke. DE-OS 3.800.468 A1.
34. Rombaut, R., Dewettinck, K., 2007. Thermocalcic aggregation of fat globule membrane fragments from acid buttermilk cheese whey. Journal of Dairy Science 90, 2665-2674.
35. Gorewit, R.C., 2001. Preparation of milk fat globule membranes in tablet or capsule form. US-PS 6444247 B1.
36. Konrad, G., Lorenz, C., Kleinschmidt, Th., 2013a. Ultrafiltration of whey buttermilk to obtain a phospholipid concentrate. International Dairy Journal 30, 39-44.
37. Morin, P., Pouliot, Y., Jiménez-Flores, R., 2006. A comparative study of the fractionation of regular buttermilk and whey buttermilk by microfiltration. Journal of Food Engineering 77, 521-528.
38. Morin, P., Jiménez-Flores, R., Pouliot, Y., 2004. Effect of temperature and pore size on the fractionation of fresh and reconstituted buttermilk by microfiltration. Journal of Dairy Science 87, 267-273.
39. Corredig, M., Roesch, R.R., Dalgleish, D.G., 2003. Production of a novel ingredient from buttermilk. Journal of Dairy Science 86, 2744-2750.
40. Roesch, R.R., Corredig, M., 2002. Production of buttermilk hydrolyzates and their characterization. Milchwissenschaft 57, 376-380.
41. Morin, P., Britten, M., Jiménez-Flores, R., Pouliot, Y., 2007b. Microfiltration of buttermilk and washed cream buttermilk for concentration of milk fat globule membrane components. Journal of Dairy Science 90, 2132-2140.
42. Sahena, F., Zaidul, I.S.M., Jinap, S., Karim, A.A., Abbas, K.A., Norulaini, N.A.N., Omar, A.K.M., 2009. Application of supercritical CO2 in lipid extraction – a review. Journal of Food Engineering 95, 240-253.
43. Fletcher, K., Catchpole, O., Grey, J.B., Pritchard, M., 2006. Beta-serum dairy products, neutral lipid-depleted and/or polar lipid-enriched dairy products, and process for their production. WO 2006/041316 A1.
44. Thompson, A.K., Singh, H., 2006. Preparation of liposomes from milk fat globule membrane phospholipids using a microfluidizer. Journal of Dairy Science 89, 410-419.
45. De Rezende Costa, M., Elias-Argote, X.E., Jiménez-Flores, R., Gigante, M.L., 2010. Use of ultrafiltration and supercritical fluid extraction to obtain a whey buttermilk powder enriched in milk fat globule membrane phospholipids. International Dairy Journal 20, 598-602.
46. Spence, A.J., Jiménez-Flores, R., Qian, M., Goddik, L., 2009a. The influence of temperature and pressure factors in supercritical fluid extraction for optimizing nonpolar lipid extraction from buttermilk powder. Journal of Dairy Science 92, 458-468.
47. Spence, A.J., Jiménez-Flores, R., Qian, M., Goddik, L., 2009b. Phospholipid enrichment in sweet and whey cream buttermilk powders using supercritical fluid extraction. Journal of Dairy Science 92, 2373-2381.
48. Spence, A.J., Jiménez-Flores, R., Qian, M., Goddik, L., 2009c. Evaluation of supercritical fluid extraction aids for optimum extraction of nonpolar lipids from buttermilk powder. Journal of Dairy Science 92, 5933-5936.

49. Rombaut, R., Dejonkheere, V., Dewettinck, K., 2007b. Filtration of milk fat globule membrane fragments from acid buttermilk cheese whey. Journal of Dairy Science 90, 1662-1673.
50. Ramachandra Rao, H.G., Lewis, M.J., Grandison, A.S., 1995. Effect of pH on flux during ultrafiltration of sweet whey and buttermilk. Journal of Dairy Research 62, 441-449.
51. Konrad, G., Lorenz, C., Kleinschmidt, Th., 2012a. Entwicklung eines Verfahrens zur Isolierung von Phospholipiden aus Molkenrahm und Nachweis des gesundheitlichen Potentials der Phospholipide. IGF-Projekt AIF 316 ZBG, Abschlussbericht.
52. Konrad, G., Lieske, B., Faber, W., 2000. A large-scale isolation of native β-lactoglobulin: Characterization of physicochemical properties and comparison with other methods. International Dairy Journal 10, 713-721.
53. Konrad, G., Kleinschmidt, Th., 2008. A new method for isolation of native β-lactalbumin from sweet whey. International Dairy Journal 15, 47-54.
54. Ye, A., Cui, J, Singh, H., 2011. Proteolysis of milk fat globule membrane proteins during in vitro gastric digestion of milk. Journal of Dairy Science 94, 2762-2770.
55. Le, T.T., Van de Wiele, T., Do, T.N.H., Debyser, G., Struijs, K., Devreese, B., Dewettink, K., Van Camp, J., 2012. Stability of milk fat globule membrane proteins toward human enzymatic gastrointestinal digestion. Journal of Dairy Science 95, 2307-2318.
56. Vanderghem, C., Francis, F., Danthine, S., Deroanne, C., Paquot, M., De Pauw, E., Blecker, C., 2011. Study on the susceptibility of the bovine milk fat globule membrane proteins to enzymatic hydrolysis and organization of some of the proteins. International Dairy Journal 21, 312-318.
57. Konrad, G., Kleinschmidt, Th., Lorenz, C., 2012b. Verfahren zur Mikropartikulierung von Molkeneiweiß. DE-OS 10 2012 216 990 A1.
58. Zhang, Z., Yang, R., Zhang, S., Zhao, H., Hua, X., 2011. Purification of lactulose syrup by using nanofiltration in diafiltration mode. Journal of Food Engineering 105, 112-118.
59. Bezelgues, J.-B., Morgan, F., Palomo, G., Crosset-Perrotin, L., Ducret, P., 2009. Milk fat globule membrane as a potential delivery system for liposoluble nutrients. Journal of Dairy Science 92, 2524-2528.
60. Catchpole, O.J., Tallon, S.J., Grey, J.B., Fletcher, K., Fletcher, A.J., 2008. Extraction of lipids from a specialist dairy stream. Journal of Supercritical Fluids 45, 314-324.
61. Morin, P., Pouliot, Y., Britten, M., 2008. Effect of buttermilk made from creams with different heat treatment histories on properties of rennet gels and model cheeses. Journal of Dairy Science 91, 871-882.
62. Wong, P.Y.Y., Kitts, D.D., 2003. A comparison of the buttermilk dry matter functional properties to nonfat dried milk, soy protein isolate, dried egg white, and egg yolk powders. Journal of Dairy Science 89, 410-419.
63. Sodini I., Morin, P., Olabi, A., Jiménez-Flores, R., 2006. Compositional and functional properties of buttermilk: a comparison between sweet, sour, and whey buttermilk. Journal of Dairy Science 89, 525-536.
64. Corredig, M., Dalgleish, D.G., 1997. Isolates from industrial buttermilk: emulsifying properties of materials derived from the milk fat globule membrane. Journal of Agricultural and Food Chemistry 45, 4595-4600.
65. Phan, T.T.Q., Asaduzzaman, M., Le, T.T., Frederick, E., Van der Meeren, P., Dewettinck, K., 2013. Composition and emulsifying properties of a milk fat globule membrane enriched material. International Dairy Journal 29, 99-106.
66. Roesch, R.R., Rincon, A., Corredig, M., 2004. Emulsifying properties of fractions prepared from commercial buttermilk by microfiltration. Journal of Dairy Science 87, 4080-4087.
67. Guggisberg, D., Chollet, M., Schreier, K., Portmann, R., Egger, L., 2012. Effects of heat treatment of cream on physical-chemical properties of model oil-in-buttermilk emulsions. International Dairy Journal 26, 88-93.
68. Konrad, G., Lorenz, C., Kleinschmidt, Th., 2013b. Emulgiereigenschaften der Phospholipide der Kuhmilch. DMW – Die Milchwirtschaft 4, 372-373.
69. Konrad, G., Lieske, B., 1992. Eine verbesserte Methode zur Bestimmung der maximalen Emulgierkapazität von Proteinlösungen. Zeitschrift für Lebensmitteluntersuchung und – Forschung 195, 320-325.
70. Deeth, H.C., 1997. The role of phospholipids in the stability of milk fat globules. Australian Journal of Dairy Technology 52, 44-46.

71. Hamzawi, L.F., 1990. Role of phospholipids and α-tocopherol as natural antioxidants in buffalo butterfat. Milchwisschenschaft 45, 95-97.
72. Newstead, D.F., 1999. Sweet cream buttermilk powders: key functional ingredients for recombined milk products. International Dairy Federation, Brussels, 9902 55-60.
73. Kasinos. M., Tran Le, T., Van der Meeren, P., 2014. Improved heat stability of recombined evaporated milk emulsion upon addition of phospholipid enriched dairy by-products. Food Hydrocolloids 34, 112-118.
74. Thompson, A.K., Couchoud, A., Singh, H., 2009. Comparison of hydrophobic and hydrophilic encapsulation using liposomes prepared from milk fat globule-derived phospholipids and soya phospholipids. Dairy Science & Technology 89, 99-113.
75. Farhang, B., Kakuda, Y., Corredig, M., 2012. Encapsulation of ascorbic acid in liposomes prepared with milk fat globule membrane-derived phospholipids. Dairy Science & Technology 92, 353-366.
76. Henry, J.V.L., Fryer, P.J., Frith, W.J., Norton, I.T., 2010. The influence of phospholipids and food proteins on the size and stability of model sub-micron emulsions. Food Hydrocolloids 24, 66-71.
77. Phan, T.T.Q., Moens, K., Le, T.T., Van der Meeren, P., Dewettinck, K., 2014. Potential of milk fat globule membrane enriched material to improve the whipping properties of recombined cream. International Dairy Journal 39, 16-23.
78. Wiking, L., Fredrick, E., Dewettinck, K., 2009. Phospholipids delay crystallization of milk fat. Milchwissenschaft 64, 376-379.
79. Fedotova, Y., Lencki, R.W., 2008. The effect of phospholipids on milk fat crystallization behaviour. Journal of the American Oil Chemists' Society 85, 205-212.
80. Le, T.T., Van Camp, J., Pascual, P.A.L., Meesen, G., Thienpont, N., Messens, K., Dewettinck, K, 2011. Physical properties and microstructure of yoghurt enriched with milk fat membrane material. International Dairy Journal 21, 798-805.
81. Métais, A., Cambert, M., Riaublanc, A., Mariette, F., 2006. Influence of fat globule membrane composition on water holding capacity and water mobility in casein rennet gel: A nuclear magnetic resonance self-diffusion and relaxation study. International Dairy Journal 16, 334-353.

3.4 Marine phospholipids – fascination of omega-3 phospholipids

Erik Løvaas

Phospholipids are one of the most fundamental prerequisites for organizing matter into living, self-reproducing entities. There is increasing attention on marine phospholipids because of their potential nutritional value and their unique biophysical properties. Central to this interest is the fact that omega-3 phospholipids are rich in eicosapentaenoic acid (EPA) and docosahexaenoic acid (DHA), two fatty acids that control and protect living organisms.

3.4.1 Introduction

Omega-3 phospholipids have their origin in the marine environment, hence the name 'marine phospholipids' (short MPL or omega-3 phospholipids). MPLs deviate from 'standard' phospholipids by having polyunsaturated fatty acid in the 2-position of the glycerol backbone. The unsaturated fatty acid is often EPA or DHA. These fatty acids are essential for human and animal health. A great body of scientific literature agrees that such fatty acids have a number of interesting biological actions, like anti-inflammatory effects (1–3), prevention of cardiovascular diseases (5–7), improved neuronal function (9), learning behaviour (10–12) and visual function (13,14), retardment of cognitive impairment in old people (13,15–19), and preventive effects of breast (20) and prostate cancer (21). There is evidence to suggest that omega-3 LCPUFAs (long-chain polyunsaturated fatty acids), DHA in particular, increase the sensitivity of cancer cells to anticancer agents and pro-oxidants (23,24).

MPLs are soluble in water as well as in oils. This allows the mixing of omega-3 fatty acids into a wide variety of products without requiring microencapsulation. This makes MPLs attractive as a multipurpose ingredient.

- MPLs spontaneously form nanoparticles (called liposomes) when mixed with water. Such nanoparticles are easily accessible to digestive enzymes, and the omega-3 fatty acids of MPL liposomes thus have better bioavailability.
- The fatty acids of MPLs have an enhanced oxidative stability compared to triglyceride omega-3 fatty acids (omega-3 oils). Enhanced oxidative stability is of great value in sensitive and expensive products like baby food, and products for clinical nutrition.

- The head groups of the omega-3 phospholipids are also of great importance, and add to the already described effects of polyunsaturated fatty acids. The most abundant head group in MPL is choline, an essential nutrient involved in several biological processes such as brain development, cell signalling, nerve impulse transmission, and lipid transport and metabolism. Despite the fact that humans can synthesize choline in small amounts, consumption through the diet is necessary for maintaining several biological functions and overall health throughout the life span (25,26). Choline is considered as an essential nutrient (27).

MPL can be obtained from fish roe, from krill, and by extraction from algae.

3.4.2 The beginning of life – fascination of omega-3 phospholipids

3.4.2.1 Chemical evolution

The importance of lipids and membranes in relation to the problem of the origin of life was recognized about three decades ago by Haldane when he said, "The critical event which may best be called the origin of life was the enclosure of several different self-reproducing polymers into a semipermeable membrane" (28). Biological membranes have to balance the need for keeping nutrients enclosed as well as the need for communication with the environment. Controlled exchange of molecules over the membrane requires fluidity. This problem was solved by introducing long-chain polyunsaturated fatty acids, like EPA and DHA, into the phospholipid membrane. The introduction was of essential importance for the development of microalgae, the first precursors for life on our planet. More will be told in a later section.

Phospholipids are the building blocks of biological membranes, and constitute the border between the internal milieu of the cell and the external environment. Phospholipids are one of the most fundamental prerequisites for organizing matter into living, self-reproducing entities. When phospholipids are mixed with water they spontaneously aggregate into microspheres (liposomes) where a phospholipid membrane encloses an internal space.

Although most of the early research on the origin of life focused on the chemical reactions leading to monomers such as amino acids and nucleobases, a few pioneers speculated on the origin of membrane structure. Goldacre (29) proposed that bilayer membranes could be formed in the ocean when lipid monolayers at the air – water interfaces were stirred by wave action. At about the same time, Alec Bangham was working at the Animal Physiology Institute a few miles south of Cambridge, England, and became interested in the properties of phospholipids. He observed that such compounds were not soluble in water, but instead assembled into microscopic globular structures that exhibited birefringence when viewed with polarized light. The birefringence was a clue that the lipids comprising the globules might have an organized internal structure, so Bangham and Horne (30) examined the globules with the first electron microscope that became available to their laboratory. They were rewarded when the image showed that the globules were in fact multilamellar lipid bilayers. This result was published in the *Journal of Molecular Biology* in 1964, and in a second paper published in the same journal a year later Bangham et al. (30) demonstrated that the bilayers in the liposomes were permeability barriers that could maintain gradients of potassium or sodium chloride for many days. These two papers have now been cited in the literature over 5,000 times because they were the first indication that certain lipids spontaneously assemble into membranous structures now called liposomes.

The major chemical components of contemporary cell membranes are lipids, primarily in the form of phospholipids. Bangham and Horne (30) originally demonstrated that phospholipid molecules self-assemble into closed vesicles whose properties resemble those of cell membranes. That is, they are osmotically active, relatively impermeable to ions and non-polar compounds like glucose, and can be given selective permeability functions by appropriate additions of other materials. These vesicles have come to be called liposomes, and are presently one of the most widely accepted model membrane systems. For example, it has been demonstrated that a mixture of lipids, light-trapping pigment and the mitochondrial coupling factor can be constituted into membranes that transduce light energy into an electrochemical proton gradient and couple it to ATP synthesis (32).

In 1999 Horrobin suggested that phospholipids and fatty acid metabolism play a major role in human evolution. His "fat utilization hypothesis" unified intriguing work from paleoanthropology, evolutionary biology, and genetic and nervous system research in a novel and coherent lipid-related context (33).

3.4.2.2 Biological evolution

The starting point for life on earth was the single-celled organisms called archaea. Archaea adapted well to life in inhospitable environments and they can still be found in extremely hot surroundings like the ones that were typical of early Earth. Archaea also thrive in salt lakes and highly acidic conditions, as well as in oxygen-starved anaerobic conditions, such as those found on the ocean bed and in hot springs. Archaea differentiated into microscopic single-cell microalgae, typically found in water systems such as oceans, lakes, rivers and streams. These microalgae are even today the prime producers of omega-3 fatty acids and MPLs. In fact, other marine sources of omega-3 oil and MPL, such as fish and krill, do not produce omega-3 fatty acids, nor MPL, by themselves, but rather accumulate these essential compounds from the prime microalgae producer.

The finding of prokaryotic microfossils being 3 to 3.5 billion years old suggests that the emergence of a primordial organism on Earth occurred 4 billion years ago or earlier. Darwin proposed already in 1871 that primordial organisms had developed because of a preceding chemical evolution. Darwin's proposal was elaborated on by Pflüger in 1885 (34), Oparin in 1924 (35), Haldane in 1928 (36) and Calvin in 1969 (37). The microalgaes are the starting point of the oceanic food chain. These organisms use photosynthesis to convert water and carbon dioxide, in the presence of sunlight, to a nutrient-rich biomass. Oxygen is produced as a by-product. Melvin Calvin, together with Andrew Benson and James Bassham, were the first to outline the basic principles of photosynthesis. Together they were awarded the Nobel Prize in Chemistry in 1961 for discovering the Calvin cycle.

The simple equation runs like this:

$$\mathbf{CO_2 + H_2O + light \rightarrow O_2 + organic\ matter}$$

It is estimated that 40% of the current world total photosynthesis occurs in the oceans, even if photosynthetic unicellular organisms account for only 0.1% of the total biomass of the earth. The rate of the photosynthesis is very fast, and the biomass produced is enormous.

3.4.2.2.1 Membrane fluidity – the need for omega-3 fatty acids

Surprisingly, there is massive photosynthetic activity in the cold Arctic waters like the oceans of the North Atlantic. It is known that a decrease in water temperature will induce production of polyunsaturated fatty acids, including EPA and DHA, while the amount of monounsaturated and saturated fatty acids is reduced at low temperatures (38–40). An increase in the unsaturation index is an adaptation mechanism that allows algae to maintain membrane fluidity at low temperatures (41,42). Algae are *the* prime producers of EPA and DHA. These fatty acids are essential for all organisms higher up in the nutrition chain. There are two reasons for this: (1) higher organisms have low ability to produce EPA and DHA in any significant amount, and (2) EPA and DHA play an essential role as regulators of a complex metabolic network, as well as building blocks for fluid phospholipid membranes. Interest is focused on DHA, since this fatty acid is of particular importance for eye and brain nutrition. DHA is highly concentrated in the brain, and is important for brain function by regulation of cell survival and neuroinflammation (43–47).

The presence of omega-3 fatty acids in cell membrane phospholipids increases membrane fluidity, modulates the opening of ion channels, and modifies the recognition and function of membrane receptors and enzymes. They also modulate several nuclear transcription factors and influence the expression of genes (48). These effects result in a further attenuation of the inflammatory response, an improved cellular antioxidant defence, and increased fat oxidation rather than deposition within the cell (49).

3.4.2.2.2 Oxidative stability

Omega-3 phospholipids are surprisingly stable to oxidation. It is known that phospholipids such as phosphatidylcholine (PC), phosphatidylethanolamine (PE) and phosphatidylinositol (PI) in themselves are known as antioxidants (50,51). Antioxidative properties of phospholipids (PL) have been demonstrated through their addition to processed vegetable oils and animal fats (52,53). Fish roe MPL containing high levels of EPA and DHA are more stable against oxidation than other marine oils containing the same or even less EPA and DHA. Various mechanisms have been proposed for the oxidative stability (54): 1) synergism between PLs and tocopherols; 2) chelation of pro-oxidant metals by the phosphate groups; 3) formation of Maillard-type products between the amino polar head and carbonyl oxidation products; and 4) action as an oxygen barrier between oil and air interfaces (55). In spite of the inherent stability of MPL it may be wise to further expand the storage time by the addition of specific antioxidants.

3.4.2.2.3 Bioavailability

Getting lipids into a water-soluble form that can be transported around the body is a major biological challenge. The obstacles that have to be overcome include substrate solubility, transport, digestion and resynthesizing etc. Many of the finer details of these processes are not well understood. However, accumulated evidence suggests that marine phospholipids are better delivery agents for n-3 fatty acids than triglycerides. To understand the situation we need to understand the essential differences in the metabolism of triglycerides versus phospholipids.

Phospholipids have the unique property of being soluble both in aqueous and lipid environments. When phospholipids are introduced into an aqueous environment they spontaneously form nanostructures (liposomes). Such structures are characterized by consisting of a membrane surrounding an internal space. Liposomes are versatile structures that may be applied both as delivery vehicles (e.g. site-specific drug delivery agents) as well as for the encapsulation of nutrients.

In an effort to increase the uptake of n-3 fatty acids there has been a drive towards identifying phospholipids that contain a high level of n-3 fatty acids. Phospholipids are more easily digested than triglycerides, and offer unique means to increase the bioavailability of the essential n-3 fatty acids.

3.4.2.3 Biological functions of EPA/DHA

DHA can be synthesized de novo in mammals from a-linolenic acid (ALA, 18:3n-3), but the synthesis is slow and DHA must therefore be obtained in the diet primarily through fish, nutraceuticals and functional foods (56). In North America the mean dietary intakes of DHA in adult (20–39 years of age) men and women are about 70 and 60 mg/d respectively (57). A recent survey recommended a daily intake of 500 mg EPA/DHA per day (58). Preformed DHA is found primarily in marine sources, while ALA is found in seeds and seed oils including flax, canola and soy (56).

Polyunsaturated fatty acids (PUFAs) are designated as 'essential' for good health as their metabolic precursors cannot be synthesized in the body and must be ingested by food intake. Increased dietary intake of omega-6 leads to the production of prostaglandins, thromboxanes and leukotrienes and interferes with the incorporation of omega-3 in cell membrane phospholipids. Omega-3 has the most potent anti-inflammatory effects. Inflammation is at the base of many chronic diseases, including coronary heart disease, diabetes, arthritis, cancer and mental health. Dietary intake of omega-3 fatty acids may prevent the development of many diseases (59).

3.4.2.3.1 Omega-3 phospholipids and the brain and eye

About 150,000 years ago a mutation in the phospholipid metabolism happened that made room for an improvement in memory and cognitive skills. The mutation paved the way for more complex connections between nerve cells. The final key mutation happened about 130,000 years ago (in humans, but not in Neanderthals) and was conceivably related to the phospholipase A2 cycle. This mutation is thought to have tremendously amplified cognitive flexibility, creativity and inventiveness by increasing the phospholipid-related connectivity between individual neurons. These mutations also form the basis for psychiatric disorders (depression schizophrenia etc) when there are dietary deficits in the range and amounts of essential fatty acids needed by the brain (60).

The key mutations that differentiate us from the Neanderthals and from the great apes are in the gene coding for proteins which regulate fat metabolism, and particularly the phospholipid metabolism of the synapses of the brain. It is known that humans and chimpanzees share close to 99% of their genetic material. Information is rapidly accumulating that the 1% difference is related to the fat metabolism of the brain, leading to a much-elevated innovative force in humans as compared to *Homo neanderthalensis*. It is important to consider that fat constitutes a large share of the brain (by weight, lipids form some 60% of the brain), and that phospholipid membranes are the key determinants of an almost inconceivably complex and beautiful neuronal network. It should also be noted that the polyunsaturated fatty acids (20:4n-6 and 22:6n-3) are key components of grey matter phospholipids, and that the turnover rate of these fatty acids is very high. The brain thus has a major transport problem bringing the essential fatty acids across the blood – brain barrier.

3.4.2.3.2 Omega-3 and the brain

Though the brain is able to make some of what it needs, it is surprisingly dependent upon raw materials from the diet, notably fatty acids (61 – 64). If the right fat is not supplied, brain structure and brain function is altered. The brain's requirement for highly specified fats, coupled with the poor dietary choices of most people living on modern diets, has created a problem in our society. Particular interest has been ascribed to omega-3 fatty acids. Recent findings show the positive effects of these fatty acids on:

- Bipolar depression. It is known that EPA and DHA gives a significant reduction in the symptoms of depression (65).
- Age-related memory loss and Alzheimer's disease. It has been shown that adults with dementia, Alzheimer's and cognitive problems have a low level of DHA in the blood, and it is suggested that deficiency of these fatty acids in the diet may be a risk factor for cognitive impairment (18,61,66,67).
- Reading and learning problems in children. Children with fatty acid deficiency have been found to have poorer reading abilities (68,69).
- Intelligence in children. Breastfed children were compared with formula-fed children who had none of the fatty acid DHA in their formula. The breastfed group had a significantly higher IQ score at 8 years of age (70).

It is highly desirable to produce food ingredients that meet the needs of the consumers by having an attractive texture and flavour profile and good oxidation stability, and at the same time having a healthy composition.

3.4.2.3.3 Pregnancy nutrition and infant formula

Important brain growth occurs during pregnancy and the first few years of life, and there are several nutritional building blocks that are vital for brain growth and visual function (61,63,71–75). The most important of these are omega-3 fatty acids (DHA) and omega-6 fatty acids (arachidonic acid, or ARA). Many expectant mothers aren't aware that what they eat makes a substantial difference to the health of their unborn child.

The DHA content of breast milk depends on the mother's diet. Most pregnant women have a low intake of omega-3 fat, which ultimately may have consequences for the unborn and the suckling child. It is thus advised that the expectant mother should eat fish several times a week to increase brain-building

DHA, or alternatively that dietary supplements should be considered when fish intake is not sufficient. Marine phospholipids should be likely candidates for such products.

There has long been a concern that lipids in infant formulas may be absorbed to a lesser extent than breast milk. It has been suggested that absorption of fish oil in preterm infants may be poor as the amount of DHA required in formulas was almost a magnitude greater than usually provided by human milk (76). Low absorption means that more lipids are left in the intestine, giving rise to adverse effects. In particular more triacylglycerols (TAG) and fatty acids will appear in the colon and lead to conditions like intestinal obstruction, an increase in colon fatty acid soap content, and altered microbial composition (77,78)

The addition of nutritional marine phospholipids may have several advantageous effects in neonates. Firstly, there is evidence that TAG absorption may be impaired when the supply of exogenous PL is insufficient for micelle formation during fat absorption (79). Secondly, nutritional phospholipids would only require the action of phospholipase A2 on one fatty acid to produce free fatty acids and 1-lysophospholipid that would be readily absorbed, utilizing more of the dietary lipid, leaving fewer lipids left in the intestine. In premature infants it has been found that DHA from egg PL is better absorbed than DHA from breast milk and TAG from single cell sources. Improved absorption of lipid and long-chain PUFA from egg phospholipids has also been shown in other studies (80,81). It has also been found that giving omega-3 eggs (presumably with high content of PL DHA) to infants of 6–12 months increased erythrocyte DHA concentration to 30–40% above that of infants fed breast milk and can be used to maintain a high omega-3 PUFA ratio. Phospholipids are also required for intestinal lipoprotein formation and lipid distribution outside the enterocytes (82). Lack of nutritional PL could lead to impaired enterocyte lipoprotein synthesis causing lipid accumulation in the cells.

3.4.2.4 Applications of MPLs

- As an omega-3 extract (oil).
- As an omega-3 ingredient (powder).

The present imbalance in the fatty acid intake represents serious trouble for the health situation in the general population, and there is a strong desire to improve the situation by introducing new products on the market with a high level of omega-3 fatty acids and low level of omega-6 fatty acids. Numerous studies have demonstrated that omega-3 fatty acids have positive effects on preventing or curing diseases like heart disease, cancer, rheumatoid arthritis, diabetes, ulcerative colitis, allergies, eczema, skin thickening, weight gain and a host of other diseases. Researchers are now also linking inadequate intake of these omega-3 fats in pregnant women to premature birth and low birth weight (71,72,83,84), and to hyperactivity in children (68).

Marine phospholipids have potential applications in diverse areas like nutraceuticals (functional food, dietary supplements, sports nutrition, maternal nutrition and baby food), pharmaceuticals (liposomes – site-specific drug delivery, clinical nutrition), cosmeceuticals (liposomes) and biochemistry (model membranes, liposomal test systems for antioxidants).

The word 'potential' is used deliberately, because many of the mentioned applications have not been tested at the present time. Marine phospholipids are new on the market, and their range of applications has yet to be determined. However, intensive development work is going on, and it is probable that we soon shall see a number of new products.

There is increasing attention on marine phospholipids because of their potential nutritional value and their unique biophysical properties. Central to this interest is the fact that enzymes that are involved in lipid metabolism (lipases) essentially act in an aqueous environment. Water-soluble lipases have better access to water-soluble substrates (like phospholipids) than to water-insoluble substrates (like triglycerides). In sum this means that omega-3 fatty acids from marine phospholipids are more easily accessible for catabolic processes than triglycerides.

Applications include use in brain nutrition (any age); clinical nutrition; pregnant women; baby food and food for children and youngsters; sport nutrition; nutrition for the elderly; additives to flour; bread and bakery products; dairy products (cheese); sauces, soups and dressings; fish products; emergency rations; snacks; chocolate bars; animal nutrition (pet food); chicken feed; aquaculture (start feed for fish) and as a starting point for subsequent processing.

Dietary fat is mainly composed of TAG, with PLs accounting for 3–6% of total fat intake (85).

The daily intake of PLs is not exactly known, however the daily intake of PC/day is estimated to be 2–8 grams (86).

3.4.3 Generations of omega-3

The stages of the omega-3 development can conveniently be divided into three 'generations'. The differences between the 'generations' reside in (1) the purity of the product, (2) the documented health effects of the product, and finally (3) the source and functionality of different omega-3 sources. Marine omega-3 phospholipids are the most recent generation.

3.4.3.1 First generation

The first generation of omega-3 products is more than 1,000 years old and is constituted of largely unrefined fish oils. In 1852 Peter Möller, a Norwegian entrepreneur and pharmacist, designed a method for producing a purified fish oil by a process involving rapid heating of fish liver by steam. In less than one hour the oil (designated as 'medical cod liver oil') was liberated and collected. The product was sold as a pharmaceutical. Although the scientific documentation was meagre at the time there was nevertheless a considerable body of anecdotal knowledge on the health effects of cod liver oil.

3.4.3.2 Second generation

The start of the second generation of omega-3 happened when the anecdotal stories were solidified by scientific documentation in the early seventies when Bang and Dyerberg observed that Eskimos had much fewer heart problems, assumedly because of high consumption of marine organisms containing much omega-3 lipid (87,88). In 1982 a demographic study found that the life expectancy level was higher in people taking more omega-3 fatty acids. The same year Bergstrom, Samuelsson and Vane earned a Nobel Prize for the discovery that omega-3 deficiency correlated with a number of diseases (89). The new understanding represented a paradigm shift when it was realized that DHA had a vital role in the nervous system (90). It became apparent by the end of the 1990s that the requirement for DHA probably is the primary reason why n-3 fatty acids are essential (91). This tentative conclusion was supported by subsequent findings indicating that DHA enhances cognition (92) and synaptic function (93), and that it is converted to lipid mediators that facilitate the resolution of acute inflammation (94), provide neuroprotection (95), and promote hippocampal development (96). Lipids suddenly became much more than a simple source of calories. In 1985 Ralph Holman discovered that heart failure could be corrected by high doses of omega-3 oils (flaxseed oil and fish oil) (see review by Holman (97)). The rise of the new era was supported by an increasing knowledge on the health effects of omega-3 fatty acids coupled with advances on purification technology that together brought new innovative products into the marketplace. A new word 'nutraceutical' was coined at that time.

3.4.3.3 Third generation

The start of the third generation of omega-3 products can arbitrarily be set to 1995, when a strategic research programme was initiated by the Norwegian Research Council. The purpose of this programme was to identify new feed components for fish fry, so that aquaculture could diversify into new arenas. Some attractive fish species had carefully been selected, and artificial feeding experiments were started. Initial problems included low survival rates of the hatched fry, slow growth rates, and many deformities among the grown fish. All these problems seemed to be related to feed composition. It was therefore imperative to identify the problem and to search for remedies. A group of Scottish researchers had already in 1989 (98) proposed that fish roe could be an ideal raw material for starter feed since roe was known to contain high amounts of omega-3 fatty acids, an essential feed component. The aims of the strategic programme were to: (1) identify a good source for small-scale production of MPLs as a feed component, (2) design methods to prevent lipid oxidation of the MPLs, and (3) perform feeding experiments on fish fry. The work was carried out at the Norwegian College of Fishery Science at the University of Tromsø. It was soon realized that fish roe was an excellent source of MPLs and DHA, and that extracts of fish roe contained efficient antioxidants, giving a naturally stabilized extract of omega-3 phospholipids. From other researchers working in adjacent fields it was demonstrated that biochemical and behavioural abnormalities were partially reversed by a dietary phospholipid supplement, especially omega-3-rich egg yolk extracts or pig brain (16). These very interesting observations were not met with a flurry of research papers, possibly because omega-3 phospholipids were available only in small amounts and at very high prices. Research focusing on nutritional aspects of omega-3 phospholipids simply lacked the impetus of having a sufficient, sustainable and economically realistic resource available. The time was not ripe for any large-scale venture into this very exciting research field. However, it was realized that omega-3 phospholipids from certain fish roes had a very attractive fatty acid composition, and that the product was superior to anything presently available on the nutraceutical market. It was also thought-provoking that MPLs, so important for the brain and vision development for fish larvae, also could be essential for brain development in humans. One key factor for carrying out big-scale nutritional research and marketing on omega-3-rich phospholipids was missing: no MPLs were available on the market. Only slightly similar products could be acquired in small amounts and at very high prices.

It was soon realized that MPLs could be obtained from different marine sources such as krill, fish meal or marine algae. All these sources had strong and weak sides. Krill, for example, could be caught in enormous quantities, but the fatty acid composition was not quite as good as wanted. The first task was to screen for available raw material and to devise scaling-up strategies going from gram quantities in the lab to ton quantities in the market.

A screening for marine raw material that contains high levels of omega-3 phospholipids was performed. A source that was sustainable, present in sufficient amounts for industrial-scale consumption, and was free from environmental polulutants was identified. The preferred source for omega-3 phospholipids turned out to be roe from the wild. The product was exceptionally rich in DHA, the omega-3 fatty acid of fundamental importance for the development of visual and brain functions.

3.4.4 Sources of MPLs

Marine phospholipids can be isolated from a variety of raw materials. Current raw materials include krill, fish roe and fish meal, which all have characteristic properties making them more or less suitable for phospholipid extraction. Important factors to consider when choosing a raw material are:

- Price of raw material.
- Level of marine phospholipids in raw material.
- Quality and freshness of raw material.
- Catch volume and stability of catch over an extended period.
- Available process technology.
- Pollutants in raw material.
- Laws and regulations (e.g. relating to 'novel food').

Marine lipids can be derived from several fish, shellfish and algae, but also from Antarctic crustacean krill (*Euphausia superba*) and from marine industry by-products such as fish roe (fully ripe internal egg masses in the ovaries of fish) (99 – 101). Fish contains between 1 and 1.5% PLs, while the amount of PLs in oil extracted from krill is typically around 40% of its total lipids (102,103). PC derivatives are the predominant phospholipids present in salmon, tuna, rainbow trout and mackerel; the second most abundant phospholipid is PE. PI, PS, lyso-PC and sphingomyelin are also present in minor quantities (104). Fish roe from herring, salmon, pollock and flying fish contain between 38 and 75% of their lipids in the form of PLs, with PC being the predominant lipid class. Notably, the main class of marine-derived PLs is PC, predominantly binding with unsaturated omega-3 PUFAs, with the most prevalent being EPA and DHA, but also stearidonic acid and docosapentaenoic acid. Marine organisms are enriched in these PUFAs by the aquatic food chain, since the main source of omega-3 PUFAs are algae that can synthetize them de novo.

3.4.4.1 Krill

Krill *(Euphausia superba)* is a small shrimp-like zooplankton (crustacean) that plays a key role in the Antarctic food web. Krill is one of the biggest biological resources on earth. The biomass of Antarctic krill is estimated at 1,350 million tons, about five times the weight of the human population (105). Krill swarms can occupy an area of 500 sq km, and hold up to 20,000 individuals per cubic metre. Biomass densities attain 10 – 16 kg/m^3. Unlike other euphausiid species, *Euphausia superba* feeds almost exclusively upon phytoplankton. This contributes to the very high abundance and potential catch of krill, since there are comparatively small losses between the primary production and the potential harvest by man. Krill occurs in groups or large swarms and occupies a niche similar to that of the herring in the North Atlantic, since large schools of pelagic fish are absent. They attain a size a size of up to 7 cm (normal range 4.5 – 5.0 cm) and feed primarily on phytoplankton or sea ice algae.

The catch limit on krill in the South Atlantic region is set at 4 million tons (106), but the annual krill catch has never amounted to more than 550,000 tons (107,108). It is still the largest crustacean fishery in the world. The catch is by trawls with small mesh size and big openings, and catch efficiency can be 5 – 20 tons/hour. The price for krill is approx. USD 700/ton delivered at the Falkland Islands. Transport from South America to Canada or Northern Europe is about USD 150/ton.

There is a considerable variation in the amount of lipid and the lipid composition of krill (109). The lipid content may vary between 12.5 and 0.5% (w/w wet mass) with about 25% as phosphatidylcholine (110). Specific production technologies give rise to a phospholipid-enriched oil, containing as much as 40–60% phospholipids, of which phosphatidylcholine is the dominating (>75%) species. The phospholipids have a high level of omega-3 fatty acids (approx. 40%), and the EPA:DHA ratio is 25:15 (which makes it similar to 18:12 oil from South America). Such oils are used as dietary supplements.

3.4.4.2 Roe

Roe has been identified as a good source for marine phospholipids (111).

Fish roe contains supplies of PUFA that are optimized nutritionally for growth of the embryo and also for growth of the resultant larva up to the stage of yolk sac absorption. Numerous analyses of the lipids of fish eggs have been published and, although the literature is fragmentary, PUFA clearly can account for more than 40% of the total fatty acids in marine fish eggs, reflecting at least in part the high concentrations of phospholipids in the egg (7–9). Fish roe is an excellent source of omega-3 phospholipids. Roe has several advantages compared to krill. The roe has a green profile, saving the environment from excessive energy usage in the catch step, and expanding the total use of the catch. The lipid composition of the roe is fairly constant over the whole development process, so that unripe roe (not well suited for consumption) is still a good source for the extraction of MPLs. In addition, the composition of the roe is under biological control. The survival of the species is critically dependent on an egg containing all the elements necessary for developing into a fish. For example, the fatty acid profile in the eggs is in conjunction with the need for a high level of DHA necessary for the development of brain and eyes. This high content of DHA is also of importance for human and animal consumers as well as for applications of MPLs. Roe is a by-product available in big quantities. The catch is carried out in accordance with advice from the Marine Stewardship Council and is collected from a sustainable and well-managed marine resource. The complete value chain from catch to customer is traced according to BRC standards. There is thus complete traceability on catch site, catch time, supplier and production.

Herring are forage fish, mostly belonging to the family Clupeidae (Fig. 1). They often move in large schools around fishing banks and near the coast. The most abundant and commercially important species belong to the genus *Clupea,* found particularly in shallow, temperate waters of the North Pacific and the North Atlantic oceans, including the Baltic Sea, as well as off the west coast of South America. Three species of *Clupea* are recognized, and provide about 90% of all herring captured in fisheries. Most abundant of all is the Atlantic herring, providing over half of all herring capture.

Figure 1.
Atlantic herring (*Clupea harengus*).

The total marine production of captured fish (excluding anchoveta) is in the order of 80 million tons/year. The main fishery nation is China (with approximately 13 million tons/year). Small Norway, with a population of only 5 million, has a catch of approximately 2.0–2.5 million tons/year. The catch of Atlantic herring *(Clupea harengus)*, which is an excellent source of roe, varied between 1.0 and 2.5 million tons/year between 1970 to 2014 (4).

Atlantic herring (*Clupea harengus*) can occur in huge schools of up to 500 million individuals. The roe percentage varies linearly from 5% to 20% over a period of 200 days; however, the catch season lasts for only 2–3 months in the winter/spring season. The adolescence biotope for the immature herring is the clean Arctic waters of the Barents Sea and Svalbard. In the spawning season the herring migrates to the Norwegian Sea. Approximately 10% of the total catch (males and females combined) is roe. It should be noted that fluctuations in catch are partially due to the natural variability of pelagic stocks such as capelin and herring. The catch is variable (Fig. 2).

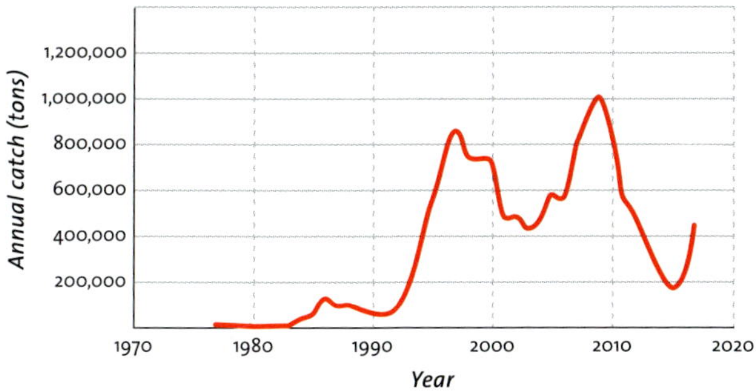

Figure 2.
Catch of Norwegian spring-spawning herring.

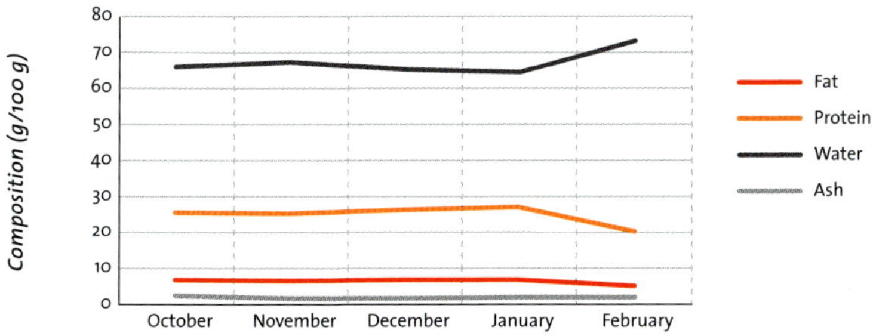

Figure 3.
Proximate composition of Norwegian spring-spawning herring.

The composition of the roe is surprisingly stable during the whole maturation process, except for the very last stage before spawning, where the water content is increasing (112) (Fig. 3). The most attractive quality for caviar production is the proportion that is classified as ripe roe, amounting to about one-third of the size of the total resource. Only a fraction of the ripe roe is used as caviar. The remaining roe is of low value but of high quality and high value for special applications, like MPL products.

The fat content of roe from Norwegian spring-spawning herring stays stable between 6.5 and 6.8% in the period October to January. At this time swelling of the eggs takes place and the fat content drops to 5.1% in February (Fig. 3).

The size and the availability of the resources are important, as is the technology for separating the roe. There are practical limitations on the amount of roe that can be processed, but the practical limitations exceed with a good margin the present market size. In Norway the practical limit presently is in the order of 10,000 tons/year. The biological resource is significantly higher than our industrial capacity. At the present time herring is the preferred biological resource, although capelin or other species may take its place in the future.

The roe composition of a number of species has been investigated by Tocher and Sargent (113); Table 1 is based on their data.

Table 1.

The roe composition of a number of fish species.

Analysis	Cod	Herring	Haddock	Whiting	Saithe	Capelin	Sand eel
Moisture	74.0	74.0	86.0	85.0	72.0	70.0	64.0
Lipid content	13.2	14.6	10.7	10.6	15.4	26.3	19.7
Polar lipids	71.7	69.0	71.3	61.2	66.5	50.7	23.4
EPA	15.3	13.7	12.6	13.3	11.5	19.0	16.7
DHA	28.6	31.4	27.6	30.3	27.7	24.6	25.5
EPA+DHA	**43.9**	**45.1**	**40.2**	**43.6**	**39.2**	**43.6**	**42.2**
DHA/EPA	**1.87**	**2.29**	**2.19**	**2.28**	**2.41**	**1.29**	**1.53**
Neutral lipids	28.3	31.0	28.7	38.8	33.5	49.3	76.6
EPA	10.9	9.7	14.0	15.0	9.7	13.1	12.4
DHA	16.0	17.1	14.5	22.3	14.8	14.1	18.8
EPA+DHA	**26.9**	**26.8**	**28.5**	**37.3**	**24.5**	**27.2**	**31.2**
DHA/EPA	**1.47**	**1.76**	**1.04**	**1.49**	**1.53**	**1.08**	**1.52**
Total lipids	100.0	100.0	100.0	100.0	100.0	100.0	100.0
EPA	12.2	10.9	10.6	11.2	9.3	15.1	13.3
DHA	22.3	24.4	21.4	24.2	21.5	19.2	20.3
EPA+DHA	**34.4**	**35.3**	**31.9**	**35.3**	**30.8**	**34.2**	**33.7**
DHA/EPA	**1.83**	**2.24**	**2.02**	**2.16**	**2.31**	**1.27**	**1.53**

The fatty acid composition of the roe is of prime importance for benchmarking against alternative MPL sources like krill and algae oil. Important parameters for the benchmarking of herring oil vs krill and algae oils are shown in Tables 2 and 3. Numbers to look for are the EPA+DHA (line 3 in Table 2), the DHA/EPA ratio (line 4 in Table 2), and the amount of DHA+EPA in the MPL fraction vs the triglyceride fraction. It is desirable to have as much as possible of EPA+DHA in the in the MPL fraction. Extracted oils from herring roe contain approximately 60% of EPA+DHA in the MPL fraction, and 40% in the triglyceride fraction. DHA is the fatty acid of prime importance, and the number should be high. The fatty acids are distributed in MPLs and triglyceride.

Table 2.

Composition and amount of triglyceride (TG) and MPL in Norwegian spring-spawning herring roe.

			Units: mg/g lipid		
		Fatty acid	**Oct**	**Dec**	**Feb**
MPL	1	EPA	43	37	47
	2	DHA	118	98	127
	3	EPA+DHA	161	135	174
	4	DHA/EPA	3	3	3
TG	1	EPA	27	31	26
	2	DHA	70	68	73
	3	EPA+DHA	96	100	98
	4	DHA/EPA	3	2	3
% DHA+EPA in MPL			63	58	64
% DHA+EPA in TG			37	42	36

Table 3.

Distribution of fatty acid classes in MPL and triglyceride (TG).

	Units: mg/g lipid	
Fatty acid class	**MPL**	**TG**
Saturated	89.0	51.2
Monounsaturated	43.1	109.6
Polyunsaturated	125.7	142.9
Omega-3	121.2	134.3
Omega-6	4.5	6.6
Omega-3/Omega-6	26.8	20.3

Table 2 displays summary data for fatty acid classes in herring roe. Of particular interest is the amount of polyunsaturated fatty acids (which should be high), and the ratio between omega-3 to omega-6 fatty acids (which also should be high to correct for the unbalanced composition of our western food habits) (114,115).

Based on a combination of several biological parameters and the production price it is possible to make a price/value estimation. Herring roe is competitive compared to both krill and algae.

With a coastline of more than 83,000 km, including fjords and islands, Norway is one of the world's leading nations regarding production from marine fisheries and aquaculture. The fisheries sector has always played a key social and economic role, nationally and regionally, and has been the basis for settlement and employment along the entire Norwegian coast. The vast marine areas under Norwegian jurisdiction are among the most productive in the world and provide ideal conditions for aquaculture production. Fishing and fish farming represented 0.7% of the gross domestic product (GDP) in 2010, with 12,000 full-time equivalents employed in the sector. In 2009, Norway ranked 11th in global capture fisheries production and seventh in aquaculture production. It produced 3.5 million tons of seafood, about 25% coming from the aquaculture industry (http://www.fao.org/fishery/facp/nor/en).

Norway, in terms of value, is the world's second-largest exporter of fish and fish products by value. Its main markets, in terms of export value, are the European Union (EU), the Russian Federation, Japan, China, Ukraine and the United States of America. With USD 8.9 billion, seafood products were the second most important export item in 2012 and made up 6% of total Norwegian merchandise exports. The biggest share is represented by oil and gas (70%). Annual per capita fish consumption amounted to an estimated 53.1 kg in 2009.

Sources of dietary EPA/DHA for humans are principally composed of fish and fish oil. The wild capture of fish peaked in the early 1990s and has been fairly constant since at about 90 million tons of fish. The yield of fish via aquaculture has grown linearly since the early 1990s and has now reached nearly the same level as the wild fish catch. As sources for EPA/DHA, anchovies are the largest single species of wild fish and much of this is obtained from the Peruvian fishery. Menhaden, cod, whiting, carp, mackerel, tuna, salmon, pollock, capelin and sardine, in roughly that order, are significant species of fish harvested. Much of the anchovy catch is consumed in the extraction fishery for fish oil and fish protein. Our best estimates of the total EPA/DHA produced is about 530,000 tons and this estimate comes from the fish industry trade association the Global Organization for EPA and DHA Omega-3s (GOED) (Fig. 4). (58)

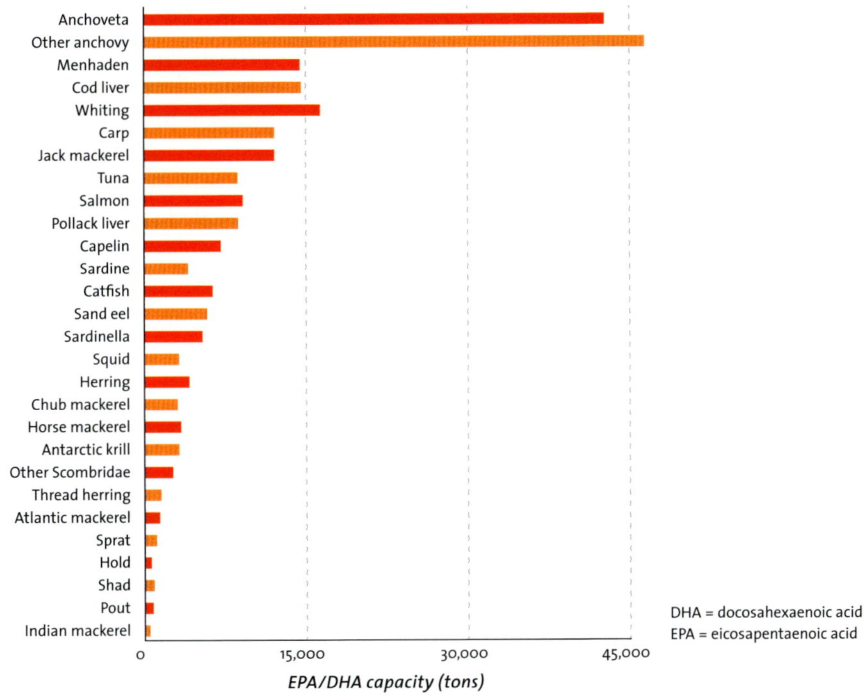

Figure 4.

Major fish species which contribute to EPA/DHA production. Data compiled by the Global Organization for EPA and DHA Omega-3s (GOED) and based on a calculation of FAO FishStat data (4).

3.4.4.3 Algae

Algae are the first link in the oceanic food chain and use photosynthesis to convert water and carbon dioxide to nutrient-rich biomass and oxygen in the presence of sunlight. In fact, other marine sources of omega-3 lipids such as fish and krill do not produce the omega-3 fatty acids EPA and DHA, but rather accumulate it from the algae they and their prey consume in their natural environment. Algae are the motor of the oceans. Their ability to capture light energy and conserve it as organic material is the basis for marine life. Algae account for 40% of world photosynthesis, even though they account for only 0.1% of the world's biomass. The photosynthesis happens in the top 20 metres of the ocean where there is enough light to keep the process going.

Considering the prime importance of microalgae for the life on earth we would like to be able to harvest the oceans for microalgae. However, it is not possible to harvest big quantities of microalgae directly from the oceans, because of the very small size of these organisms and the lack of technologies to catch them. Instead, work is going on to bring the production into land-based production units. Considerable work is taking place to establish economic processes. Since CO_2 is an essential input factor, the process units are placed close to locations that produce CO_2 as a by-product of ongoing industrial processes. Bioreactors are located close to industries that have large CO_2 emissions. CO_2 is immediately absorbed by algae and converted into organically bound CO_2 in the form of organic material including triglycerides, glycolipids and MPLs. The type of lipids can be controlled by selecting the right colour of light as well as the right light intensity.

Biochemical and fatty acid composition of a given species of phytoplankton can be modified according to its growth phase (116,117), culture system (118) and growing conditions, i.e. nutrients, temperature, irradiance and light quality (119–121). Groundbreaking work is carried out by a number of research institutions (22,122–126) to make use of CO_2 for the production of organic matter. The work of Da Costa et al. (8) and Marchetti et al. (22) is indicating the potential for land-based industrial production of tailor-made microalgae. Certain strains of algae like *Tisochrysis lutea* are well suited for lipid production (8) (Table 4A). The same strain can also be stimulated by light to produce significant levels of omega-3 phospholipids (22) (Table 4B). Of particular interest is the finding that MPL-DHA (22:6n-3) is very high compared to the MPL-EPA (20:5n-3). This is of great importance when evaluating the nutraceutical and pharmaceutical potential of the product. A high DHA/EPA ratio is preferred.

Tables 4A and 4B.

Component analysis of the algae *Tisochrysis lutea*. The data are compiled from two different publications. (A) Proximate composition is from Table 1 in (8), while (B) Fatty acid profile is from Table 4 in (22). Unfortunately this publication has no data on the relative amount of neutral lipids (NL), glycolipids (GL) and phospholipids (PL).

	g/100 g	
Component	**T**	**T+**
Lipids	43	62
Proteins	26	11
Carbohydrate	20	15
Ash	11	12
Sum	100	100

Fatty acid	NL	GL	PL
20:5n-3	0.3	0.3	0.2
22:6n-3	4.3	2.6	17.3
Σn-6	5.3	2.5	1.7
Σn-3	13.9	36.5	28.4
n-3/n-6	2.7	14.7	18.1
ΣSFA	48.5	34.1	37.0
ΣMUFA	28.3	18.7	16.6
ΣPUFA	19.3	39.6	30.3
SUM	96.1	92.4	83.9

At present, algal oil represents less than 2% of human EPA/DHA consumption, but this source has grown appreciably due to several socially desirable properties including its environmental friendliness, the absence of ocean-borne contaminants, its vegetarian nature, as well as the possibility to manufacture under kosher or halal conditions. However, the fermentation and refining of algal oils is at present more expensive than the production of fish oils and is likely to remain so for some time. Thus, microorganism-based production of EPA/DHA-enriched oils can accommodate greater demand for these nutrients, albeit at a higher price.

3.4.5 Conclusion

Phospholipids are one of the most fundamental prerequisites for organizing matter into living, self-reproducing entities. There is increasing attention on marine phospholipids because of their potential nutritional value and their unique biophysical properties. Central to this interest is the fact that omega-3 phospholipids are rich in EPA and DHA, two fatty acids that control and protect living organisms. The preferred source for omega-3 phospholipids is probably fish roe. This natural resource has a green profile; it has phospholipids of a constant composition and with a high level of DHA and EPA. From an economic viewpoint fish roe is very attractive.

References

1. James, M.J., Gibson, R.A., Cleland, L.G., 2000. Dietary polyunsaturated fatty acids and inflammatory mediator production. Am. J. Clin. Nutr. 71(1 Suppl), 343-348.
2. Ross, R., 1999. Atherosclerosis is an inflammatory disease. Am. Heart J. 138(5 Pt 2), 419-420.
3. Adam, O., 2003. Dietary fatty acids and immune reactions in synovial tissue. Eur. J. Med. Res. 8(8), 381-387.
4. Ababouch, L., 2016. FAO. The state of the world fisheries and aquaculture. In Contributing to food security and nutrition for all. Food and Agriculture Organization of the United Nations: Rome, 200.
5. Siscovick, D.S., et al., 2000. Dietary intake of long-chain n-3 polyunsaturated fatty acids and the risk of primary cardiac arrest. Am. J. Clin. Nutr. 71(1), 208-212.
6. Marckmann, P., Grønbæk, M., 1999. Fish consumption and coronary heart disease mortality. A systematic review of prospective cohort studies. Eur. J. Clin. Nutr. 53(8), 585-590.
7. Anonymous, 1999. Dietary supplementation with n-3 polyunsaturated fatty acids and vitamin E after myocardial infarction: results of the GISSI-Prevenzione trial. Gruppo Italiano per lo Studio della Sopravvivenza nell'Infarto miocardico. Lancet 354(9177), 447-455.
8. Da Costa, F., et al., 2016. Influence of one selected Tisochrysis lutea strain rich in lipids on Crassostrea gigas larval development and biochemical composition. Aquacult. Nutr. 22, 813-836.
9. Loudes, C., et al., 1983. Release of immunoreactive TRH in serum-free cultures of mouse hypothalamic cells. Brain Res. 285(2), 231-234.
10. Wainwright, P., 2003. Nutrition and behaviour: the role of n-3 fatty acids in cognitive function. Br. J. Nutr. 83(4), 337-339.
11. Zimmer, L., et al., 2000. Chronic n-3 polyunsaturated fatty acid deficiency alters dopamine vesicle density in the rat frontal cortex. Neurosci. Lett. 284(1-2), 25-28.
12. Uauy, R., et al., 1996. Role of essential fatty acids in the function of the developing nervous system. Lipids 31 Suppl., 167-176.
13. Horrocks, L.A., Yeo, Y.K., 1999. Health benefits of docosahexaenoic acid (DHA). Pharmacol. Res. 40, 211-225.
14. Jacobson, S.W., 1999. Assessment of long-chain polyunsaturated fatty acid nutritional supplementation on infant neurobehavioral development and visual activity. Lipids 34(2), 151-160.
15. Kalmijn, S., et al., 1997. Polyunsaturated fatty acids, antioxidants, and cognitive function in very old men. Am. J. Epidemiol. 145, 33-41.
16. Bourre, J.M., 2004. Roles of unsaturated fatty acids (especially omega-3 fatty acids) in the brain at various ages and during ageing. J. Nutr. Health. Aging 8(3), 163-174.
17. Haag, M., 2003. Essential fatty acids and the brain. Can. J. Psychiatry 48(3), 195-203.
18. Markesbery, W.R., 1997. Oxidative stress hypothesis in Alzheimer's disease. Free Radic. Biol. Med. 23, 134-147.
19. Youdim, K.A., Martin, A., Joseph, J.A., 2000. Essential fatty acids and the brain: Possible health implications. Int. J. Dev. Neurosci. 18(4-5), 383-399.
20. Bartsch, H., Nair, J., Owen, R.W., 1999. Dietary polyunsaturated fatty acids and cancers of the breast and colorectum: emerging evidence for their role as risk modifiers. Carcinogenesis 20(12), 2209-2218.
21. Leitzmann, M.F., et al., 2004. Dietary intake of n-3 and n-6 fatty acids and the risk of prostate cancer. Am. J. Clin. Nutr. 80(1), 204-216.
22. Marchetti, J., et al., 2018. The combined effects of blue light and dilution rate on lipid class and fatty acid composition of Tisochrysis lutea. J. Appl. Phycol. 30(3), 1483-1494.
23. Timmer-Bosscha, H., et al., 1998. Differential effects of all-trans-retinoic acid, docosahexaenoic acid, and hexadecylphosphocholine on cisplatin-induced cytotoxicity and apoptosis in a cisplantin-sensitive and resistant human embryonal carcinoma cell line. Cancer Chemother. Pharmacol. 41(6), 469-476.
24. Germain, E., et al., 1998. Enhancement of doxorubicin cytotoxicity by polyunsaturated fatty acids in the human breast tumor cell line MDA-MB-231: relationship to lipid peroxidation. Int. J. Cancer 75(4), 578-583.
25. Blusztajn, J.K., 1998. Choline, a vital amine. Science 281(5378), 794-5.
26. Zeisel, S.H., da Costa, K.A., 2009. Choline: an essential nutrient for public health. Nutr. Rev. 67(11), 615-23.

4 Modification and Fractionation of Lecithin

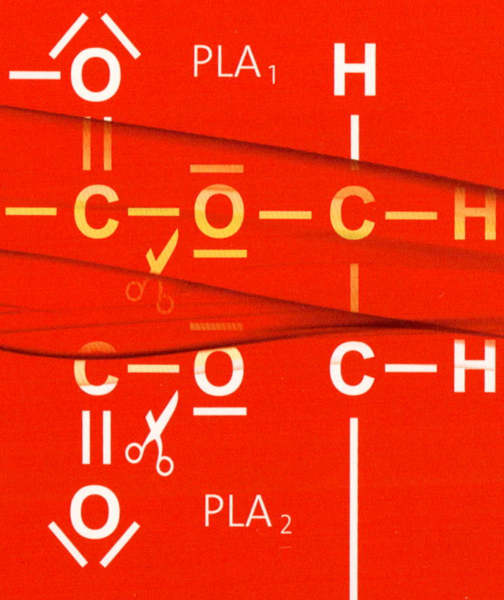

4.1 Enzymatic and chemical modification of lecithin

Jingbo Li, Zheng Guo

4.1.1 Summary

With increasing exploration of plant and microbial enzyme sources, and the advances in genetic engineering and directed evolution technology to improve their intrinsic nature, a wide variety of lipases and phospholipases have been developed and applied as potent tools for either modification of natural lecithin or synthesis of new and less abundant phospholipids (PLs). In addition to a summary of classic methodology and strategy, this work reviews the latest progress in the research on phospholipases and lipases and their applications in industrial degumming, modification of lecithin for function and nutrition, production of structured PLs for targeting applications, as well as design and synthesis of new phosphatidyl derivatives as prodrugs or for pharmaceutical applications. Focus is directed on discussion of the effects of key enzymatic reaction parameters, strategy and approach design for optimal production of targeted phospholipid products. The catalytic characteristics and specificity of phospholipases and their applications in practical reaction systems are also highlighted.

4.1.2 Enzymatic modification

Biochemical modification or synthesis provides alternative and possible cheaper routes for the modification or manufacture of phospholipids (PLs) for different applications, as compared to traditional extraction methods (1). The main biochemical modification options are enzymatic modification and whole-cell modification. The whole-cell modification approach utilizes a series of enzymes in living cells to reach the target, which has less efficacy because the substrates are also used metabolically to keep the cells alive. However, such whole-cell enzyme catalysis is able to systematically carry out continuous reactions, and the reaction systems can be self-repaired (2). In contrast, enzymatic modification methods are usually more straightforward and effective than the whole-cell modifications. Therefore in this chapter the main topic discussed is the enzymatic modification of PLs.

Several phospholipases are able to catalyse specific reactions on specific chemical bonds in a phospholipid molecule. An overall picture of such reactions and enzymes using phosphatidylcholine (PC) as a model substrate is summarized in Fig. 1 below (from Guo et al. (1)). Meaningful modifications of PLs can be achieved based on these reactions and enzymes.

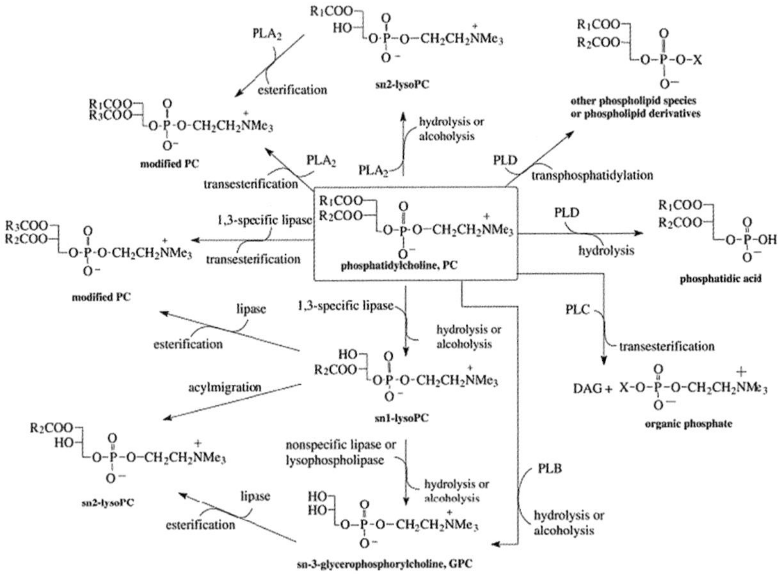

Figure 1.
Phospholipase/lipase-catalysed transformations of phosphatidylcholine as a model for phospholipids (PLs) (1).

4.1.2.1 Lipases, phospholipases and other enzymes related to PLs

The first group of enzymes that can be used for the modification/synthesis of PLs is phospholipases. These enzymes catalyse cleavage of ester or ether bonds in glycerophospholipids as well as sphingophospholipids, and their sites of action are indicated in Fig. 2 (from Guo et al. (1)). Phospholipase A1 (PLA1, phosphatidylcholine 1-acylhydrolase, EC 3.1.1.32) and A2 (PLA2, phosphatidylcholine 2-acylhydrolase, EC 3.1.1.4) both are acyl hydrolases, which specifically hydrolyse 1- and 2-acyl ester bonds of PLs, respectively (Fig. 2). Phospholipase B or PLB (EC 3.1.1.5) is able to hydrolyse both 1- and 2- positional acyl ester bonds. Lysophospholipase (EC 3.1.1.5) is the preferred enzyme to hydrolyse both 1- and 2-monoacyl PLs to diacylglycerol PLs. Phospholipase C or PLC (phosphatidylcholine cholinephosphohydrolase, EC 3.1.4.3) and D (phosphatidylcholine phosphatidohydrolase, EC 3.1.4.4) show similar activity to phosphodiesterases in cleaving the phosphorus-oxygen bond between the glycerol backbone and the phosphate, and between the phosphate and the head group, respectively. Interestingly, phospholipase C can also hydrolyse phosphate ester bonds in sphingomyelin and ether PLs at the similar sites. At the same time, phospholipase A2 and lysophospholipase show hydrolytic activity on the acyl ester bond of ether PLs.

Sphingomyelinase (sphingomyelin phosphodiesterase, EC 3.1.4.12) catalyses the hydrolysis of sphingomyelin to ceramide and phosphocholine. The ceramide can then be hydrolysed to free sphingoid bases and fatty acids by ceramidase. The cleavage of the bond between choline and phosphate, which yields ceramide 1-phosphate and choline, is completed by phospholipase D and sphingomyelinase D (Fig. 2, upper right).

Besides phospholipases, there are many other enzymes that can be used for the modification of PLs. One of the big groups of such enzymes is lipases, frequently used for modification of the acyl groups in PLs. For example, lipase Cal-A was found recently to be able to concentrate n-6 polyunsaturated fatty acids in monoglycerides (3). Different lipases show different preferences to either fatty acid chain length and number of double bonds, or other region-specificity (summarized by Xu (4)). These kinds of properties of lipases also can be used for producing structured PLs. Therefore, in the current work, both phospholipases and lipases are discussed.

Phospholipase A1 (PLA1)

Phospholipase A1 constitutes a very diverse subgroup of phospholipases with 1-acyl hydrolase activity. A number of enzymes with PLA1 activity have been explored and characterized from different sources, among which a kind of integral membrane protein from *E. coli* was well studied (1). However, only a limited number of PLA1s have been purified and cloned. Phosphatidylserine-specific PLA1s (PS-PLA1) from rat platelets, PLA1 from vespid venom and a phosphatidic acid-preferable PLA2 from bovine and human testes are relatively well developed, and these enzymes are also members of the triacylglycerol lipase family, showing dramatic sequence similarity to human hepatic and pancreatic lipases and the guinea pig pancreatic lipase-related protein 2 (5).

Other PLA1s were gradually explored since 2007. A phospholipase A1 from the venom of the social wasp *Polybia paulista* was purified, sequenced and structurally characterized. This PLA1 is a 33,961.40 Dalton protein, referred to as a lipase of the GX class, lipoprotein lipase superfamily, pancreatic lipase (ab20.3) homologous family and RP2 subgroup of phospholipase. The 3D structure of the enzyme molecule revealed 23% α-helix, 28% β-sheet and 49% coil. This protein shares the common α/β fold to many lipases (6).

A metal ion-independent PLA1 of *Streptomyces albidoflavus* isolated from Japanese soil has been purified and characterized (7). The recombinant enzyme expressed using *Streptomyces lividans* cells preferably hydrolysed phosphatidic acid and phosphatidylserine. The substrate specificity was also dependent on the reaction pH. The pH-controlled substrate specificity is a great advantage.

A novel cold-adapted PLA1 from *Serratia* sp. xjF1 was cloned, expressed and characterized (8). The gene product of one open reading frame encoding 320 amino acids with a molecular weight of 33.8 kiloDaltons (kDa) was identified as a PLA1, which exhibited the highest homology to PLA1 of *Serratia marcescens* (71%). The other open reading frame encoded a protein of 251 amino acids, which did not show any enzymatic activity but was found to be an enhancer for expressing the PLA1 in *E. coli.* This PLA1 showed highest activity at 35°C, and 70% of its maximum activity is maintained at 10°C. The low temperature active PLA1 could potentially be used for handling sensitive PLs with polyunsaturated fatty acids. In addition, PLA1-producing bacteria were screened from intestinal contents of marine animals, and a bacteria isolate from a flatfish was found to have PLA1 activity at low temperatures (9).

Phospholipase A2 (PLA2)

Phospholipase A2 (PLA2) preferentially catalyses the release of fatty acid from the *sn*-2 position of PLs (1). PLA2s are normally divided into three categories: 1) secreted PLA2; 2) cytosolic PLA2; and 3) Ca^{2+}-independent PLA2 (5). Secreted PLA2s have a low molecular weight and contain a large amount of disulphide bridges working with a certain amount of Ca^{2+} ions for optimum catalytic activity. Cytosolic PLA2s preferably hydrolyse PLs containing sn-2 arachidonic acid. Ca^{2+} binding to an N-terminal C2 domain and phosphorylation on serine residues is required to fully activate the activity of these enzymes (5).

Ca^{2+}-independent PLA2s of 85 – 88 kDa molecular weight have a lipase motif and a serine catalytic triad, and exhibit both lysophospholipase and transacylase activities. Their molecules contain seven to eight ankyrin repeats (one of the most common sequence motifs), contain the consensus lipase motif G-X-S-X-G, and were detected mainly in human tissues (5).

Phospholipase B (PLB) and lysophospholipase
Several phospholipase B enzymes (PLBs) from fungi, including from *Cryptococcus neoformans, Saccharomyces cerevisiae* and *Penicillium notatum,* have been discovered, characterized and/or cloned (1). Many fungal species appear to produce PLBs that hydrolyse both acyl groups in the PL molecule, resulting in only minimal accumulation of lysophospholipid product. Hence, these enzymes often also have lysophospholipase activity, removing the remaining acyl moiety on lyso-PLs (10). Therefore, both phospholipase B and lysophospholipase can catalyse the hydrolysis of PLs and lyso-PLs, with more lyso-PLs accumulating from PLB-catalysed reactions (1).

Some fungal PLBs have been shown to exert transacylase activity, since they are able to convert lyso-PLs and free fatty acids into PLs (10). Certain small lyso-PLA enzymes contain the lipase motif of G-X-S-X-G and the catalytic triad of Ser-Asp-His. They are independent of Ca^{2+}, Mg^{2+} and EDTA and have an optimal pH between 7 and 9 (1).

Phospholipase C (PLC)
Phospholipase C enzymes (PLCs) are members of a large superfamily to enzymes, namely phosphoinositide-specific phospholipase C (PI-PLCs). PI-PLC molecules occur in a broad spectrum of organisms including bacteria, simple eukaryotes, plants and animals. However, the structures and catalytic mechanisms of PI-PLCs from bacteria and from mammals are significantly different. Bacterial PI-PLCs consist of a single domain of 30 to 35 kDa, while the eukaryotic enzymes are much larger, around 60 to 150 kDa, and are organized in several distinct domains. Both the eukaryotic and the prokaryotic enzymes need cofactors, calcium and zinc respectively, for optimum catalytic activity. However, the difference is that the utilization of a catalytic calcium ion stabilizes the transition state electrostatically in eukaryotic enzymes, while this role in bacterial PI-PLCs is fulfilled by an analogously positioned arginine. The two different PI-PLCs are from a common ancestor molecule with divergent evolution (1,5).

Phospholipase D (PLD)
Phospholipase D (PLD) is a ubiquitous enzyme found in bacteria, fungi, plants and mammals (5). PLD hydrolyses the phosphodiester bond of the glycerolipid phosphatidylcholine, resulting in the production of phosphatidic acid and free choline (11). Phosphatidic acid (PA) is a compound heavily involved in signal transduction. The PLD superfamily includes enzymes that are involved in phospholipid metabolism, nucleases, toxins and virus envelope proteins of unknown function (5). Sequence comparison discloses that PLDs from different sources share high close homologues and sequence motifs (1).

The conserved histidine, lysine and aspartate residues in PLD enzymes form a catalytic triad responsible for formation or hydrolysis of phosphodiester bonds (12,13). Almost all PLDs are Ca^{2+} dependent, but differ in their calcium concentration levels for activation. For instance, while mammalian PLDs need around a 1 mM Ca^{2+} concentration, plant PLDs require higher levels and the yeast PLDs need much lower levels of Ca^{2+} (1). PLD is able to carry out both hydrolysis of the phosphodiester bonds, yielding PA, as well as the transphosphatidylation reaction (head group exchange), and uniquely it can carry out both reactions in the presence of a bulk water phase (14). The transphosphatidylation – head group exchange reaction commands the main interest for industry applications since it allows for different functional PLs to be formed.

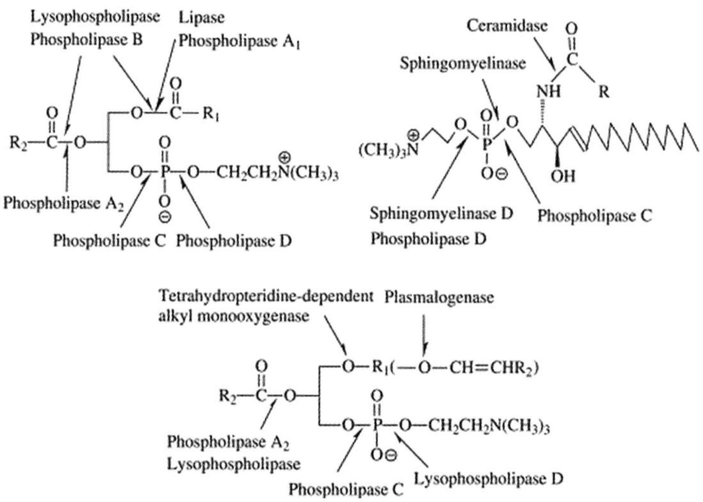

Figure 2.

Positional specificity of phospholipases and other PL-cleaving enzymes. Adapted from Guo et al. (1).

4.1.2.2 Enzymatic production of lyso-PLs

Lyso-PLs are partially hydrolysed PL molecules containing one fatty acid moiety, either on sn-1 or sn-2 position, as catalysed by PLA1, PLA2 or PLB. Several methods are widely used for lyso-PL production except for hydrolysis; three such examples are alcoholysis, esterification of glycerophosphorylcholine, and thermodynamic transacylation of 2-acyl lyso-PLs to 1-acyl lyso-PLs (1). Hydrolysis and alcoholysis are in principle the same because both water and alcohols provide the hydroxyl group to accept the fatty acid moiety. Commercialization of lyso-PL production by enzymatic routes has been implemented by for instance Degussa, Loders Croklaan, Cereol, Central Soya and Taiyo Kagaku. In addition, research is actively ongoing to explore more possibilities.

4.1.2.2.1 Hydrolysis

Hydrolysis of PLs can be achieved by cleavage of their fatty acid moiety/moieties, or catalysed by either enzymes or chemicals in the presence of water as shown in Scheme 1. In an enzymatic reaction, the specific enzyme species used determines the types of the final products such as 1-lyso or 2-lyso PLs or fully deacylated PLs (glycerophosphoryl molecular species). Parameters such as substrate concentration, water activity and enzyme loading will influence the reaction rate and final product yield.

Lipase preparations from *Mucor* and *Rhizopus* were found to be the most efficient among 19 commercial lipase preparations for hydrolysing phosphatidylcholine into its lyso forms (15). Later on, this lipase was immobilized covalently onto effectively functionalized silica nanoparticles and the resulting hydrolytic activities in an aqueous medium were significantly high, almost in the same range as the free enzyme. The immobilized lipase was also more temperature resistant than the free form (16).

Another type of immobilized lipase, Lipozyme, was able to hydrolyse the fatty acyl ester bonds of phosphatidylcholine in organic media (17). Eight commercial lipase preparations were examined for this purpose, but only enzymes from *Humicola lanuginosa, Rhizopus delemar* and *Candida rugosa* displayed appreciable hydrolytic activity to phosphatidylcholine.

Lipase preparations from *Aspergillus niger* and *Penicillium cyclopium,* phospholipases from *A. niger,* and porcine pancreas were compared in the hydrolysis of soybean PLs (18). The researchers found that lipase preparations which contained both phospholipase and lysophospholipase activities were more efficient than simpler phospholipase preparations in the hydrolysis of soybean PLs. This may imply synergism between phospholipase and lysophospholipase activities.

Milk PLs were hydrolysed by three fungal phospholipases and porcine pancreatic phospholipase A2, and the most efficient enzyme was *Fusarium venenatum* phospholipase A1 (19).

Hydrolysis of phosphatidylcholine was investigated in two different solvent systems using three different enzymes (20). It was found that both immobilized phospholipase A1 and Novozym 435 were able to hydrolyse phosphatidylcholine into its lyso form in ethanol. The highest degree of hydrolysis (70%) was reached with the enzyme Lipase PS after 48 h at 60°C in water.

4.1.2.2.2 Alcoholysis

Scheme 2 shows the enzymatic alcoholysis of PLs to produce lyso-PLs and fatty acid esters. This reaction is similar to the hydrolysis. In an alcoholysis reaction, water activity is a critical parameter because short-chain alcohols would compete with water molecules for binding with the enzymes and this could lead to deactivation of the enzymes. For example, addition of 4% water increased the conversion of PLs dramatically when ethanol, 2-propanol and 1-butanol were used (21). More water was required to reach the maximum activity in water-miscible alcohols than in more hydrophobic homologues. With 10% water in the reaction system, equimolar proportions of soy PLs and individual alcohols were reacted with the aid of Lipozyme IM 20 at 55°C for 24 h, and a 69–78% molar yield of lyso-PLs was obtained (22). Alcohol chain lengths up to 18 did not decrease the yield of ester and lyso-PLs, suggesting the enzyme used was versatile with the various alcohols.

Recently, several lipases including Novozym 435 (from *Candida antarctica*), Lipozyme TLIM (from *Thermomyces lanuginosus*) and Lipozyme RMIM (from *Rhizomucor miehei*) were used to prepare lysophosphatidylcholine in both hexane-mediated and solvent-free systems (23). The three lipases exhibited higher conversion than 93% in both systems. Novozym 435 showed the best catalytic efficiency. Hexane-mediated reaction worked better than solvent-free media. It should be noted that water was supplemented in both reaction systems.

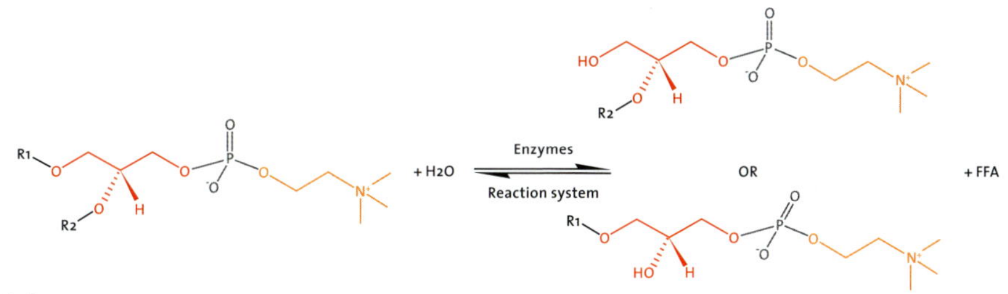

Scheme 1:

Hydrolysis of PLs to produce lyso-PLs and free fatty acids.

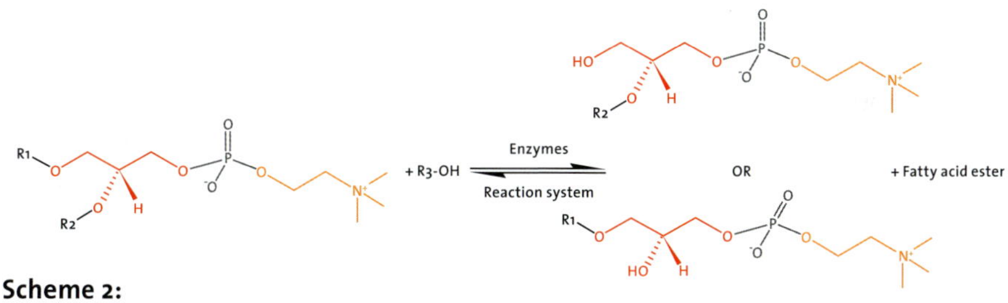

Scheme 2:

Alcoholysis of PLs to produce lyso-PLs and fatty acid esters.

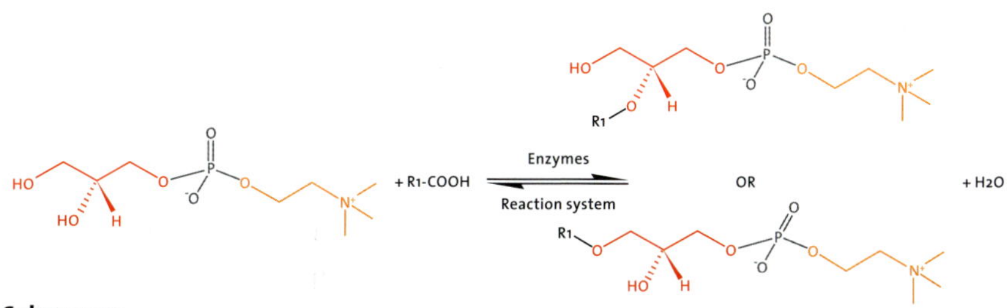

Scheme 3:

Esterification of glycerophosphorylcholine to produce lysophosphatidylcholine and water.

4.1.2.2.3 Esterification

Lyso-PLs can be synthesized from glycerophosphorylcholine due to the two free and active hydroxyl groups on its glycerol backbone. Two types of lyso-PLs could be formed by using corresponding position-specified lipases such as 1,3-specific lipases and 2-specific lipases. Scheme 3 illustrates the esterification reaction. The reaction is reversible and apparently could be controlled by increasing either the loading of reactants or the removal of products. The conversion of 71% of glycerophosphorylcholine was achieved by esterifying with fatty acid anhydrides (24). Five different lipases were compared for lysophosphatidylcholine synthesis from glycerophosphorylcholine, and Lipozyme IM (*Mucor miehei*) was found to be much better than Novozym 435 (*Candida antarctica*), Lipase AP-6 (*Aspergillus*), Lipase F-AP-15 (*Rhizopus japonicus*) and Lipase N (*Rhizopus niveus).*

Lysophosphatidylcholine and phosphatidylcholine yields of 88% and 12% from glycerophosphorylcholine were achieved when palmitic acid was used (25). In situ water activity was enhanced by adding a small amount of water-mimicking solvents such as dimethylformamide to the reaction media, and this significantly enhanced the reaction rate and the product yield (to 90%) (26). Novozym 435 was effective (95% conversion) in the synthesis of lysophosphatidylcholine via transesterification reaction, in both a solvent-free system and a t-butanol mediated system (27). Almost no phosphatidylcholine was formed in the first 24 h. However, subsequently the formation of phosphatidylcholine increased slowly over time, indicating the optimal time should be focused in order to obtain purer lysophosphatidylcholine. High excess of vinyl laurate over glycerophosphorylcholine was critical in this reaction. Induction of product precipitation by decreasing the reaction temperature could drive the equilibrium towards lysophosphatidylcholine formation. It was shown that vinyl esters with different chain lengths required different optimal reaction temperatures.

It has been suggested that the acid form of fatty acid was less effective for direct esterification with glycerophosphorylcholine than fatty acid derivatives, such as fatty acid anhydrates or fatty acid vinyl esters. However, simultaneous water removal increased the yield of lysophosphatidylcholine dramatically in the esterification reaction (28). The lowest vacuum of 0.5 mm Hg did not result in a higher lysophosphatidylcholine yield than 1 mm Hg, suggesting a too high vacuum risks deactivating the enzyme by dehydration. Vacuum levels between 100 and 300 mm Hg gave rise to a lower reaction rate and yield of lysophosphatidylcholine compared with vacuum levels between 1 and 50 mm Hg. It should be noted that the optimal vacuum level has a close relationship with the reaction temperature in this system. Therefore, these two parameters should be considered at the same time.

4.1.2.2.4 Acyl migration

Acyl migration is an often-encountered problem in selective synthesis of regiospecific PLs, i.e. intramolecular transfer of one fatty acid moiety from one hydroxyl group to the adjacent one (1). This reaction is non-enzymatic and can be catalysed by acids or bases or other chemicals (1). For example, it was found that carriers used for enzyme immobilization can promote acyl migration (29). Acyl migration could be suppressed by blocking one of the hydroxyl groups by for instance acylation of glycerophosphorylcholine (30). sn-2 position lyso-PLs are less stable than sn-1 position ones and the equilibrium between these two forms is approximately 1:9 (31).

4.1.2.3 Enzymatic synthesis of structured PLs

In order to meet different applications of PLs, modification of existing fatty acid species and/or polar head groups on selected PL molecules could endow special functional properties. The term 'structured phospholipid' generally indicates PL molecules with different functional fatty acids region-specifically incorporated (32). The incorporation of docosahexaenoic acid (DHA, omega-3) may have potential medical applications such as in promoting cell differentiation in leukaemia (33) and in enhancing survival of tumour-bearing mice (34).

Enzymatic modification of PLs is based on selective/positional recognition of the PL molecule by phospholipase A1, A2 and 1,3-specific lipases (1). Two reaction routes could realize the production of structured PLs, i.e. a direct or one-step process, and a two-step process. In the direct process, interesterification between PLs and fatty acids or their ester forms is carried out. The two-step process includes, first, hydrolysis of PLs in order to gain free hydroxyl groups on the glycerol backbone, then to esterify or transesterify them in a targeted manner. Generally, the one-step route is simpler and easier to be performed, and the two-step route results in higher-purity products (29).

Conjugated linoleic acid (CLA) was incorporated with PLs via enzymatic acidolysis catalysed by several different enzymes including Lipozyme RM IM, Lipase F-AP15, Novozym 435, Lipozyme TL IM and phospholipase A2. Only Lipozyme RM IM and Lipozyme TL IM were found effective, and the maximum incorporation achieved by the latter enzyme was 16% with soy PL in 72 h (35). Immobilized phospholipase A1 was successfully used to incorporate CLA with phosphatidylcholine via acidolysis reaction, with the maximum yield being 85.8% incorporation of CLA (36). The optimal conditions were explored using response surface methodology and found to be 15% enzyme loading at 55°C for a 1:4 (PC/CLA) substrate molar ratio. Meanwhile, the same group of authors reported an even higher incorporation of CLA (90%) was achieved at 50°C, with phospholipase A1 loading of 15% and the same substrate molar ratio within 24 h (37). It seems that PLA1 was more effective for incorporation of CLA into phosphatidylcholine than were the other enzymes.

Acidolysis catalysed by Lipozyme RM IM for the production of structured PLs between phosphatidylcholine and caprylic acid in the solvent-free system obtained an incorporation of 46% at the optimal conditions (38). Commercial immobilized lipase Lipozyme TL IM was used in a hexane-mediated system for incorporation of caprylic acid with phosphatidylcholine and the reaction conditions were optimized (39). An incorporation of caprylic acid up to 46% and recovery of PC up to 60% can be obtained under the optimal conditions, i.e. enzyme dosage 29%, reaction time 50 h, temperature 54°C, substrate ratio 15 mol/mol caprylic acid/PC, and 5 mL of hexane per 3 g substrate.

PLA2 was immobilized on a non-ionic weakly polar macroreticular resin and further tested for incorporation of caprylic acid with phosphatidylcholine in a solvent-free system (40). Highest incorporation during acidolysis was 36% and lysophosphatidylcholine was found to be the major by-product. Novozym 435 incorporated 28% of α-Linolenic acid into the *sn*-1 position of egg yolk phosphatidylcholine via a one-step transesterification process, and PLA2 yielded 25% incorporation of α-Linolenic acid into the *sn*-2 position (41).

Polyunsaturated fatty acids are of great interest to incorporate with PLs because of the possibilities for use of these products in medical applications. Phospholipase A1 from *Thermomyces lanuginosus/Fusarium oxysporum* was employed to incorporate the n-3 (omega-3) fatty acids eicosapentaenoic (EPA; C20:5), docosapentaenoic (DPA; C22:5) and docosahexaenoic (DHA; C22:6) with phosphatidylcholine in a solvent-free system. The reaction conditions were optimized and after only 6 h of reaction at 55°C, 21.0 mol% n-3 polyunsaturated fatty acids (sum of EPA, DPA and DHA) was incorporated into PC at a loading of the enzyme solution of 10% of the total weight of substrates. Longer reaction time up to 24 h led to 28.0 mol% incorporation of n-3 polyunsaturated fatty acids. However, phosphatidylcholine content decreased as the reaction time exceeded 6 h (42).

The effect of water activity in this reaction system was further investigated by using the immobilized phospholipase A1 from *Thermomyces lanuginosus/Fusarium oxysporum* for the same purpose (43). Water activities greater than 0.53

could increase the activity of the enzyme but increased water led to hydrolysis of phosphatidylcholine. A compromise has to be made between the increased enzyme activity and the decreased product yield.

Most recently, the effects of immobilization methods for phospholipase A1 from *Thermomyces lanuginosus/Fusarium oxysporum* for the same purpose were extensively investigated (44). Lewatit VP OC 1600 was selected as a carrier and the highest incorporation (57.4 mol%) of n-3 polyunsaturated fatty acids into phosphatidylcholine was obtained at 24 h and the yield of PC was 16.7 mol%. Immobilized lipase from *Rhizomucor miehei* was examined to incorporate n-3 polyunsaturated fatty acids into soybean lecithin via transesterification reaction (45). After 48 h, incorporation was highest in the presence of Mg^{2+}, urea and solvent. After 72 h, it was highest with Mg^{2+} and urea in the presence or absence of solvent (56.8 and 45.7%, respectively). Exploration of new enzymes, reduction of cost, and identification of higher-value applications of such structured PLs should be carried out in order to implement the technologies at the large industrial level.

The aforementioned modifications focus mainly on exchange of fatty acids on the *sn*-1 and *sn*-2 positions of PLs. Alternatively, some other compounds can also be incorporated with PL molecules to endow other properties such as antioxidant and antimicrobial activities. For example, cinnamic acid and benzoic acid were incorporated with phosphatidylcholine by chemo-enzymatic multi-reactions (46). The synthesis of such compounds involves a typical two-step process as mentioned above. First phosphatidylcholine was enzymatically converted into lysophosphatidylcholine by ethanolysis. Phenolic compounds were then esterified with the prepared lysophosphatidylcholine to get the final products. Some of the synthesized structured phosphatidylcholine showed strong antimicrobial activities to both Gram-negative and positive bacterial strains. The cinnamic acid-phosphatidylcholine molecule exhibited almost the same EC_{50} values against bacteria as did BHT and α-tocopherol. This work expands the research field of the structured PLs. It implies that many other functional compounds bearing reactive groups like carbonyl, carboxylic, hydroxyl etc. could be incorporated with PLs in order to obtain products with ever more versatile application potential.

4.1.2.4 Phospholipase complexes and enzymatic degumming

Degumming is the first step in the process of refining vegetable oils, and it removes PLs and mucilaginous gums (47). The purpose of degumming is to remove PLs in order to decrease the phosphorus content to less than 10 mg/kg of oil. There are two main routes available for vegetable oil refining, namely chemical refining and physical refining. Water and acid degumming are the methods most commonly used in industry. Compared to chemical refining, physical refining results in much less waste water and discharge (48). However, in water degumming, only hydratable PLs can be removed while the remaining phosphorus content is still high (48). Alternatively, enzymatic degumming offers environmentally friendly opportunity in terms of converting non-hydratable PLs into hydratable species. As shown in Fig. 2, PLA1, PLA2, PLB and PLC all are applicable for this purpose because their hydrolysis of fatty acids chains or phosphates on PLs results in easier removal of phosphorus.

The first industrial enzymatic degumming process, Lurgi's EnzyMax® process, was launched in Budapest in 1992 (49). Lecitase® 10L (porcine phospholipase A2) was the enzyme used in the EnzyMax® process. However, the production capacity of Lecitase® 10L is limited as it is extracted from porcine pancreas, which is insufficient to meet the increasing market demand. Meanwhile, the optimal pH for activity of Lecitase® 10L is too high to make it a fully practical option. Additionally, utilization of this enzyme leads to restrictions against kosher food certification (50). Therefore, microbial-based phospholipases need to be developed to solve these problems.

Lecitase® Novo and Lecitase® Ultra (PLA1) developed by Novozymes A/S, Rohalase MPL (PLA2) developed by AB Enzymes, Purifine® (PLC), made by Verenium Corporation, and LysoMax® made by Genencor, a division of Danisco A/S, were gradually commercialized until 2011 (51). Phospholipases from different sources are still under development. For instance, a novel PLC from *Bacillus licheniformis* MTCC 7445 was characterized and applied in a degumming process (52). Heterologous overexpression of phospholipase B from *Pseudomonas fluorescens* BIT-18 in *Pichia pastoris* achieved two times higher enzyme activity than the wild type, and the enzyme can decrease the phosphorus content of soybean oil from 125.1 mg/kg to 4.96 mg/kg, which is dramatically promising (53). A novel PLB from *Thermotoga lettingae* TMO has been cloned and functionally overexpressed in *Escherichia coli,* and purified to homogeneity. The recombinant PLB showed a half-life of 4 h at 90°C and its degumming efficiencies for rapeseed, soybean, peanut and sunflower oils were very high, i.e. starting from 225.2, 189.3, 85.6 and 70.4 mg/kg and moving to 4.9, 4.7, 3.2 and 2.2 mg/kg within 5 h, respectively (54). With the current fast-moving development of genetic engineering and protein engineering, more effective phospholipases could be produced to realize different purposes.

As an application technology, optimization of the process conditions for degumming is commonly required to achieve the best performance. Table 1 shows the optimal conditions for degumming of various vegetable oils using different phospholipases. It can be seen that the optimal pH value is around 5 and the temperature close to 50°C for all the enzymes investigated. In addition, a time period of 5 h is enough to achieve satisfactory phosphorus removal.

Table 1.

Summary of optimal conditions for degumming.

Enzyme	Feedstock	Enzyme dosage mg/kg	Temperature °C	pH	Time h	Phosphorus content mg/kg	Ref.
Lecitase® Ultra; EC 3.1.1.3	Rapeseed oil	39.6	48.3	4.9	6	3.1	Yang et al. (48)
Lecitase® Ultra	Rapeseed oil	30	50	5.0	5	10	Yang et al. (55)
Lecitase® Ultra	Rice bran oil	50	49.2	5	4	8.9	Jahani et al. (56)
Lecitase® Ultra	Soybean oil	40	48 ± 2	4.8–5.1	6	Less than 10	Yang et al. (57)
PLB from *Pseudomonas fluorescens* BIT-18	Soybean oil	500 U/kg	40	4.7	4–6	4.9	Jiang et al. (47)
PLB from *Pseudomonas fluorescens* BIT-18	Soybean oil Rapeseed oil Peanut oil	500 U/kg	40	4.7	5	4.6 4.5 3.0	Jiang et al. (58)
ecitase® Ultra	Crude rice bran oil	–	50	–	2	10	Manjula et al. (59)
PLC from a genetically modified *Bacillus cereus*	Camellia oil	400	53	5.4	2.2	15.1	Jiang et al. (60)
Magnetically immobilized Lecitase® Ultra	Soybean oil	100	56	–	6.3	10.4	Yu et al. (61)

In order to realize an economical process to utilize phospholipases as degumming reagents, immobilization of such enzymes is worth studying. Lecitase® Ultra was immobilized on three different supports – calcium alginate, calcium alginate-chitosan (CAC) and calcium alginate-gelatin – and cross-linked with glutaraldehyde (62). The loss of enzyme activity was around 35–40% compared to free enzymes. However, the thermostability was improved and the pH/activity profile range was considerably broadened. Especially, the recoveries of immobilized PLA1–CAC and activity of PLA1 were 80.2 and 78.2%, respectively. Similar observations were reported even when a different carrier was used (63). These works suggest that not only the enzymes' reusability but also other properties can be improved by the immobilization process. Subsequently Lecitase® Ultra was immobilized on magnetic nanoparticles and the ability to degum soybean oil was explored and optimized (61). However, recovery of the immobilized Lecitase® Ultra was not mentioned (61). Thereafter, the same phospholipase was immobilized onto magnetic Fe3O4/SiOx-g-P nanoparticles and after ten cycles through the soybean oil degumming process at 55°C, the immobilized PLA1 still possessed more than 80% of its initial activity (63). Immobilization is definitely a promising technique to reduce the cost of enzymes by reusing them. The cost of the carriers also needs to be considered, as well as any by-products or pollutants produced during immobilization processes. An overall life cycle assessment should be performed in order to determine the best technology.

4.1.2.5 Phospholipase-mediated exchange of head groups for synthesis of novel PLs

The polar head groups on PLs can be either hydrolysed or exchanged via reactions catalysed by PLD. This can be used for 1) enrichment of particular phospholipid species; 2) testing the ability of PLD to catalyse the transphosphatidylation of acceptors with different hydroxyl groups and varied structures or to optimize reaction conditions; 3) transforming drugs such as genipin, ascorbic acid, arbutin or others; and 4) preparing novel derivatives with potentially pharmaceutical values, such as alkyl phosphate ester and plasmalogen (1).

Ethanol, ethanolamine, serine, choline, inositol, glycerol, glucose and water all contain the active hydroxyl group and therefore can react with PLs catalysed by PLD. The difference between the reactions of alcohols and water is that the former reaction is called transphosphatidylation while the latter is called hydrolysis. Generally, water-miscible alcohols are more easily incorporated whereas the alcohols with longer chain length are more difficult to incorporate. Apparently there is a problem to control the transphosphatidylation and hydrolysis reactions. In an aqueous/organic two-phase system containing hexane and Tris-HCl buffer, the formation of phosphatidylglycerol and phosphatidylserine was very close at around 51% and the phosphatidic acid content was 6% and 14% (64), indicating this solvent system works well for transphosphatidylation. Therefore, biphasic systems containing organic solvent and buffer are widely used for transphosphatidylation reactions (65,66). But organic solvents are not easy to operate, especially at an industry level, and they are toxic. Therefore, a novel reaction system for transphosphatidylation without involvement of organic solvents was discovered (67). In this system, $CaSO_4$ was used to adsorb lecithin in an aqueous buffer solution (pH 5.6), and 80% phosphatidylserine was obtained after reaction for 24 h. In another study (68), PLD was immobilized on VA-Epoxy Biosynth and the reaction was performed in a complete aqueous system. The formation of PA became insignificant and the yield of phosphatidylglycerol was 60%. In comparison with the biphasic system, the purely aqueous system had a lower reaction rate. Therefore, in order to eliminate utilization of organic solvents, improved immobilization of PLD or improved use of absorbents like CaSO4 should be further explored.

Recently, new 'green' solvents such as ionic liquids (69), deep eutectic solvents (70) and 2-methyltetrahydrofuran (71) were introduced into the transphosphatidylation reaction system. PLD did not lose its activity after being soaked in 1-butyl-3-methylimidazolium hexafluoroborate for 1 hour, and a microemulsion of toluene and IL/buffer 90/10 (V/V) proved to be a very promising system for carrying out PC transphosphatidylation and greatly inhibiting the formation of phosphatidic acid (69). So-called 'biobased' solvents including limonene and p-cymene were adopted for transphosphatidylation reactions to produce phosphatidylserine. The reaction system contained 3 mL of such solvent, 0.05 mmol substrate, 0.15 mmol L-serine, 60 units PLD and 0.5% water. The results showed no hydrolysis reaction and 95% yield of phosphatidylserine after 12 h (72). Biobased cholinium amino acid ionic liquids were recently introduced as reaction solvents, and their phosphatidylserine yield reached 86.5% while 75% of the enzyme activity was retained after ten batch reactions (73). To even further explore the most effective solvent system, enzymes themselves are of course another way to go in order to achieve high reaction rate and final yield.

A new PLD constitutively secreted by *Ochrobactrum* sp. is ASAG-PL1. The molecular weight of the enzyme was ~37 kDa by SDS-PAGE, with optimal temperature of 40°C and pH of 7.0. More than 90% of the activity was retained after 30 days at 4°C and pH 7.0 (74). Another versatile PLD was produced by *Streptomyces* sp. isolated from soil- and waste-water-contaminated palm oil. The purified PLD showed convertibilities of 100% and 94.7% in biphasic system and aqueous system, respectively (75). Cell-surface display, which allows enzymes to be displayed on the surface of microorganisms, such as *Escherichia coli, Saccharomyces cerevisiae* and *Pichia pastoris,* and to function as immobilized enzymes, shows additional advantages. Phospholipase D from *Streptomyces chromofuscus* was surface displayed on *Pichia pastoris* and the resultant PLD showed good stability over broad ranges of temperature (20–65°C) and pH (4.0–11.0). The thermostability, acid stability and organic solvent tolerance of the displayed PLD were also significantly enhanced. In addition to an initial 67.5% convertibility of phosphatidylcholine into phosphatidylserine, 50% convertibility was retained after seven cycles (75). Cell-surface display technology does not need purification of enzymes, which enhances its cost-effectiveness.

4.1.2.6 Industrial reactors for the enzymatic modification

Enzyme reactors are often classified by operation (batch or continuous) and mixing performance (from well mixed to plug flow). A stirred tank reactor can be operated in batch or continuous mode, and are often considered well-mixed reactors in which the molecules are maintained in a constant state of agitation (Fig. 3, A and B).

Most published works on lipase applications in lipid modification has been conducted in batch reactors, including tubes or flasks with shaking or magnetic stirring. In these reactors, the effectiveness factor varies with the time of the reaction because the substrate concentration decreases with time. One of the problems of this type of reactor is the additional step for recovery of the enzymes after reaction. In order to tackle this problem, enzymes were immobilized or fixed into a reactor, and the new reactor type called packed-bed reactor was invented (Fig. 3C). Additionally, the packed-bed reactor allows for continuous operation, which can be treated as plug flow in the ideal state, meaning that no back-mixing occurs. The effectiveness factor is high near the inlet and low near the outlet, due to the decrease in substrate concentration as the reactants pass through the reaction. The fluidized-bed reactor is generally used for multiphase reactions (Fig. 3D). This reactor is not commonly used in lipid modifications.

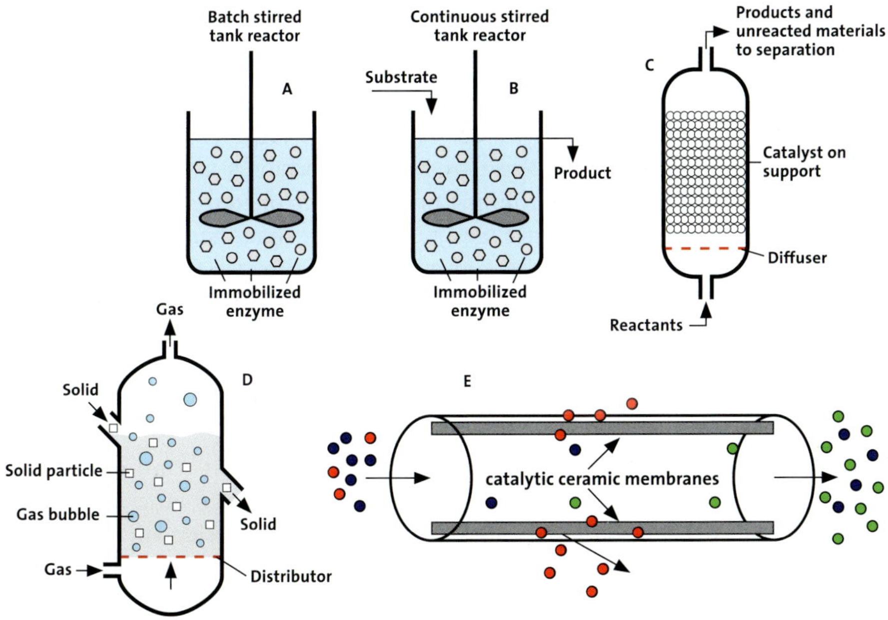

Figure 3.

Different types of reactors. A: Batch stirred tank reactor. B: Continuous stirred tank reactor. C: Packed-bed reactor. D: Fluidized-bed reactor. E: Membrane reactor.

Membranes are thin barriers that can perform various degrees of separation using differences in concentration, pressure and electrical potential gradients between the two compartments they separate (76). Membrane reactors are defined as the reaction systems in which membranes are implemented (Fig. 3E). The selective removal of either products or inhibitors by specific membranes would lead to higher productivity or yield. Many lipid modification reactions are reversible and therefore online separation of the products would shift the equilibrium forwards. Meanwhile, different types of inhibitors would decrease enzyme activity. The in situ removal of such inhibitors will increase the product yields as well. To date, there have not been many reports regarding utilization of these reactors for phospholipid modifications. Commercialization of such technologies could become a future focus of effort. For those who are interested, a very good in-depth discussion of these reactors can be found in an earlier publication (76).

4.1.3 Chemical modification

4.1.3.1 Chemical hydrolysis

Hydrolysis of PLs can be achieved by chemical catalysis in acid or alkaline conditions. Acid-catalysed hydrolysis can be organic or inorganic, is generally performed at higher temperatures, and takes longer time than alkali-mediated reactions. The main disadvantages of a chemical hydrolysis are the non-specificity of the reaction and undesirable dark colour formation. Metal oxide-catalysed hydrolysis in alcohols, on the other hand, can be performed at lower temperatures and consequently yields light-coloured products (1).

4.1.3.2 Acylation

The purpose of acylation is to block the 'zwitterionic' group of phosphatidylethanolamine and thus improve the oil/water emulsifying property of lecithin, or to ease purification of phosphatidylethanolamine (77). Acid anhydrides are preferable acylating agents, and non-acylatable amines such as trimethylamine or pyridine are preferable catalysts for this reaction (1).

4.1.3.3 Hydroxylation

Hydroxylation is the insertion of two hydroxyl groups at the carbon – carbon double bonds of unsaturated fatty acids of lipids under acidic conditions. The resulting hydroxylated lecithin has a lower iodine value, light colour and higher hydrophilic capacity (1). The acidic conditions can be achieved using lactic acid (1) and other solid acids (78). Full conversion with 80% selectivity at 57°C over 24 h was achieved by using Amberlyst 15, and this catalyst can be reused five times (78).

4.1.3.4 Hydrogenation

Unsaturated fatty acids can be converted into saturated ones, in order to increase product oxidative stability. Catalysts such as palladium on carbon, platinum, nickel, sulphur-poisoned nickel, copper and chromium complexes all can work (79). Hydrogenation was also performed in supercritical carbon dioxide and hydrogen catalysed by nickel and found promising (King et al., 2001).

4.1.4 Conclusions

The ranges of enzymatic and chemical modifications of PLs and the applications of phospholipases were reviewed in this chapter. It is apparent that additional studies are ongoing and that more and more promising processes, new enzymes, solvent systems, high-value products et cetera have been emerging. From the enzyme perspective, advanced technologies in gene manipulation can be employed in order to obtain designed enzymes. Microbes are a huge community on this planet, and we should make full use of these precious living organisms to discover more relevant enzymes. Production, purification and immobilization of enzymes have to be optimized with consideration of economic feasibility since the high price of enzymes is still an obstacle for their utilization in large-scale industry. Green chemistry, sustainability and food safety have been emphasized in recent years. Therefore, these should also be considered when designing reaction systems for PL modification. For example, toxic organic solvents should be avoided and thus novel solvent systems are urgently required. Although many research works have been very promising on the lab scale, applications at industrial scale are not as fruitful as in the lab. The development of reactors, processes, large-scale operations and other engineering issues should catch up in order to better apply these emergent technologies in real industry.

References

1. Guo, Z., Vikbjerg, A.F., Xu, X., 2005. Enzymatic modification of phospholipids for functional applications and human nutrition. Biotechnol. Adv.
2. De Carvalho, C.C.C.R., 2011. Enzymatic and whole cell catalysis: Finding new strategies for old processes. Biotechnol. Adv. 29, 75-83.
3. He, Y., Li, J., Kodali, S., Chen, B., Guo, Z., 2016. The near-ideal catalytic property of Candida antarctica lipase A to highly concentrate n-3 polyunsaturated fatty acids in monoacylglycerols via one-step ethanolysis of triacylglycerols. Bioresour. Technol. 219, 466-478.
4. Xu, X., 2000a. Production of specific-structured triacylglycerols by lipase-catalyzed reactions: a review. Eur. J. Lipid Sci. Technol. 102, 287-303.
5. De Maria, L., Vind, J., Oxenbøll, K.M., Svendsen, A., Patkar, S., 2007. Phospholipases and their industrial applications. Appl. Microbiol. Biotechnol. 74, 290-300.
6. Santos, L.D., Santos, K.S., de Souza, B.M., Arcuri, H. a., Cunha-Neto, E., Castro, F.M., Kalil, J.E., Palma, M.S., 2007. Purification, sequencing and structural characterization of the phospholipase A1 from the venom of the social wasp Polybia paulista (Hymenoptera, Vespidae). Toxicon 50, 923-937.
7. Sugimori, D., Kano, K., Matsumoto, Y., 2012. Purification, characterization, molecular cloning and extracellular production of a phospholipase A1 from Streptomyces albidoflavus NA297. FEBS Open Bio 2, 318-327.
8. Fu, J., Huang, H., Meng, K., Yuan, T., Yao, B., Shi, Y., Ouyang, P., 2008. A novel cold-adapted phospholipase A1 from Serratia sp. xjF1: Gene cloning, expression and characterization. Enzyme Microb. Technol. 42, 187-194.

9. Nishihara, M., Kamata, M., Koyama, T., Yazawa, K., 2008. New phospholipase A1-producing bacteria from a marine fish. Mar. Biotechnol. 10, 382-7.
10. Köhler, G.A., Brenot, A., Haas-Stapleton, E., Agabian, N., Deva, R., Nigam, S., 2006. Phospholipase A2 and Phospholipase B activities in fungi. Biochim. Biophys. Acta – Mol. Cell Biol. Lipids 1761, 1391-1399.
11. Jenkins, G.M., Frohman, M.A., 2005. Phospholipase D: A lipid centric review. Cell. Mol. Life Sci. 62, 2305-2316.
12. Frohman, M.A., Morris, A.J., 1999. Phospholipase D structure and regulation. Chem. Phys. Lipids 98, 127-140.
13. Waite, M., 1999. The PLD superfamily: insights into catalysis. Biochim. Biophys. Acta – Mol. Cell Biol. Lipids 1439, 187-197.
14. Doig, S.D., Diks, R.M.M., 2003a. Toolbox for modification of the lecithin headgroup. Eur. J. Lipid Sci. Technol. 105, 368-376.
15. Haraldsson, G.G., Thorarensen, A., 1999. Preparation of phospholipids highly enriched with n-3 polyunsaturated fatty acids by lipase. J. Am. Oil Chem. Soc. 76, 1143-1149.
16. Kim, M. Il, Ham, H.O., Oh, S.D., Park, H.G., Chang, H.N., Choi, S.H., 2006. Immobilization of Mucor javanicus lipase on effectively functionalized silica nanoparticles. J. Mol. Catal. B Enzym. 39, 62-68.
17. Haas, M.J., Cichowicz, D.J., Phillips, J., Moreau, R., 1993. The hydrolysis of phosphatidylcholine by an immobilized lipase: Optimization of hydrolysis in organic solvents. J. Am. Oil Chem. Soc. 70, 111-117.
18. Mustranta, A., Forssell, P., Poutanen, K., 1995. Comparison of lipases and phospholipases in the hydrolysis of phospholipids. Process Biochem. 30, 393-401.
19. Lilbæk, H.M., Fatum, T.M., Ipsen, R., Sørensen, N.K., 2007. Modification of milk and whey surface properties by enzymatic hydrolysis of milk phospholipids. J. Agric. Food Chem. 55, 2970-2978.
20. Baeza-Jiménez, R., López-Martínez, L.X., Otero, C., Kim, I.H., García, H.S., 2013. Enzyme-catalysed hydrolysis of phosphatidylcholine for the production of lysophosphatidylcholine. J. Chem. Technol. Biotechnol. 88, 1859-1863.
21. Sarney, D.B., Fregapane, G., Vulfson, E.N., 1994. Lipase-catalyzed synthesis of lysophospholipids in a continuous bioreactor. J. Am. Oil Chem. Soc. 71, 93-96.
22. Ghosh, M., Bhattacharyya, D.K., 1997. Enzymatic alcoholysis reaction of soy phospholipids. J. Am. Oil Chem. Soc. 74, 597-599.
23. Yang, G., Yang, R., Hu, J., 2015. Lysophosphatidylcholine Synthesis by Lipase-catalyzed Ethanolysis. J. Oleo Sci. 64, 443-447.
24. Mazur, A.W., Hiler, G.D., Lee, S.S.C., Armstrong, M.P., Wendel, J.D., 1991. Regio- and stereoselective enzymatic esterification of glycerol and its derivatives. Chem. Phys. Lipids 60, 189-199.
25. Kim, J., Kim, B.-G., 1998. Lipase-Catalyzed Synthesis of Lysophosphatidylcholinea. Ann. N. Y. Acad. Sci. 864, 341-344.
26. Kim, J., Kim, B.-G., 2000. Lipase-catalyzed synthesis of lysophosphatidylcholine using organic cosolvent for in situ water activity control. J. Am. Oil Chem. Soc. 77, 791-797.
27. Virto, C., Adlercreutz, P., 2000a. Lysophosphatidylcholine synthesis with Candida antarctica lipase B (Novozym 435). Enzyme Microb. Technol. 26, 630-635.
28. Hong, S.I., Kim, Y., Kim, C.T., Kim, I.H., 2011. Enzymatic synthesis of lysophosphatidylcholine containing CLA from sn-glycero-3-phosphatidylcholine (GPC) under vacuum. Food Chem. 129, 1-6.
29. Adlercreutz, P., Virto, C., Persson, M., Vaz, S., Adlercreutz, D., Svensson, I., Wehtje, E., 2001. Enzymatic conversions of polar lipids. Principles, problems and solutions. J. Mol. Catal. B Enzym. 11, 173-178.
30. Virto, C., Adlercreutz, P., 2000b. Two-enzyme system for the synthesis of 1-lauroyl-rac-glycerophosphate (lysophosphatidic acid) and 1-lauroyl-dihydroxyacetonephosphate. Chem. Phys. Lipids 104, 175-184.

31. Adlercreutz, D., Budde, H., Wehtje, E., 2002. Synthesis of phosphatidylcholine with defined fatty acid in the sn-1 position by lipase-catalyzed esterification and transesterification reaction. Biotechnol. Bioeng. 78, 403-411.
32. Peng, L., Xu, X., Mu, H., Høy, C.E., Adler-Nissen, J., 2002. Production of structured phospholipids by lipase-catalyzed acidolysis: Optimization using response surface methodology. Enzyme Microb. Technol. 31, 523-532.
33. Sakai, K., Okuyama, H., Yura, J., Takeyama, H., Shinagawa, N., Tsuruga, N., Kato, K., Miura, K., Kawase, K., Tsujimura, T., Naruse, T., Koike, A., 1992. Composition and turnover of phospholipids and neutral lipids in human breast cancer and reference tissues. Carcinogenesis 13, 579-584.
34. Jenski, L.J., Zerouga, M., Stillwell, W., 1995. ω-3 Fatty Acid-Containing Liposomes in Cancer Therapy. Exp. Biol. Med. 210, 227-233.
35. Hossen, M., Hernandez, E., 2005. Enzyme-catalyzed synthesis of structured phospholipids with conjugated linoleic acid. Eur. J. Lipid Sci. Technol. 107, 730-736.
36. Baeza-Jiménez, R., González-Rodríguez, J., Kim, I.-H., García, H.S., Otero, C., 2012. Use of immobilized phospholipase A1-catalyzed acidolysis for the production of structured phosphatidylcholine with an elevated conjugated linoleic acid content. Grasas y Aceites 63, 44-52.
37. Baeza-Jiménez, R., Noriega-Rodríguez, J.A., García, H.S., Otero, C., 2012. Structured phosphatidylcholine with elevated content of conjugated linoleic acid: Optimization by response surface methodology. Eur. J. Lipid Sci. Technol. 114, 1261-1267.
38. Vikbjerg, A.F., Mu, H., Xu, X., 2005. Parameters affecting incorporation and by-product formation during the production of structured phospholipids by lipase-catalyzed acidolysis in solvent-free system. J. Mol. Catal. B Enzym. 36, 14-21.
39. Vikbjerg, A.F., Mu, H., Xu, X., 2008. Lipase-Catalyzed Acyl Exchange of Soybean Phosphatidylcholine in n-Hexane: A Critical Evaluation of Both Acyl Incorporation and Product Recovery. Biotechnol. Prog. 21, 397-404.
40. Vikbjerg, A.F., Mu, H., Xu, X., 2007. Synthesis of structured phospholipids by immobilized phospholipase A2 catalyzed acidolysis. J. Biotechnol. 128, 545-554.
41. Chojnacka, A., Gładkowski, W., Kiełbowicz, G., Wawrzeńczyk, C., 2009. Enzymatic enrichment of egg-yolk phosphatidylcholine with α-linolenic acid. Biotechnol. Lett. 31, 705-709.
42. Kim, I.-H., Garcia, H.S., Hill, C.G., 2007. Phospholipase A1-catalyzed synthesis of phospholipids enriched in n–3 polyunsaturated fatty acid residues. Enzyme Microb. Technol. 40, 1130-1135.
43. Kim, I.H., Garcia, H.S., Hill, C.G., 2010. Synthesis of structured phosphatidylcholine containing n-3 PUFA residues via acidolysis mediated by immobilized phospholipase A1. JAOCS, J. Am. Oil Chem. Soc. 87, 1293-1299.
44. Zhao, T., No, D.S., Kim, B.H., Garcia, H.S., Kim, Y., Kim, I.H., 2014. Immobilized phospholipase A1-catalyzed modification of phosphatidylcholine with n-3 polyunsaturated fatty acid. Food Chem. 157, 132-140.
45. Marsaoui, N., Naghmouchi, K., Baah, J., Raies, A., Laplante, S., 2015. Incorportation of Ethyl Esters of EPA and DHA in Soybean Lecithin Using Rhizomucor miehei Lipase: Effect of Additives and Solvent-Free Conditions. Appl. Biochem. Biotechnol. 176, 938-946.
46. Balakrishna, M., Kaki, S.S., Karuna, M.S.L., Sarada, S., Kumar, C.G., Prasad, R.B.N., 2017. Synthesis and in vitro antioxidant and antimicrobial studies of novel structured phosphatidylcholines with phenolic acids. Food Chem. 221, 664-672.
47. Jiang, F.Y., Wang, J.M., Ju, L.C., Kaleem, I., Dai, D.Z., Li, C., 2011b. Optimization of degumming process for soybean oil by phospholipase B. J. Chem. Technol. Biotechnol. 86, 1081-1087.
48. Yang, B., Wang, Y.H., Yang, J.G., 2006a. Optimization of enzymatic degumming process for rapeseed oil. JAOCS, J. Am. Oil Chem. Soc. 83, 653-658.
49. Dijkstra, A.J., 2010. Enzymatic degumming. Eur. J. Lipid Sci. Technol. 112, 1178-1189.
50. Clausen, K., 2001. Enzymatic oil-degumming by a novel microbial phospholipase. Eur. J. Lipid Sci. Technol. 103, 333-340.
51. Dijkstra, A.J., 2011. Enzymatic degumming. Lipid Technol. 23, 36-38.

52. Bora, L., 2013. Characterization of novel phospholipase C from Bacillus licheniformis MTCC 7445 and its application in degumming of vegetable oils. Appl. Biochem. Microbiol. 49, 555-561.
53. Huang, S., Liang, M., Xu, Y., Rasool, A., Li, C., 2014. Characteristics and vegetable oils degumming of recombinant phospholipase B. Chem. Eng. J. 237, 23-28.
54. Wei, T., Xu, C., Yu, X., Jia, W., Yang, K., Jia, C., Mao, D., 2015. Characterization of a novel thermophilic phospholipase B from Thermotoga lettingae TMO: applicability in enzymatic degumming of vegetable oils. J. Ind. Microbiol. Biotechnol. 42, 515-522.
55. Yang, J.G., Wang, Y.H., Yang, B., Mainda, G., Guo, Y., 2006b. Degumming of vegetable oil by a new microbial lipase. Food Technol. Biotechnol. 44, 101-104.
56. Jahani, M., Alizadeh, M., Pirozifard, M., Qudsevali, A., 2008. Optimization of enzymatic degumming process for rice bran oil using response surface methodology. LWT – Food Sci. Technol. 41, 1892-1898.
57. Yang, B., Zhou, R., Yang, J.G., Wang, Y.H., Wang, W.F., 2008. Insight into the enzymatic degumming process of soybean oil. JAOCS, J. Am. Oil Chem. Soc. 85, 421-425.
58. Jiang, F., Wang, J., Kaleem, I., Dai, D., Zhou, X., Li, C., 2011a. Degumming of vegetable oils by a novel phospholipase B from Pseudomonas fluorescens BIT-18. Bioresour. Technol. 102, 8052-8056.
59. Manjula, S., Jose, A., Divakar, S., Subramanian, R., 2011. Degumming rice bran oil using phospholipase-A1. Eur. J. Lipid Sci. Technol. 113, 658-664.
60. Jiang, X., Chang, M., Jin, Q., Wang, X., 2014. Optimization of the degumming process for camellia oil by the use of phospholipase C in pilot-scale system. J. Food Sci. Technol. 52, 3634-3644.
61. Yu, D., Ma, Y., Jiang, L., Elfalleh, W., Shi, M., Hu, L., 2013a. Optimization of magnetic immobilized phospholipase A1 degumming process for soybean oil using response surface methodology. Eur. Food Res. Technol. 237, 811-817.
62. Yu, D., Jiang, L., Li, Z., Shi, J., Xue, J., Kakuda, Y., 2012. Immobilization of phospholipase A 1 and its application in soybean oil degumming. JAOCS, J. Am. Oil Chem. Soc. 89, 649-656.
63. Yu, D., Ma, Y., Xue, S.J., Jiang, L., Shi, J., 2013b. Characterization of immobilized phospholipase A1 on magnetic nanoparticles for oil degumming application. LWT – Food Sci. Technol. 50, 519-525.
64. Doig, S.D., Diks, R.M.M., 2003b. Toolbox for modification of the lecithin headgroup. European Journal of Lipid Science and Technology, 105, 368-376
65. Hosokawa, M., Shimatani, T., Kanada, T., Inoue, Y., Takahashi, K., 2000. Conversion to docosahexaenoic acid-containing phosphatidylserine from squid skin lecithin by phospholipase D-mediated transphosphatidylation. J. Agric. Food Chem. 48, 4550-4554.
66. Piazza, G.J., Marmer, W.N., 2007. Conversion of phosphatidylcholine to phosphatidylglycerol with phospholipase D and glycerol. JAOCS, J. Am. Oil Chem. Soc. 84, 645-651.
67. Iwasaki, Y., Mizumoto, Y., Okada, T., Yamamoto, T., Tsutsumi, K., Yamane, T., 2003. An aqueous suspension system for phospholipase D-mediated synthesis of PS without toxic organic solvent. J. Am. Oil Chem. Soc. 80, 653-657.
68. Dittrich, N., Ulbrich-Hofmann, R., 2001. Transphosphatidylation by immobilized phospholipase D in aqueous media. Biotechnol. Appl. Biochem. 34, 189-94.
69. D'Arrigo, P., Cerioli, L., Chiappe, C., Panzeri, W., Tessaro, D., Mele, A., 2012. Improvements in the enzymatic synthesis of phosphatidylserine employing ionic liquids. J. Mol. Catal. B Enzym. 84, 132-135.
70. Yang, S.L., Duan, Z.Q., 2016. Insight into enzymatic synthesis of phosphatidylserine in deep eutectic solvents. Catal. Commun. 82, 16-19.
71. Duan, Z.Q., Hu, F., 2013. Efficient synthesis of phosphatidylserine in 2-methyltetrahydrofuran. J. Biotechnol. 163, 45-49.
72. Bi, Y.H., Duan, Z.Q., Du, W.Y., Wang, Z.Y., 2015a. Improved synthesis of phosphatidylserine using bio-based solvents, limonene and p-cymene. Biotechnol. Lett. 37, 115-119.
73. Bi, Y.H., Duan, Z.Q., Li, X.Q., Wang, Z.Y., Zhao, X.R., 2015b. Introducing biobased ionic liquids as the nonaqueous media for enzymatic synthesis of phosphatidylserine. J. Agric. Food Chem. 63, 1558-1561.
74. Hu, F., Wang, H., Duan, Z.Q., Yao, R.S., 2013. A novel phospholipase D constitutively secreted by Ochrobactrum sp. ASAG-PL1 capable of enzymatic synthesis of phosphatidylserine. Biotechnol. Lett. 35, 1317-1321.

75. Choojit, S., Bornscheuer, U.T., Upaichit, A., H-Kittikun, A., 2016. Efficient phosphatidylserine synthesis by a phospholipase D from Streptomyces sp. SC734 isolated from soil-contaminated palm oil. Eur. J. Lipid Sci. Technol. 118, 803-813.
76. Xu, X., 2000b. Modification of Oils and Fats by Lipase-Catalyzed Interesterification: Aspects of Process Engineering, in: Enzymes in Lipid Modification. 190-215.
77. Nasir, M.I., Bernards, M.A., Charpentier, P.A., 2007. Acetylation of Soybean Lecithin and Identification of Components for Solubility in Supercritical Carbon Dioxide. J. Agric. Food Chem. 55, 1961-1969.
78. Rios, L., Echeverri, D., Cardeño, F., 2013. Hydroxylation of vegetable oils using acidic resins as catalysts. Ind. Crops Prod. 43, 183-187.
79. Coenen, J.W.E., 1976. Hydrogenation of Edible oils. J. Am. Oil Chem. Soc. 53, 382-389.
80. King, J.W., Holliday, R.L., List, G.R., Snyder, J.M., 2001. Hydrogenation of vegetable oils using mixtures of supercritical carbon dioxide and hydrogen. J. Am. Oil Chem. Soc. 78, 107-113.

4.2 Fractionation of lecithin

Michael Schneider

As it has already been described in previous chapters, lecithin – as defined in the EU under E322 (1) – is a complex mixture of polar lipids (phospho- and glycolipids) containing some neutral lipids, carbohydrates and several other minor components. Vegetable lecithins are as such by-products from vegetable oil refining (degumming) (2), whereas animal lecithins are extracted from animal raw materials like for example egg yolk.

Lecithins are the main starting materials to obtain specific phospholipid concentrates or isolates.

4.2.1 Fluid (crude) vegetable lecithins

The preferred route to crude vegetable phospholipids is to separate them as a by-product of the physical refining process from crude vegetable oils. The first step of refining, called degumming, uses the fact that phospholipids tend to hydrate and to form high molecular aggregates in the presence of water. These aggregates are insoluble in oil and can be removed as so-called wet gums by centrifugation. Drying the gums gives lecithin with approximately 45 to 50% polar lipids, predominantly phospholipids. As polar lipid aggregates entrap a certain amount of oil, the dried gums (crude lecithin) are of fluid to semi-fluid viscosity.

The literature reports many different variations of the degumming process (2) all aiming to reduce the phosphorus level in the crude oil to a minimum. Low residual phosphorus is of importance for the stability of food-grade oils and for further processing steps, particularly for hydrogenation and for further use for biodiesel production.

Degumming conditions have a strong influence on crude vegetable lecithin composition and quality.

Based on such crude lecithins the industry has developed a complex system of different standardizing, refining, fractionation, purification and modification steps to serve the different industrial needs for products having application-optimized properties.

Table 1.
Composition of a commercial fluid soybean lecithin (3; modified).

	%
Phospholipids	50
Phosphatidylcholine (PC)	15
Phosphatidylethanolamine (PE)	11
Phosphatidylinositol (PI)	10
Phosphatidic acid (PA)	4
N-acylphosphatidylethanolamine (NAPE)	2
Other phospholipids (e.g. lyso-)	3
Glycolipids	6
Carbohydrates	4
Free fatty acids	2
Triglycerides (oil)	37
Moisture	1
Fatty acid composition	
Palmitic acid	16
Stearic acid	4
Oleic acid	17
Linoleic acid	55
Linolenic acid	7

The following simplified scheme illustrates the industrial processing basics of vegetable lecithins/phospholipids, including fractionation, enzyme and chemical modification:

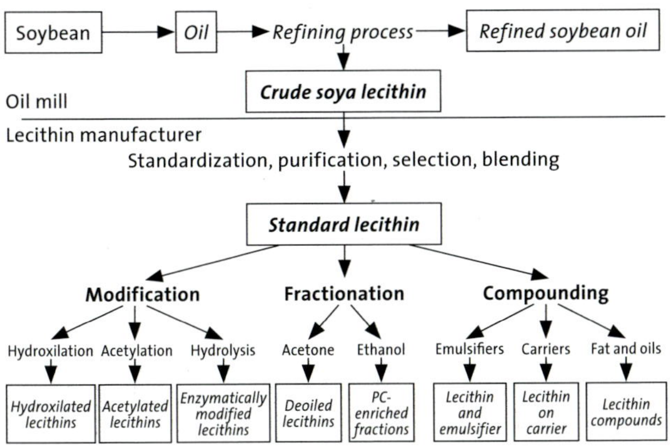

Figure 1.
Vegetable lecithin processing (13).

Fluid (crude) vegetable lecithins (approximately 450,000 tons annually) represent the biggest market share of all phospholipid-containing products. In general they serve as non-toxic, highly functional food and feed emulsifiers, dispersants, structure modifiers and stabilizers. Soy lecithins dominate this segment, with more than 80% market share.

During the last decades the quality of fluid vegetable lecithins has been dramatically improved, especially with regards to colour and impurities (so-called hexane insolubles (HI), soybean fines – mainly protein – resulting from the oil extraction process). Even careful filtration of crude oils has not been able to remove the last small traces of protein from lecithin to allow an exemption from allergy labelling of commercial lecithins.

4.2.2 Membrane degumming

It has long been known that non-polar solvents like hexane do not dissolve polar lipids in a molecular dispersed manner, contrary to neutral lipids. Polar lipids, like phospholipids, form high-molecular-weight lamellar or vesicular structures (4).

Based on this observation a new degumming process has been developed (5). The hexane solution from oil seed extraction (miscella) contains a mixture of high-molecular-weight polar lipid aggregates dispersed in a homogeneous solution of neutral lipids. Passing such dispersions through ultrafiltration modules with cut-offs of approximately 20,000 Dalton leads to a transport of the neutral lipid solution across the membrane (permeate), whereas the polar lipid aggregates are retained and concentrated in the retentate.

Operating such ultrafiltration processes the right way, water degumming can be avoided. Lecithin is separated in almost exactly the same quality and composition as with the standard water degumming process. The advantage for the oil mill is an extremely low residual phosphorus in the degummed oil, lower than with conventional degumming procedures.

Modern membrane systems are solvent-resistant and allow long standing times. However, the limiting problem is membrane fouling caused even by minimal protein impurities in the extracted miscella, even when it has gone through very careful filtration steps.

4.2.3 Solvent-based fractionation

Commercial solvent technologies are limited to a small selection of solvents, which are appropriate to the image and use of a natural, functional ingredient for the various finished goods, mainly for food and feed use. The solvents of choice are:

- Hexane.
- Acetone.
- Ethanol.
- Compressed gases.
- Methanol, isopropanol.

4.2.4 Membrane deoiling

Exhaustive membrane separation from hexane solutions of crude lecithins (or even of miscella solutions from seed extraction) can lead to a complete removal of neutral lipids resulting in oil-free polar lipid products (deoiled lecithins). The first commercial product (oil-free powder lecithin) was made by dissolving water-degummed lecithin in hexane followed by membrane separation (6).

4.2.5 Acetone deoiling

However, the most current method of choice for the production of oil-free vegetable lecithins is not membrane separation but processing with acetone.

Vegetable crude lecithins contain approximately 30–40% neutral lipids, predominantly triacylglycerols ("triglycerides"). In order to improve the handling of the viscous crude lecithins and to improve dispersibility properties, industry makes use of the fact that polar lipids are almost insoluble in acetone, whereas neutral lipids dissolve easily (7).

Batch or continuous acetone extraction leads to products in powder or granulated form containing a residual of just 2–3% neutral lipids. The products are slightly hygroscopic, but flow freely especially, after the addition of minor amounts of flow agents like tricalcium phosphate or silica.

Because acetone also removes most of the natural tocopherol traces in lecithin (approx. 1,000 ppm in crude soybean lecithin), the shelf life of deoiled lecithin products is usually limited to 18 months.

A drawback of using a highly reactive solvent like acetone for lecithin treatment is the formation of small amounts of reaction products. Acetone can be catalysed by polar lipids to dimerize into diacetone alcohol, and subsequently – by splitting off water – to finally form mesityl oxide, a strong and unpleasant-smelling compound.

The formation of mesityl oxide is strongly dependent on the contact time between acetone and phospholipids. Whenever acetone has not been completely removed immediately after deoiling its typical cat-urine-like smell can be recognized, a typical sign of improper processing.

Another problem of using acetone for deoiling is the fact that acetone has a slight tendency to form Schiff base-type adducts with aminophospholipids like phosphatidylethanolamine (8). This risk of side reactions, together with environmental concerns and problems with acceptance of even trace solvent levels in a product for food and nutraceutical use, has stimulated lots of attempts to develop an alternative to acetone for deoiling. One such is carbon dioxide.

4.2.6 Carbon dioxide deoiling

Compressed carbon dioxide shows good solubility for neutral lipids under certain conditions of moderate temperature and high pressure (so-called super-critical conditions) and does not dissolve polar lipids at all – ideal conditions for the complete removal of neutral lipids from polar lipids (deoiling).

However, several decades with many disappointing results passed until one German company found the way (9) to overcome the problem of dramatically increasing viscosity in the course of reducing the oil content of crude lecithin combined with a limited transport of oil into the compressed gas. The poor diffusion issue could finally be solved with a clever spray nebulizing approach creating – like in spray drying – small lecithin droplets which could be deoiled more or less instantaneously.

Another major problem is less to separate the oil from the polar lipids, but to remove the oil-free powder continuously from the pressurized equipment, sometimes operated at more than 500 bar and fighting with the Joule – Thompson effect of rapid cooling during the expansion. All these issues didn't restrain a Chinese company together with the German equipment manufacturer Uhde from realizing this interesting process. CO_2-deoiled soy lecithin powder is now available from China.

Table 2.

Typical composition of a commercial deoiled soybean lecithin.

	wt %
Phospholipids	80
Phosphatidylcholine (PC)	23
Phosphatidylethanolamine (PE)	22
Phosphatidylinositol (PI)	15
Phosphatidic acid (PA)	8
N-acylphosphatidylethanolamine (NAPE)	6
Lysophospholipids	3
Others	3
Glycolipids	10
Carbohydrates	5
Neutral lipids	3
Sterols	traces
Moisture	1

Alternative and successful approaches with mixtures of propane and CO_2 or even pure propane (10) have clear advantages in terms of pressure (and energy as well as investment costs). But they have not been implemented in industrial processing so far.

Advantages of acetone-free deoiling are manifold: low temperature, oxygen-free process, no reaction between acetone and aminophospholipids, no solvent residues, environmentally friendly.

4.2.7 Alcohol fractionation

The industrial production of vegetable lecithin 'fractions' uses ethanol or ethanol/water systems predominantly. The principle is that phosphatidylcholine (PC) is preferentially soluble in ethanol, phosphatidylethanolamine (PE) distributes more or less evenly between the soluble and the insoluble phase, and the more acidic phospholipids like phosphatidylinositol (PI) and phosphatidic acid (PA) remain practically insoluble.

The oil part of fluid lecithins partitions between both the soluble and the insoluble phase, leading to fluid fractions.

The commercial interest in both these fractions comes from the difference in their technological functionality (the soluble fraction is more an O/W, the insoluble fraction a W/O emulsifier). But the interest of the industry in PC-enriched products for dietetic purposes is a driving force as well.

Alcohol fractionation is carried out both before and after deoiling. Table 3 shows the changes in phosphatidylcholine concentration after fractionation of both a liquid and a deoiled soybean lecithin powder using different types of alcohol or alcohol/water mixtures (11).

Table 3.

Alcohol fractionation of soybean lecithin (13).

	Triacylglycerols %	Phosphatidylcholine %	Yield %
Crude lecithin	39	15	
Soluble fraction:			
Methanol	25	39	23
Ethanol (100%)	60	22	26
Ethanol (95%)	48	33	15
Ethanol (90%)	23	46	12
Isopropanol	54	19	49
Insoluble fraction:			
Ethanol (95%)	38	7	85
Deoiled lecithin	2.5	24	
Soluble fraction:			
Ethanol (100%)	2.5	52	28
Insoluble fraction:			
Ethanol (100%)	2.5	12	72

The PC-enriched fraction of deoiled powder is of waxy consistency whereas the insoluble part is a powder after solvent removal.

The preferred process for fluid lecithin fractionation uses a continuous approach with either countercurrent or concurrent extraction. Alcohol extraction of deoiled lecithin is a batch process.

What has been described for soybean lecithin here can in principle be applied to other crude vegetable phospholipid products (lecithins) as well. The results are similar, but today not used commercially.

I have previously reviewed this topic in detail (11). Generally speaking, chromatographic separation is the method of choice to prepare polar lipids of high purity. It is almost impossible to give a short summary of the numerous methods published in the scientific and patent literature. Many of them are used for the quantitative analysis of tissue extracts, less for the production of commercial quantities of phospho- or glycolipids.

The best results in separation are usually achieved by combining different separation techniques, such as solvent extraction of crude lipid mixtures first, followed by adsorption processes. Amongst the long list of different adsorbents are aluminium oxide, silica, Florisil (magnesium silicate), DEAE (diethylaminoethyl) cellulose and ion-exchanging resins.

Alumina and silica play an outstanding role for industrial use, especially for phosphatidylcholine concentrates or isolates. Both materials are applied in low- and medium-pressure chromatography to produce large quantities of soybean and egg PC products.

A typical alumina separation is as follows: deoiled soy bean lecithin is fractionated with ethanol first to produce a soluble fraction with about 50% PC. This solution is then passed down a column filled with aluminium oxide under low pressure. Almost all of the PC accompanying phospholipids is bound to the alumina and the effluent consists of almost pure choline phospholipids (PC and lyso-PC).

The biggest advantage of this process is the extremely high loading capacity of aluminium oxide for the non-PC phospholipids in comparison with other adsorbents and methods. Drawbacks of the method however are the impossibility to recycle the adsorbent or to remove lysophosphatidylcholine (LPC) from the purified material and the limitation to PC purification.

Applying alumina chromatography in ethanol to egg phospholipids shows that sphingomyelin (SPM, a choline-containing compound) has the same elution behaviour as PC and is included in the PC/LPC effluent.

Table 4 gives a short overview of different chromatography methods as published in the literature and patents.

Table 4.
Chromatographic systems for phospholipid fractionation/purification (13).

Adsorbent	Solvent system
Aluminium oxide	Ethanol
	Chloroform/methanol
Silica gel	Hexane
	Ethanol
	Hexane/isopropanol/water
	Chloroform/methanol
	Acetone/chloroform/methanol
	Chloroform/acetone
Florisil	Chloroform/methanol
Amberlite	Ethanol
	Ethyl acetate/ethanol
DEAE cellulose	Chloroform/methanol/acetic acid

In comparison to the alumina separation mentioned, silica chromatography seems to be the more flexible method with respect to solvent systems, separation characteristics and reconditioning of the column. But more sophisticated equipment and much higher quantities of adsorbent (at much higher prices) are needed.

Silica can also be used for the selective elution of different polar lipid classes from a given lecithin. The choice of solvent systems are chloroform/methanol and methanol/acetone mixtures. Separation is mainly done stepwise and in some cases by gradient elution. For more details about the wide range of possibilities for chromatographic separation refer to the literature on this subject (11).

Today's industrial-scale chromatography is mainly focused on the preparation of PC from soya and egg. These products are of great commercial interest and are available even in tons quantities.

Preparation of other phospholipids like PE and PI is of growing interest. The chosen separation technique is large-scale HPLC, which is also used for the purification process of semi-synthetic and synthetic phospholipids. Limitations are expensive equipment, low capacity and consequently high production costs.

It has to be considered that even purified PC obtained from a natural raw material with their source-specific different fatty acids (e.g. egg or soya) is still a mixture of various molecular species differing in fatty acid composition (12).

So far literature gives little indication of the possibility to separate phospholipids on the basis of the degree of saturation. Only for analytical purposes the use of argentation chromatography, that is using silver nitrate impregnated silica as an adsorbent, is described. The higher the degree of unsaturation, the more firmly the phospholipids are bound to the stationary phase and can be separated from each other.

A more recent development in this area is to use reverse-phase HPLC. However, it is not known if these processes are already being used commercially. Synthesizing the desired molecule seems to be the more appropriate and efficient route.

References

1. E322 – Lecithins. 2012. Commission Regulation (EU) No 231/2012. L 83/103.
2. Flider, F.J., 1985. The manufacture of soybean lecithins. In Szuhaj, B.F., List, G.R. (Eds), Lecithins. J. Am. Oil Chem. Soc. 21.
3. Van Nieuwenhuyzen, W., 1976. Lecithin production and properties. J. Am. Oil Chem. Soc. 53, 425.
4. SenGupta, A.K., 1986. Novel developments in refining of edible oils. Fette, Seifen, Anstrichmittel 88, 79.
5. Cargill (2001) Method for removing phospholipids from vegetable oil miscella, method for conditioning a polymeric microfiltration membrane and membrane. US Patent 6,207,209.
6. Archer-Daniels-Midland Co. (2000) Process for producing deoiled phospholipids. US Patent 61405190.
7. Gober, K.H., Guenther, B.R., Lunebach, E.M., 1995. Isolation and analysis of phospholipids and phospholipid derivatives. In Cevc, G. (Ed.), Phospholipids Handbook. Marcel Dekker 1.
8. Kuksis, A., Ravandi, A., Schneider, M., 2005. Covalent binding of acetone to aminophospholipids in vitro and in vivo. Ann. N. Y. Acad. Sci. 1043, 417.
9. Meyer, L., 1994. Liquid high pressure spraying extraction process. European Patent 0785814.
10. Peter, S., 1994. Process for removing oil from raw lecithin. International Patent PCT 94/01004.
11. Schneider, M., 1989. Fractionation and purification of lecithin. In Szuhaj, B.F. (Ed.), Lecithins, Sources, Manufacture and Uses. J. Am. Oil Chem. Soc., 109-130.
12. Kuksis, A., 1985. Animal lecithins. In Szuhaj, B.F., List, G.R. (Eds), Lecithins. J. Am. Oil Chem. Soc. Champaign IL, 105.
13. Schneider, M., 2008. Major sources, composition and processing. In Gunstone, F.D. (Ed.), Phospholipid Technology and Applications. The Oily Press.

5 Quality Control and Legal Principles of Lecithin

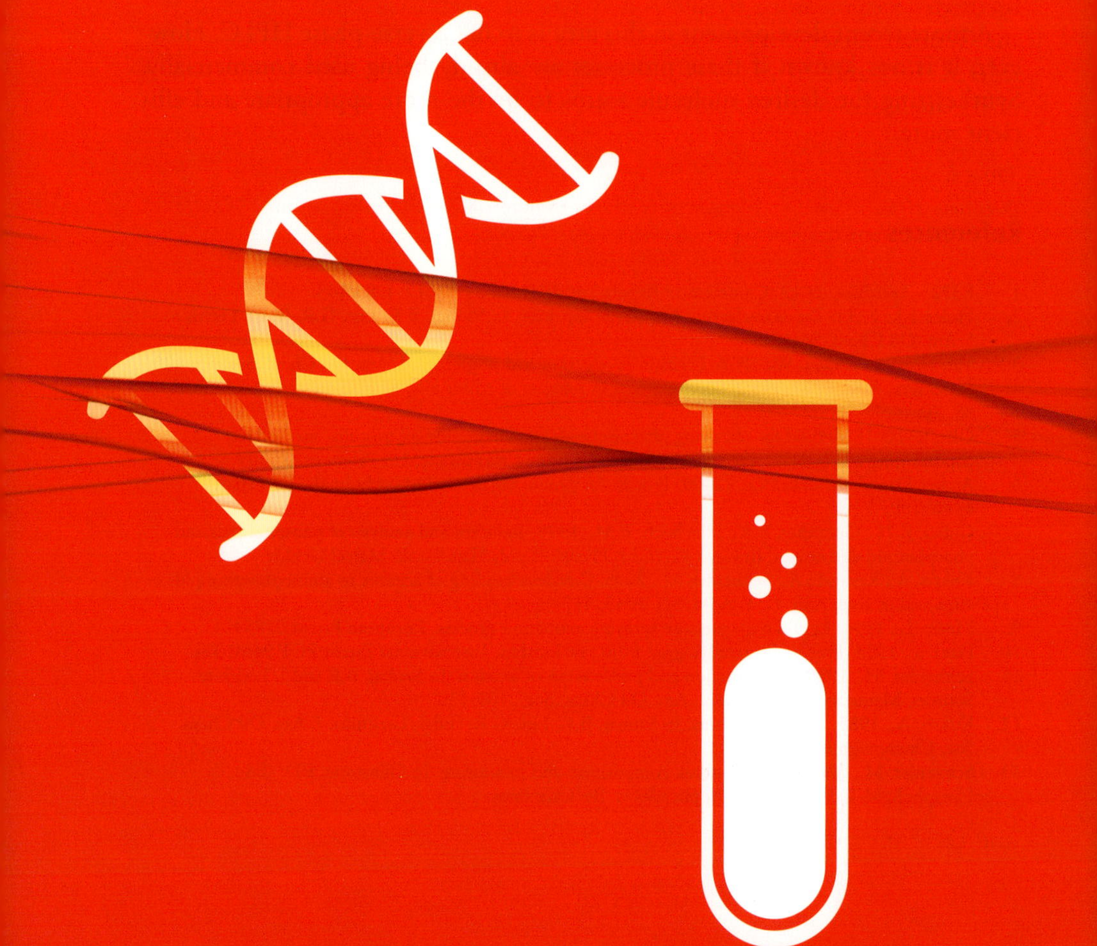

5.1 Routine analytical control methods for lecithin

Willem van Nieuwenhuyzen, Waldemar Buxmann

Chemical analysis of lecithin is an essential part of quality control during processing and enables fulfilling the quality requirements in terms of chemical-physical properties and thus the technical functionality of the finished lecithin products. The same methods can be used for all commercially available lecithin products, mostly for soy-, sunflower- or rapeseed-derived lecithin. The methods recommended by the American Oil Chemists' Society (AOCS) and the German Society for Fat Science (DGF) are widely accepted and can be considered the standard analytical method for lecithin. Table 1 gives an overview of the analytical methods according to AOCS and DGF.

Table 1.

Overview of the standard analytical methods for commercial lecithin by American Oil Chemists' Society (AOCS) and German Society for Fat Science (DGF).

Method	AOCS	DGF
Acetone-insoluble matter	Ja 4-46	F-I 5
Acid value	Ja 6-55	F-I 3
Peroxide value	Ja 8-87	F-I 3b
Hexane-insoluble matter	Ja 3-87	–
Toluene-insoluble matter	–	F-I 4b
Moisture (Karl Fischer)	Ja 2b-87	F-I 4
Gardner colour	Ja 9-87	C-IV 4a
Viscosity	Ja 10-87	F-I 2a
TLC	Ja 7-86	
HPLC-UV	Ja 7b-91	
HPLC-LSD	Ja 7c-07	F-I-6a
HPTLC	–	F-I-6

In contrast, in the CIS countries, GOST methods are mostly used. These differ in some cases significantly and are not clearly correlatable with the AOCS and DGF methods.

5.1.1 Chemical and microbiological analysis

The following methods are routinely used for quality control of lecithin.

Acetone-insoluble (AI) matter
The techno-functional performance of lecithin is mainly based on the surface-active part, the phospholipids. So, the total phospholipid content can be considered as the key parameter of lecithin. The amount of acetone-insoluble matter (% AI) is the approximate indication for the amount of all polar lipids, including phospholipids, glycolipids and carbohydrates. While the non-polar lipids including oil (triglycerides but also mono- and diglycerides) and fatty acids dissolve in acetone. The % AI is a rough indication for the 'active substance' (1,2).

Acid value (AV)
Acid value describes the acidity as mg KOH/g of sample which is necessary to neutralize free fatty acids but also acidic phospholipids. The AV represents the acidity contributed by phospholipids (in vegetable lecithin often 18 – 24 mg KOH/g) and (added) free fatty acids. AV can serve as an indicator for the stability of liquid lecithin in terms of phase separation. Furthermore, high AV could be an indication of product damage due to improper storage of the raw material (seeds) or inadequate processing conditions. Enzymatic and acid/base hydrolysis also leads to AV increase (3,4).

Peroxide value (POV)
The peroxide value (POV) specifies the content of functional peroxide groups of a fat or oil. It serves as an indicator for the fat rancidity and therefore for the freshness of lecithin. The POV of lecithin produced from fresh or optimally stored oil seeds is usually below 2 mEq/kg. The higher POV in lecithin is mostly caused by the residual hydrogen peroxide from the bleaching process step. This is in contrast to POV values in oil as a result of oxidation. POV is determined by dissolving a lecithin sample in an acetic-chloroform solution. After the addition of saturated potassium iodide the solution is titrated with sodium thiosulphate in the presence of a starch indicator (2,3).

Hexane-insoluble (HI) or toluene-insoluble (TI) matter
This method determines the content of impurities (residual fibre, husk, filtering aid etc.). The level of HI/TI matter in commercial lecithin is defined by the Food Chemicals Codex and by the European Union as E322 and is not to exceed 0.3%. Whereas crude soy and rapeseed lecithin rarely exceeds 0.3%, sunflower lecithin usually shows higher TI, mostly because of the presence of husk residues and wax-like substances, and needs to be purified. In

international trade usually the AOCS method using hexane as analytical solvent is applied, while European regulation advises toluene as solvent (DGF method). The porosity of the used glass filters is of greater importance on the result than the type of solvent used. Usually G-3 filters with a pore size of 16–40 µm are used (2,3).

Moisture

The water content of lecithin products is usually less than 1.0%. Moisture is usually determined by the Karl Fischer method. A less accurate moisture level can also be determined by azeotropic toluene distillation or by gravimetric method = drying in an oven at 105°C (loss on drying) (2).

Colour

The colour of the standard lecithin, which can be described as yellowish to brownish, is a result of the combination of pigments including carotenoids and melanoids. The colour of lecithin is measured by comparison with the Gardner colour scale or Iodine colour scale. The analysis can be done using either undiluted (pure) lecithin or a solution with toluene, usually in a 1:10 ratio. The colour range of most clear lecithins is generally in the range of Gardner values 12–18 in the undiluted products or 10–13 in the solution with toluene (2,4).

Viscosity

Liquid lecithin exhibits non-Newtonian flow characteristics – the viscosity is not a constant ratio of shear stress to shear rate. The viscosity of lecithin depends on the acetone-insoluble content (AI), moisture, content of divalent ions and acid value. High AI and moisture content increase the viscosity, whereas an increased acid value decreases it. The viscosity is measured using an LVT viscometer at 25°C. Before measurement, the lecithin sample should be homogenized in order to eliminate some pseudo-plastic effects. The measured result is a useful indication of the flow properties of lecithin during pumping and in dosing equipment (2). The viscosity of commercial lecithin (crude and standardized) usually ranges between 20,000 and 7,000 mPas.

Microbiology

Lecithin with low microbiological load is required for food and pharmaceutical uses. Due to the low water content both liquid but also deoiled lecithin can be considered as rather uncritical in terms of microbiological risk. Due to a high water content of 30–50%, there might be a bit higher risk of microbiological contamination in lecithin wet gums. Therefore, it is highly recommended to dry the wet gum right after separation from the crude oil. A low addition of for example 0.1% hydrogen peroxide 35% solution to the gums may reduce the total microbiological count even further.

The lecithin production should be carried out in closed equipment, complying with state-of-the-art GMP (Good manufacturing praxis), QA and HACCP standards (2).

The methods for microbiological examination are mostly laboratory-specific. The microbiological analyses usually include the following parameters: total plate count, coliforms, *Salmonella,* mould and yeasts (5).

5.1.2 Regulatory specifications

According to most legislative definitions for their use as food ingredients, lecithins are mixtures of phosphatides (= phospholipids) derived from vegetable and animal origin. In the USA lecithins are classified in the Food Chemicals Codex (FCC III) (6). The US Food and Drug Administration (FDA) regulates under Title 21, Part 184 direct food substances affirmed as Generally Recognized as Safe (GRAS) (7). Standard lecithin, fractions and acetylated lecithin are listed GRAS in §184.1400, enzyme-modified lecithin in §184.1063 and hydroxylated lecithin in §172.814. Table 2 lists the various purity specifications.

The EU-approved food additive number E322 covers all standard, physically fractionated, bleached and enzymatically hydrolysed lecithins for use in the food industry (8). Hydroxylated and acetylated lecithins do not have an EU-approved food additive status (3).

EU legal specifications are similar or close to the Codex Alimentarius recommended specifications and methods of analysis. The Codex numbers are recommendations to nations for worldwide use. However, national laws are binding. Regulations for use as health foods or supplements, or in pharmaceutical products, may require further specific purity specifications (2,3).

Table 2.

Legal purity specifications of food-grade lecithin.

Purity	FAO/WHO Codex Alimentarius	European Union E322	Food Chemicals Codex
Acetone insolubles, %	> 60	> 60 Hydrolysed > 56	> 50
Hexane insolubles, % Toluene insolubles, %	< 0.3	< 0.3	< 0.3
Moisture, % Drying loss, %	< 2.0	< 2.0	< 1.5
Acid value, mg KOH/g	< 36	< 35 Hydrolysed < 45	< 36
Peroxide value, mEq/kg	< 10	< 10	< 100
Arsenic, ppm	< 3	< 3	
Lead, ppm	< 10	< 5	< 1
Mercury, ppm		< 1	
Heavy metals (as Pb), ppm	< 40	< 10	

References

1. Official Methods and Recommended Practices of the AOCS, 1998. 5th Edition.
2. Van Nieuwenhuyzen, W., 2015. Production and Utilization of Natural Phospholipids. In Ahmad, M., Xu, X., Polar Lipids, Biology, Chemistry and Technology. AOCS Press, Urbana, Illinois.
3. Van Nieuwenhuyzen, W., Tomás M.C., 2008. Update of vegetable lecithin and phospholipid technologies. Eur. J. Lipid Sci. Tech. 5, 472-486.
4. Lantz, R., 1989. Industrial Methods of Analysis. In Szuhaj, B. Lecithins, Sources, Manufacture & Uses. AOCS Champaign, Illinois.
5. Pardun, H., 1988. Die Pflanzenlecithine. Verlag für chemische Industrie H. Ziolkowsky, Augsburg, 492-540.
6. Food Chemicals Codex (FCC III).
7. US Food and Drug Administration (FDA), regulates under Title 21, Part 184.
8. Implementing Regulation (EU) 2017/2325.

5.2 Analytical methods for the characterization of lecithins

Gerhard Helmerich

A large number of different analytical techniques for the characterization of lecithins have been described. Apart from conventional methods of determining characteristic parameters such as fat values and the fatty acid profile, the literature describes methods of quantifying phospholipids using chromatography and spectrometry that are also of interest. In addition to checking purity, tests focus on determining the contents of the phospholipid fraction and identifying individual phospholipids. Separation via HPLC has been established as the standard method, although thin-layer chromatography (TLC) and analysis using ^{31}P NMR are also widespread. This chapter contains a summary of the results published by Helmerich and Köhler in 2004 (1).

5.2.1 Determination methods using chromatography

Two types of chromatography, HPLC and TLC, are particularly important methods used to characterize phospholipids.

5.2.1.1 High-performance liquid chromatography (HPLC)

For over twenty years HPLC has been used to separate phospholipids, and many separation systems have been developed. In medical analysis the ratio between phosphatidylcholine and sphingomyelin in the amniotic fluid is determined by HPLC and permits an assessment of fetal lung maturity (2). In food chemistry, it is used to determine the phosphatidylcholine (PC) content in high-quality lecithin, the distribution of PE/PI/PS/PC in foodstuffs or to determine the composition of egg yolk (3).

The stationary phase used in the separation of phospholipids may be a normal-phase substance such as silica gel, diol phases and amino phases, as well as reversed phases, although silica gel is the most frequently used (4). UV detectors with wavelengths between 203 and 210 nm are generally used to detect the phospholipids; less frequently used are flame ionization (FI) (5), evaporative light scattering (ELS) (6) or refractive index (RI) detectors (3,7). Examples of post-column derivatization can also be found (8) or mass spectrometry fast atom bombardment (FAB) detection (9).

In the case of polar stationary phases, a variety of solvent mixes of chloroform, methanol, acetonitrile, isopropanol, hexane and water are used for the mobile phase. Usually a small amount of an aqueous buffer (e.g. potassium phosphate, ammonium hydroxide, ammonium formate or sodium acetate) is added. In the case of non-polar phases, solvent mixes of isopropanol, methanol, water and a very small amount of hexane are used. Classic eluent mixes are 2-propanol/hexane/water, chloroform/methanol/water or acetonitrile/methanol/water in varying concentrations (3).

5.2.1.2 Thin-layer chromatography (TLC)

Thin-layer chromatography is a low-cost method capable of rapidly separating phospholipid preparations directly into their individual components. TLC is preferred to HPLC whenever complex mixes of phospholipids are to be separated. A deciding factor in favour of TLC is that, unlike using HPLC, there are no restrictions regarding the solvent. For example the UV activity, viscosity or volatility of a solvent plays no role in this case. Thin-layer chromatography on a semi-preparative scale is also used to isolate individual phospholipids.

The list of separating systems is very long. Aloisi et al. (10) provide an overview of the most regularly used solvents. Silica gel is used as the stationary phase, and molybdatophosphoric acid as the detection agent, thereby making both polar and non-polar lipids visible. The phospholipids are separated by using a mixture of chloroform/methanol/water (65+25+4, v+v+v). In another study Vuorela et al. (11) described a rapid method to quantify the phospholipids in lecithins. For this the lecithins are separated into polar and non-polar lipids using methyl acetate/chloroform/1-propanol/methanol/0.25% aqueous potassium chloride solution (m/v) (25+25+25+10+9, v+v+v+v+v). The detection agent is copper (II) sulphate – phosphoric acid solution and the photo file derived from the scan of the thin-layer plate is transformed into x/y data sets with the aid of an image-processing program.

5.2.2 Determination methods using spectrometry

Characterization analysis of lecithins can be performed with three types of spectrometry: IR spectroscopy, ^{31}P NMR spectrometry and mass spectrometry.

5.2.2.1 Infrared spectroscopy

Fourier transform infrared spectroscopy may be used to determine the quality and quantity of three lecithin components found in vegetable oil – PC, PE and PI (12).

5.2.2.2 ^{31}P nuclear magnetic resonance (NMR) spectrometry

Because the ^{31}P nuclide is highly sensitive, this method is ideal for quantification purposes. The reproducibility of the measurement depends on the temperature, type and concentration of inorganic cations, the distribution of a variety of surfactants and the pH value (13).

^{31}P NMR spectrometry is particularly susceptible to cations which can coordinately react with the functional groups of the phosphatidyl moiety of the phospholipids. For example, magnesium cations at concentrations above 30 ppm mask the phosphorus signals (14,15). Apart from alkali and alkaline-earth metals, other polyvalent cations ($Fe^{2+/3+}$, Al^{3+}) can also make the resonance signal weaker. The result is very wide, poorly differentiated peaks which do not permit reproducible quantification because the phospholipids all show in a field width of no more than 6 ppm units (16).

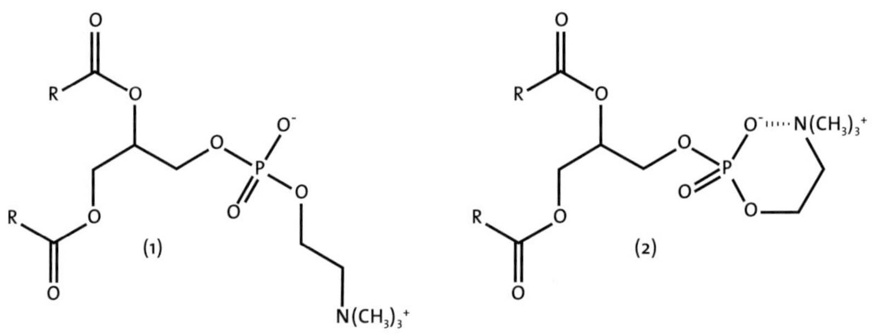

Figure 1.
Intramolecular ring formation (2) results in chemically non-equivalent compounds in the case of phosphatidylcholine (1). From Henderson et al. (15).

To remove the unwanted inorganic cations, stable chelate complexes are required that are capable of binding both mono- and polyvalent ions. For this purpose the sample must be washed with a dication salt of EDTA. The cation chosen is an element that does not weaken the resonance signal and does not trigger a competing reaction with the newly formed EDTA complex. One advantage of the washing process is that introducing the monovalent cation prevents intramolecular ring formation, as shown in Fig. 1 in the case of PC. Thus the measurement is not hampered by the presence of several signals for one phospholipid (15).

5.2.2.3 Mass spectrometry

In order to determine the molecular composition of the phospholipid classes contained in raw lecithin, mass spectrometry is used for detection. To characterize lecithins, the phospholipid classes are separated using HPLC and then determined using electrospray ionization (ESI) or atmospheric pressure chemical ionization (APCI) mass spectrometry (17).

5.2.3 Characterization of lecithins

Lecithins were characterized using liquid and deoiled variations of soy, rapeseed and sunflower lecithin.

5.2.3.1 Sensory qualities

The liquid raw lecithins examined were dark brown, thick, viscous liquids. In common with the deoiled lecithins they were observed to range in smell from neutral to faintly hay-like and slightly fatty. This accords with data about soy lecithin published by Stephan and Steinhart (18). Deoiled rapeseed lecithin smelled very strongly of sotolon, a lactone with an aroma reminiscent of maple syrup. Both the liquid and deoiled lecithin used in the experiments had a nutty, grass-like, slightly bitter, abrasive taste.

5.2.3.2 Basic assay results

Dry matter, water content, residue on ignition, phosphate content and nitrogen content (Dumas method) were determined as laid down by the Working Group on Grain Research, standard methods for grain, flour and bread (1994, unpublished). The results obtained are shown in Table 1.

Table 1.

Figures obtained in basic assays of the lecithins examined in % (m/m) with n = 3 assays.

Assay	Liquid soy % m/m	Deoiled soy % m/m	Liquid rapeseed % m/m	Deoiled rapeseed % m/m	Liquid sunflower % m/m	Deoiled sunflower % m/m
Dry matter	99.4	98.2	99.1	98.2	99.7	98.4
Residue on ignition	7.9	8.5	6.4	9.4	6.0	9.2
Phosphate	7.2	7.7	5.8	8.5	5.5	8.4
Phosphorus	2.3	2.5	1.9	2.8	1.8	2.7
Nitrogen	0.75	1.09	0.96	1.24	0.81	1.18
FFA	2.2	4.1	3.3	3.1	2.0	2.4
AI/AS	78/22	97/3	68/32	96/4	65/35	97/3

AI: acetone insolubles (polar lipids)
AS: acetone soluble (non-polar lipids)
FFA: free fatty acids

5.2.4 Characteristic lecithin parameters

5.2.4.1 Acetone insolubles and acetone solubles

Following several cycles of deoiling raw lecithins with cold acetone, differential weighing resulted in the distribution of polar (acetone-insoluble) and non-polar (acetone-soluble) lipids as shown in Fig. 2.

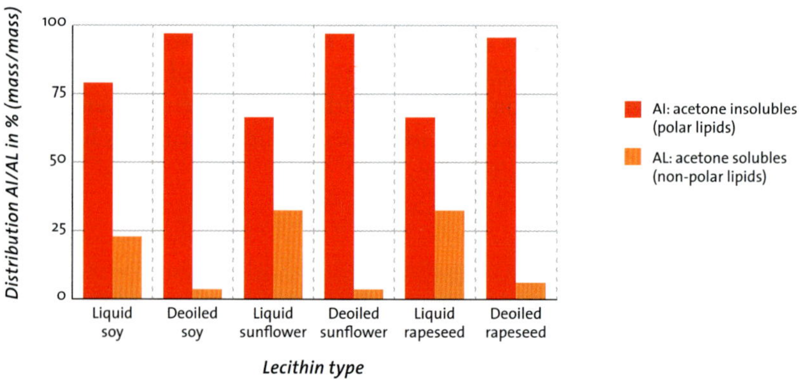

Figure 2.

Distribution of acetone-insolubles and acetone-solubles in lecithins.

5.2.4.2 Total fatty acid distribution

To determine the distribution of fatty acids, the lecithin samples were heated in boron trifluoride-methanol solution, whereby both the esterified and free fatty acids were converted into fatty acid methyl esters (FAME). The distribution of fatty acids in the samples is shown in Table 2.

Table 2.

Distribution of fatty acids found in various lecithins in % (m/m), and the ratios of saturated to unsaturated fatty acids (S/U).

Lecithin	C16:0 % m/m	C18:0 % m/m	C18:1 % m/m	C18:2 % m/m	C18:3 % m/m	S/U
Liquid soy	32	26	40	2	–	1.4
Deoiled soy	43	26	29	2	–	2.2
Deoiled rapeseed	52	–	44	2	2	1.1
Deoiled sunflower	39	14	44	–	3	1.1

C16:0 palmitic acid, C18:0 stearic acid, C18:1 oleic acid, C18:2 linoleic acid, C18:3 alpha-linolenic acid

The lowest molecular mass fatty acid present in all the lecithins examined was palmitic acid (C16:0). In terms of proportion this was the most common acid in all four lecithins, reaching a proportion of up to 52% (m/m). Small amounts of linoleic and alpha-linolenic acids, i.e. 2–3% (m/m), were found in some but not all lecithins examined. As anticipated, the ratio of saturated to unsaturated fatty acids (S/U) was higher in deoiled soy lecithin than in liquid soy lecithin. Deoiled rapeseed lecithin contained no stearic acid. Deoiled sunflower lecithin displayed an S/U ratio that was only half that of liquid soy lecithin.

5.2.4.3 Fat indices

The fat indices found in the raw lecithin samples are summarized in Table 3. In the process of determining the lecithins' acid values the phosphatidyl derivative of the phospholipids is likewise deprotonated by the sodium hydroxide solution. Thus, when the lecithins' acid value is determined, the point of change is not the point at which the free fatty acids are neutralized. In liquid lecithins the iodine value and the saponification value were, as expected, higher than in deoiled lecithins. This may be attributed to progressive ageing during storage.

Table 3.
Fat values determined in lecithins.

Lecithin	Acid value mgK^OH/g	Saponification value mgKOH/g	Iodine value $gI_2/100g$	Peroxide value $meqO_2/kg$
Liquid soy	23.5	188	110	5.5
Deoiled soy	30.0	170	95	3.7
Liquid sunflower	25.6	193	105	3.1
Deoiled sunflower	33.4	175	90	1.3
Liquid rapeseed	27.1	198	122	2.0
Deoiled rapeseed	35.1	185	98	1.5

5.2.4.4 Quantification of the classes of compounds in acetone-insoluble matter

The acetone-insoluble fraction of the lecithins is composed of the more polar lipids – phospholipids, glycolipids, sphingolipids and free fatty acids. The following section presents data from the quantification of these classes of compounds. Three methods were used to identify the phospholipid classes: TLC, HPLC and ^{31}P NMR spectrometry. TLC was used to quantify the sphingolipids, and the free fatty acid content was determined by titration. The glycolipid content in % (m/m) was calculated as the difference using the formula

$$\text{Gl. 1 Glycolipids (\%)} = AI - \Sigma(PL) - \Sigma(LPL) - \Sigma(SL) - \Sigma(FFA)$$

Where AI: acetone-insoluble matter, Σ(PL): phospholipid content, Σ(LPL): lysophospholipid content, Σ(SL): sphingolipid content, Σ(FFA): free fatty acid content.

5.2.4.5 Thin-layer chromatography

Phospholipids, glycolipids and sphingolipids were separated by thin-layer chromatography using silica gel according to Helmerich and Koehler (19). With the method described it is possible to achieve reproducible separation of traditional lecithin preparations and determine the quantities of phospholipid classes they contain within a relatively short time and without deploying expensive equipment. Fig. 3 shows a typical thin-layer chromatogram of the lecithin components found in deoiled rapeseed lecithin. Contrasted with this is the quantifiable chromatogram derived from converting the spots' varying intensities into x/y data sets.

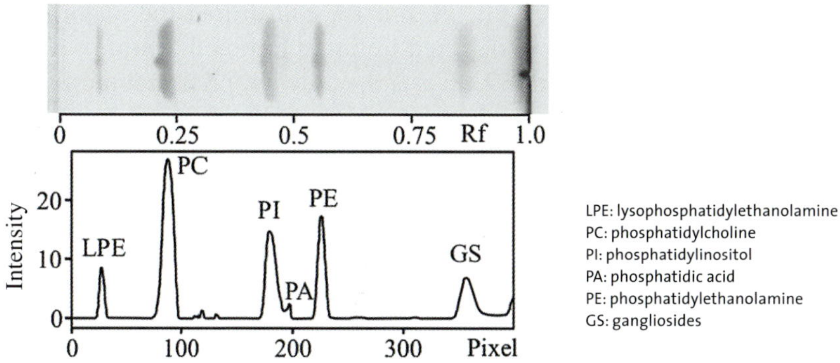

Figure 3.

Conversion of thin-layer chromatogram of separation of deoiled rapeseed lecithin into x/y data sets with the aid of the image processing program Image J.

Thin-layer chromatography succeeded in fully splitting up the lecithins into their separate components. It was possible to quantify all the phospholipids contained (PC, PI, PA, PE, PS) and almost all the lysophosphatidic acid (LPA), lysophosphatidylinositol (LPI) and lysophosphatidylethanolamine (LPE). The presence of lysophosphatidylcholine (LPC) could be detected but not quantified, because the reference substance displayed extreme band broadening when the calibration line was drawn. The sphingolipids (ceramides, gangliosides) were also quantifiable. The phospholipid and sphingolipid contents were calculated by means of calibration lines that were verified by recovery testing.

The detection limits of the method were between 200 and 500 ng for phospholipids and between 300 and 1400 ng for their lyso compounds. Based on a feed volume of 5 µL, a solvent volume of 100 µL and a phospholipid content of 5% in the sample, it can be calculated that 80–280 µg of a lecithin sample will be sufficient to determine the quantities of polar and non-polar lipids it contains. The concentrations of the phospholipid components in the acetone-insoluble matter of the lecithins tested are summarized in Table 4.

In all lecithins examined, PC, PE, PI and PA constituted the major part of the polar lipid fraction. PS was detected in liquid rapeseed lecithin. In liquid rapeseed and sunflower lecithin LPE was the only lysophospholipid found above the detection limit. The concentration of sphingolipids was 4.7 – 11.8% (m/m), whereby more gangliosides than ceramides were found in all the lecithins examined. Thus the lecithins examined contained 3.2 – 9.5% (m/m) gangliosides and 1.0 – 2.5% (m/m) ceramides.

Table 4.

Concentrations of components in acetone-insoluble matter of the lecithins tested using TLC.

Compound	Liquid soy % m/m	Deoiled soy % m/m	Liquid rapeseed % m/m	Deoiled rapeseed % m/m	Liquid sunflower % m/m	Deoiled sunflower % m/m
Phospholipids, min.	**55.7**	**82.4**	**48.4**	**68.3**	**57.8**	**79.0**
PC	17.5	24.0	19.5	22.1	19.6	28.9
PE	17.3	26.7	9.2	10.8	10.5	15.3
PI	16.0	20.3	14.4	28.7	18.1	25.4
PA	4.9	11.4	4.0	6.7	9.7	9.4
PS	< 0.8	< 1.7	1.3	< 1.4	< 0.7	< 2.0
Lysophospholipids, min.	**n.d.**	**n.d.**	**2.6**	**n.d.**	**0.9**	**n.d.**
LPC	< 1.0	< 2.0	< 0.9	< 1.8	< 0.8	< 2.4
LPE	< 0.5	< 1.0	2.6	< 0.9	0.9	< 1.2
LPI	< 2.2	< 4.4	< 2.0	< 3.9	< 1.8	< 5.3
LPA	< 2.3	< 4.7	< 2.2	< 4.2	< 1.9	< 5.7
Sphingolipids	**4.7**	**6.6**	**9.8**	**11.8**	**4.4**	**6.4**
Gangliosides	3.2	4.4	7.3	9.5	3.4	5.3
Ceramides	1.5	2.2	2.5	2.3	1.0	1.2
Free fatty acids	**2.2**	**4.1**	**2.0**	**2.4**	**3.3**	**3.1**
Acetone-insoluble matter	**78.0**	**97.0**	**68.0**	**96.0**	**67.0**	**97.0**
Glycolipids	**15.4**	**3.9**	**5.2**	**13.5**	**0.6**	**8.5**

PC: phosphatidylcholine, PE: phosphatidylethanolamine, PI: phosphatidylinositol, PA: phosphatidic acid, PS: phosphatidylserine, LPC: lysophosphatidylcholine, LPE: lysophosphatidylethanolamine, LPI: lysophosphatidylinositol, LPA: lysophosphatidic acid

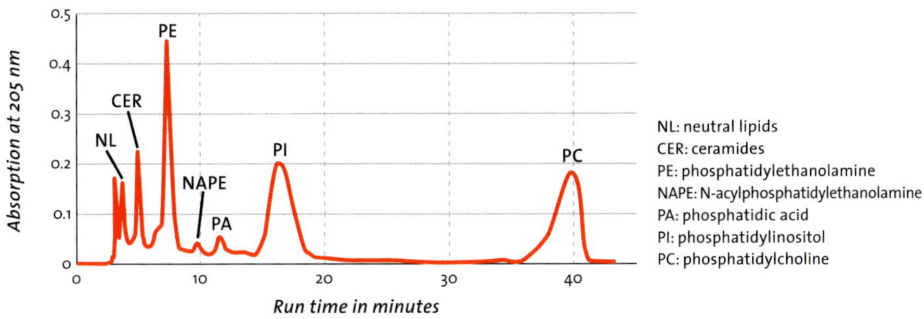

Figure 4.

HPLC separation of deoiled sunflower lecithin. Stationary phase: silica gel. Mobile phase: hexane/2-propanol/0.1% formic acid (m/v) (8+8+1 v+v+v).

5.2.4.6 HPLC

Phospholipids were separated and quantified using the HPLC technique described by Helmerich and Koehler (19). The eluent developed was required to meet two important criteria. In addition to separating individual classes of phospholipids, the desired option was one capable of coupling the process with mass spectrometry. This was achieved by use of a readily volatile buffer (0.1% formic acid). The individual phospholipids were quantified by calibration with an external standard. A typical HPLC separation of deoiled sunflower lecithin is shown in Fig. 4.

It was shown that the HPLC method described was able to separate and quantify all the phospholipids in the sample. LPE was the only lyso compound eluted, because the elution system was optimized to extract phospholipids rather than their lyso forms.

Fig. 5 shows a typical mass spectrum for PE from liquid sunflower lecithin. With the aid of mass spectrometry in the negative APCI mode it was possible to detect sufficient quantities of molecular ions to definitely attribute the separate peaks to the corresponding phospholipids. PE and PA produced the most molecular ion species, compared to the other phospholipid classes present.

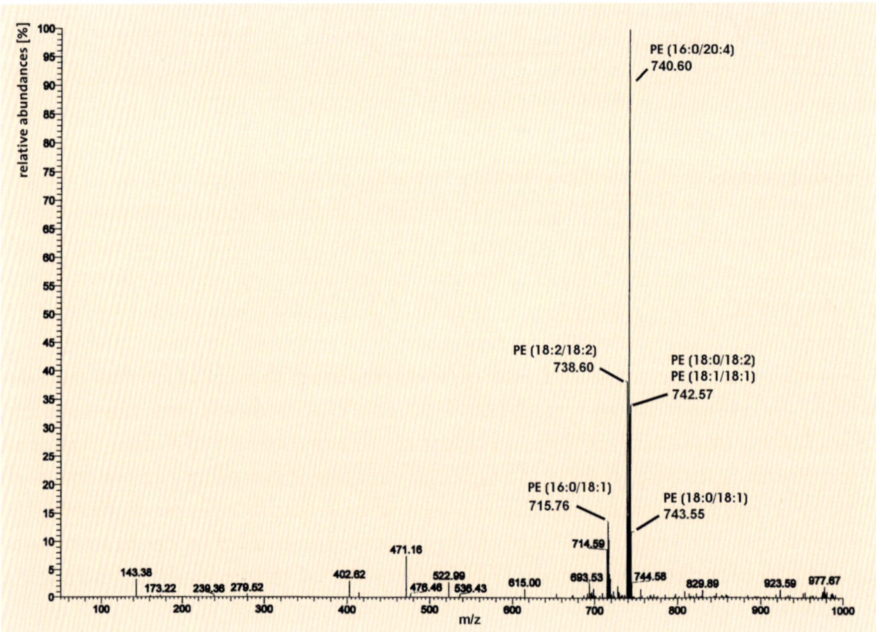

Figure 5.

APCI mass spectrum of phosphatidylethanolamine (PE) (M-H) from liquid sunflower lecithin. Measurement of molecular species as negatively charged ions.

Table 5.

Concentrations of components in acetone-insoluble matter in the lecithins tested using HPLC.

Compound	Liquid soy % m/m	Deoiled soy % m/m	Liquid rapeseed % m/m	Deoiled rapeseed % m/m	Liquid sunflower % m/m	Deoiled sunflower % m/m
Phospholipids, min.	**57.8**	**79.4**	**33.1**	**41.6**	**49.2**	**72.4**
PC	32.4	43.1	15.6	19.8	27.5	41.1
PE	7.9	11.8	3.5	5.8	3.5	5.3
PI	13.8	19.6	10.7	11.8	15.4	21.5
PA	3.7	4.9	3.2	3.9	2.8	4.5
PS	< 0.2	< 0.2	0.1	0.3	< 0.2	< 0.2
Lysophospholipids	**n.d.**	**n.d.**	**1.7**	**1.2**	**n.d.**	**n.d.**
LPE	< 0.1	< 0.1	1.7	1.2	< 0.1	< 0.1
Sphingolipids	**4.7**	**6.6**	**9.8**	**11.8**	**4.4**	**6.4**
Gangliosides	3.2	4.4	7.3	9.5	3.4	5.3
Ceramides	1.5	2.2	2.5	2.3	1.0	1.2
Free fatty acids	**2.2**	**4.1**	**2.0**	**2.4**	**3.3**	**3.1**
Acetone-insoluble matter	**78.0**	**97.0**	**68.0**	**96.0**	**67.0**	**97.0**
Glycolipids	**13.3**	**6.9**	**21.4**	**39.0**	**10.1**	**15.1**

PC: phosphatidylcholine, PE: phosphatidylethanolamine, PI: phosphatidylinositol, PA: phosphatidic acid, PS: phosphatidylserine, LPE: lysophosphatidylethanolamine

The limits of detection for the HPLC method ranged from 20 to 170 µg/mL depending on the class of phospholipid. It was possible to quantify LPE up to a concentration of 10 µg/mL. The concentrations of the phospholipid components in the acetone-insoluble matter in the lecithins tested are summarized in Table 5.

In all the lecithins tested PC constituted the major part of the phospholipid fraction, followed by PI and PE. In liquid and deoiled rapeseed lecithin PS and LPE were detected.

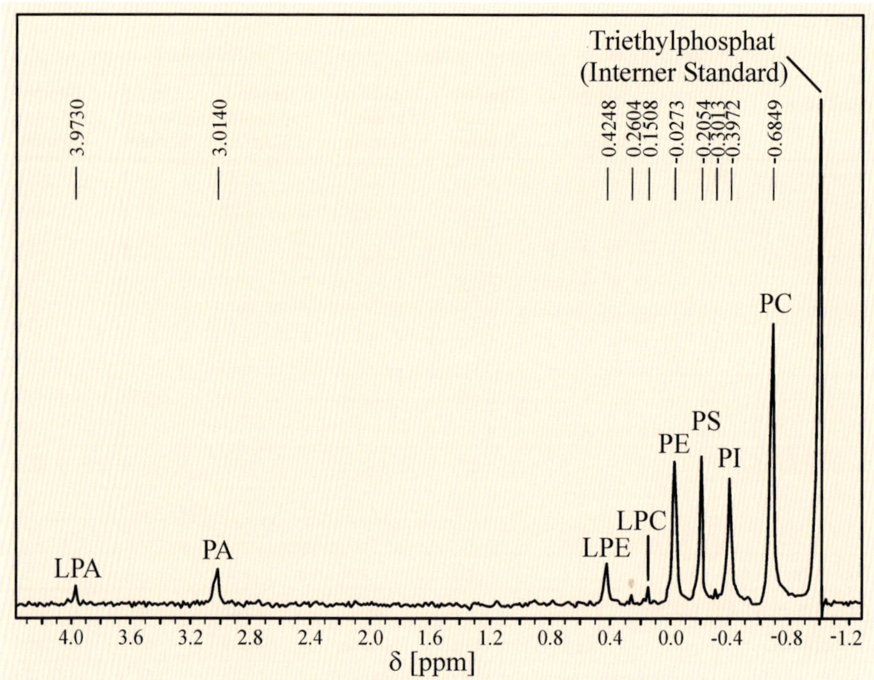

LPA: lysophosphatidic acid, PA: phosphatidic acid, LPE: lysophosphatidylethanolamine, LPC: lysophosphatidylcholine, PE: phosphatidylethanolamine, PS: phosphatidylserine, PI: phosphatidylinositol, PC: phosphatidylcholine, **δ**: chemical shift (ppm), Triethylphosphat: triethyl phosphate, Interner: internal

Figure 6.

^{31}P NMR spectrum of liquid rapeseed lecithin.

5.2.4.7 ^{31}P NMR spectrometry

Individual phospholipids were quantified as described by Helmerich and Koehler (19). To make a qualitative allocation of phospholipid classes the spectra were calibrated against the internal standard triethyl phosphate, to a chemical shift δ of -1.0 ppm. The separate peaks were identified via the corresponding reference substances. In order to achieve a complete separation of the phospholipid classes and their lyso compounds, the solvent wash described by Meneses and Glonek (20) was optimized from pH 6.0 to pH 10.5. Amounts were quantified through calibration lines showing the peak areas of individual phospholipid classes relative to the peak area of the internal standard. Recovery tests produced values in the range of 105–120%. A typical ^{31}P NMR spectrum of liquid rapeseed lecithin is shown in Fig. 6.

Table 6.

Concentrations of components in acetone-insoluble matter in the lecithins tested using ^{31}P NMR spectrometry.

Compound	Liquid soy % m/m	Deoiled soy % m/m	Liquid rapeseed % m/m	Deoiled rapeseed % m/m	Liquid sunflower % m/m	Deoiled sunflower % m/m
Phospholipids	**53.3**	**61.3**	**41.6**	**59.3**	**44.9**	**70.2**
PC	23.0	23.2	17.0	24.5	18.1	30.2
PE	12.3	14.1	8.4	12.2	7.6	11.3
PI	11.4	13.4	10.6	12.5	12.6	20.9
PA	6.6	6.7	5.6	7.2	6.6	7.8
PS	1.8	1.9	< 1.2	2.9	< 1.3	< 1.3
Lysophospholipids	**4.6**	**4.9**	**5.7**	**11.8**	**n.d.**	**n.d.**
LPC	< 2.0	< 2.0	< 2.0	< 2.0	< 2.0	< 2.4
LPE	4.6	4.9	5.7	11.8	< 2.3	< 2.4
LPI	< 2.6	< 2.3	< 2.2	< 2.4	< 2.3	< 3.6
LPA	< 4.0	< 3.5	< 3.3	< 3.7	< 3.5	< 2.0
Sphingolipids	**4.7**	**6.6**	**9.8**	**11.8**	**4.4**	**6.4**
Gangliosides	3.2	4.4	7.3	9.5	3.4	5.3
Ceramides	1.5	2.2	2.5	2.3	1.0	1.2
Free fatty acids	**2.2**	**4.1**	**2.0**	**2.4**	**3.3**	**3.1**
Acetone-insoluble matter	**78.0**	**97.0**	**68.0**	**96.0**	**67.0**	**97.0**
Glycolipids	**13.2**	**20.1**	**8.9**	**22.5**	**14.4**	**17.3**

PC: phosphatidylcholine, PE: phosphatidylethanolamine, PI: phosphatidylinositol, PA: phosphatidic acid, PS: phosphatidylserine, LPC: lysophosphatidylcholine, LPE: lysophosphatidylethanolamine, LPI: lysophosphatidylinositol, LPA: lysophosphatidic acid

All of the phospholipids in the test (PC, PE, PI, PA, PS) and all the lysophospholipids were quantifiable using ^{31}P NMR spectrometry. Due to their high polarity and thus good solubility in the aqueous washing phase, it was necessary to add at least 10 mg/mL of the reference substances for the lyso compounds. Therefore the detection limits were high at between 1.1 and 3.3 mg/ml for phospholipids and between 2.5 and 5.0 mg/mL for the corresponding lyso compounds. The quantities of phospholipid components in the acetone-insoluble matter in the lecithins tested by ^{31}P NMR spectrometry are summarized in Table 6.

In all the lecithins tested phosphatidylcholine constituted the major part of the phospholipid fraction. Almost all the samples had nearly identical quantities of PE and PI. The one exception was sunflower lecithin, where the phosphatidylinositol content was considerably higher than that of PE. PS was detected in liquid soy lecithin, deoiled soy lecithin and deoiled rapeseed lecithin. Of the lysophospholipids, only LPE was detected in liquid and deoiled rapeseed lecithin and in liquid and deoiled soy lecithin.

5.2.5 Comparison of methods for quantifying phospholipids

An uncomplicated method was developed for performing thin-layer chromatography in order to determine the quantities of phospholipid components and the non-polar lipids in lecithin preparations. The method employs standard laboratory equipment, a PC, a conventional flatbed scanner and analysis software that is available free of charge.

An HPLC method was developed in which lecithins can be split into their main components in a relatively short time using a simple isocratic separation process. Due to the volatility of the buffer employed (0.1% formic acid) it is possible to identify molecular subgroups by LC/MS in addition to quantifying the phospholipids. Freeze-drying can remove the solvent without leaving any residue.

^{31}P NMR spectrometry is a fast, selective method of quantifying the phospholipid contents of lecithins. Unlike TLC and HPLC processes, all lyso compounds also can be quantified, albeit with very low sensitivity. Calculating the phospholipid contents via nitrogen and phosphorus concentrations proved to be a rapid, reliable method of determining the overall content and the content of nitrogenous phospholipids in lecithins.

For eight different lecithins, the phospholipid classes were quantified using TLC, HPLC and ^{31}P NMR spectrometry (Table 7). The total phospholipid content thus determined and the quantities of separate classes of phospholipids in the lecithins tested were comparable, but depended first and foremost on the quantification method used. The key reasons for variations in results were the selectivity of the detectors and the sensitivity of the methods used.

^{31}P NMR spectroscopy gave the highest level of selectivity, because it succeeded in detecting all phospholipids and their lyso compounds. Non-polar lipids, sphingo- and glycolipids, phospholipids and their lyso compounds were quantifiable using TLC. Thin-layer chromatography failed only in the detection of LPC, whereas HPLC resulted in the elution of none of the lyso compounds apart from LPE. Variations in selectivity are due to differences in detection methods. The detection of phospholipids by non-selective oxidizing reagents such as copper sulphate – phosphoric acid solution (in TLC) or by measuring the sample in a UV range with the non-selective wavelength of 206 nm (in HPLC) relies not on the presence of the phosphorus atom in the molecule, but on the detection of many other functional groups. In a UV range of 203–210 nm, for example, amino groups (PE, sphingolipids), quaternary ammonium salts (PC, sphingomyelin), carboxylic ester, carboxyl groups, hydroxyl groups and double bonds (unsaturated fatty acids) are detected. When heated, copper (II) sulphate – phosphoric acid results in a wide range of decomposition and oxidation products, which also makes selective detection impossible (21). A specific phosphorus reagent for TLC has not yet been found. The molybdatophosphoric acid reagent often used to detect phosphorus is also a non-selective oxidant. By contrast, measurements using ^{31}P NMR spectrometry are very selective, because the phospholipid molecule is detected solely by reason of its phosphorus atom.

HPLC was the most sensitive method with detection limits of 20–170 µg/mL, followed by TLC (Table 8). In the case of ^{31}P NMR spectrometry the limits are ten to 80 times higher. The non-selective detection made the HPLC and TLC methods more sensitive and thus resulted in extremely low detection limits. ^{31}P NMR spectrometry revealed definite drawbacks in terms of sensitivity, because the washing process resulted in the samples being considerably diluted up to a total volume of 3 mL. Another finding was that, to a certain degree, the phospholipids and their lyso compounds are soluble in methanol and thus cleansing the sample causes a transition from the deutero-chloroform phase into the aqueous methanol phase (15). Thin-layer chromatography was the method offering the best level of reproducibility. The average coefficients of variation were 5.5% for TLC, 6.8% for HPLC and 12.8% for ^{31}P NMR spectrometry. The reason why results varied, at times very considerably, was because the material in the samples was not homogeneous and, being derived from a natural product, was subject to natural variations. Coefficients of variation higher than 10% were found only in the quantification of minor components.

In summary, it may be concluded that it was possible to quantify the phospholipids in the lecithin using all three of the methods developed. However, the method of choice was ^{31}P NMR spectrometry. It is quick and simple to perform and, in conjunction with its high level of selectivity, this method of quantification delivered rapid, precise results. The TLC method is a cheap, simple alternative, because it supplies meaningful results in a short time using a minimal amount of equipment. A tandem of HPLC and mass spectrometry detection permitted additional statements in respect of the phospholipids with defined fatty acid esters contained in the lecithins.

Table 7.

R_f values, retention times and chemical shift δ for the lipids examined by TLC, HPLC and ^{31}P NMR.

Lipid	R_f value TLC	Retention time (mins) HPLC	δ (ppm) ^{31}P NMR
PA	0.59	11:38	+2.70 to +3.80
PC	0.44	39:58	-0.70 to -0.60
PE	0.63	05:43	±0.00 to 0.20
PI	0.54	18:25	-0.40 to -0.20
PS	0.50	12:23	-0.20 to -0.05
NAPE	0.61	09:51	n.d.
LPA	0.32	n.d.	+3.70 to +4.30
LPC	0.24	n.d.	+0.20 to +0.30
LPE	0.35	08:28	+0.30 to +0.50
LPI	0.28	n.d.	+0.55 to +0.60
SM	0.39	45:31	-0.10 to ±0.00
SL	0.68	n.d.	n.d.
CER	0.76	04:11	n.d.
GANG	0.83	n.d.	n.d.
NPL	0.90	03:35	n.d.

PA: phosphatidic acid, PC: phosphatidylcholine, PE: phosphatidylethanolamine,
PI: phosphatidylinositol, PS: phosphatidylserine, NAPE: N-acylphosphatidylethanolamine,
LPA: lysophosphatidic acid, LPC: lysophosphatidylcholine, LPE: lysophosphatidylethanolamine,
LPI: lysophosphatidylinositol, SM: sphingomyelin, SL: sulpholipids, CER: ceramides,
GANG: gangliosides, NPL: non-polar lipids, n.d.: not detected

Table 8.

Limits of detection (µg/mL) and coefficients of variation (%) of phospholipids detected using TLC, HPLC and ^{31}P NMR spectroscopy.

	Limit of detection µg/mL			Coefficient of variation %		
Phospholipid	TLC	HPLC	^{31}P NMR	TLC	HPLC	^{31}P NMR
PA	80	140	1,100	6.0	6.5	21.1
PC	100	20	1,400	5.3	5.8	8.2
PE	80	60	1,000	5.3	6.6	12.3
PI	40	170	1,500	5.2	6.0	11.4
PS	100	40	3,300	5.3	4.2	18.9
LPA	280	n.d.	5,000	n.q.	n.d.	n.q.
LPC	150	n.d.	2,500	n.q.	n.d.	n.q.
LPE	60	10	3,300	6.1	4.8	4.6
LPI	260	n.d.	3,300	n.d.	n.d.	n.q.

PA: phosphatidic acid, PC: phosphatidylcholine, PE: phosphatidylethanolamine, PI: phosphatidylinositol, PS: phosphatidylserine, LPA: lysophosphatidic acid, LPC: lysophosphatidylcholine, LPE: lysophosphatidylethanolamine, LPI: lysophosphatidylinositol, n.q.: not quantifiable, n.d.: not detected

References

1. Helmerich, G., Koehler, P., 2004. Struktur-Wirkungsbeziehungen von Phospholipiden bei Backwaren. Verlag Dr. Hut, ISBN 3899631196, 9783899631197.
2. Dilena, B.A., Ku, F., Doyle, I., Whiting, M.J., 1997. Six Alternative Methods to the Lecithin/Sphingomyelin Ratios in Amniotic Fluid for Assessing Fetal Lung Maturity. Anal. Clin. Biochem. 34, 106-108.
3. Aitzetmüller, K., 1984. HPLC and Phospholipids. Part I: General Considerations. Fette, Seifen, Anstrichmittel 86, 318-322.
4. Sheeley, R.M., Hurst, W.J., Sheeley, D.M., Martin, R.A., 1987. HPLC Phospholipid Separation Mechanisms on Silica, Amino and Diol Columns. J. Liq. Chromatogr. 10, 3173-3181.
5. Kiuchi, K., Ohta, T., Ebine, H., 1977. High-Performance Liquid Chromatographic Separation and Quantitative Analysis of Synthetic Phospholipids. J. Chromatogr. 133, 226-230.
6. Becart, J., Chevalier, C., Biesse, J.P., 1990. Quantitative Analysis of Phospholipids by HPLC With a Light Scattering Evaporating Detector – Application to Raw Materials for Cosmetic Use. J. High Resolut. Chromatogr. 13, 126-129.
7. Amari, J.V., Brown, P.R., Turcotte, J.G., 1992. Detection for HPLC Separations of Glycerophospholipids, Part 1. American Laboratory 24, 23-29.
8. Postle, A.D. 1987. Method for the Sensitive Analysis of Individual Molecular Species of Phosphatidylcholine by High-Performance Liquid Chromatography Using Post-Column Fluorescence Detection. J. Chromatogr. 415, 241-251.
9. Dasgupta, A., Ayanoglu, E., Wegmann-Szente, A., Tomer, K.B., Djerassi, C., 1986. Phospholipid Studies of Marine Organisms. Part 12. Mass Spectral Behaviour and HPLC of Some Unusual Molecular Phospholipid Species. Chem. Phys. Lipids 41, 335-347.
10. Aloisi, J.D., Sherma, J., Fried, B.,1990. Comparison of Mobile Phases for Separation and Quantification of Lipids by One-Dimensional TLC on Preadsorbent High Performance Silica Gel Plates. J. Liq. Chromatogr. 13, 3949-3961.

11. Vuorela, P., Vuorela, H., Suppula, H., Hiltunen, R., 1996. Development of a Quantitative TLC Assay for Phospholipid Products. J. Planar. Chromatogr. 9, 254-259.
12. Nzai, J.M., Proctor, A., 1998. Determination of Phospholipids in Vegetable Oil by Fourier Transform Infrared Spectroscopy. J. Am. Oil Chem. Soc. 75, 1281-1289.
13. Gupta, R., Muralidhara, H.S., Davis, H.T., 2001. Structure and Phase Behaviour of Phospholipid-Based Micelles in Nonaqueous Media. Langmuir 17, 5176-5183.
14. Merchant, T.E., Glonek, T., 1992. ^{31}P NMR of Tissue Phospholipids: Competition for Mg2+, Ca2+, Na+ and K+ Cations. Lipids 27, 551-559.
15. Henderson, T.O., Glonek, T., Myers, T.C., 1974. Phosphorus-31 Nuclear Magnetic Resonance Spectroscopy of Phospholipids. Biochemistry 13, 623-628.
16. London, E., Feigenson, G.W., 1979. Phosphorus NMR Analysis of Phospholipids in Detergents. J. Lipid Res. 20, 408-412.
17. Jensen, N.J., Gross, M.L., 1988. A Comparison of Mass Spectrometry Methods for Structural Determination and Analysis of Phospholipids. Mass 7, 41-69.
18. Stephan, A., Steinhart, H., 1999. Quantification and Sensory Studies of Character Impact Odorants of Different Soybean Lecithins. J. Agric. Food. Chem. 47, 4357-4364.
19. Helmerich, G., Koehler, P., 2003. Comparison of methods for the quantitative determination of phospholipids in lecithins and flour improvers. J. Agric. Food. Chem. Nov. 5; 51(23), 6645-6651.
20. Meneses, P., Glonek, T., 1988. High Resolution Phosphorus-31 NMR of Extracted Phospholipids. J. Lipid Res. 29, 679-688.
21. Seel, F., Schuh, J., 1987. The Reaction of Copper Sulfate with Carboxylic Acids and Their Derivates in Aqueous Solutions Under Hydrothermal Conditions. Z Naturforsch, Teil B: Chem. Sci. 42, 157-162.

5.3 GMO analysis

Hans-Ulrich Waiblinger

5.3.1 Introduction

Genetically modified organisms (GMO)

For genetic modification, genes coding for certain information (e.g. coding for a defined protein) is transferred from a donor species to a species naturally not containing this gene. This process is frequently termed genetic engineering.

Frequently, bacterial genes are transferred to plants, e.g. to obtain a tolerance against herbicides.

Organisms in which the genetic material (DNA) has been altered in a way that does not occur naturally can be defined as genetically modified organisms (GMO).

Current use of genetic engineering in soybean cultivation

Worldwide, more than 100 genetically modified plants (GM plants, GMP) are commercially cultivated (1,2).

Different genetically modified lines (events) of important crops, such as corn, soybean and rapeseed, are commercially grown, especially in North and South America on large acreages. For example, in 2014 GMPs were cultivated on 82% and 25% of the overall acreages of soybeans and corn, respectively.

In Brazil, the most important country for the cultivation of non-GM soybeans so far, GM soy has increased to more than 90% (2).

Herbicide tolerance is the most dominant trait of GM soybeans in worldwide commercial cultivation, especially to the herbicides glyphosate (Roundup), glufosinate (Liberty Link) and 2,4-D. In recent years, GM soybeans with new traits have been grown: notably, traits that increase levels of oleic acid (C18:1, omega-9) and reduce levels of linoleic acid (C18:2, omega-6) and alpha-linolenic acid (C18:3, omega-3) contribute to reduced formation of trans-fatty acids during food production. Another GM soybean already commercially grown exhibits an increased level of an omega-3 fatty acid (stearidonic acid, C20:4, omega-3) in order to obtain benefits for health.

Authorization within the European Union

Within the European Union (EU), no GM soybean event has been authorized for cultivation to date (6/2017). However, 12 different GM soybean events have been authorized for food and feed purposes (as of 06/2017). These authorizations are restricted to import and processing (4).

Regulatory requirements within the EU

The EU requirements for authorization and labelling of genetic modifications in food and feed are laid down in the regulations (EC) No. 1829/2003 and 1830/2003 (5). Food and feed produced from GMOs that has not passed the authorization procedure according to regulation (EC) No. 1829/2003 must not be marketed within the EU. Even minimal (trace) amounts of unauthorized GMOs and their compounds are not permitted in food. As far as feed is concerned, under defined conditions, the presence of unauthorized GMPs is tolerated at low levels (< 0.1%) (6).

Food and feed produced from GMOs can be marketed in the EU if labelling provisions are followed. No labelling is required for food and food ingredients produced from authorized GMOs in proportions below 0.9%. However, this exemption can only be made if it can be demonstrated that the presence of GMOs is adventitious and technically unavoidable.

Legal background and GMO analysis

GMO analysis is therefore used on the one hand to check if labelling requirements are fulfilled and, on the other hand, for the detection of unauthorized GMO products.

For the control of labelling provisions and the 0.9% threshold, quantitative methods are required. This threshold also refers to the ingredients. This means that for reliable results, analysis on the ingredient level – if possible at the place of production – is recommended.

The European GMO labelling regulations are extensive because any ingredient 'produced from GMO' is covered, irrespective of the degree of processing and analytical detectability.

Distinguishing 'technically unavoidable' from 'adventitious'

If GM constituents are detectable in food, "operators must be in a position to supply evidence to satisfy the competent authorities that they have taken appropriate steps to avoid the presence of such material" (article 12 paragraph 3 of regulation (EC) No 1829/2003).

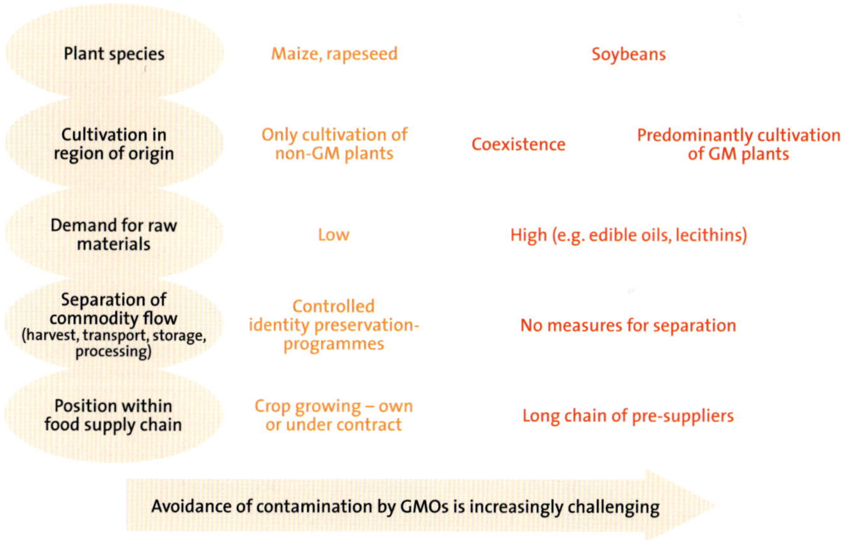

Figure 1.

Technically unavoidable factors that influence whether or not GMO material is present in foods.

Experts of official German food control laboratories described an approach for the interpretation of the terms 'technically unavoidable' and 'adventitious' (7). A practical approach for interpretation was proposed that takes into account quality control measures and documentation, as well as present findings in comparable products. For example, GMO contaminations repeatedly found in one commodity (e.g. different lots of soy protein concentrates containing 0.2% genetically modified soybeans) may not be regarded as adventitious but as technically unavoidable. On the other hand, a singular finding of 0.5% of

GM soybean in the context of several negative or trace results obtained by a well-elaborated self-monitoring system may be regarded as adventitious.

There are different causes of the presence of GM material in foodstuffs. Admixture or contamination can occur during the whole food chain, from cultivation (seed) to harvest, transport and storage, and processing.

Evaluation of the question if the presence of GM material could have been technically unavoidable depends on several factors which cannot clearly be separated from each other (see also Fig. 1).

Separation of commodity flow and traceability

Starting from harvest until final use in food production, many soybean and maize products pass through a considerable number of steps and geographic locations (almost worldwide) in the course of manufacture.

If identity, i.e. relation to traditional (non-GM) raw material, has been evaluated and documented by means of traceability and identity preservation (IP) systems, contamination of a sample by GM material usually should be regarded as technically unavoidable.

Traceability systems that (for example) have to be installed according to article 18 of regulation (EC) No. 178/2002 by food operators can facilitate inspection. Frequently, traceability along several steps of the supply chain or along different production steps within one plant is required. Hence, provision of the minimum traceability documentation regulated by law (one step forward, one back) may not be sufficient for a meaningful assessment.

Separation of commodity flow

Furthermore, if no particular separation of commodity flow of traditional and GM products is implemented, contaminations by GM material can be regarded as technically avoidable.

If, according to the circumstances of production and technical prerequisites, no complete separation of commodities is possible, it has to be decided on a case-by-case basis whether and how contamination by GM material can be avoided and controlled by quality assurance and documentation measures. One possible measure within a production plant could be for example to rinse pipelines using non-GM material following the processing of GM material.

Demand for raw materials, availability
For some food ingredients such as soybean oils, soy lecithins, maize starches or maltodextrins, there is a high demand for industrial food production. Possibly, the amounts of certified non-GM material available from reliable suppliers or producers do not suffice to satisfy the demands of food production. Therefore, the availability of traditional raw materials on the global market can be one consideration for the assessment of technical (un-)avoidability.

Practical approach for official control laboratories
The results of GMO analysis of food derived (for example) from soybean can be used to assist with the consideration of whether or not contamination by GM plant ingredients is technically unavoidable. For this purpose, 95 percentile values, expressed in per cent genetically modified (GM) soybean or maize and specified in categories of products, were calculated from the results of the German food control from 2011 to 2013 (8). These 95 percentile values represent the GMO percentages that have not been exceeded by 95% of all samples within one category. In the period 2011 to 2013, the highest 95 percentile value was 0.15% for soy protein isolates. According to these results, it seemed to be feasible to obtain soy protein isolates in the range of (or below) this value (8).

5.3.2 Strategies for GMO detection

Detection by the phenotype via introduced proteins

In living systems, most typically new proteins are produced (expressed) due to the introduced genetic information. Immunological methods using specific antibodies directed against new proteins can be used for specific detection.

A number of immunoassays (enzyme-linked immunosorbent assays (ELISA) and lateral-flow devices) were developed and are commercially available (9). For example, tests for the detection of different crystal proteins (Cry) from insect-tolerant lines and for the detection of the enzymes 5-enol-pyruvyl-shikimate-3-phosphate-synthase (EPSPS) and phosphinothricin acetyltransferase (PAT) in herbicide-tolerant lines are used.

ELISA tests to detect insect-tolerant maize MON810 (Cry1Ab protein) and herbicide-tolerant Roundup Ready™ soybean (EPSPS protein) have successfully been validated in collaborative trials; for the analysis of flours a sensitivity of 0.1 to 0.3% could be achieved (9).

Especially during harvest (i.e. under field conditions), rapid tests based on lateral-flow strip devices are used. Fig. 2 illustrates an example of one such commercially available lateral-flow sandwich test, which can be used for the simultaneous detection of the following proteins:

- CP4 EPSPS (e.g. Roundup Ready soybean).
- Cry3Bb (e.g. MON863 maize).
- Cry1Ab (e.g. Bt11 and MON810 maize).

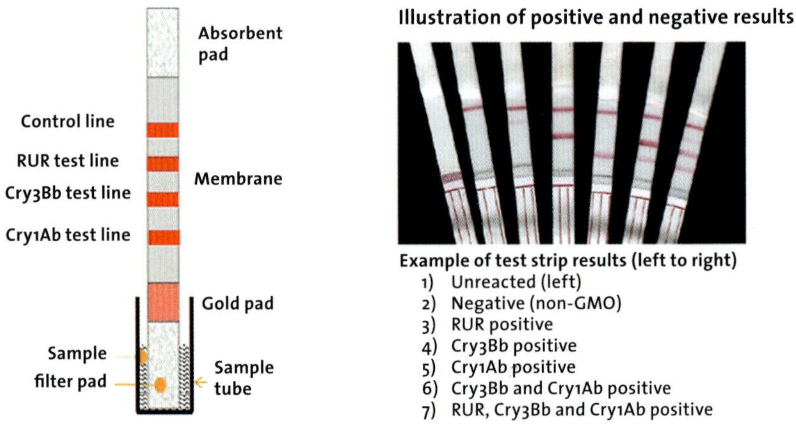

Figure 2.

Protein-based rapid tests. Example of a commercial lateral-flow test for the simultaneous detection of different crystal proteins (Cry) and the EPSPS protein of the Roundup Ready® (RRS) soybean.

A specific monoclonal antibody is coupled with colloidal gold ('gold pad'). The test procedure requires that the colloidal antibody – gold conjugate is dissolved by the sample extract solution and is directed by capillary effect of the carrier material along areas with immobilized specific antibodies, to the adsorber material. The formation of a sandwich antigen – antibody complex at the predicted areas of the lateral-flow strip indicates the presence of the individual proteins. The limits of detection of the tests range from 0.1 to 1% of each GMP (9).

However, the use of such tests is limited due to the following facts:

- the amount of expressed protein can vary, depending on (for example) cultivar and location;
- the amount of expressed proteins can strongly depend on the type of tissue. For example, the Cry protein is present in leaves and not in kernels;
- in complex food matrices unspecific reactions and matrix effects may occur;
- in heated foodstuffs due to denaturation of proteins (i.e. the antigens) binding of specific antibodies may be reduced.

Therefore, such rapid tests primarily serve as a fast and cheap initial control of raw materials and kernels during harvest.

Detection by other phenotypic characteristics

In the case of modified composition of such ingredients as fatty acids, chromatographic methods such as gas chromatography (GC) or LC/LC-MS (liquid chromatography – mass spectrometry) may be used. However, their parameters are subject to natural fluctuations so large that sensitive and quantitative GMO analysis based on these phenotypic parameters is usually not possible.

Detection by the genotype

The most sensitive and quantitative detection of genetic modifications is predominantly done by genotype. Introduced and modified DNA sequences are detected by molecular biological methods, especially based on polymerase chain reaction (PCR). For routine detection of genetic modifications in plants, two different strategies are followed (10 – 13):

Screening methods

In addition to the gene responsible for the new trait ('structural gene'), plasmid vector sequences, reporter and selection genes (e.g. antibiotic resistance genes) and regulatory sequences of promoters as well as terminators are often transferred from a donor to the genetically modified plant. In the past, the so-called 35S promoter from cauliflower mosaic virus (P35S) and the NOS terminator from *Agrobacterium tumefaciens* (T-nos) were frequently used in GMPs. For example, such constitutive promoters as P35S were frequently used to obtain high expression rates of the structural genes. These sequences are common to different GMPs and can therefore be used for screening purposes (see Fig. 3 and 4) (10 – 13).

In cases of positive screening results for P35S and T-nos sequence respectively, natural contaminations by cauliflower mosaic virus (CaMV) and *Agrobacterium tumefaciens* have to be considered. For example, members of the Brassicaceae family including rapeseed are frequently infected by CaMV. By means of detection of further DNA sequences specific to CaMV, such contamination can be detected.

Screening methods may often be the strategy of choice for the detection of unauthorized GMPs, if detailed specific sequence information and/or suitable control samples are not available to the laboratory.

Construct-specific methods

In order to detect genetic modifications, so-called construct-specific methods are used. Overlapping sequences from DNA constructs are targeted that have been used for genetic modification. As an example, the link of the P35S sequence (origin CaMV) to the herbicide resistance gene bar (from soil bacterium *Streptomyces hygroscopicus*) (see Fig. 3) does not occur in nature.

Fig. 4 shows the map of the plasmid vector that has been used for transformation of the soybean Roundup Ready® (event GTS 40-3-2) (14). A construct-specific method for the detection of this GM soybean is based on the detection of a sequence region spanning the overlap between CP4 EPSPS gene from *Agrobacterium sp. 4* and chloroplast transit peptide sequence from petunia (CP4 EPSPS – CTP 4) (15).

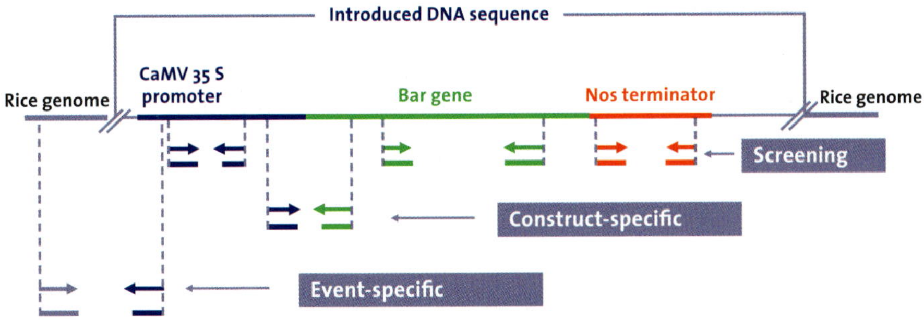

Figure 3.

Principle of the DNA-based detection of genetic modifications in plants. Screening methods target DNA sequences frequently occurring in GMPs, examples being the CaMV 35S promoter, the bar gene or the nos terminator. For a construct-specific detection, overlapping sequences from DNA constructs are targeted that have been used for genetic modification (e.g. the P35S sequence of the herbicide-resistance bar gene. For the event-specific detection, the integration site of the introduced sequence into the plant genome is used.

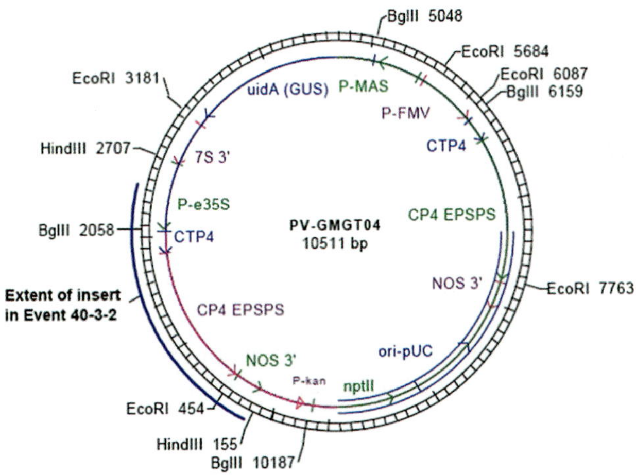

Figure 4.

Map of the plasmid vector that was used for transformation of the soybean Roundup Ready® (event GTS 40-3-2) (14). P-e35S = 35S promoter from cauliflower mosaic virus (P35S), NOS 3' = nopalin synthase 3' transcription terminator from *Agrobacterium tumefaciens* (T-nos), CTP4 = chloroplast transit peptide sequence from *Petunia hybrida;* CP4 EPSPS = enol-pyruvyl-shikimate-phosphate synthase gene from *Agrobacterium sp. 4*.

Event-specific methods

The individual GMP (i.e. transformation event, e.g. event GTS 40-3-2 soybean) is finally identified by the detection of sequences in the region of the integration site of the transformation vector into the plant genome; for example of the junction between the P35S sequence on the plasmid vector and the plant genome (see Fig. 3).

5.3.3 Sampling

Products mostly of plant origin are relevant for possible genetic modifications. Several factors argue for the analysis of genetic modification at an early stage of the production chain, e.g. after harvest. At this stage, DNA can be extracted from the products in large yield and with good quality. By contrast, the poor quality and miniscule amounts of DNA of the relevant species in highly processed products and in food composed of many different ingredients makes sensitive detection very difficult or outright impossible. In addition, threshold values for labelling usually refer to the ingredient. Therefore, for verification, analysis has to be carried out on the ingredient level.

Along the production chain from harvest to processing, kernels of corn or legumes are often aggregated into large lots. Production and retail stages close to harvest may exhibit considerable inhomogeneities in relation to GM kernels. Therefore, sampling is a very important step of GMO analysis.

On the one hand, the number of sampling points and increments has to be adapted to the lot size. On the other hand, an adequate lab sample requires a minimum number of particles. In a typical technical specification 10,000 kernels/particles is given as the target value (16). With this laboratory size, a sampling error of about 20% can be assumed for samples with GMO proportions of 1% (95% probability level) (16). In addition, this particle number allows the detection of GM constituents down to 0.03% with 95% probability in homogeneous lots. Due to the differing thousand kernel weights, the required sizes of the laboratory samples vary from 40 grams for rapeseed to about 2,000 grams (2 kilograms) for soybeans (16).

The further steps of sample preparation within the laboratory must be adapted to this minimum size requirement. In molecular biological analysis, usually the size of the analytical sample is below one gram. This may be reasonable if particle size of the homogenized sample material is very small and therefore the particle number is high enough to represent the original sample (10,000 kernels and more) in a small portion of the analytical sample. Otherwise, the sample size has to be increased; many protocols therefore use an analytical sample weight of 2 grams.

5.3.4 Molecular biological DNA detection methods

Molecular biological DNA-based methods directly detect DNA sequences introduced by genetic engineering. DNA is very stable against heat or high pressure; in addition, DNA is present in any type of tissue. Thus, a broad spectrum of products from raw to processed material can be analysed. However, DNA can be degraded (e.g. depurinated or hydrolysed), for example due to low pH (e.g. juices) or hydrolytic processes (e.g. soy sauces).

Fragment length of DNA sequences used for specific detection should therefore be selected in a way that amplification is still possible even if only short DNA fragments (e.g. below 300 base pairs) are present in a sample (18,19).

DNA extraction

A large number of DNA extraction methods have been described that facilitate GMO analysis (20). Frequently, hexadecyltrimethylammonium bromide (CTAB) methods are used for the extraction of DNA from food and feed of plant origin. Extraction methods based on the purification of DNA on silica matrices are also very commonly used (21). Various protocols adapted for difficult matrices have been validated and standardized in Germany for the extraction of DNA from soy lecithin, maize starch and honey pollen (22–24).

Guidelines for the validation of DNA extraction methods have been described, especially in the scope of analysis of genetically modified sequences in food and feed products by real-time polymerase chain reaction (real-time PCR) (25).

According to these guidelines, performance testing of extracted DNA is primarily done by real-time PCR determination of the quantity and quality of amplifiable, species-specific DNA. Accurate estimation of the quantity of such DNA allows for prediction as to whether the extracted DNA is suitable for a sensitive GMO analysis (see below, Practical limits of detection) (18,21).

Detection by polymerase chain reaction (PCR)

Methods based on PCR are currently the most commonly used for the detection of genetic modifications. So-called end-point PCR and real-time PCR methods have been tested in many ring trials and proficiency tests. Methods are described in international norms and platforms (3,10,15,25).

Endpoint PCR (qualitative PCR)

In 1995, the first PCR method for the detection of the GM tomato ('Flavr Savr' tomato) was published (26). The first methods to screen for the P35S, the T-nos and the *nptII* sequences were described (9), as well as methods for the construct-specific detection of Bt176 maize and Roundup Ready® soybean (RRS) (27,28). These methods were based on qualitative PCR (= endpoint PCR). For the detection of amplified DNA fragments, DNA separation in agarose gels is used, followed by staining of the DNA with ethidium bromide.

For endpoint PCR detection in heated and other processed foodstuffs, relatively small sizes of the amplified DNA fragments are chosen; usually below 200 base pairs (bp) and ideally between 60 and 150 bp (18,19,25).

Fig. 5 shows an example of the detection of RRS in soy lecithin. Even from soy lecithins, soy-specific (left, lanes 2 to 4) and GM soy-specific (right, lanes 3 and 4) DNA fragments were amplifiable.

However, if no amplicon is visible (right, lanes 2 and 5), no statement is possible if the sample contains GM proportions that have to be labelled in terms of existing thresholds, e.g. 0.9% as laid down in regulation (EC) No. 1829/2003 (5). Additionally, in the case of DNA from soy lecithins (left, lanes 2 to 4), no evidence can be given if a sensitive detection of RRS at the 0.9% level and below is possible at all.

This can only be made by means of quantitative methods which quantify the amount of amplifiable soybean DNA species (see below).

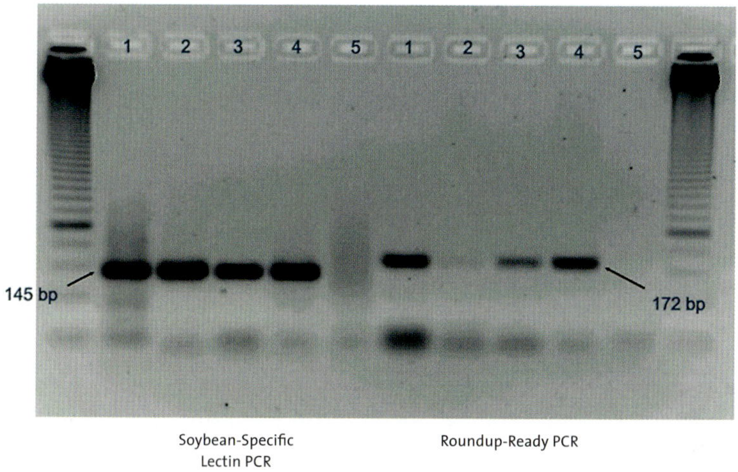

Figure 5.

Qualitative (= endpoint) PCR for the detection of GM soy in soy lecithin. Agarose gel electrophoresis after amplification of a 145 bp sequence of the soybean-specific lectin gene (= amplification control of extracted DNA) (left) and a Roundup Ready® soy (RRS) construct-specific sequence (172 bp) (right). Left and right border: 50 bp ladder; 1 = positive control (1% RRS, matrix soy flour), 2–5 = raw lecithin samples.

5.3.5 Quality assurance

General

The requirements for performing and evaluating PCR analyses of genetic modifications are described in international norms (18). General requirements for technical organization of the laboratory include the forward flow principle, at least three separate areas for extraction / PCR set-up / post PCR, and instrumental and personnel requirements (e.g. clothing). Results of the analysis are based on two separate test portions, beginning with sample weight (18).

Controls

Furthermore, controls have to be analysed together with the samples. Positive controls (e.g. genomic DNA containing the target sequence) are included, at low concentrations to ensure sensitivity of detection. In order to check the suitability of the extracted DNA for further GMO analysis, the sample DNA is analysed for the presence of a species-specific DNA sequence. Inhibition controls are performed to test for inhibitors of PCR reactions, e.g. by analysis of different dilutions of the sample DNA or by spiking of the target sequence. Negative controls (extraction blank instead of sample DNA) and PCR reagent controls (water instead of sample DNA) also are mandatory (18).

Evaluation

According to the requirements (18), the identity of the amplicons has to be verified by more than just comparison of the fragment length in gel electrophoresis. For this purpose, probe hybridization using blotting techniques, restriction analysis or sequencing of the fragment can be used. Such more or less laborious verification techniques are not required if real-time PCR is applied. Additional specification is achieved by the hybridization of specific fluorescence-labelled probes.

Detection and quantification using real-time PCR

Today, the most commonly used method for GMO analysis is real-time PCR. Compared to endpoint PCR, quantification of DNA target sequences is possible, contaminations can be avoided by simultaneous amplification and detection without the need for opening the tubes, and the verification of the detected target sequences is included (see above).

Principle of quantification

Fig. 6 illustrates the principle of quantification of genetic modifications by real-time PCR for the Roundup Ready® soybean (RRS). The result (in 'GMO per cent') is calculated on the basis of the quantitative ratio of the results of the species-specific gene(s) (e.g. a soy lectin gene) and a transgene-specific sequence (e.g. the RRS construct).

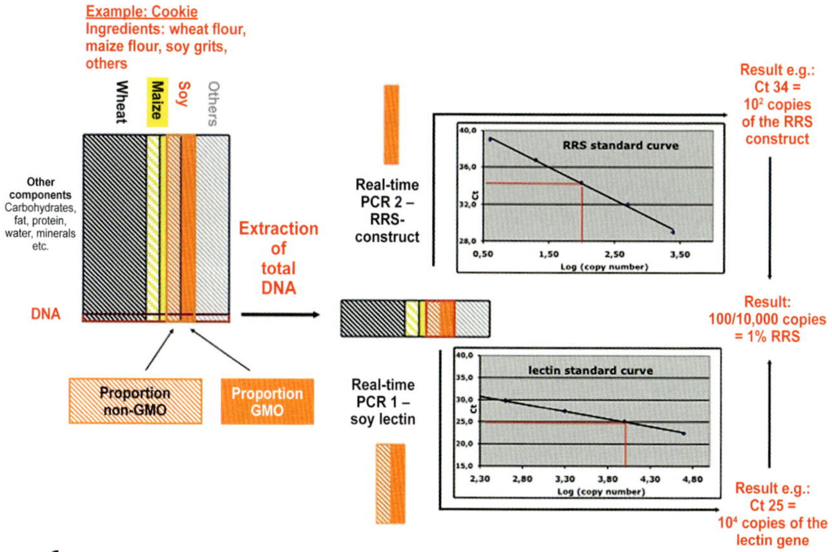

Figure 6.

Principle of quantification of genetic modifications by real-time PCR. Example: Roundup Ready® soybean (RRS). After the extraction of the total DNA of the food (example: a cookie), the DNA is analysed in parallel by two separate real-time PCR reactions. In real-time PCR 1, as a reference, the amount of DNA 'copies' (= genome equivalents) of a species-specific soy lectin sequence is quantified using an external standard curve. In real-time PCR 2, the amount of DNA copies of a RRS-specific DNA construct is quantified. The quantification of lectin and RRS target is done assuming that in a '100% RRS' standard material, lectin and RRS target are present in a ratio of 1:1 and as 'single copies' (meaning, one copy per genome). For this purpose, DNA from reference materials (e.g. flours) is used that contain defined proportions of RRS (e.g. 5% RRS w/w). The reference DNA is analysed in different dilution steps (i.e. copy numbers) resulting in standard curves for the lectin and RRS sequence (Ct against $\log_{\text{copy number}}$). Using the standard curves, the Ct values obtained for the sample DNA can be assigned to copy numbers. Finally, the ratio of the copy numbers of the RRS and the lectin sequence is calculated (15).

Quality control

The requirements mentioned above for endpoint PCR also have to be considered for real-time PCR (18), which also requires further controls: different concentrations of an external standard (usually different DNA dilutions prepared from one standard material) are analysed as positive control and for quantitation. Quality control samples prepared from materials with known GMO percentage should be analysed in parallel.

The measured value 'haploid genome copies'

Haploid genome copies are the main quantity measured by real-time PCR. For genomic DNA, the number of genome equivalents in an extract of a defined material can be estimated from literature values of haploid genome weights (30). However, such calculations are based on different assumptions. Reference material may be produced from material which is heterozygous in terms of the transgene (31). Furthermore, the integration number of the transgenic sequence per genome may vary depending on the GMP. Finally, different tissue types (endosperm, embryo) may exhibit different ploidies (diploid, triploid and other multiples of haploid), and therefore different proportions of integrated haploid transgenic genomes in relation to the total number of haploid genomes (32). In conclusion, determination of the *absolute* number of haploid genome copies from standard materials (e.g. soybean flours) certified for GMO proportions in weight per cent can be regarded as estimations.

The introduction of the digital PCR technique has now offered the opportunity to absolutely quantify DNA copy numbers without the need for reference material (33). However, reference materials are still necessary to transform the results of GMO quantification obtained in 'DNA percentages' into weight percentages.

By the use of uniform reference materials, reproducible results with good precision data can be obtained between different laboratories.

Especially for whole kernels and seeds, quantification via haploid genome copies may result in over- or underestimations of the 'true' GMO percentage, typically expressed as per cent of the seed kernels. This may be the case if for example heterozygous seed is present and only one out of three genomes contains the transgenic sequence.

Therefore, for the estimation of the proportion of transgenic kernels in seeds, qualitative methods combined with statistic evaluation are used (so-called subsampling methods) (34).

Authorization procedure and event-specific methods

Within the European Union, applicants for authorization of GM organisms for food and feed purposes have to provide control samples as well as event-specific methods for the GMO detection and quantification. The methods are tested and validated by the European Network of GMO Laboratories (ENGL). For this purpose, the ENGL has defined acceptance criteria for validation results (35). After an authorization is granted, its methods are published in the EU Database of Reference Methods for GMO Analysis (10). Reference materials are available from the Institute of Reference Materials and Measurement of the European Commission (36). A considerable number of real-time PCR methods for the event-specific quantification of GMP have been validated so far (37).

5.3.6 Appendix. Specific characteristics of GMO analysis

Practical limits of detection and quantification

The genetic modifications are quantified in relation to a species-specific reference gene. Results are given in percentages. Limits of detection (LOD) and limits of quantification (LOQ) can be determined or at least estimated for the transgene-specific PCR method (as DNA copy number). According to long-term experience as well as the results of validation studies, efficient real-time PCR methods show LODs in the range of five to ten copies (genome equivalents) (38,39).

In routine analysis, apart from the absolute LODs and LOQs, the practical LOD/LOQ or practical limit of detection (LOD_{pract}) is very important. It can be used as an additional criterion of the suitability of the extraction method and of the individual sample DNA extract, respectively.

For calculation of the LOD_{pract}, the known absolute limit of detection (LOD_{abs}) of the method for detection of the GM target sequence (e.g. ten copies) and the copy number of the reference target sequence determined for the sample are taken and used for the following ratio:

$$\mathbf{LOD_{pract}\ (\%)} = \frac{\textit{LODabs (GM target sequence copy number)}}{\textbf{measured copy number reference gene}} \times \mathbf{100}$$

LODs given as GMO percentages can be determined using material with defined GMO proportions (e.g. 0.1%). Thus, relative LODs of 0.01 to 0.1% in flour material can usually be achieved within method validation. However, many samples contain only miniscule DNA of the relevant species, too little to allow for a sensitive GMO analysis.

For example, if a negative result for the GM target sequence is obtained and the LOD_{pract} for the given sample is above 0.9%, an analytical verification of the sample in terms of the labelling requirements for authorized GM food and feed according to the European Union legislation is not possible. In this case, only the analysis of the (relatively) DNA-rich raw material can give meaningful results.

Therefore, in the case of negative results, laboratories have to provide a practical LOD that is directly related to the sample analysed and not to the original method validation using DNA-rich flour materials (18).

Table 1.

Ratio of practical limit of detection (LOD_{pract}) and PCR amplifiable copy number of the reference gene in GMO analysis (40).

Copy number of reference gene estimated by PCR reference gene	LOD_{abs} of the GMO detection method (nominal)	LOD_{pract} %
100,000	10 copies	0.01
10,000	10 copies	0.1
1,000	10 copies	1

In Table 1, the relation between the amount of species DNA (reference gene) present in a sample and the practical LOD is shown. Table 2 gives the range of practical LODs obtained in routine analysis of important food ingredients, including soy lecithins.

Table 2.

Practical limits of detection (LOD_{pract}) obtained for important food ingredients.

Ingredient	Species-specific DNA (observed range) copy number of reference gene	LOD_{pract} %
Soy lecithin	0 – approx. 10,000	100 – 0.05
Maize starch, native	0 – approx. 7,000	100 – 0.1
Maize starch, modified	0 – approx. 500	100 – 1
Glucose syrup, maltodextrins	< 20	> 25

Detection of unauthorized GMPs by screening

The unambiguous identification of unauthorized GMPs may be difficult in many cases: usually, detailed sequence information about introduced DNA and suitable reference material is not available. Therefore, special screening strategies have to be used for detection.

Screening

Indications of the presence of unauthorized GMPs can be obtained using screening methods that detect sequences from the introduced genetic construct. Publicly available sequence databases and information platforms (1,2) can help when establishing such screening methods. If positive controls can be made available, at least qualitative statements and indications can be given by the laboratory. However, an exact quantification and identification of the transformation event will not be possible if reference material is lacking.

A combination of different targets used for screening, including construct-specific sequences, should be used for a best possible specification of the result. Inter-laboratory validated methods for the detection of target sequences frequently occurring in GMPs have been published in national and international norms and standards (10,38,39,41).

A strategy to screen for authorized and unauthorized genetically modified plants using five different target sequences has been presented (13). A combination of real-time PCR detection methods was selected, including for the P35S sequence, the T-nos sequence, the CTP2-CP4 EPSPS construct sequence, a sequence of the bar gene from *Streptomyces hygroscopicus* and the synthetic phosphinothricin-acetyltransferase gene (pat) (see also "Detection by the genotype"). The target sequences can be detected simultaneously by multiplex real-time PCR (42). For most of the relevant crop species (e.g. maize, canola, cotton), this screening strategy covers nearly all GM events described in publicly available databases. At least one of the targeted sequences is present in each such GMP.

However, especially for soybean plant species several new GM events do not contain these sequence elements any more. Table 3 lists GM soybean events for which sequence information from databases is available.

Table 3.

Screening table for the detection of GM plants (3, 13, 43). Extract: plant species soybean. '–' = event does not contain the screening element, '+' = event does contain the screening element. Shaded: event contains none of the elements and is not detected by the proposed screening.

Event	Unique identifier (first event mentioned)	Plant	P35S	T-nos	CTP2-CP4EPSPS	bar	35S-pat	Cry1Ab/Ac	p-FMV	p-nos
260-05, G94-1, G94-19, G168	DD-026005-3	Soybean	+	+	–	–	–	–	–	–
305423	DP-305423-1	Soybean	–	–	–	–	–	–	–	–
356043	DP-356043-5	Soybean	+	–	–	–	–	–	–	–
A2704-12	ACS-GM005-3	Soybean	+	–	–	–	+	–	–	–
A2704-21, A5547-35	ACS-GM004-2	Soybean	+	–	–	–	+	–	–	–
A5547-127 (LibertyLink)	ACS-GM006-4	Soybean	+	–	–	–	+	–	–	–
BPS-CV127-9	BPS-CV127-9	Soybean	–	–	–	–	–	–	–	–
DAS-44406-6	DAS-44406-6	Soybean	–	–	–	–	–	–	–	–
DAS-68416-4	DAS-68416-4	Soybean	–	–	–	–	–	–	–	–
DAS-81419-2	DAS-81419-2	Soybean	–	–	–	–	–	–	–	–
FG72	MST-FG072-2	Soybean	–	+	–	–	–	–	–	–
GTS 40-3-2 (Roundup Ready)	MON-04032-6	Soybean	+	+	–	–	–	–	–	–
GU262 (LibertyLink)	ACS-GM003-1	Soybean	+	–	–	–	+	–	–	–
MON87701	MON-87701-2	Soybean	–	–	–	–	–	+	–	–
MON87705	MON-87705-6	Soybean	–	–	+	–	–	–	+	–
MON87708	MON-87708-9	Soybean	–	–	–	–	–	–	–	–
MON87751	MON-87751-7	Soybean	–	–	–	–	–	+	–	–
MON87754	MON-87754-1	Soybean	–	–	–	–	–	–	–	–
MON87769	MON-87769-7	Soybean	–	–	–	–	–	–	–	–
MON89788	MON-89788-1	Soybean	–	–	+	–	–	–	+	–
SYHT0H2	SYN-000H2-5	Soybean	+	+	–	–	+	–	–	–
W62, W98 (Liberty Link)	ACS-GM001-8	Soybean	+	+	–	+	–	–	–	–

The columns of this table specify the presence of screening target sequences. Additional target sequences for screening are listed (Cry1Ab/Ac, p-FMV and P-nos). For all the target sequences, official inter-laboratory validated screening methods are available (41). In Table 3, GM soybean events which cannot be detected by any of the screening targets are shaded in grey. For the future detection of these events other approaches will have to be developed. One option, multiplex event-specific PCR methods that would cover these GM soybean events not currently detected by common screening methods, could become an effective and economical complement to current testing methods (44–46).

References

1. Center for Environmental Risk Assessment (CERA), GM crop database. Retrieved from http://www.cera-gmc.org/GMCropDatabase
2. International Service for the Acquisition of Agri-Biotech Applications (ISAAA). Retrieved from http://www.isaaa.org/
3. Federal Office of Consumer Protection and Food Safety (BVL) (D) and RIKILT Wageningen UR (NL): Euginius, The European GMO reference database. Retrieved from http://www.euginius.eu/
4. European Commission: Community Register of GM Food and Feed. Retrieved from http://ec.europa.eu/food/dyna/gm_register/index_en
5. Regulation (EC) No 1829/2003 of the European Parliament and the Council of 22 September 2003 on genetically modified food and feed (Official Journal No. L 268/1, 18.10.2003), and Regulation (EC) No. 1830/2003 concerning the traceability and labelling of genetically modified organisms and the traceability of food and feed products produced from genetically modified organisms and amending Directive 2001/18/EC of the European Parliament and the council of 22 September 2003 on genetically modified food and feed (Official Journal No. L 268/24, 18.10.2003).
6. Commission regulation (EU) No. 619/2011 of 24 June 2011 laying down the methods of sampling and analysis for the official control of feed as regards presence of genetically modified material for which an authorisation procedure is pending or the authorisation of which has expired (Official Journal No. L 166/9, 25.06.2011).
7. Waiblinger, H.U., Graf, N., Mäde, D., Woll, K., Busch, U., Holland, B., Pilsl, H., Schulze, M., Tschirdewahn, B., Brünen-Nieweler, C., Hempel, G., Weidner, M., Winterstein, A.R., 2007. "Technically Unavoidable" in Terms of Genetically Modified Organisms - an Approach for Food Control. J. Verbr. Lebensm. 2, 126-129.
8. Waiblinger, H.U., Busch, U., Denker, S., Döpping, C., Eichner, C., Graf, N., Josefowitz, P., Wirries, F.M., Krujatz, I., Mäde, D., Näumann, G., Pecoraro, S., Reiting, R., Tschirdewahn, B., Woll, K., 2014. Results of GMO analysis in food derived from soybean and maize obtained by official German control laboratories in the years 2011-2013 and their use for the interpretation of current results. Deutsche Lebensmittel-Rundschau 110, 460-464.
9. Stave, J.W., 2002. Protein immunoassay methods for detection of biotech crops: Applications, limitations, and practical considerations. J. AOAC Int. 85, 780-786.
10. European Commission, Joint Research Centre: EU Database of Reference Methods for GMO Analysis. Retrieved from http://gmo-crl.jrc.ec.europa.eu/gmomethods/
11. Holst-Jensen, A., Ronning, S.B., Lovseth, A., Berdal, K.G., 2003. PCR technology for screening and quantification of genetically modified organisms. Anal. Bioanal. Chem. 375, 985-993.
12. Fraiture, M.A., Herman, P., Taverniers, I., De Loose, M., Deforce, D., Roosens, N.H., 2015. Current and New Approaches in GMO Detection: Challenges and Solutions. BioMed Res. Int. 15, 392872.
13. Waiblinger, H.U., Grohmann, L., Mankertz, J., Engelbert, D., Pietsch, K., 2010. A practical approach to screen for authorised and unauthorised genetically modified plants. Anal. Bioanal. Chem. 396, 2065-2072.
14. Padgette, S.R., Kolacz, K.H., Delannay, X., Re, D.B., LaValee, B.J., Tinius, C.N., Rhodes, W.K., Otero, Y.I., Barry, G.F., Eichholtz, D.A., Peschke, V.M., Nida, D.L., Taylor, N.B., Kishore, G.M., 1995. Development, identification, and characterization of a glyphosate-tolerant soybean line. Crop Sci. 35, 1451-1461.
15. ISO 21570, 2005. Foodstuffs - methods of analysis for the detection of genetically modified organisms and derived products - quantitative nucleic acid based methods. ISO 21570, 1-103.
16. EN/TS 15568, 2006. Foodstuffs - methods of analysis for the detection of genetically modified organisms and derived products - Sampling strategies. Beuth, Berlin.
17. Hübner, P., Waiblinger, H.U., Pietsch, K., Brodmann, P., 2001. Validation of PCR methods for the quantification of genetically modified plants in food. J. AOAC Int. 84 (6), 1855-1864.
18. ISO 24276, 2006. Foodstuffs - methods of analysis for the detection of genetically modified organisms and derived products – general requirements and definitions. ISO 24276. Beuth, Berlin.
19. Engel, K.H., Moreano, F., Ehlert, A., Busch, U., 2006. Quantification of DNA from genetically modified organisms in composite and processed foods. Trends Food Sci. Technol. 17, 490-497.
20. Terry, C., Harris, N., Parkes, H., 2002. Detection of GM crops and their derivatives: Critical steps in sample preparation and extraction. J. AOAC Int. 85 (3), 768-774.

21. ISO 21571, 2005. Foodstuffs - methods of analysis for the detection of genetically modified organisms and derived products - Nucleic acid extraction. ISO 21571. Beuth, Berlin.
22. Waiblinger, H.U., Ernst, B., Graf, N., Pietsch, K., 2007. Ring trial validation of a method for the extraction of DNA from soy lecithins. J. Verbr. Lebensm. 2, 113-115.
23. Waiblinger, H.U., Schilinger, M., Hess, N., 2012. In-house and ring trial validation of a method for the extraction of DNA from maize starch. J. Verbr. Lebensm. 7, 163-170.
24. Waiblinger, H.U., Ohmenhaeuser, M., Meissner, S., Schillinger, M., Pietsch, K., Goerlich, O., Mankertz, J., Lieske, K., Broll, H., 2012. In-house and interlaboratory validation of a method for the extraction of DNA from pollen in honey. J. Verbr. Lebensm. 7, 243-254.
25. ISO 21569, 2005. Foodstuffs - methods of analysis for the detection of genetically modified organisms and derived products - qualitative nucleic acid based methods.
26. Meyer, R., 1995. Detection of genetically modified plants by means of PCR: Example FLAVR SAVR tomato. Zeitschrift für Lebensmitteluntersuchung und Forschung 201, 583-586.
27. Pietsch, K., Waiblinger, H.U., Brodmann, P., Wurz, A., 1997. Screening method for the identification of "genetically modified" food of plant origin. Deutsche Lebensmittel-Rundschau 93 (2), 35-38.
28. Wurz, A., Willmund, R., 1997. Identification of transgenic glyphosate-resistant soybeans. BgVV-Hefte 01/1997, 115-117.
29. Arumuganathan, K., Earle, E.D., 1991. Estimation of nuclear DNA content of plants by flow cytometry. Plant Mol. Biol Rep 9, 208-218.
30. Kew, Royal Botanic Gardens. Plant DNA C-values Database. Retrieved from http://www.rbgkew.org.uk/cval/
31. European Commission. Institute for reference materials and measurements, 2006. ERM Application note 4. Retrieved from http://www.erm-crm.org/html/ERM_products/application_notes/application_note_4/index.htm
32. Trifa, Y., Zhang, D., 2004. DNA content in embryo and endosperm maize kernel (Zea mays L.): Impact on GMO quantification. J. Agric. Food. Chem. 52, 1044-1048.
33. Pinheiro, L., Coleman, V., Hindson, C., Herrmann, J., Hindson, B., Bhat, S., Emslie, K., 2012. Evaluation of a Droplet Digital Polymerase Chain Reaction Format for DNA Copy Number Quantification. Anal. Chem. 84, 1003–1011.
34. European Commission, JRC Technical Reports, 2015. European Network of GMO Laboratories, Working Group "Seed Testing" (WG-ST). Working Group Report.
35. European Network of GMO Laboratories (ENGL), 2015. Definition of minimum performance requirements for analytical methods of GMO testing. Retrieved from http://gmo-crl.jrc.ec.europa.eu/guidancedocs.htm
36. European Commission, JRC. Institute for Reference Materials and Measurements. European Reference Materials (ERM). Retrieved from http://www.erm-crm.org/html/homepage.htm
37. European Commission, JRC. European Union Reference Laboratory for Food and Feed. Status of dossiers. Retrieved from http://gmo-crl.jrc.ec.europa.eu/StatusOfDossiers.aspx
38. Waiblinger, H.U., Ernst, B., Anderson, A., Pietsch, K., 2007. Validation and collaborative study of a P35S and T-nos screening duplex real-time PCR screening method to detect genetically modified organisms in food products. Eur. Food Res. Technol. 26, 1221-1228.
39. Grohmann, L., Brünen-Nieweler, C., Nemeth, A., Waiblinger, H.U., 2009. Collaborative trial validation studies of real-time PCR based screening methods for detection of the bar gene and the ctp2-cp4epsps construct. J. Agric. Food Chem. 57 (19), 8913-8920.
40. Waiblinger, H.U., Grohmann, L., 2014. Guidelines for validation of DNA extraction methods applied in subsequent PCR analysis of food and feed products for the presence of genetically modified material. J. Verbraucherschutz Lebensmittelsicherh. 9, 183-190.
41. Official Collection of Test Methods. German food and feed law - food analysis, article 64. Beuth, Berlin.
42. Huber, I., Block, A., Sebah, D., Debode, F., Morisset, D., Grohmann, L., Berben, G., Stebih, D., Milavec, M., Zel, J., Busch, U., 2013. Development and validation of duplex, triplex, and pentaplex real-time PCR screening assays for the detection of genetically modified organisms in food and feed. J. Agric. Food Chem. 61 (43), 10293-10301.

43. German Federal Office for Food Safety and Consumer protection, 2015. Screening table for the detection of GM plants. Retrieved from http://www.bvl.bund.de/DE/09_Untersuchungen/01_Aufgaben/02_Referenzlaboratorien/02_Nationale_Referenzlaboratorien_NRL/02_NRL_GVO/untersuchungen_NRLs_GVO_node.html
44. Koeppel, R., Sendic, A., Waiblinger, H.U., 2014. Two quantitative multiplex real-time PCR systems for the efficient GMO screening of food products. Eur. Food Res. Technol. 239 (4), 653-659.
45. Koeppel, R., Bucher, T., Meuwly, A., Moor, D., 2014. Multiplex real-time PCR Method for the detection and quantification of DNA from the four transgenic soy traits DP-356043-5, DP-305423-1, MON 87701 and BPS CV127- 9 and lectin. Eur. Food Res. Technol. 239(2), 347-355.
46. Koeppel, R., Bucher, T., Frei, A., Waiblinger, H.U., 2015. Droplet digital PCR versus multiplex real-time PCR. Method for the detection and quantification of DNA from the four transgenic soy traits MON87769, MON87708, MON87705, FG72 and lectin. Eur. Food Res. Technol. 241(4), 521-527.

6 Application of Lecithin in Food

6.1 Interactions of phospholipid fractions with the surfaces of solids

Dana Middendorf, Ute Bindrich

The production of foods may involve combining a wide variety of ingredients into greatly differing substance systems. The resulting dispersed, multiphase systems fall into three broad categories: foams, emulsions and suspensions. They each consist of a continuous 'outer' phase and a dispersed 'inner' phase which may be solid, liquid or gaseous (see Table 1).

Table 1.

Selected examples of dispersed multiphase systems.

Food	Dispersed system	Dispersed phase	Continuous phase
Whipped cream, ice cream, beaten egg white	Foam	Gaseous	Liquid
Mayonnaise, milk, dressing	Emulsion	Liquid	Liquid
Hazelnut spread, ice cream	Suspension	Solid	Liquid
Meringue, bread, marshmallows	Solid foam	Gaseous	Solid
Butter, margarine	Solid emulsion	Liquid	Solid
Chocolate	Solid suspension	Solid	Solid
Muesli, sugar, starch	Loose bulk	Solid	Gaseous
Spray (mist)	Aerosol	Liquid	Gaseous

If the relevant quality attributes of foods are to be maintained over a considerable time, their structures must have a certain degree of stability. In this connection, phospholipids play a significant role. Mixtures of phospholipid fractions are often known as lecithin. In the case of emulsions, complexes of proteins and phospholipid fractions adsorb to the surface of the dispersed droplets, reduce the surface tension and increase the repulsive electrostatic and steric forces, thus preventing coalescence and phase separation. In the case of suspensions, the stability is largely determined by processes on the boundary of the dispersed phase. Here, too, adsorption of phospholipids at the surface of the particles has a stabilizing effect, influencing the interactions between particles that result in agglomeration and phase separation (see also Fig. 1).

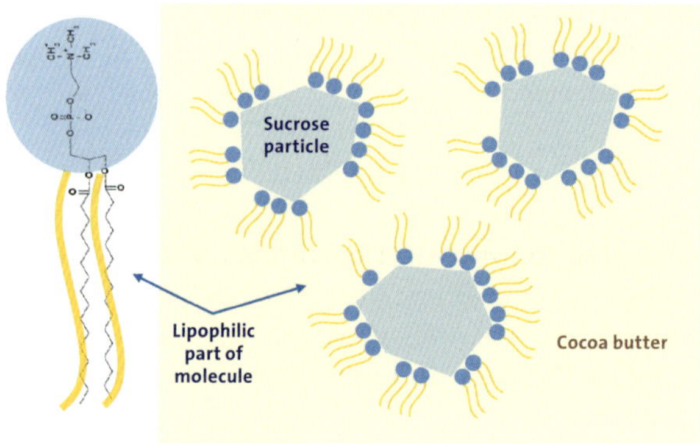

Figure 1.

Diagram showing the functionality of emulsifiers in lipophilic suspensions, taking phosphatidylcholine as an example.

Examples of suspensions in foods include fondant, nougat and also chocolate. In these systems the dispersed phase consists mainly of sugar, cocoa and/or dried milk particles. The nature of these particles and their interactions with each other and with the continuous phase is therefore extremely important for maintaining the quality of the foods.

As already described, the interaction of phospholipids with the surfaces of solids plays an important role in stabilizing suspensions. The atomic force microscope (AFM) offers a diversity of possibilities for characterizing the surfaces of particles (1,2).

In the AFM, the surface of a sample is passed line by line under a probe tip on the end of a cantilever with the aid of a piezoelectric scanner. The cantilever is deflected according to the forces interacting between the probe tip and the surface of the sample – on a vertical plane (attractive or repulsive interaction) or on a horizontal plane (lateral forces, e.g. friction). The deflection of the cantilever is recorded using a laser beam focused on the end of the cantilever and reflected onto a photodiode divided into four segments (3). The deflection of the cantilever on the vertical plane, and thus the difference between the upper and lower segments, results in a topographic signal. The structure and principle of the AFM enable three-dimensional detection of surface characteristics down to the Ångström range (molecular resolution). Besides recording surface topography (imaging modes) it is possible to determine other properties such as viscoelasticity, friction and also the surface charge, surface polarity and layer thicknesses. Spectroscopic measurements of forces are also possible (4,5).

6.1.1 Interactions with sucrose

Dana Middendorf

6.1.1.1 Structure of the boundary layer

Sucrose is used in foods for other reasons besides its sweetness. In bakery products and confectionery, especially, sugar plays an important role in forming the structure of the food matrix and in chocolate, for example, it constitutes the dispersed phase to which the surface-active substances are adsorbed.

Sucrose itself has a crystalline structure in the shape of layers. The AFM can detect these molecular structures in detail. The fractured surface of a sucrose crystal and a height profile of its topography are shown in Fig. 2. In this particular case, the height of a layer of the crystal is 23 Å (see Fig. 2 b). That corresponds approximately to the height of two unit cells of sucrose and illustrates the molecular resolution of these images (6).

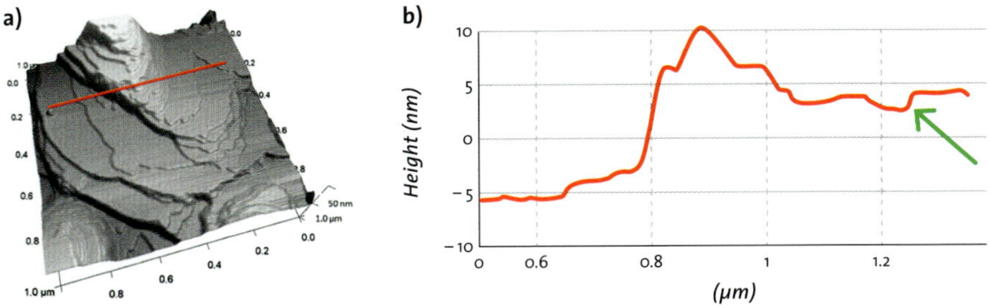

Figure 2.

a) Topography of the fractured surface of a sucrose crystal;
b) Height profile of the topography at the marked position in a).

Changes in the surface of the sugar particles brought about by the adsorption of phospholipids were investigated using model suspensions of cocoa butter and sugar. As in the production of chocolate, 0.5% soy lecithin based on the sucrose content was added to the cocoa butter phase. The sugar was then ground in a ball mill and the model suspension subjected to a conching process during which the emulsifier was adsorbed to the newly formed fractured surfaces (7,8). A suspension without the addition of the emulsifier was made up as a reference.

As a first step, the adsorption of the phospholipids to the interface between the sugar and the cocoa butter was determined with the aid of confocal laser scanning microscopy (CLSM) (9). To permit this, the phospholipids were coloured selectively with a fluorescent dye and activated with a laser beam (helium-neon laser) so that they emitted light with a specific wavelength. This emitted light was recorded pixel by pixel in the observation plane. However, the dye did not react with the sugar particles, which therefore appeared as grey-black spots. A CLSM image of phospholipids adsorbed to sugar particles is shown in Fig. 3.

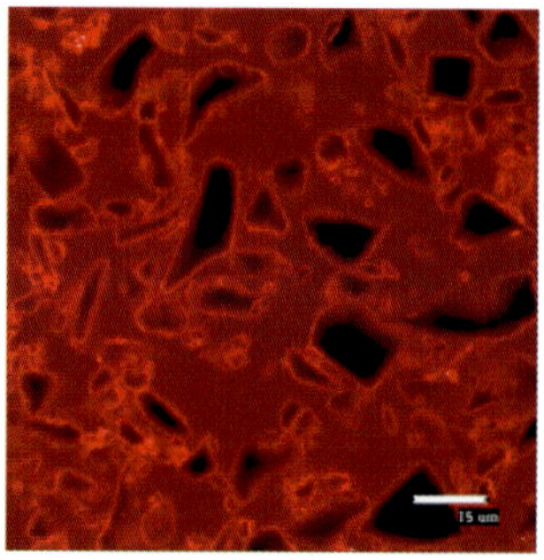

Figure 3.

CLSM image of phospholipids (red) adsorbed to ground sugar particles (grey-black).

This boundary layer was then characterized in greater detail by AFM. Fig. 4 shows that the surface of the sugar particles has undergone radical changes through adsorption of the phospholipids from soy lecithin. The surface no longer contains step-like structures like those previously observed on the fractured surface of a sugar particle (see Fig. 2). The surface topographies now appear less sharply contoured and look much smoother.

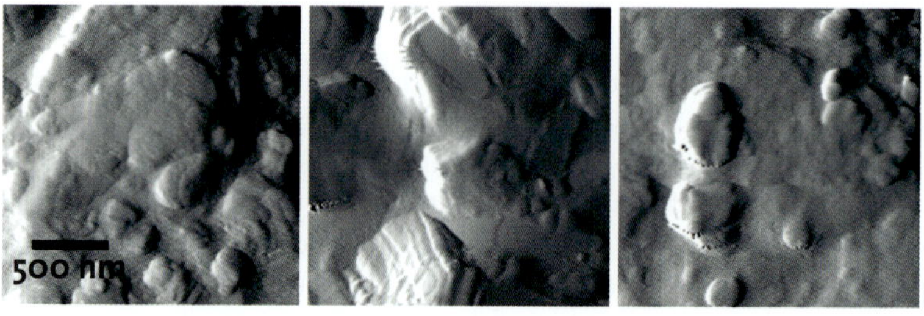

Figure 4.

Typical AFM surface topographies of sucrose ground in a ball mill and in the presence of soy lecithin; after (10).

So-called force – distance curves, which can also be recorded with the aid of AFM, can be used as a further means of characterizing the changes in particle surfaces brought about by the adsorption of emulsifiers (10). To do this, the probe tip is brought closer to the surface of the sample until contact is achieved. The direction of movement is then reversed and the tip drawn away from the sample. The adhesive force occurring between the probe tip and the sample in the course of this process can then be depicted as a function of the distance between the surface of the sample and the tip (see Fig. 5 left). Furthermore, force – distance curves can be strung together in the form of a scan and combined into a chart (see Fig. 5, top right). The values for local adhesion are then converted into colour values, and these are likewise combined to form a chart. In this way it is possible to produce a diagram showing the local degree of adhesion at the surface of the sample. Finally, the force charts can be used to calculate histograms in which the frequency of certain forces in a defined area of the surface is depicted (see Fig. 5 bottom right).

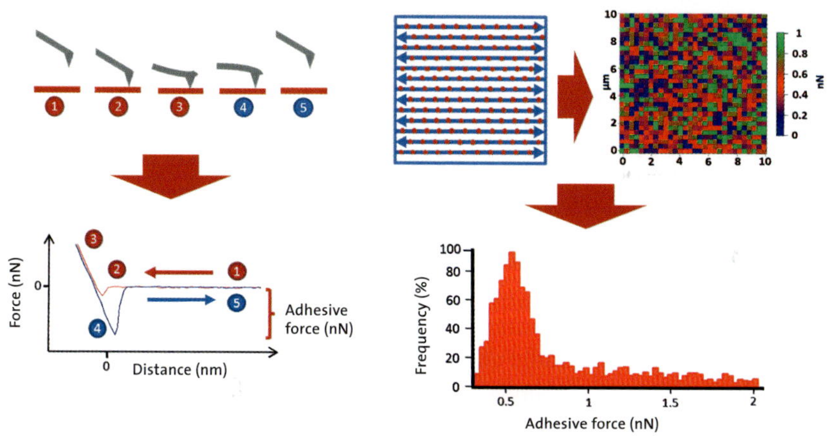

Figure 5.
Force – distance curve, measured by AFM; for explanations see text.

Fig. 6 now shows some frequency distributions typical of the condition of the surface of sugar particles ground in a ball mill without adsorbed phospholipids. The adhesive forces between the probe tip and the surface are in the range of 5 to 50 nN; at 10 to 40 nN (e.g. blue bar), the distribution of the forces in the surface area shown on the left is slightly narrower than in the area on the right, where it is 15 to > 50 nN (e.g. red bar).

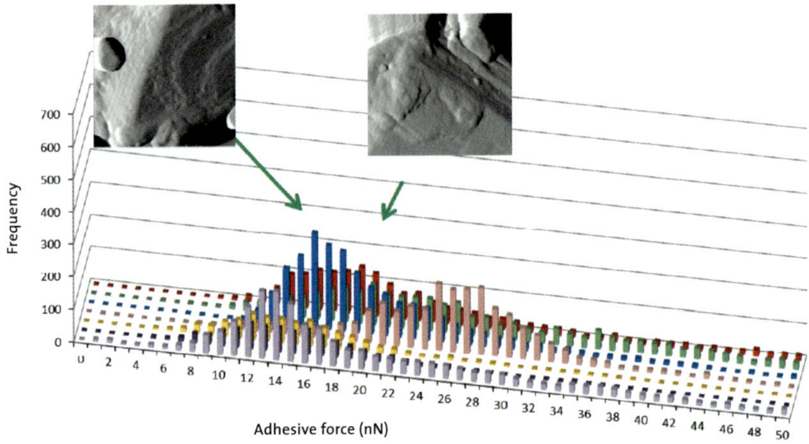

Figure 6.

Histograms from the force charts of AFM measurements of sucrose particles ground in a ball mill; size of scan: 1 × 1 µm.

Fig. 7 shows the histograms derived from the force charts of the surfaces of the sugar particles coated with soy lecithin. When compared directly with the surface of the uncoated particles, the relevant area of the forces is now considerably wider (up to > 100 nN). Moreover, it can be divided into several sections: the bottom section represents forces of 10 to 30 nN, the middle section those between 20 and 60 nN, and the forces in the top section are from 40 to about 100 nN, in some cases even greater. The narrower distribution in the bottom section corresponds to that found on uncoated particle surfaces. The position of the distributions in the middle and top sections shows that the adhesive properties of the surfaces of the sugar particles have changed substantially through adsorption of the emulsifier. The fact that it is possible to differentiate between these areas may be interpreted as indicating the different phospholipid fractions and fatty acid groups with differing degrees of adhesion between the probe tip and the surface of the sample.

The coating of the surfaces of the particles with emulsifiers can be calculated from the difference between the quantity of emulsifier added and the concentration of the emulsifier still detectable in the lipophilic phase after conching of the suspension. This difference is set in relation to the surface of the particles, which can be determined by laser diffraction. For coating of the surface of the sugar particles with soy lecithin, this calculation results in a value of 2.8 mg lecithin per m^2 of surface area, corresponding to 98% of the quantity of lecithin used.

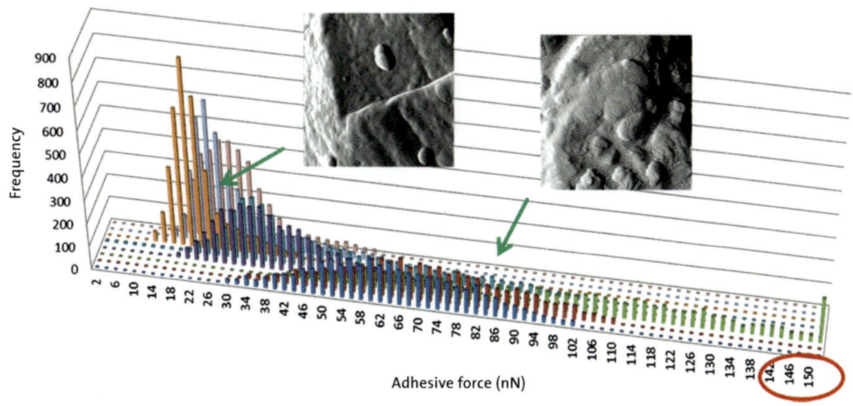

Figure 7.

Histograms from the force charts of AFM measurements of sucrose particles ground in a ball mill (ball diameter 1.5 cm) and coated with soy lecithin; size of scan 2 × 2 µm.

Besides the coating of the surface, the thickness of the layer of emulsifier can be determined. To achieve this, the force – distance curves recorded by AFM are used again. On the basis of these curves it is possible to determine the depth to which the probe tip has penetrated a surface coated with the emulsifier and then estimate the thickness of the coating.

The depth of penetration results from the difference between the force – distance curves for approach and withdrawal of the tip of the probe while this is in contact with the surface (see red double-arrow in Fig. 8, left). In the case of a particle surface coated with emulsifier, the cantilever is therefore immersed deeper on its way out (blue line) than on the way in (red line). This results in a difference of, for example, 34 ± 2 nm as a measurement of the thickness of the coating of soy lecithin (10,11). The standard deviation can be calculated from the difference measured on an uncoated surface (see Fig. 8, right).

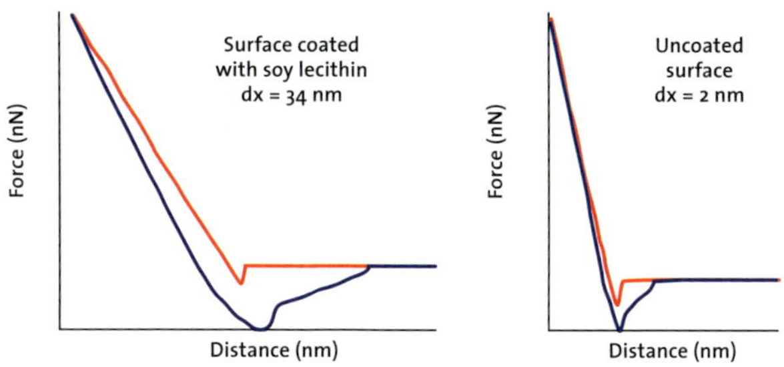

Figure 8.

Force – distance curves from an AFM force chart of a particle surface coated with soy lecithin (left) and an uncoated particle surface (right).

6.1.1.2 Mechanisms of phospholipid adsorption

The process of adsorption of soy lecithin in a lipophilic suspension was investigated more closely on the molecular level with the aid of phosphatidylcholine (PC), the main component of lecithin. To do so, a fractured surface of sugar particles was prepared for AFM measurement in a phase consisting of liquid cocoa butter. The local changes in the condition of the surface were then determined on an area of 10 × 10 µm before and after adsorption of the PC. The AFM topography of the surfaces and the corresponding force charts are shown in Fig. 9 a) to c).

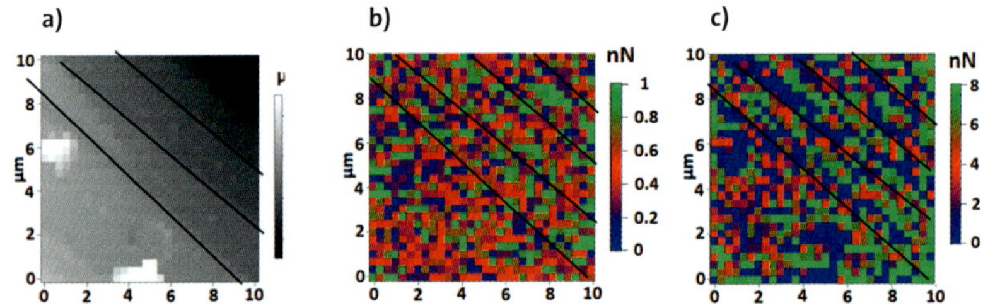

Figure 9.

Changes in surface adhesion brought about by application of the emulsifier PC to a sucrose surface (see also Fig. 2 a); the steps in the surface are indicated by diagonal lines. a) Topography, b) chart of adhesion before and b) after the addition of PC (AFM).

Fig. 9 a) first shows the topography of the surface of the sample. The black lines drawn on it indicate the steps in the structure already described (see Fig. 2 a). The corresponding adhesion of this surface is to be seen in Fig. 9 b). The position of the steps is again indicated by black lines but does not correlate with the distribution of the forces on this surface. The spectrum of the adhesive forces in the chart ranges from 0 to 1 nN. Fig. 9 c) shows the condition of the surface after adsorption of the PC to the surface of the sucrose particles. Here, too, the step-like pattern is indicated by the black lines. However, a connection can now be seen with the colour pattern of this chart, and the force range of the spectrum is wider. It now extends from 0 to 8 nN. A comparison of the charts shows that the changes in the adhesive force tend to be in the flat areas of the surface, whereas the forces in the area of the steps in the crystal have changed only slightly or not at all. A consideration on the molecular level shows why it is quite plausible that the emulsifier adsorbs to the flat areas rather than the steps. Fig. 10 shows the three-dimensional structure of the sucrose in the unit cell of the crystal, which is made up of two sucrose molecules. The atoms of the molecule are indicated by different colours (light grey: hydrogen, red: oxygen, dark grey: carbon). The illustration makes it clear that the surface of the particle is made up largely of the oxygen atoms of the molecule.

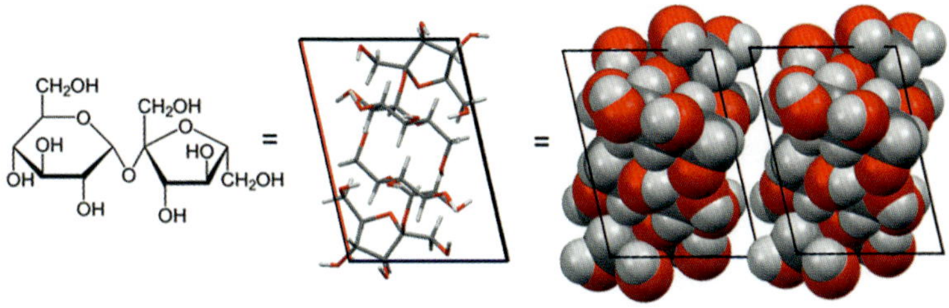

Figure 10.

Molecular structure of sucrose. Left: structural formula; centre: two sucrose molecules arranged in the unit cell; right: three-dimensional representation of the atoms as a ball model (Mercury 3.0, CCDC, Cambridge, UK; data base after Brown and Levy (12), accessible online at http://www.ccp14.ac.uk/ccp/ccp14/ftp-mirror/platon-spek/pub/special/.cif; last accessed on 17 May 2018).

Because of the free electron pairs of the oxygen atoms, interaction with the positively charged head group of the PC is advantageous (see Fig. 11 a). A different view of the three-dimensional model of the sucrose molecule shows that the positively polarized carbon and hydrogen atoms tend to be on the 'edges' of the crystal (see Fig 11 b). Interaction with the positively charged head group of the PC is not very likely; it might rather be classified as repulsive. Moreover, adsorption of PC in the vicinity of the edges is much more difficult for steric reasons.

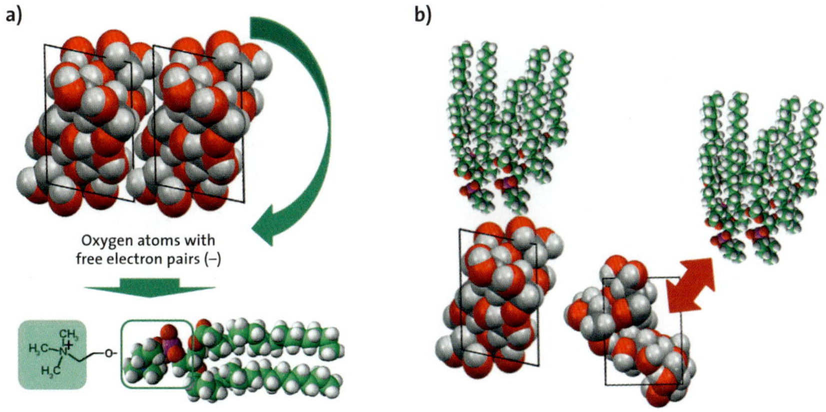

Figure 11.

a) Interaction of the PC head group with the surface of the sucrose molecule; b) comparison of interactions on the surface with interaction at a broken edge. Illustration of the lecithin molecule by kind permission of Avanti Polar Lipids, Inc.; for explanations, see text.

In further tests using AFM, the time frame was now investigated in which PC is adsorbed to the surface of the sucrose particles. For this purpose, a fractured surface of a sugar particle was prepared in a liquid phase of cocoa butter, and several force charts of one and the same area were recorded (13). PC was then injected into the liquid cocoa butter in the measuring cell. Distribution was controlled entirely by diffusion; active mixing was not possible because of the experimental set-up and the measuring principle of the AFM. Force charts were then measured in the same area.

Fig. 12 shows the changes in the distribution of surface adhesion in a surface area of 10 × 10 μm as a function of time. Initial changes in the distribution of the forces took place approximately two hours after injection of the PC into the AFM measuring cell. The frequencies in the range of forces below 2 – 3 nN decreased; the spectrum was initially widened to a range of about 8 nN. Over the course of time the distributions widened further, until an equilibrium was reached after about 10 – 11 h. This equilibrium was characterized by a continuous process of adsorption and desorption of the PC molecules; the distribution of the adhesive forces shown in Fig. 12 remained unchanged, but local changes in adhesion to the surface of the sucrose particles were observed (see force charts in Fig. 13).

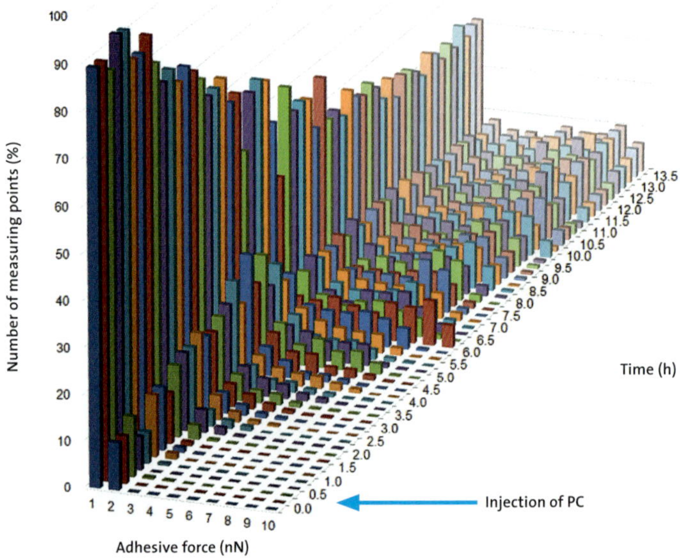

Figure 12.

Changes in the distribution of adhesion to the surface of a sucrose crystal (AFM) after injection of a solution of phosphatidylcholine and cocoa butter as a function of time; times of injection are marked with an arrow.

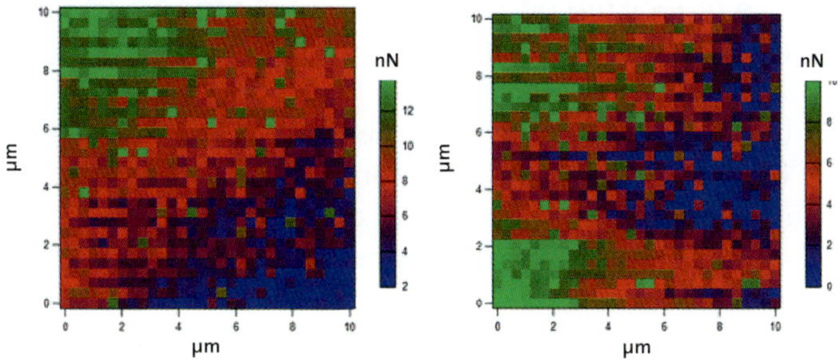

Figure 13.

Two AFM force charts recorded in a state of equilibrium after a period of 12 h; the time between two measurements is approx. 15 min. For explanations see text.

It is very probable that these simultaneous processes of adsorption and desorption of the PC also take place during conching of liquid chocolate masses, for example. But in this case a much shorter time up to the establishment of a state of equilibrium is to be expected because of active mixing of the suspension. This assumption was confirmed by CLSM images of sugar particles with marked phospholipids after different conching times (see Fig. 14). Immediately after grinding, sugar particles in an unconched suspension a) showed very irregular coating of the surface with the phospholipids. As the conching process continued, the phospholipid film became increasingly regular, as shown especially in Fig. 14 d).

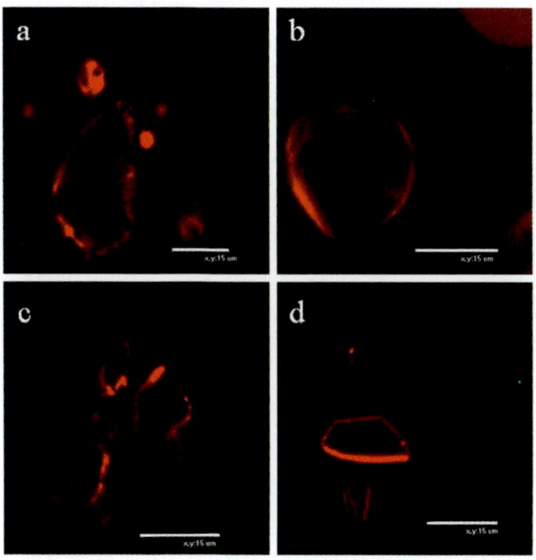

Figure 14.

CLSM image of finely ground sugar particles with coloured phospholipids after different conching times of a suspension of sugar and cocoa butter; a) without conching, b) after 1 h, c) after 2 h and d) after 2:45 h.

6.1.1.3 Influence of lecithin on the macroscopic properties of the lipophilic suspension: Fat immobilization and flow properties

On the macroscopic level, too, the changes to the surface of the particles can be characterized on the basis of their interaction with the surrounding fatty phase. This is done by determining the proportion of fat immobilized, which depends on the shape of the particles and in particular the different polarities on the particle surfaces as well as the particle size. It is this surface polarity, especially, that is decisively influenced by adsorption of the emulsifier molecules.

A distinction can be made between two types of fat or liquid immobilization (14): immobilization on the particle surface and as the part of the liquid immobilized as a volume. The part immobilized on the surface of the particles can be attributed to molecular or energetic interactions between solid particles and the liquid phase or between the molecules of the liquid phase themselves. These interactions cause the formation of a fairly large number of layers of liquid molecules which may be regarded as immobilized on the surface of the particles. On the other hand, the part of the liquid immobilized as a volume can be attributed to the cavities or depressions in particles and internal gaps (dead spaces) in particle agglomerates.

An indirect measurement for both types of immobilization is the percentage of fat that can be separated off from the suspension by centrifugation. The larger this percentage, the less fat is immobilized on particles and in dead spaces. Assuming a uniform particle size distribution and specific surface of the sucrose particles, the coating of emulsifier on the boundary surfaces of the solid particles has an influence on the percentage of immobilized cocoa butter and thus especially on the flow properties of the suspension (15,16,17).

In order to examine this influence more closely, the proportion of the cocoa butter that could be separated off from the suspension by centrifugation was determined (18). For the reference suspension without added lecithin a value of 0.33 ± 0.01 g cocoa butter / m^2 of the particle surface was established. When soy lecithin was used, a higher level of fat release was detected, the quantity being $0.28 \pm 0.01 g / m^2$. The proportion of immobilized fat had therefore been reduced by adsorption of the emulsifier molecules; more cocoa butter was released. As already described, the tendency of the sugar particles to agglomerate is reduced by adsorption of the emulsifier molecules (see Fig. 1). This means, in particular, that the cocoa butter previously enclosed in the dead spaces of these agglomerates is released.

This additionally released cocoa butter is now available for promoting the flow of the suspension and therefore also influences its flow properties (19). For the reference suspension without added lecithin a yield value of 1.29 Pa was determined; for the suspension with soy lecithin the value was 1.16 Pa. The model suspension without added lecithin therefore had the higher yield value. In this case there were stronger interactions between the particles, which were reduced to some extent by the addition of the emulsifier, thus also lowering the yield value. The emulsifier adsorbs to the surface of the sucrose particles; this helps the particles to glide past each other and largely prevents the formation of agglomerates. Because of the reduced interaction between the particles, the point at which the suspension begins to flow is therefore reached more easily than in the case of the suspension without the emulsifier (14,19).

6.1.1.4 Differences in functionality between soy and sunflower lecithin

Since soy lecithin is very often produced from genetically modified soy plants, there is increasing discussion of replacing it with lecithin from alternative sources such as sunflowers (20,21). However, lecithins consist of a mixture of phospholipids with differing polarities, the distribution of the phospholipids differing according to the plant species. As an example, Fig. 15 shows the differences between phospholipid distributions in soy and sunflower lecithin.

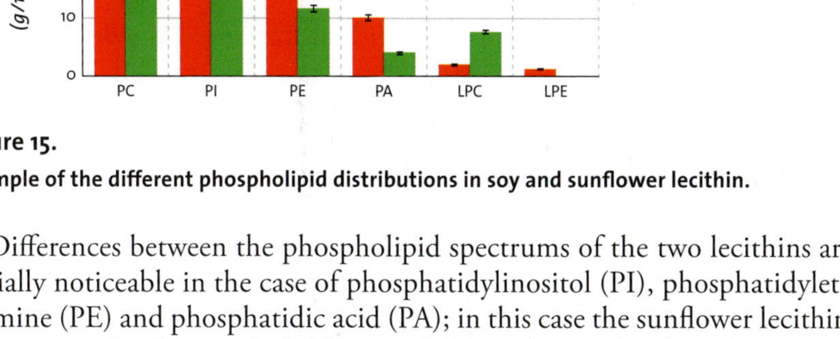

Figure 15.
Example of the different phospholipid distributions in soy and sunflower lecithin.

Differences between the phospholipid spectrums of the two lecithins are especially noticeable in the case of phosphatidylinositol (PI), phosphatidylethanolamine (PE) and phosphatidic acid (PA); in this case the sunflower lecithin has twice as much PI but only half as much PE as the soy lecithin. Because of the different phospholipid head groups, some of which differ greatly in their polarity, these two lecithins may be expected to have very different functionalities in respect of their adsorption to the surfaces of sugar particles (10).

This can already be seen from the nature of the surface coatings made visible by AFM (see Fig. 16). The particle surfaces coated with sunflower lecithin show much more sharply defined structures; in some cases, the crystalline surfaces are still easily recognizable, characterized by step-like structures (see arrows in Fig. 16 below).

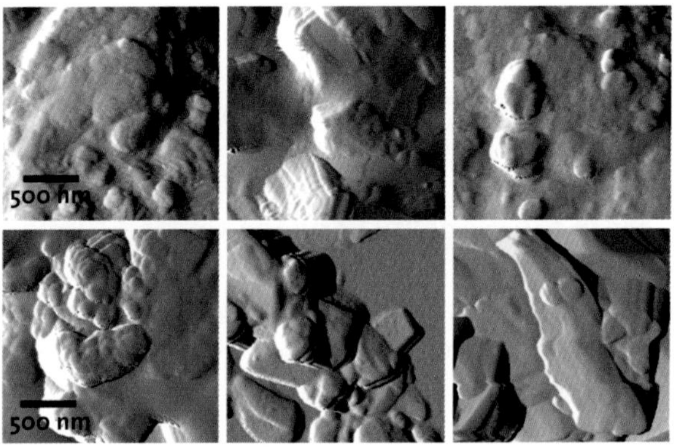

Figure 16.

Characteristic AFM surface topographies of sucrose ground in a ball mill (Ø 1.5 cm) in the presence of soy lecithin (top) and sunflower lecithin (bottom); for explanations of the markings, see text.

In contrast to this, the topographic AFM images of the surface coated with soy lecithin suggest a heavier coating of lecithin. The surfaces of the sugar particles appear smoother and show very few structures indicating an uncoated surface (see also Section 6.1.1).

The different functionalities of the two lecithins are also reflected in the different distributions of the adhesive forces (see Fig. 17 and Fig. 18; for explanations see Section 6.1.1.1). Fig. 17 shows the histogram of the force charts of particle surfaces coated with soy lecithin. Compared with the particles coated with sunflower lecithin (Fig. 18), the area of the forces is wider and can be divided into two main sections. The lower section represents forces from 10–40 nN, the upper section ranges from 30–100 nN, in some cases even beyond. Like soy lecithin, sunflower lecithin is also found to have a fairly wide force distribution (5–60 nN), with greater frequencies in the range of the smaller forces between 5 and 25 nN. This suggests that the surface is less heavily coated. Whether and to what extent the different phospholipid or fatty acid compositions of the two lecithins play a role has not yet been determined in detail.

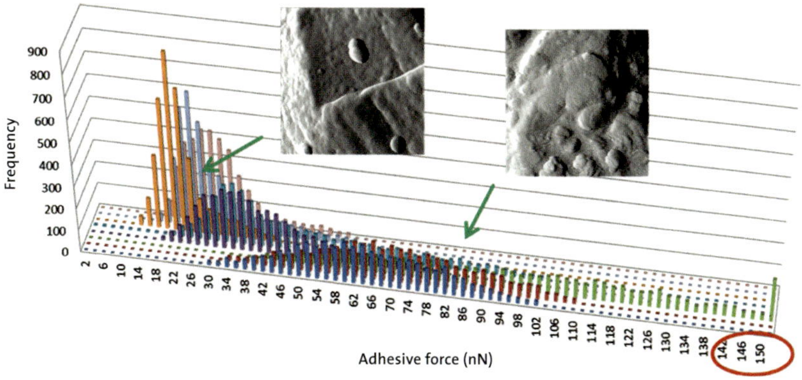

Figure 17.

Histograms from the force charts of AFM measurements of sucrose particles ground in a ball mill (ball diameter 1.5 cm) and coated with soy lecithin; size of scan: 2 × 2 µm.

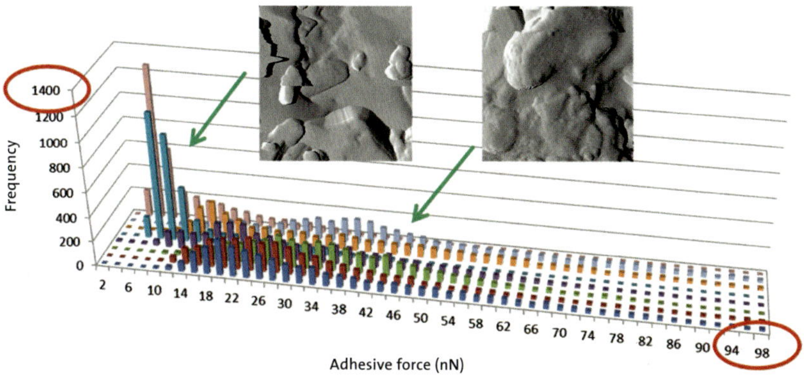

Figure 18.

Histograms from the force charts of AFM measurements of sucrose particles ground in a ball mill (ball diameter 1.5 cm) and coated with sunflower lecithin; size of scan: 2 × 2 µm.

As already described, the data acquired by AFM permit conclusions as to the thickness of the emulsifier coating (see Section 6.1.1.1). The force measurements carried out on the particle surfaces coated with lecithin were evaluated in respect of plastic deformation. This resulted in the layer thicknesses listed in Table 2.

and roller/drum-dried milk powder (RDMP) or spray-dried skimmed milk powder (SDMP) can be depicted very satisfactorily with the aid of scanning electron microscopy (see Fig. 20).

Figure 20.

Cocoa powder (left); roller-dried milk powder (centre); spray-dried milk powder (right).

Although the particles have greatly differing surface characteristics, there are no major differences in the surface energy of sugar, cocoa and dried milk particles (see Fig. 21). In all the particles the dispersive part (DP) of the surface energy predominates and is three to 4.5 times greater than the polar part (PP). There are, however, considerable differences between roller-dried milk powder (RDMP) and spray-dried skimmed milk powder (SDMP). The roller-dried powder has both lower surface energy and a larger polar part. That is due to the different mechanical and thermal effects of the drying processes.

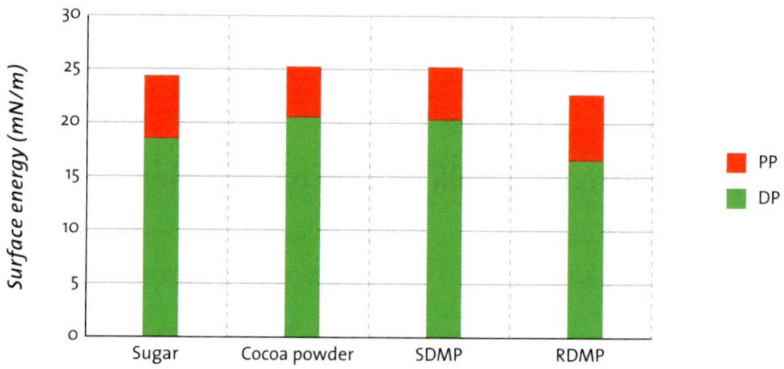

Figure 21.

Polar parts (PP) and dispersive parts (DP) of the surface energy of the solid particles.

These different surface characteristics are particularly noticeable in connection with fat immobilization. The fat immobilizing properties correlate with the surface energy (ratio of DP to PP); however, this does not apply to spray-dried skimmed milk powder because of modification of the surface in the production process (see Fig. 22).

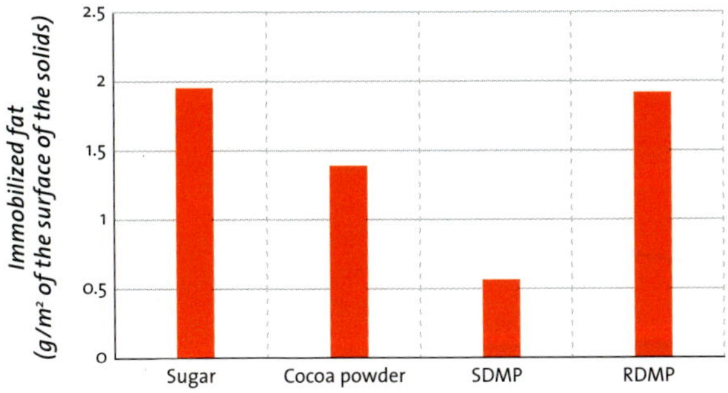

Figure 22.
Proportion of immobilized fat on the corresponding particle surfaces.

After coating of the particle surfaces with lecithin, fat is released due to modification of the boundary surfaces. However, the value for this is not constant; it depends largely on the lecithin added (see Fig. 23).

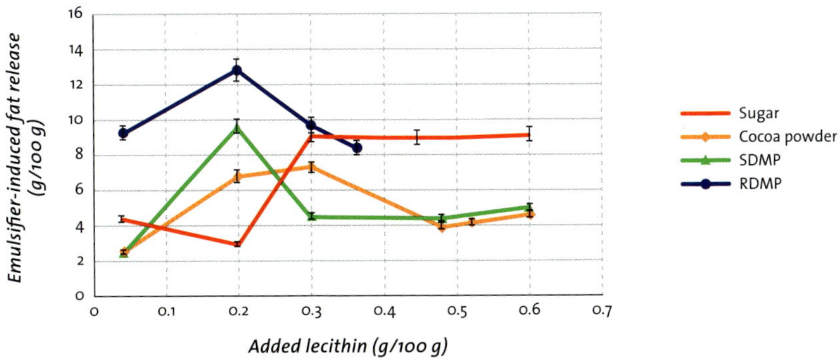

Figure 23.
Emulsifier-induced fat release from the solid particles as a function of the lecithin added.

Whereas, in the case of sugar, at least 0.3 g/100 g is required in order to achieve appreciable fat release, 0.2 g/100 g is sufficient for cocoa, SDMP and RDMP. This shows that there are clear qualitative differences in coating with the emulsifier between sugar and the other particulate constituents. This cannot be explained by their surface energy.

Coating of the particle surfaces with phospholipids is shown as a graph in Fig. 24. In the case of sugar and SDMP, coating with phospholipids is comparatively light and does not depend to any appreciable extent on the amount of lecithin added. Coating of the surfaces is particularly heavy in the case of cocoa powder.

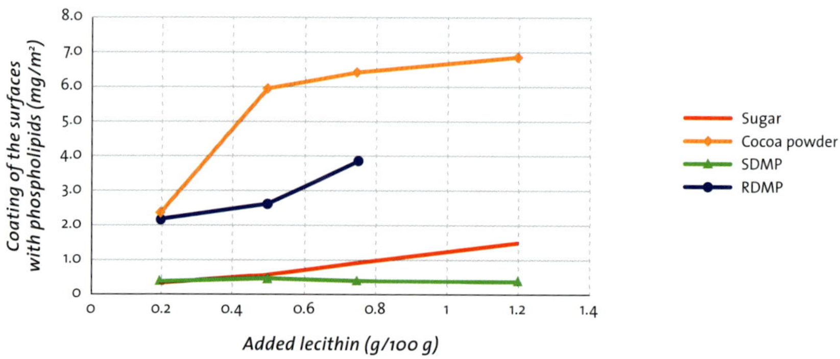

Figure 24.
Coating of the surfaces of the solid particles with phospholipids as a function of the lecithin added.

In the case of the milk powder particles, the variations in coating of the surfaces are due to the different surface characteristics. This does not only apply in terms of quantity; the ratios of the individual phospholipid (PL) fractions also differ appreciably (see Fig. 25).

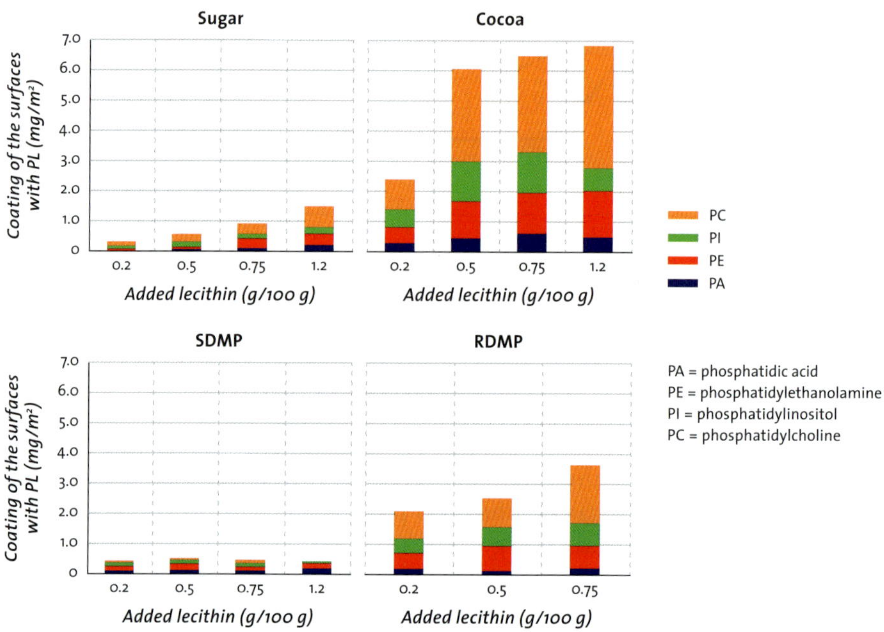

Figure 25.
Coating of the surfaces of the solid particles with the individual phospholipid fractions as a function of the lecithin added (g/100 g).

As expected, phosphatidylcholine (PC) is the fraction with the highest level of surface coating. The more lecithin is added, the more PC displaces phosphatidylinositol from the particle surfaces, whereas the adsorption of phosphatidic acid remains relatively unaffected. Here, too, the exception is spray-dried skimmed milk powder, where coating of the surface with PL is very slight and practically no PC adsorbs to the surface of the particles.

The interaction of the phospholipids with the different solid particles can be demonstrated by confocal laser scanning microscopy (CLSM) (see Fig. 26). This technique also makes the substantial quantitative differences visible. Whereas cocoa particles (top row) and roller-dried milk powder (centre row) are seen to contain phospholipids not only on the surface (particles coloured throughout), the coating of the surface of the spray-dried milk powder is incomplete (bottom row).

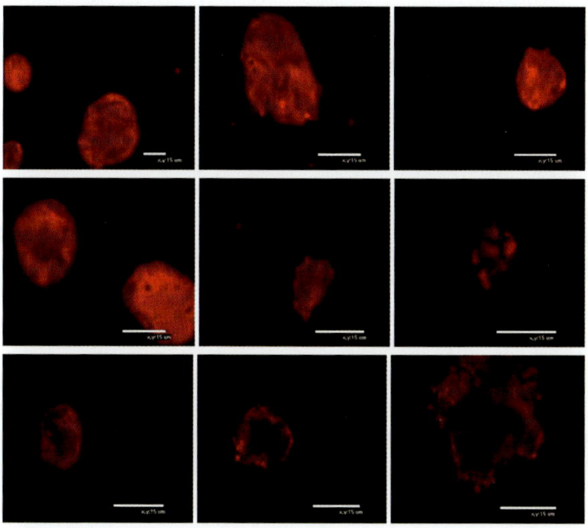

Figure 26.

CLSM images of cocoa particles coated with phospholipids (top row), roller-dried milk powder (centre row) and spray-dried milk powder (bottom row).

References

1. Binnig, G., Quate, C., Gerber, C., 1986. Atomic force microscope. Physical Review Letters 56(9), 930-933. Retrieved from http://www.ncbi.nlm.nih.gov/pubmed/10033323
2. Eaton, P., West, P., 2010. Atomic Force Microscopy. Oxford: Oxford University Press.
3. Putman, C., 1992. A detailed analysis of the optical beam deflection technique for use in atomic force microscopy. J. Appl. Phys. 72(1), 6.
4. Butt, H.-J., Cappella, B., Kappl, M., 2005. Force measurements with the atomic force microscope: Technique, interpretation and applications. Surf. Science Reports 59(1-6), 1-152.
5. Berger, R., Butt, H.-J., Retschke, M.B., Weber, S.A.L., 2009. Electrical Modes in Scanning Probe Microscopy. Macromol. Rapid Commun. 30(14), 1167-1178.
6. Beevers, C.A., McDonald, T.R.R., Robertson, J.H., Stern, F., 1952. The crystal structure of sucrose. Acta Cryst. 5(5), 689-690.
7. Afoakwa, E.O., 2010. Chocolate Science and Technology. Oxford: John Wiley & Sons.
8. Beckett, S., 2009. Industrial chocolate manufacture and use. Beckett, S.T. (Ed.) (4th Edition). New York: John Wiley & Sons.
9. Dürrenberger, M.B., Handschin, S., Conde-Petit, B., Escher, F., 2001. Visualization of Food Structure by Confocal Laser Scanning Microscopy (CLSM). Lebensmittel-Wissenschaft & Technologie 34(1), 11-17.
10. Middendorf, D., Bindrich, U., Mischnick, P., Juadjur, A., Franke, K., Heinz, V., 2016. Atomic Force Microscopy study on the effect of different lecithins in cocoa-butter based suspensions. Colloids Surf. A 499, 60-68.
11. Kindlein, M., Elts, E., Briesen, H., 2018. Phospholipids in chocolate: Structural insights and mechanistic explanations of rheological behaviour by coarse-grained molecular dynamics simulations. J. Food Eng. 228, 118-127.
12. Brown, G.M., Levy, H.A., 1973. Further refinement of the structure of sucrose based on neutron-diffraction data. Acta Cryst. B 29(4), 790-797.
13. Middendorf, D., 2015. Rasterkraftmikroskopische Untersuchungen zur Adsorption von Emulgatoren an Saccharosepartikeloberflächen in kakaobutterbasierten Suspensionen. Technical University Braunschweig.
14. Windhab, E., 2000. Fluid immobilization – a structure-related key mechanism for the viscous flow behaviour of concentrated suspension systems. Appl. Rheol. 10(3), 134-144.
15. Afoakwa, E.O., Paterson, A., Fowler, M., 2007. Factors influencing rheological and textural qualities in chocolate – a review. Trends Food Sci. Technol. 18(6), 290-298.
16. Beckett, S., 2000. Controlling the Flow Properties of Liquid Chocolate. In The Science of Chocolate, 66-84. Cambridge: The Royal Society of Chemistry.
17. Bolenz, S., Holm, M., Langkrär, C., 2014. Improving particle size distribution and flow properties of milk chocolate produced by ball mill and blending. Eur. Food Res. Technol., 139-147.
18. Franke, K., Heinzelmann, K., 2008. Structure improvement of milk powder for chocolate processing. Int. Dairy J. 18(9), 928-931.
19. Franke, K., Bindrich, U., 2009. Influence of solid particle surface properties on adsorption of surface active components in lipophilic suspensions. In 5th International Symposium on Food Rheology and Structure, 770-771, Zurich.
20. Nguyen, M.T., Van De Walle, D., Petit, C., Beheydt, B., Depypere, F., Dewettinck, K., 2014. Mapping the chemical variability of vegetable lecithins. J. Am. Oil Chem. Soc. 91(7), 1093-1101.
21. Van Nieuwenhuyzen, W., Tomás, M.C., 2008. Update on vegetable lecithin and phospholipid technologies. Eur. J. Lipid Sci. Technol. 110(5), 472-486.
22. Buxmann, W., Bindrich, U., Strijowski, U., Heinz, V., Knorr, D., Franke, K., 2010. Influencing emulsifying properties of egg yolk by enzymatic modification by phospholipase D from Streptomyces chromofuscus Part 1: Technological properties of incubated egg yolk. Colloids Surf. B, Biointerfaces 76(1), 192-8.

6.2 Lecithins in chocolate

Gunther Arnold, Yvonne Schneider, Harald Rohm

Chocolate is a suspension containing sugar and cocoa, and also milk powder in the case of milk chocolate, as the dispersed phase, and cocoa butter (plus free milk fat in milk chocolate) as the dispersion medium (continuous phase).

Generally speaking, the basic steps in chocolate manufacture are mixing, grinding, conching, tempering and moulding. After the relevant parts of the cocoa beans have been ground into a cocoa liquor and mixed with sugar, milk powder and other ingredients according to the recipe, the mass is finely ground to a maximum particle size of 15–35 µm in two-roll and five-roll refiners (1). The mass is then sheared intensively in a conche to achieve the desired flavour and texture. At the end of the conching process, more cocoa butter is added along with surfactants[1], depending on the recipe. If these ingredients are added earlier they may inhibit the removal of water and steam-volatile substances (e.g. acetic acid) and thus have a negative effect on flavour formation (1,2). Subsequent tempering (pre-crystallization) serves to form discrete fat crystals in the stable ß(V) modification. As a result, the cocoa butter in the moulded product generally crystallizes in the ß(V) modification during cooling and storage, giving the chocolate the desired gloss and melting and breaking properties. Moreover, when allowed to set in this modification the chocolate is less prone to fat bloom and therefore has a longer shelf life (1).

6.2.1 Rheology

During chocolate processing, the continuous phase is in a completely or partially molten state (formation of fat crystals during tempering). In this state, chocolate exhibits the pseudoplastic behaviour typical of concentrated suspensions, characterized by a reduction in apparent viscosity with increasing shear rate (shear thinning) (Fig. 1).

[1] Under the food laws the most widely used surfactant, lecithin, is classified as an emulsifier and therefore declared as such. In this context, however, the emphasis is on techno-functional effects, namely modification of the surface properties of the dispersed phase, through the adsorption of surfactants (3) and the resulting influence on interaction between the hydrophilic solid particles and the non-polar continuous phase.

The rheological properties of chocolate are extremely important for its processing and handling performance (transportation, dosing, moulding, coating), and also for its mouth feel and melting characteristics. These are influenced by the volume fraction and particle size distribution of the dispersed phase, the intensity of conching and the moisture content. Since these parameters can only be varied to a limited extent, the use of surfactants such as lecithin is crucial for achieving optimum adjustment of flow behaviour.

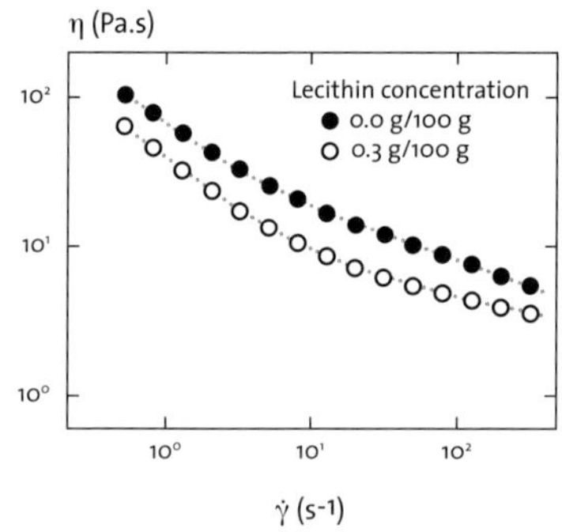

Figure 1.

Apparent viscosity (η) as a function of shear rate ($\dot{\gamma}$) for a dark chocolate mass (fat content: 34 g/100 g) in relation to the lecithin concentration.

To describe the rheological properties of chocolate masses, rheological models (Casson model, Herschel – Bulkley model, Windhab model) are used or values are derived directly from flow or viscosity curves. The objective of all approaches is to obtain comparative values that characterize the rheological behaviour of chocolate during transportation and dosing and also in an unstressed state (e.g. spreading of a melted coating). Although some authors (4,5) recommend using values taken directly from the measurement curve, the models are still used in the current literature (6-10). Moreover, the determination of model parameters is a requirement for the numerical simulation of processing steps. From the functional relationship between shear rate ($\dot{\gamma}$) and shear stress (τ) it is then possible to extrapolate parameters for describing the yield stress and apparent viscosity outside the measuring range.

6.2.1.1 Mechanisms of action

The addition of lecithin lowers yield stress and apparent viscosity of liquid chocolate. At a lecithin concentration of $C_{Lec} = 0.3 - 0.6$ g/100 g the surfactant achieves maximum lowering of the yield stress while simultaneously reducing apparent viscosity (Fig. 2). A further increase in the lecithin concentration has little effect on the apparent viscosity and may cause the yield stress to rise again, as in the right chart of Fig. 2 (6-8,10-12).

The reason for this functionality is an accumulation of the surfactant components of the lecithin at the interface between hydrophilic solids and the lipophilic liquid phase of the suspension (11,13-15). This is coupled with an increase in the lipophilicity of the surfaces of the solids and facilitates wetting of the dispersed phase with lipophilic dispersant. Some authors (16-19) see a possible reason for the effect of lecithin in the adsorption of lecithin components to the surface of the solids which causes the release of fat immobilized on the solid material and thus an increase in the free-flowing dispersant. According to Dedinaite et al. (20), the adsorbed surfactants may act as a lubricant film, lessening the friction between the particles and thus reducing the forces necessary for shearing. Harris (13) already had speculated that the surfactants counteract a direct contact between the surfaces of the solids. Sedimentation tests with sugar/oil suspensions and investigations using atomic force microscopy have shown that phospholipid fractions and native lecithin reduce attractive forces between the surfaces of sugar particles (21,22,23). We may deduce from this finding that lecithin inhibits the formation of agglomerates, and that during shearing it promotes the dissolution of agglomerates into their primary particles and the release of continuous phase immobilized in the agglomerates. These lecithin effects would facilitate the movement of particles in relation to each other and reduce the apparent viscosity (24).

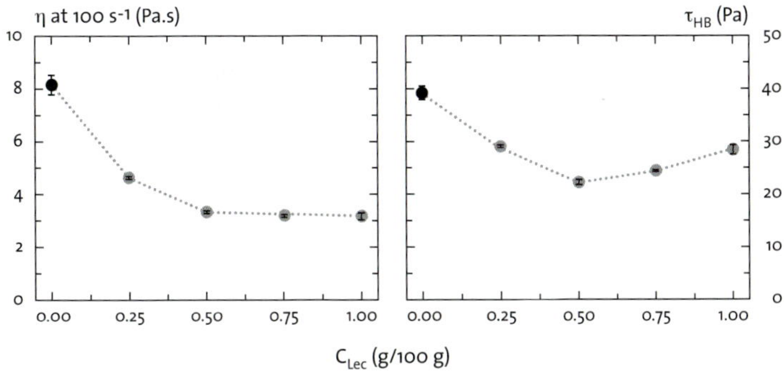

Figure 2.

Apparent viscosity (ηs) at $\dot{\gamma}$ = 100 s^{-1} (left) and Herschel – Bulkley yield stress (τ_{HB}) of a dark chocolate mass as a function of the lecithin concentration (C_{Lec}).

The subsequent rise in the yield stress as the lecithin concentration increases (Fig. 2, right) can be explained by the achievement of complete monomolecular loading of the particle surfaces. Excess surfactant molecules apparently form partial double layers on the particle surfaces or aggregates (micelles) in the dispersion medium; these have a structure-forming effect and promote interaction between the particles (11,25,26).

6.2.1.2 Influence of lecithin composition

Lecithin preparations vary in respect of their phospholipid and glycolipid content and the percentages of the individual surfactant fractions. When comparing different lecithin preparations it is advisable to consider the actual proportion of surfactants they contain in order to prevent misinterpretations of their functionality.

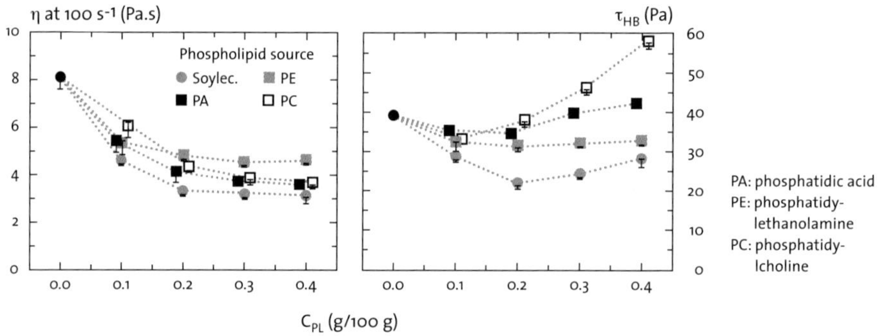

Figure 3.

Apparent viscosity (η_s) at $\dot{\gamma}$ = 100 s^{-1} (left) and Herschel – Bulkley yield stress (τ_{HB}) (right) of a dark chocolate mass as a function of the type and concentration (shown as the phospholipid concentration (CPL)) of the phospholipid source.

According to Nieuwenhuyzen (27,28), the phospholipid composition may affect the techno-functional properties of lecithin, and enriching lecithin with specific phospholipid fractions may lower yield stress and apparent viscosity to different degrees. It was found that lecithin enriched with phosphatidylcholine lowered the viscosity of dark chocolate masses further than lecithin enriched with phosphatidylinositol or phosphatidylethanolamine, but was less effective in lowering the yield stress (27,28). Tests with enzymatically produced phosphatidyl fractions and soy lecithin in dark chocolate masses show that the phospholipid fractions lower yield stress and apparent viscosity to a lesser degree, and that at higher dosages they cause a much more pronounced rise in the yield stress than does native lecithin (Fig. 3). Possible reasons for the differences in effect between the various phospholipid fractions and the native lecithin preparation are synergisms between the surfactants present in the lecithin and also differences in solubility and adsorption properties.

Most of the studies of the effect of lecithin in chocolate are restricted to the use of soy lecithin with about 60 g/100 g acetone-insoluble matter and a phospholipid content of about 40–45 g / 100 g. Lecithins derived from sunflower and rapeseed are also commercially available, but their phospholipid composition may differ slightly from that of soy lecithin (29). However, Fig. 4 shows that there is little difference between soy, rapeseed and sunflower lecithin in respect of maximum reduction of the yield stress and apparent viscosity. These lecithins are therefore a viable alternative to soy lecithin (30).

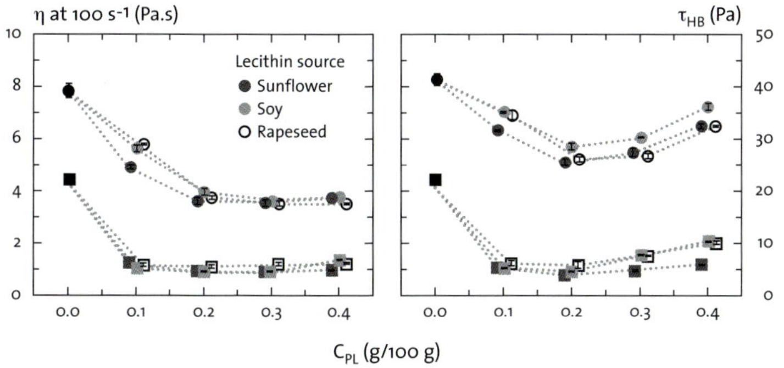

Source: Arnold et al., *Süßwaren* (31).

Figure 4.

Apparent viscosity (ηs) at $\dot{\gamma}$ = 100 s⁻¹ (left) and Herschel – Bulkley yield stress (τ_{HB}) (right) of dark (circles) and milk (squares) chocolate mass as a function of the type of lecithin used and the lecithin concentration in the chocolate (shown as phospholipid concentration, C_{PL}).

6.2.1.3 Other surfactants, synergisms

Besides lecithin, there are other surfactants that can be used for adjusting the flow properties of chocolate. They include ammonium phosphatides (E442), usually produced from rapeseed oil by interesterification, phosphorylation and neutralization with ammonium. These compounds are practically tasteless and odourless (11,32). In chocolate, ammonium phosphatides reduce the rheological parameters to much the same extent as lecithin, but up to a concentration of 0.6 g/100 g they do not cause the yield stress to rise again (11,33).

Another alternative is polyglycerol polyricinoleate (PGPR, E476), obtained chemically by condensation of ricinoleic acid, production of polyglycerol and partial esterification of polyricinoleic acid (34). PGPR is more effective than lecithin in lowering the yield stress and does not cause this parameter to increase again at higher concentrations. On the other hand, its effect on viscosity in the higher shear rate range is slightly less effective than that of lecithin (12,35).

Mixing different surfactants makes it possible to utilize synergistic effects. Schantz and Rohm (12) demonstrated that the optimum mixture ratios of lecithin to PGPR differ according to the target value and the type of chocolate. They found the optimum lecithin-to-PGPR ratio for minimizing the yield stress to be 30:70, whereas the maximum possible reduction of apparent viscosity in the upper shear rate range was achieved at a ratio of 50:50 for dark chocolate and 75:25 for milk chocolate.

6.2.2 Texture

Besides odour, flavour and melt, the texture of chocolate is a quality attribute perceived by customers and readily apparent in its breaking properties and mouth feel. The addition of lecithin reduces the hardness of the chocolate, presumably because lecithin facilitates wetting of the dispersed phase and its surfactant constituents promote the dissolution of agglomerates (36). Tests on dark and milk chocolate samples have shown that rapeseed and sunflower lecithins reduce the hardness to much the same degree as soy lecithin (Fig. 5) (31).

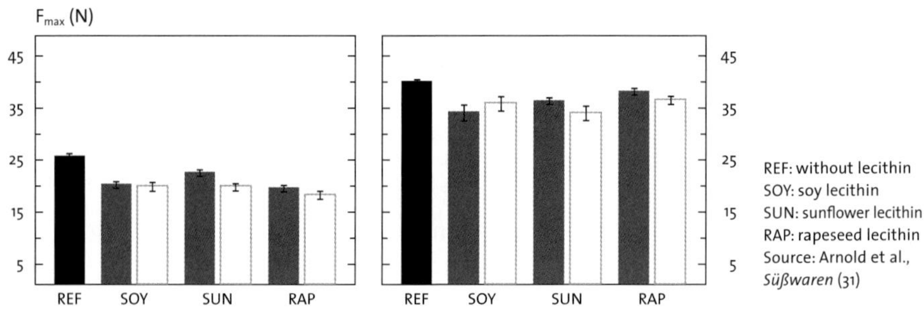

Figure 5.

Hardness (shown as maximum penetration force F_{max}) of milk chocolate (left) and dark chocolate (right) as a function of the concentration (filled – 0.3 g/100 g; unfilled – 0.6 g/100 g) and source of the lecithin.

The use of lecithin may have an adverse effect on the shelf life of products with a filling containing alcohol, since surfactants increase the tendency of water and alcohol to migrate in chocolate and compounds containing cocoa (37,38). Consequently, chocolate containing lecithin swells and softens sooner under the effects of such fillings, which in turn impairs the enjoyment and saleability of the products (39,40). Tests carried out by Schantz et al. (33) also show that, compared to lecithin, PGPR and ammonium phosphatides delay the structural breakdown of dark and milk chocolate caused by alcoholic fillings. In the case of chocolate products with fillings containing fat, on the other hand, Schantz et al. (33) and Choi et al. (41) observed no significant effect of lecithin on the migration of fat and reduction of the hardness of the samples.

6.2.3 Sensory considerations

Both the mouth feel and perception of the taste of chocolate depend on its rheological properties (1). Lecithin may therefore influence the sensory properties of chocolate by changing its rheology and because of its own taste and odour. Attributes typically used to describe the sensory properties of lecithin are hay-like, straw-like, grassy, rusty, nutty, earthy, malty and mushroom-like (42).

At low lecithin concentrations ($C_{Lec} \approx 0.3$ g/100 g), no sensory differences between a lecithin-free reference and the sample containing lecithin were perceived, but upwards of a lecithin concentration of 0.6 g/100 g, significant differences were detected, depending on the type of chocolate (30,31). The same investigations showed that it is possible to substitute rapeseed or sunflower lecithin for soy lecithin in dark or milk chocolate without significantly affecting its popularity. A further increase in the lecithin concentration to 1 – 2 g/100 g may have an adverse effect on acceptance. This is presumably because the characteristic lecithin taste is more conspicuous, and consumers perceive it as an off-taste (9,43).

References

1. Beckett, S.T., 2008. The Science of Chocolate, 2nd Edition. Royal Society of Chemistry, Cambridge.
2. Ziegleder, G., 2009. Flavour development in cocoa and chocolate, in: Beckett, S.T. (Ed.), Industrial Chocolate Manufacture and Use. Wiley-Blackwell, Chichester, 169-191.
3. Rousset, Ph., Sellappan, P., Daoud, P., 2002. Effect of emulsifiers on surface properties of sucrose by inverse gas chromatography. Journal of Chromatography A 969, 97-101.
4. Aeschlimann, J.-M., Beckett, S.T., 2000. International inter-laboratory trials to determine the factors affecting the measurement of chocolate viscosity. Journal of Texture Studies 31, 541-576.
5. Servais, C., Ranc, H., Roberts, I.D., 2004. Determination of chocolate viscosity. Journal of Texture Studies 34, 467-497.
6. Afoakwa, E.O., Paterson, A., Fowler, M., 2008a. Effects of particle size distribution and composition on rheological properties of dark chocolate. European Food Research and Technology 226, 1259-1268.
7. De Graef, V., Depypere, F., Minnaert, M., Dewettinck, K., 2011. Chocolate yield stress as measured by oscillatory rheology. Food Research International 44, 2660-2665.
8. Karnjanolarn, R., McCarthy, K.L., 2006. Rheology of different formulations of milk chocolate and the effect on coating thickness. Journal of Texture Studies 37, 668-680.
9. Nebesny, E., Zyzelewicz, D., 2005. Effect of lecithin concentration on properties of sucrose-free chocolate masses sweetened with isomalt. European Food Research and Technology 220, 131-135.
10. Sokmen, A., Gunes, G., 2006. Influence of some bulk sweeteners on rheological properties of chocolate. LWT – Food Science and Technology 39, 1053-1058.
11. Beckett, S.T., 2009. Chocolate flow properties, in: Beckett, S.T. (Ed.), Industrial Chocolate Manufacture and Use. Wiley-Blackwell, Chichester, 224-246.
12. Schantz, B., Rohm, H., 2005. Influence of lecithin-PGPR blends on the rheological properties of chocolate. LWT – Food Science and Technology 38, 41-45.
13. Harris, T.L., 1968. Surface active lipids in chocolate. SCI Monograph 32, 108-122.
14. Nieuwenhuyzen, W. v., Szuhaj, B.F., 1998. Effects of lecithins and proteins on the stability of emulsions. Lipid/Fett 100, 282-291.
15. Schuster, G., 1985. Emulgatoren für Lebensmittel. Akademie-Verlag, Berlin.
16. Aylward, F., 1960. Chocolate. In: Texture in Foods. Society of Chemical Industry, 75-88.
17. Moran, D.P.J., 1969. The influence of some surfactants on the viscosity of confectionery fat/sugar mixtures. Revue Internationale de la Chocolaterie 24, 478-482.
18. Dimick, P.S., Hoskin, J.M., 1981. Chemico-physical aspects of chocolate processing – a review. Canadian Institute of Food Science and Technology Journal 14, 269-282.
19. Franke, K., Heinzelmann, K., Bindrich, U., 2011. Kakaobutterbindung an die Feststoffe und die Wirkung der Emulgatoren in der aufgeschmolzenen Schokoladenmasse. Süsswaren Technik und Wirtschaft 56, 14-17.
20. Dedinaite, A., Claesson, P.M., Campbell, B., Mays, H., 1998. Interactions between modified mica surfaces in triglyceride media. Langmuir 14, 5546-5554.
21. Arnold, G., Schade, E., Schneider, Y., Friedrichs, J., Babick, F., Werner, C., Rohm, H., 2014. Influence of individual phospholipids on the physical properties of oil-based suspensions. Journal of the American Oil Chemists' Society 91, 71-77.
22. Arnold, G., Schuldt, S., Schneider, Y., Friedrichs, J., Babick, F., Werner, C., Rohm, H., 2013. The impact of lecithin on rheology, sedimentation and particle interactions in oil-based dispersions. Colloids and Surfaces A: Physicochemical and Engineering Aspects 418, 147-156.
23. Johansson, D., Bergenståhl, B., 1992. The influence of food emulsifiers on fat and sugar dispersions in oils. 1. Adsorption, sedimentation. Journal of the American Oil Chemists' Society 69, 705-717.
24. Windhab, E.J., 2000. Fluid immobilization – a structure-related key mechanism for the viscous flow behaviour of concentrated suspension systems. Applied Rheology 10, 134-144.
25. Schantz, B., 2003. Zur Wirkung oberflächenaktiver Substanzen in Schokolade auf Verarbeitungseigenschaften und Endprodukt (Dissertation). Technische Universität Dresden, Dresden.
26. Vernier, F.C., 1997. Influence of Emulsifiers on the Rheology of Chocolate and Suspensions of Cocoa or Sugar Particles in Oil (Dissertation). University of Reading.
27. Nieuwenhuyzen, W. v., 1995. Lecithin fractions for novel foods. Food Technology International Europe 40, 47-50.

28. Nieuwenhuyzen, W. v., 1997. Funktionalität von Lecithinen. Lipid/Fett 99, 10-14.
29. Schneider, M., 2008. Major sources, composition and processing, in: Gunstone, F.D. (Ed.), Phospholipid Technology and Applications. Oily Press, Bridgwater, 21-40.
30. Böhme, B., Symmank, C., Rohm, H., 2016. Physical and sensory properties of chocolate made with lecithin of different origin. European Journal of Lipid Science and Technology 118, 1839-1845.
31. Arnold, G., Könk, U., Böhme, B., Schneider, Y., Rohm, H., 2012. Lecithine verschiedener Herkunft in Schokolade. Süsswaren Technik und Wirtschaft 57, 15-17.
32. Gaunt, I.F., Grasso, P., Gangolli, S.D., 1967. Short-term toxicity study of emulsifier YN in rats. Food and Cosmetics Toxicology 5, 623-629.
33. Schantz, B., Linke, L., Rohm, H., 2003. Effects of different emulsifiers on rheological and physical properties of chocolate. Proceedings ISFRS 3, 329-333.
34. Wilson, R., van Schie, B.J., Howes, D., 1998. Overview of the preparation, use and biological studies on polyglycerol polyricinoleate (PGPR). Food and Chemical Toxicology 36, 711-718.
35. Aguilar, C.A., Ziegler, G.R., 1995. Viscosity of molten milk chocolate with lactose from spray-dried whole-milk powders. Journal of Food Science 60, 120-124.
36. Afoakwa, E.O., Paterson, A., Fowler, M., Vieira, J., 2008b. Particle size distribution and compositional effects on textural properties and appearance of dark chocolates. Journal of Food Engineering 87, 181-190.
37. Ghosh, V., Ziegler, G.R., Anantheswaran, R.C., 2002. Fat, moisture, and ethanol migration through chocolates and confectionary coatings. Critical Reviews in Food Science and Nutrition 42, 583-626.
38. Ghosh, V., Ziegler, G.R., Anantheswaran, R.C., 2005. Moisture migration through chocolate-flavored confectionery coatings. Journal of Food Engineering 66, 177-186.
39. Robinson, L., 1971. Veränderungen in krustenlosen Pralinen mit alkoholhaltiger flüssiger Füllung I. Fette, Seifen, Anstrichmittel 73, 521-526.
40. Zürcher, K., 1971. Veränderungen in krustenlosen Pralinen mit alkoholhaltiger flüssiger Füllung II: Ergänzende Untersuchungen an unter Betriebsbedingungen hergestellten Weinbrandbohnen. Fette, Seifen, Anstrichmittel 73, 527-529.
41. Choi, Y.J., McCarthy, K.L., McCarthy, M.J., 2005. Oil migration in a chocolate confectionery system evaluated by magnetic resonance imaging. Journal of Food Science 70, E312-E317.
42. Stephan, A., Steinhart, H., 1999. Identification of character impact odorants of different soybean lecithins. J. Agric. Food Chem. 47, 2854-2859.
43. Prawira, M., Barringer, S.A., 2009. Effects of conching time and ingredients on preference of milk chocolate. Journal of Food Processing and Preservation 33, 571-589.

6.3 The effects of lecithins on bakery products

Gerhard Helmerich

6.3.1 Influencing the baking properties of wheat flour by adding lecithin

In baking technology it is preferable to use deoiled lecithins in powder form since they are easier to dose than viscous native lecithins. The technical effect of the lecithins differs according to the type of bakery product concerned. In the case of yeast-leavened wheat doughs, lecithins are known to improve the gas retention capacity of the dough and thus its fermentation stability, and this in turn results in a larger volume and an even texture (1,2). For the complex requirements of freezing the most suitable choice is enzymatically hydrolysed lecithins. Being hydrophilic, they inhibit the growth of large ice crystals and thus protect the yeast cells and the gluten network against mechanical damage (3).

In pastry goods, lecithin ensures better distribution of the basic raw materials, good emulsification of the fat constituents, better processing properties (2) and more even browning of the baked products due to the Maillard reaction of the amino groups of certain phospholipids (PE, PS) with reducing carbohydrates. In the case of wafers the addition of lecithin ensures clean release from the irons and enhances the texture of the wafers themselves (4). For prolonging the shelf life of baked goods, hydrolysed lecithins are of particular interest (5). Like the mono- and diglycerides, they form inclusion compounds with the amylose.

The paper by Working in 1924 (6) was the first publication to focus on the effect of raw lecithins on dough and gluten. Numerous studies, listed in Table 1, confirmed that some lecithins (e.g. liquid and deoiled soy lecithin) have a positive effect on baked volume by changing the viscosity of the dough. Lecithins improve the oven rise values and fermentation behaviour of doughs; the crumb of the finished bread is softer and the texture finer than in loaves without such an addition.

Lecithins increase the kneading tolerance of doughs. The addition of lecithins enables bakeries to adjust flours of different qualities to a quality that results in better and reproducible results of the baking process and uniform baked goods.

A further effect observed when lecithins are used in bakery products is a reduction of the time required for final proofing, whereas the DATEM emulsifier and monoglycerides, for example, lengthen this step in the process (7). The reason for this is interactions between the emulsifiers and the gluten, which prolong the final proof and result in a longer fermentation time (7,8).

The larger volume of carbon dioxide resulting from fermentation expands correspondingly during the baking process and results in greater oven rise. The use of lecithins and certain lecithin fractions influences further parameters that fall under the heading of 'texture': the size of the pores, the firmness and elasticity of the crumb and changes in the latter during storage (9,10).

According to Pomeranz et al. (11), the use of hydrolysed lecithins lengthens the dough development time. According to Munz and Brabender (12), the use of 1% soy lecithin increases the stability of the dough and simultaneously reduces its extensibility. Monoglycerides were reported to have the opposite effect (12). The use of triacylglycerides (triglycerides) reduced both the stability of the dough and its extensibility (14).

6.3.2 The baking activity of different lecithins

6.3.2.1 Preliminary remarks

This section contains a brief summary of the results published by Helmerich and Köhler in 2004 (15). The results given were generated in micro-baking trials, and the main conclusions drawn from them were then verified in normal baking trials.

As a first step, the optimum usage level of the eight lecithins considered was to be determined with the aid of micro-baking trials in order to subsequently investigate their efficacy in the optimum range. In conjunction with the quantification data (described in Chapter 5.2), it was hoped that the information thus acquired would help to clarify the structure – effect relationships of phospholipids in baked goods.

Besides the volume of the bread, other measurement parameters such as the shape of the loaves, the pore size of the crumb and the consistency of the crust were included in the assessment of baking activity. Furthermore, the influence of the lecithins on the properties of the crumb was investigated as a function of the length of storage.

The parameter chosen for assessing the baking efficacy of the lecithins used was the loaf volume in per cent as compared to loaves without the additive, since the change in loaf volume as a measure of the mechanical strength, fermentation tolerance and gas retention capacity of a dough is the most important parameter for rating the quality of baked goods (16).

Table 1.

Studies relating to the baking activity of phospholipids and lecithins.

Author	Year	Results
Working (6)	1924	Phosphatides impair the gluten and the baking properties of wheat flour.
Sullivan (17)	1940	Lecithin improves handling of the dough but has hardly any effect on the baking quality of wheat flour.
Pratt (18)	1946	Lecithin increases kneading tolerance but has hardly any effect on the baking activity of wheat flour.
Olcott and Mecham (19)	1947	During paste formation the proportion of unbound flour lipids falls by one-third, mainly phospholipids being bound to the gluten.
Coppock et al. (20)	1954	The addition of acetone-insoluble phosphatides to defatted flour increases the loaf volume.
Adler and Pomeranz (21)	1959	Lecithins improve the loaf volume and crumb of bread containing soy flour.
Grosskreutz (22)	1961	Model of the structure of the lipoproteins in wheat gluten: protein chains bind to the phospholipid bilayer.
Zentner (23)	1964	Phosphatidyl derivatives of the phospholipids in wheat flour bind to the proteins and thus help to stabilize the dough.
Pomeranz (24)	1966	A lecithin fraction with a PC/PE ratio of 2 to 1 (% m/m) shows very good baking activity.
Pomeranz et al. (11)	1970	Individually added phospholipids enlarge the volume of the loaves; PI shows the best effect.
Wassermann (25)	1983	The baking activity of lysolecithin is comparable to that of the emulsifier DATEM.
Jodlbauer (26,27)	1987 a, b	Patenting of lecithin fractions in which the ratio of PC to PE is smaller or greater than 1.
Adams et al. (16)	1991	From a usage level > 0.75% lecithin in wheat flour, negative effects occur (poor fermentation stability, tough crumb, soapy taste).
Mettler et al. (7,8)	1991 a, b	A correlation can be seen between oven rise, fermentation behaviour and larger loaf volume through the use of lecithins.
Jodlbauer et al. (9)	1992	Certain components of lecithin 'cannibalize' each other in the baking process; this again emphasizes the outstanding importance of the ratio of PC to PE.
Schneider (28)	1997	Patent for the use of lecithins for adjusting to uniform flour properties.

6.3.2.2 Micro-baking trials: Determining the optimum usage levels for lecithins

To determine the optimum amounts of lecithins to be added, liquid and deoiled soy, sunflower and rapeseed lecithin and also egg and oat lecithin were added in concentrations of 0.5 to 3.0%, and the volume of the loaves made in this way was measured. The increases in loaf volume thus determined are shown in Fig. 1, in comparison with loaves without added lecithin.

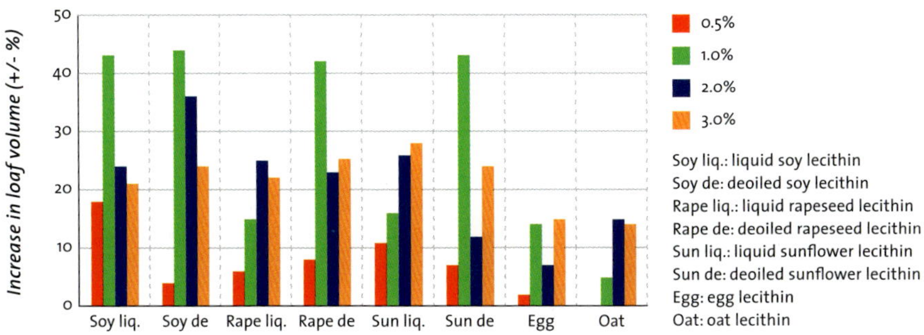

Figure 1.

Increase in volume, in %, of loaves to which lecithins derived from different raw materials have been added at 0.5 to 3.0% in comparison with loaves without added lecithin.

In the concentration range from 0.5 to 3.0% (m/10 g flour), all eight lecithins used increased the loaf volume compared to loaves without added lecithin. Only oat lecithin at a concentration of 0.5% failed to increase the loaf volume. The greatest increase in loaf volume resulted from the use of liquid and deoiled soy lecithin, deoiled sunflower lecithin and deoiled rapeseed lecithin at a concentration of 1.0%. These four lecithins all achieved a loaf volume increase of 42–44%. The smallest increase in loaf volume resulted from the use of egg and oat lecithin, the maximum increase for these lecithins being 15%. On the basis of the above results and in keeping with the data from the literature (16), the optimum usage level for lecithin in further trials for baking efficacy was assumed to be 0.2 to 1.0%.

6.3.2.3 Determining the lecithins' baking efficacy in the optimum range

The baking efficacy of the lecithins tested was investigated more closely at an addition of 0.2 to 1.0% (m/10 g flour) and the increase in loaf volume measured in comparison with loaves without added lecithin. The increases in loaf volume thus determined are shown in Fig. 2. It emerged that liquid, deoiled soy lecithin and deoiled rapeseed lecithin achieved the greatest increase in loaf volume of all the lecithins tested at a usage level of 0.8–1.0%.

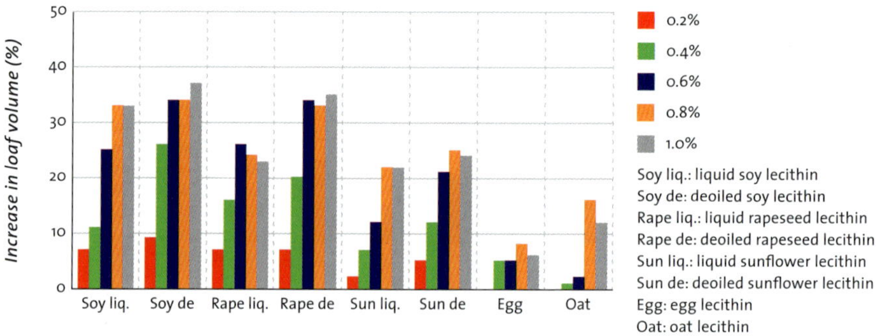

Figure 2.

Increase in loaf volume in % at an addition of 0.2 to 1.0% lecithin in comparison with loaves without added lecithin.

Deoiled sunflower lecithin resulted in a much smaller increase in loaf volume, similar to that of liquid rapeseed and sunflower lecithin. Of all the lecithins tested, egg and oat lecithin had by far the least baking efficacy, with a maximum volume increase of 16%. At an addition of 0.4%, the three deoiled lecithins showed the clearest differences in baking activity. Of these, deoiled soy lecithin achieved by far the biggest increase in loaf volume (+26%), followed by deoiled rapeseed lecithin (+20%) and deoiled sunflower lecithin (+12%).

6.3.2.4 Further measurement parameters

Besides the measurement of loaf volume, the assessment included the crumb (size, regularity of the texture), the shape of the loaves (height, diameter) and the crust (consistency, colour). Some examples of these parameters are shown in Table 2.

Table 2.

Further measurement parameters for the baking activity of the lecithins tested at usage levels of 0.2 to 1.0%.

Lecithin	Loaf shape m/m	Crust	Crumb texture
No addition	Oval, misshapen (*h* 29, *d* 49)	Crisp, golden brown	Irregular
Soy, liquid	Round (*h* 34, *d* 53)	Crisp, golden brown	Fine, regular
Soy, deoiled	Round (*h* 33, *d* 52)	Dry, brittle, golden brown	Irregular
Sunflower, deoiled	Round (*h* 34, *d* 51)	Crisp, golden brown	Irregular
Rapeseed, deoiled	Round (*h* 33, *d* 51)	Dry, brittle, golden brown	Irregular

h: height; d: diameter

The addition of lecithins at a concentration of 0.2 to 1.0% (m/10 g flour) resulted in a larger loaf diameter and much greater height than was the case in loaves without added lecithin. The loaves were regular and round in shape, the crust crisp and golden brown, except with deoiled rapeseed and soy lecithin, which resulted in dry, brittle crusts. Liquid soy lecithin produced a fine, regular crumb texture, of the kind desirable in toast loaves for example. The addition of deoiled soy, rapeseed and sunflower lecithin resulted in an irregular crumb containing both large and small pores. The higher the dosage of lecithin, the finer was the texture. The cross sections of the 10 g loaves in Fig. 3 show that the differences between the lecithins in respect of crumb texture were slight, so the sections that follow no longer go into the details of the measurement parameters described here for rating the results of baking. Loaf volume remains the most important criterion for assessing baking activity.

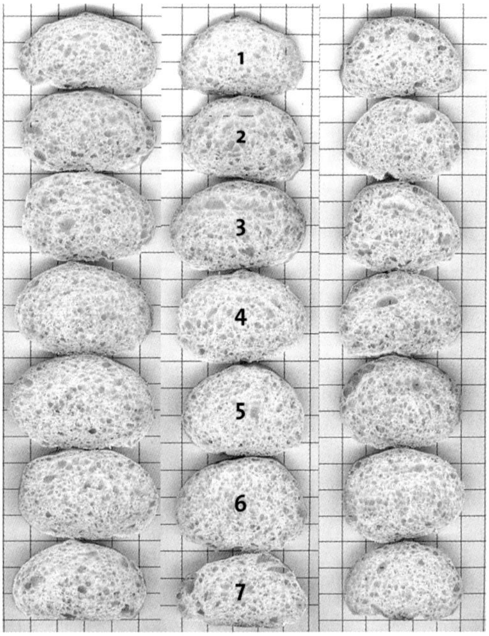

Figure 3.

Cross sections of 10 g loaves with 0.2 to 1.0% added lecithin. Loaves 1 and 7 without added lecithin, loaves 2–6 with 0.2 to 1.0% lecithin added. □ = 1 cm².

6.3.2.5 Influence of the lecithins on firmness of the crumb

In order to measure the firmness of the crumb as a function of the length of storage it was necessary to produce larger loaves. Pan loaves proved to be ideal for crumb measurement, since these had a uniform texture and therefore permitted reproducible measurements. The pan loaves were made with the addition of 0.4% (m/10 g flour) deoiled soy, rapeseed and sunflower lecithin. To investigate the changes in the crumb during storage, the loaves were kept in plastic bags at 20°C and 60% relative humidity. Every 24 hours over a three-day period, the force *F* necessary for compressing cylindrical pieces cut out of the crumb by 0.7 cm, with a punch, was measured in (N).

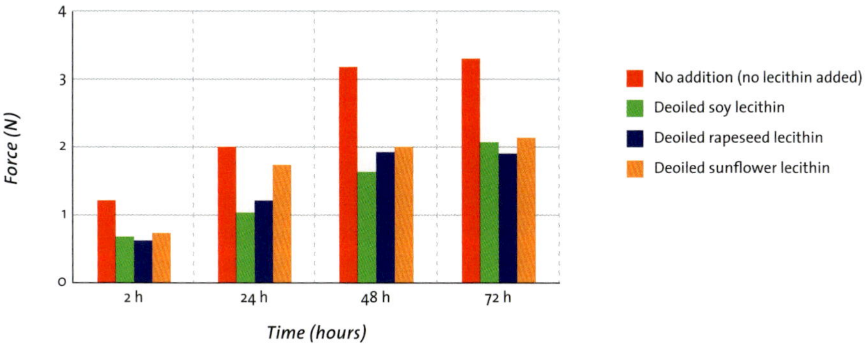

Figure 4.
Dependence of crumb firmness on the lecithin added and the length of storage of pan bread.

As can be seen from Fig. 4, in the course of storage both the pan bread without added lecithin and the three pan loaves with lecithin showed an increase in the force *F* that was necessary to compress the crumb. This is due to progressive staling of the bread. Whereas the force increased 2.5-fold from 1.28 N to 3.28 N in 72 hours in the case of the bread without added lecithin, in the case of the bread with added lecithin the force *F* required was already 47 – 51% lower after just 2 hours in the fresh loaves. Even after 72 hours the crumb was still noticeably softer in all three lecithin loaves than in the loaves without added lecithin. This effect of delaying crumb hardening was similarly good with all three lecithins. Only after 24 hours of storage did a clear deviation in softness between the three lecithins become apparent. At this point, compared with the crumb of the bread without lecithin the force needed for the soy crumb was 47% lower, for the rapeseed crumb 37% lower, and for the sunflower lecithin crumb 15% lower.

6.3.2.6 Summary of micro-baking trials

With the aid of micro-baking trials it was possible to rate the baking efficacy of the eight lecithins tested. All the lecithins used enlarged the volume of the loaves. In some cases the differences in their baking activity were very considerable. The optimum usage level of the lecithins was between 0.2 and 1.0% (m/10 g flour). Liquid, deoiled soy lecithin and deoiled rapeseed lecithin proved to have the best baking activity. At 33%, the most obvious increase in loaf volume compared to loaves without added lecithin was found at a lecithin concentration of 0.8%. The greatest differences in baking activity were found at a lecithin concentration of 0.4%. Deoiled soy lecithin increased the loaf volume by 26%, followed by deoiled rapeseed lecithin (20%) and liquid soy lecithin (11%). In contrast to deoiled rapeseed and soy lecithin, deoiled sunflower lecithin only achieved a slight enlargement in the loaf volume (7%) in spite of its high polar lipid content. Egg and oat lecithin brought about the least enlargement in loaf volume. In spite of its low polar lipid content, liquid soy lecithin proved to have similarly high baking activity to that of deoiled soy and rapeseed lecithin at higher concentrations (33% volume increase at a lecithin concentration of 0.8 to 1.0%).

Measurements of crumb firmness showed plainly that adding lecithins made the crumb softer than in loaves without added lecithin. Staling of the bread during storage was delayed considerably by the addition of lecithins.

6.3.3 Role of the phosphoric acid derivative in baking activity

6.3.3.1 Preliminary remarks

In order to determine the baking activity of phospholipids from lecithins as a function of the phosphoric acid derivative, isolated phospholipid classes phosphatidylcholine (PC), phosphatidylethanolamine (PE), phosphatidylinositol (PI), phosphatidic acid (PA), phosphatidylserine (PS), lysophosphatidylcholine (LPC) and lysophosphatidylethanolamine (LPE) were tested for their baking efficacy in micro-baking trials. On the basis of the baking activity of each commercially available lecithin and its content of the individual phospholipid classes, reconstituted samples were made up out of the different phospholipid classes. Both the individual phospholipid classes and the reconstituted samples were added in the amounts present in 0.2 to 1.0% lecithin (m/10 g flour).

In the trials for the baking activity of the phospholipid classes and the reconstituted samples made up from them, at least one duplicate determination was carried out using the 10 g mixer.

6.3.3.2 The baking activity of individual phospholipid classes

The main phospholipid classes PC, PE, PI and PA were added in concentrations of 0.02 – 0.1% during production of the bread. The volume of these test loaves was determined; the results for the soy lecithin components are diagrammed in Fig. 5, those for the sunflower lecithin components in Fig. 6. This concentration range corresponded to the amounts in which the individual phospholipid classes were present in the commercially available lecithin products at a lecithin addition of 0.2 to 1.0%. The baking activity of minor components of the phospholipid fraction of PS and LPE commercial lecithins was also determined at a concentration of 0.02 to 0.1% (see Fig. 7).

A comparison of the results of the micro-baking trials conducted with the individual phospholipid classes from deoiled sunflower and soy lecithin showed that PI was the most baking-active phospholipid in both cases. The PI effect was very potent: even a mere 2 mg of PI added to 10 g flour resulted in a 12% (ml/ml) enlargement of the loaf volume, compared to loaves without added phospholipids. But as the amount of PI added was further increased, the loaf volume actually decreased; in fact at an addition of 0.1% (m/m) PI the loaf volume was even smaller than with loaves without the addition.

In the case of the other phospholipid classes clear differences in baking efficacy became apparent, depending on the vegetable source chosen. Whereas PC from deoiled soy lecithin resulted in a loaf volume enlargement of up to 10%, when PC from deoiled sunflower lecithin was added the loaf volume remained practically unchanged. Similarly, while PE from deoiled soy lecithin reduced the loaf volume by up to 15%, PE from deoiled sunflower lecithin also had very little effect on loaf volume. At an addition level of 0.06%, PA from deoiled sunflower lecithin enlarged the loaf volume by 7%, yet PA from deoiled soy lecithin reduced the loaf volume by 4%.

The influence of the minor components of the lecithins on loaf volume is shown in the form of a diagram in Fig. 7. The addition of PS reduced the loaf volume by up to 7%. LPE also showed low baking activity, similar to that of PA from deoiled sunflower lecithin.

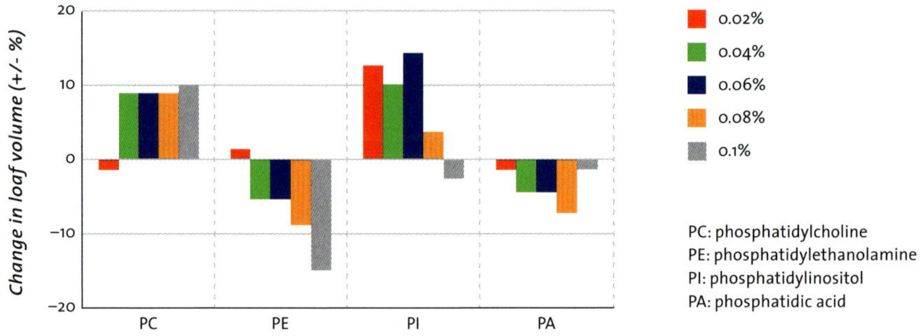

Figure 5.

Change in loaf volume resulting from the addition of the isolated phospholipid classes from deoiled soy lecithin at a concentration of 0.02 to 0.1%.

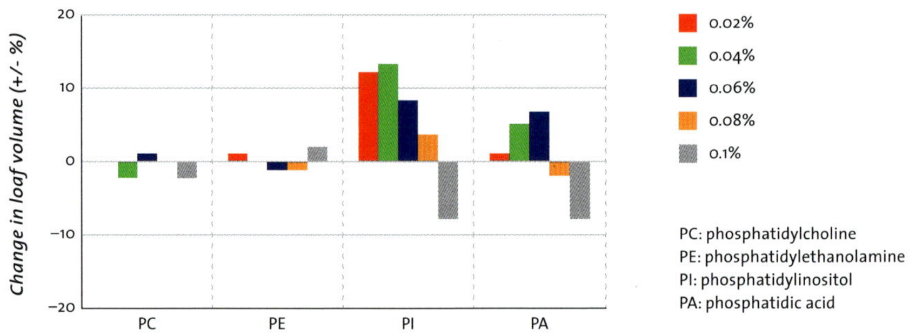

Figure 6.

Change in loaf volume resulting from the addition of the isolated phospholipid classes from deoiled sunflower lecithin at a concentration of 0.02 to 0.1%.

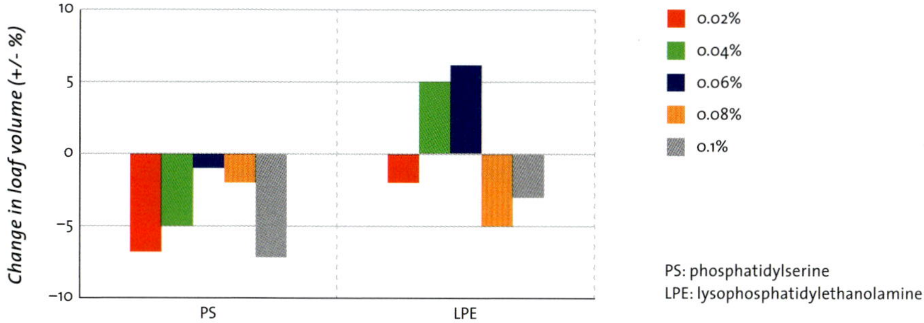

Figure 7.

Change in loaf volume resulting from the addition of the minor components PS and LPE at a concentration of 0.02 to 0.1%.

The tests described show clearly that the baking efficacy of the various phospholipid classes is not determined by their phosphatidyl derivative alone. A comparison of the volume yields of the phospholipid classes from deoiled sunflower lecithin in Fig. 8 with the volume increases achieved with the use of sunflower lecithin at a concentration of 0.2 to 1.0% in Fig. 2 shows that no simple correlation exists between the baking activity of a lecithin and the baking activity of its individual phospholipid classes. This proves that no single phospholipid is responsible for the baking activity of the lecithins.

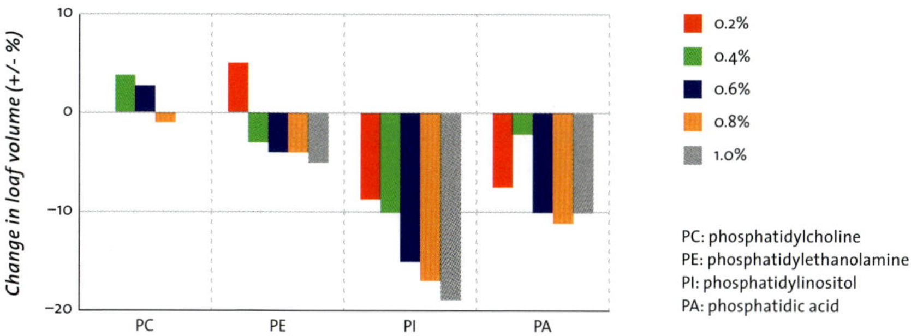

Figure 8.

Change in loaf volume resulting from the addition of the isolated phospholipid classes from deoiled sunflower lecithin at a concentration of 0.2 to 1.0%.

6.3.3.3 Baking activity of reconstituted samples from isolated phospholipid classes

After it was shown, per the previous sections, that lecithins influence loaf volume differently depending on their composition, but that the individual phospholipid classes do not achieve the baking activity of the whole commercials, the next step was to examine the baking activity of the individual phospholipid classes in the compositions in which they occurred in the deoiled lecithins used (soy, rapeseed, sunflower). To do this, two different reconstituted samples were first made up with the following main components of the phospholipid fraction: PC, PE, PI and PA (see Table 3).

The composition of reconstituted sample 1 corresponded to that of deoiled soy and rapeseed lecithin, in which the main classes of the phospholipid fraction are present in almost the same amounts. The composition of reconstituted sample 2 corresponded to that of deoiled sunflower lecithin (see Fig. 9).

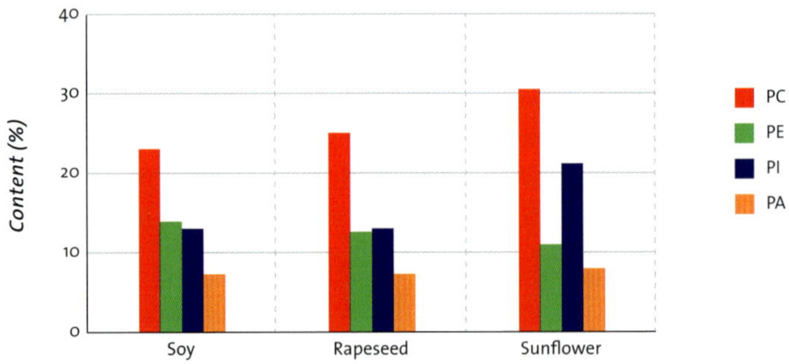

Figure 9.
PC, PE, PI and PA content of deoiled soy, rapeseed and sunflower lecithin determined by ^{31}P NMR spectrometry (see also Table 6, Chapter 5.2).

The decisive difference between the two reconstituted samples was the higher PC and PI content of the deoiled sunflower lecithin. Samples 1 and 2 were added in quantities that contained the four phospholipid classes PC, PE, PI and PA at total added concentrations of 0.2 to 1.0% lecithin. The quantities added and the composition of the two samples are shown in Table 3. Fig. 10 compares the volume yields achieved with the two reconstituted samples with the volume yields of deoiled soy, rapeseed and sunflower lecithin. The amounts of the samples added corresponded to the equivalent amount of lecithin, i.e. 0.2 to 1.0% (m/10 g flour).

Table 3.

Composition and added quantity of reconstituted sample 1 (soy/rapeseed) and sample 2 (sunflower) in micro-baking trials.

Reconstituted sample		Composition %		Quantity added %
1 (soy/rapeseed)	PC	23.9	41.6	0.11 = 0.2 lecithin
	PE	13.2	23.1	0.23 = 0.4 lecithin
	PI	13.0	22.8	0.34 = 0.6 lecithin
	PA	7.0	12.5	0.46 = 0.8 lecithin
	Σ	*57.1*	*100.0*	0.57 = 1.0 lecithin
2 (sunflower)	PC	30.2	43.0	0.14 = 0.2 lecithin
	PE	11.3	16.1	0.28 = 0.4 lecithin
	PI	20.9	29.8	0.42 = 0.6 lecithin
	PA	7.8	11.1	0.56 = 0.8 lecithin
	Σ	*70.2*	*100.0*	0.70 = 1.0 lecithin

PC: phosphatidylcholine; PE: phosphatidylethanolamine; PI: phosphatidylinositol;
PA: phosphatidic acid; **Σ**: total phospholipid content in % (m/m)

A clear difference emerged between the baking activities of the reconstituted samples 1 and 2. Although the quantity of sample 1 was less than that of sample 2 in absolute terms, sample 1 brought about a much better volume yield than sample 2. It corresponded to the volume yield of deoiled sunflower lecithin but failed to achieve the baking activity of intact, deoiled soy and rapeseed lecithin. These micro-baking trials showed that reconstituted samples had better baking activity than the phospholipid classes added singly. They also showed that the composition of the reconstituted sample was decisive for its baking activity.

Since deoiled sunflower lecithin has a higher PC and PI content than deoiled soy and rapeseed lecithin, and also contains neither PS nor LPE (see Fig. 11), further reconstituted samples were needed in order to explain the differences in baking activity between the different types of lecithins. Since sample 1 showed better baking activity than sample 2 because of its composition, it was used as a basis for further reconstituted samples in the subsequent baking trials; the composition of these samples is shown in Table 4.

Sample 1 was first baked without phosphatidylcholine (mixture PE/PI/PA: reconstituted sample 3, see Table 4). This was intended to demonstrate the importance of phosphatidylcholine as the main component of the phospholipid fraction in the baking activity of the lecithins. The missing phospholipid classes were also added to the basic mixture PC/PE/PI/PA (reconstituted sample 1) and to reconstituted sample 3. This was to show whether the missing components of the phospholipid classes (PS, LPE) have an influence on baking efficacy. Therefore reconstituted sample 3 (PE/PI/PA) was broadened with the individual minor components PS and LPE to make samples 4, 5 and 6, respectively (PE/PI/PA/PS as sample 4, PE/PI/PA/LPE as sample 5 and PE/PI/PA/PS/LPE as sample 6, see Table 4). Additionally, sample 1 was broadened with PS for comparison (PC/PE/PI/PA/PS as sample 7).

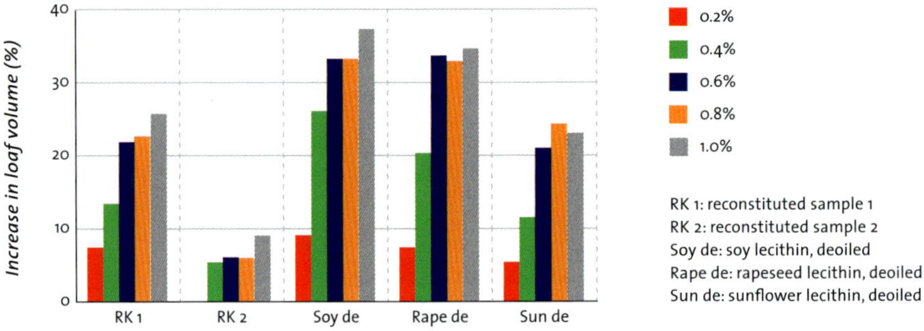

Figure 10.

Increase in loaf volume after the addition of reconstituted samples 1 and 2 and the corresponding lecithins at an equivalent level of 0.2 to 1.0% compared with loaves without additions.

The composition and mixture ratios of the reconstituted samples used are shown in Table 4. Fig. 12 shows the volume yields achieved with the reconstituted samples 3 to 7 listed in Table 4.

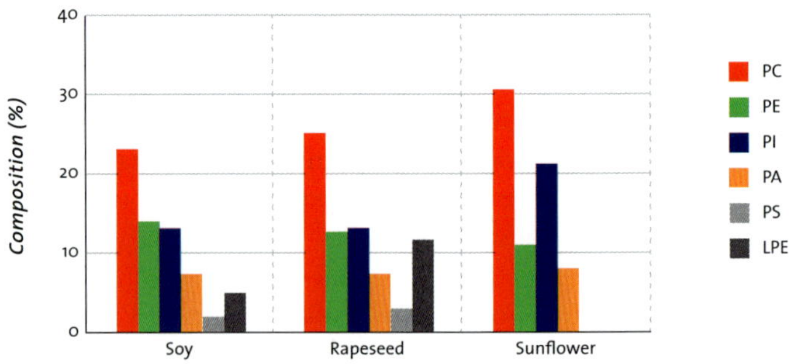

Figure 11.

Composition of the phospholipid fractions of deoiled soy, rapeseed and sunflower lecithin according to quantification by ^{31}P NMR spectrometry (see also Table 6, Chapter 5.2).

The micro-baking trial with the addition of reconstituted sample 4 showed that baking activity improved after the addition of PS and the removal of PC from the mixture. On the other hand, the addition of LPE impaired the results of baking. Clearly the presence of PC and LPE had a negative effect on the baking activity of the samples tested. The optimum reconstituted composition therefore proved to be a mixture of PE, PI, PA and PS in the ratio stated in Table 4.

Table 4.

Composition and added quantities of reconstituted samples 3–7 on the basis of reconstituted sample 1 (soy/rapeseed) in micro-baking trials.

Reconstituted sample		Composition %		Quantity added %
3 (PE/PI/PA)	PE	13.2	39.8	0.07 = 0.2 lecithin
	PI	13.0	39.2	0.13 = 0.4 lecithin
	PA	7.0	21.0	0.20 = 0.6 lecithin
	Σ	*33.2*	*100.0*	0.27 = 0.8 lecithin
				0.33 = 1.0 lecithin
4 (PE/PI/PA/PS)	PE	13.2	37.1	0.07 = 0.2 lecithin
	PI	13.0	36.5	0.14 = 0.4 lecithin
	PA	7.0	19.7	0.21 = 0.6 lecithin
	PS	2.4	6.7	0.28 = 0.8 lecithin
	Σ	*35.6*	*100.0*	0.36 = 1.0 lecithin
5 (PE/PI/PA/LPE)	PE	13.2	31.7	0.08 = 0.2 lecithin
	PI	13.0	31.3	0.17 = 0.4 lecithin
	PA	7.0	16.8	0.25 = 0.6 lecithin
	LPE	8.4	20.2	0.33 = 0.8 lecithin
	Σ	*41.6*	*100.0*	0.42 = 1.0 lecithin
6 (PE/PI/PA/PS/LPE)	PE	13.2	30.0	0.09 = 0.2 lecithin
	PI	13.0	29.5	0.18 = 0.4 lecithin
	PA	7.0	15.9	0.26 = 0.6 lecithin
	PS	2.4	5.5	0.35 = 0.8 lecithin
	LPE	8.4	19.1	0.44 = 1.0 lecithin
	Σ	*44.0*	*100.0*	
7 (PC/PE/PI/PA/PS)	PC	20.1	36.1	0.11 = 0.2 lecithin
	PE	13.2	23.7	0.22 = 0.4 lecithin
	PI	13.0	23.3	0.33 = 0.6 lecithin
	PA	7.0	12.6	0.45 = 0.8 lecithin
	PS	2.4	4.3	0.56 = 1.0 lecithin
	Σ	*55.7*	*100.0*	

PC: phosphatidylcholine; PE: phosphatidylethanolamine; PI: phosphatidylinositol; PA: phosphatidic acid; PS: phosphatidylserine; LPE: lysophosphatidylethanolamine; **Σ**: total phospholipid content in % (m/m)

A comparison of the volume yield of sample 4 (PE/PI/PA/PS) with the yields from soy, rapeseed and sunflower lecithin showed good conformity of the baking efficacy of the optimum reconstituted product with the baking activity of deoiled soy and rapeseed lecithin. The baking activity of deoiled sunflower lecithin was surpassed by that of reconstituted sample 4 (see Fig. 13).

In summary, on the basis of the micro-baking trials using reconstituted phospholipid samples in the quantities and ratios corresponding to the deoiled soy, rapeseed and sunflower lecithins tested it was possible to show structure – activity relationships of the phospholipid classes as a function of their phosphatidyl derivative. These trials clearly establish that it is not the overall phospholipid content of the deoiled lecithin that is decisive for the lecithin's baking activity, but its quantitative phospholipid derivative distribution.

Sample 4 (PE/PI/PA/PS) resulted in the best volume yield of all the reconstituted samples tested. It enlarged the loaf volume by as much as 31%, compared to loaves without the addition, and therefore showed even better baking activity than sample 1 (soy/rapeseed). Compared to sample 1, the addition of PS in sample 7 reduced the volume of the loaf.

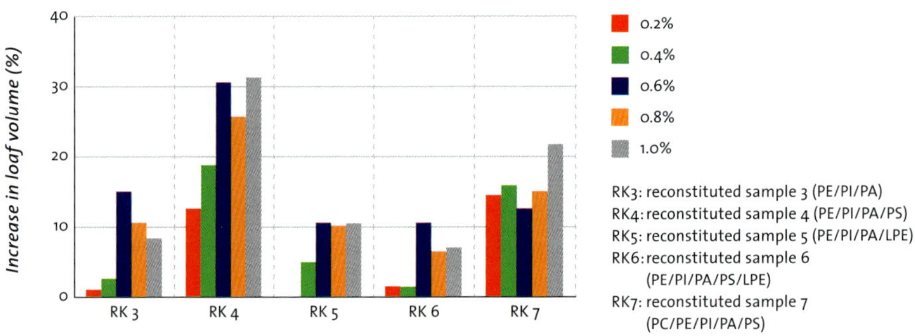

Figure 12.

Increase in loaf volume following the addition of the modified variants of reconstituted sample 1 (soy/rapeseed) at a quantity equivalent to 0.2 to 1.0% lecithin, as compared to loaves with no addition.

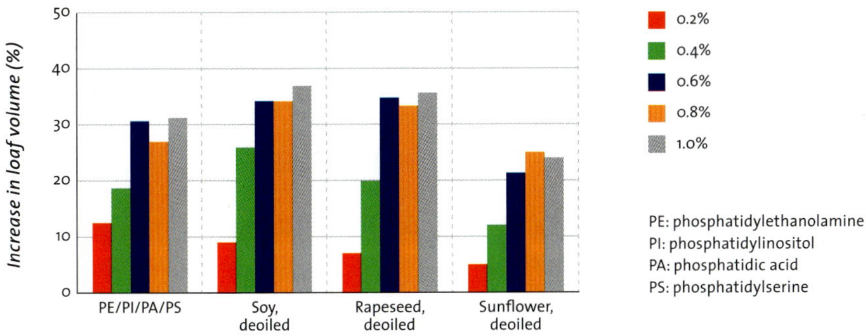

Figure 13.

Increase in loaf volume following the addition of the mixture PE/PI/PA/PS and deoiled soy, rapeseed and sunflower lecithin as compared to an equivalent addition of 0.2 to 1.0% lecithin and loaves with no addition.

For example, reconstituted sample 2, sunflower lecithin, with the major components PC/PE/PI/PA, proved to have baking activity far inferior to that of the corresponding sample soy/rapeseed in spite of a larger addition to the loaves. Micro-baking trials with the reconstituted samples showed that the reason for this is the greater PC and PI content. Another important reason for the poorer volume yield of deoiled sunflower lecithin as compared to deoiled rapeseed and soy lecithin was that the sunflower lecithin lacked PS. It became clear that this phospholipid class has a decisive influence on baking activity in spite of the small amount present in the lecithins. On the other hand, the addition of LPE impaired the baking activity. For this reason, deoiled rapeseed lecithin, which has a higher LPE content than deoiled soy lecithin, had less baking activity than the latter.

6.3.4 Baking activity of phosphatidylcholine with defined fatty acids

6.3.4.1 Preliminary remarks

After exploration of the individual phospholipid classes, the micro-baking trials showed that, assuming the amount added was equal, the baking activity of the phospholipid classes depended on the source of the lecithin from which the phospholipids had been isolated. At an addition of 0.02–0.1%, for example, PC derived from deoiled soy lecithin enlarged the loaf volume by up to 10% compared to loaves without such an addition, whereas PC from deoiled sunflower lecithin showed no baking activity when the same amount was added. The conclusion from this result was that the baking activity of a phospholipid class depends on the nature of its fatty acid radical groups (tails) as well as its phosphatidyl derivative group (head group). Taking PC as an example, this section will discuss how the chain length, the degree of saturation and the number of fatty acid radicals influences baking activity.

6.3.4.2 Dependence of baking activity on the nature of the fatty acid radicals

In order to test the baking activity of phosphatidylcholine in relation to the nature of its fatty acid radicals, we used molecular species of PC with four different fatty acid radicals: 1,2-dihexanoyl-glycero-3-phosphatidylcholine, 1,2-didodecanoyl-glycero-3-phosphatidylcholine, 1,2-dioctadecanoyl-glycero-3-phosphatidylcholine and 1,2-di-cis-9-octadecenoyl-glycero-3-phosphatidylcholine. To ensure a recognizable difference in baking activity, the samples were used at a concentration of 0.5% (basis: 10 g flour). The loaf volumes achieved in the micro-baking trials are shown in Fig. 14.

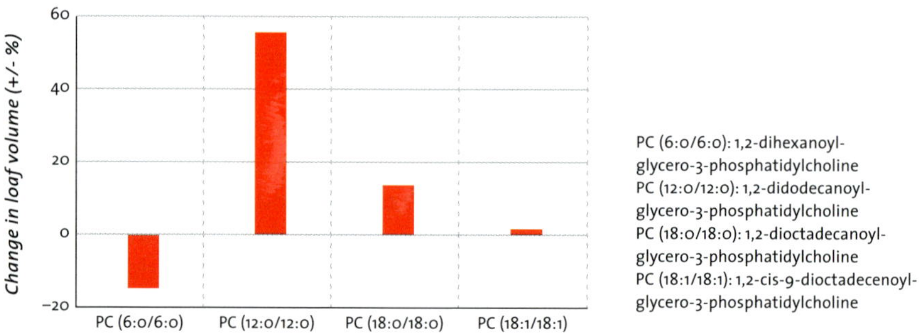

Figure 14.

Change in loaf volume brought about by the addition of 0.5% PC with different fatty acid radicals as compared to loaves with no addition.

The addition of PC resulted in considerable differences in the volume yield, depending on the nature of the fatty acid radicals. Whereas the short-chain variant PC (6:0/6:0) had a negative effect on the volume yield (–14%), the medium-chain compound PC (12:0/12:0) showed excellent baking activity (+54%), exceeding even the maximum volume yield achieved with deoiled soy lecithin. The PC with long-chain fatty acid radicals, PC (18:0/18:0), also enlarged the loaf volume as compared to loaves with no addition, although it had much less effect than PC (12:0/12:0) (+13%). Compared to the saturated compound with two stearic acid radicals, the PC with two oleic acid radicals brought practically no improvement in baking activity (+1%). The 10 g loaves are shown in Fig. 15.

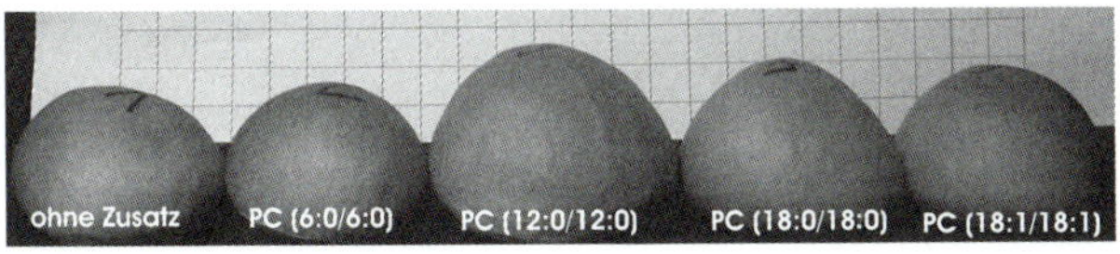

PC (6:0/6:0): 1,2-dihexanoyl-glycero-3-phosphatidylcholine
PC (12:0/12:0): 1,2-didodecanoyl-glycero-3-phosphatidylcholine
PC (18:0/18:0): 1,2-dioctadecanoyl-glycero-3-phosphatidylcholine
PC (18:1/18:1): 1,2-cis-9-dioctadecenoyl-glycero-3-phosphatidylcholine

Figure 15.

Ten-gram loaves with and without the addition of 0.5% phosphatidylcholine with different fatty acid radicals. Ohne Zusatz: no addition.

6.3.4.3 Dependence of baking activity on the number of fatty acid radicals

The objective of testing the baking activity of phosphatidylcholine as a function of the number of fatty acid radicals was to determine whether PC or LPC results in the better volume yield. For this purpose, PC and LPC derived from egg lecithin were added to loaves at concentrations from 0.02 to 0.8%. The corresponding volume yields are shown in Fig. 16.

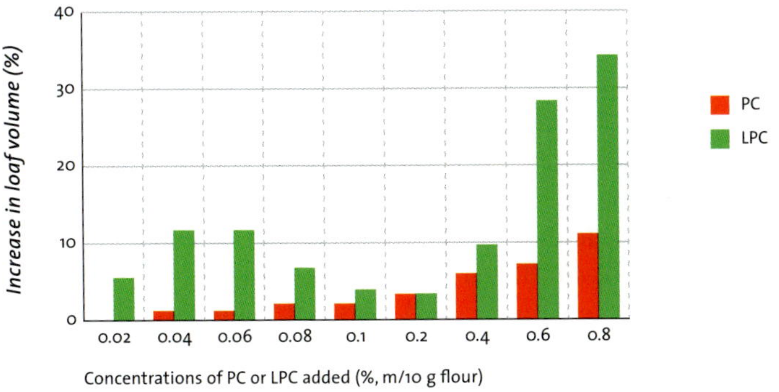

Figure 16.

Increase in loaf volume in % (ml/ml) achieved through the addition of 0.02 to 0.8% (m/10 g flour) phosphatidylcholine (PC) and lysophosphatidylcholine (LPC) derived from egg lecithin.

There was a considerable volume difference between loaves to which PC or LPC had been added. Whereas LPC increased the loaf volume by as much as 34%, the maximum volume increase with PC was only 11%. While the volume yield with PC did not increase noticeably until a concentration of 0.4% (for a +6% volume increase), for LPC at a concentration of 0.04% the volume increase was already 12%. LPC therefore showed much better baking activity than PC.

6.3.4.4 Dependence of baking activity on the distribution of the fatty acid radicals

After it had been demonstrated using phosphatidylcholine with defined fatty acid radicals that the baking activity of this phospholipid class depends on the chain length, the degree of saturation and the number of fatty acid radicals, these results were compared with the volume yields achieved in the baking trials with phosphatidylcholines from different sources, and with the deoiled lecithins. As a basis for assessment the fatty acid distributions of the samples added, determined by gas chromatography of the fatty acid methyl esters, were used for comparison (see Table 2 in Chapter 5.2). In order to assess the baking activity as a function of the number of fatty acid radicals, the volume yields of the deoiled lecithins were compared against their compositions as determined by ^{31}P NMR spectrometry.

The results shown in Fig. 17 demonstrated clearly that the baking activity of PC depends to a large extent on the type of lecithin from which it was isolated. Whereas PC derived from sunflower and egg lecithin showed practically no baking activity at a concentration of 0.02 to 0.1% (m/10 g flour), with PC from deoiled soy lecithin added at the same concentrations the volume increased by as much as 10% (as compared to loaves with no PC added). The fatty acid distributions of PC from deoiled soy and egg lecithin (see Fig. 18) also showed considerable differences. Whereas the stearic acid content was almost equal in PC from sunflower and egg lecithin, PC from deoiled soy lecithin contained 7% more palmitic acid, and at 15% nearly twice as much oleic acid, as PC from egg lecithin. Since PC from egg lecithin also contained 20% arachidic acid, its S/U (saturated/unsaturated) ratio was much higher than PC from deoiled soy lecithin. For PC from egg lecithin it was 4.9, for PC from deoiled soy lecithin just 2.1.

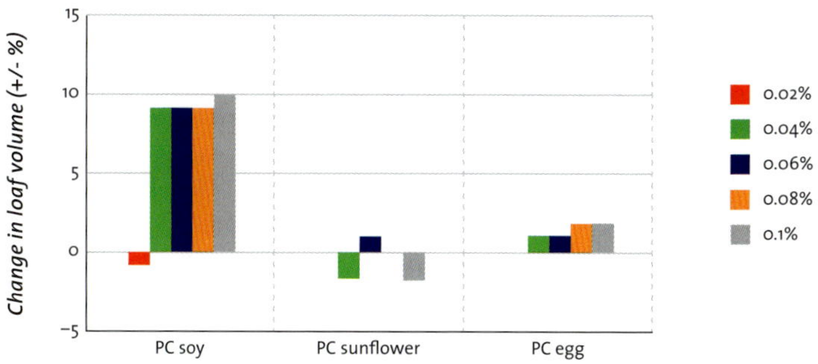

Figure 17.

Change in loaf volume after the addition of 0.02 to 0.1% (m/10 g flour) phosphatidylcholine from either deoiled soy lecithin (PC soy), deoiled sunflower lecithin (PC sunflower) or egg lecithin (PC egg), compared against loaves with no addition.

By comparing the volume yields of the deoiled lecithins against their respective fatty acid distributions, it could be determined whether the degree of saturation or the ratio of saturated to unsaturated fatty acids ultimately had an effect on the baking activity of the lecithins. Fig. 19 shows the volume yields of the deoiled lecithins used at an addition of 0.2 to 1.0% (m/10 g flour).

Unlike the different PCs, for the deoiled lecithins comparison of the volume yields against their fatty acid distributions (see Fig. 20) showed no correlations. Although deoiled rapeseed and soy lecithin have similarly good baking activity, the S/U ratio of deoiled rapeseed lecithin (1.1) does not correlate with that of deoiled soy lecithin (2.1); rather, it corresponds to that of deoiled sunflower lecithin (1.1). But the latter has much poorer baking activity than deoiled rapeseed lecithin.

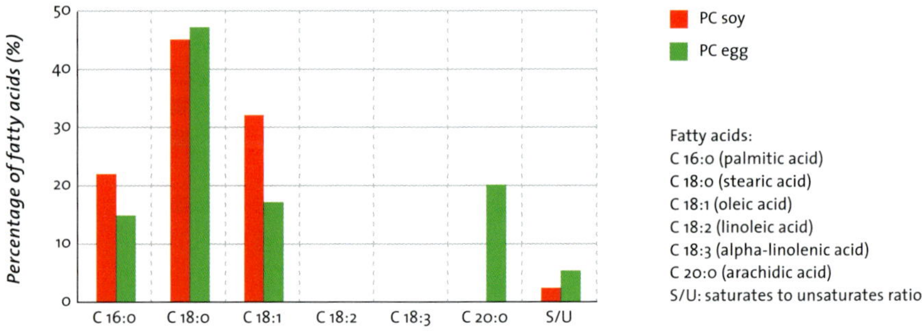

Figure 18.

Distributions of fatty acids (measured as methyl esters) as % in PC derived from soy lecithin (PC soy) or egg lecithin (PC egg) and the resulting ratio of saturated to unsaturated fatty acids (S/U).

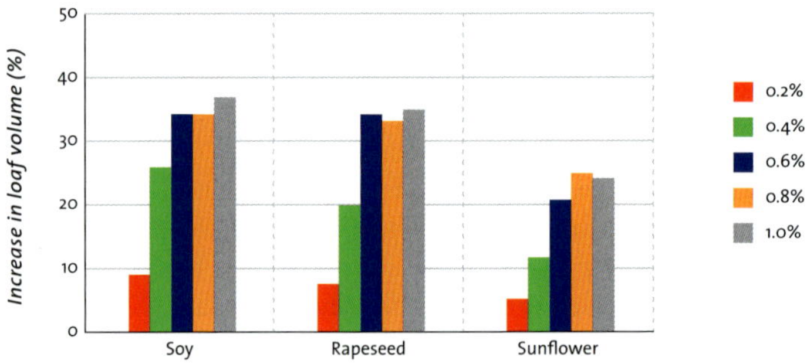

Figure 19.

Increase in loaf volume in % through the addition of deoiled lecithins derived from soy, rapeseed and sunflower at a concentration of 0.2 to 1.0% (m/10 g flour).

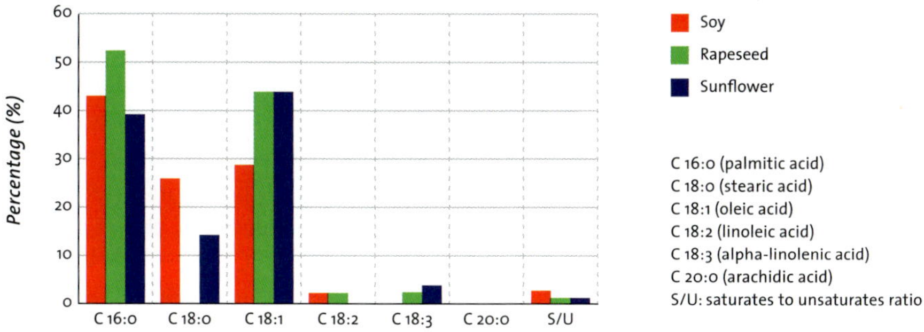

Figure 20.

Fatty acid distributions of different lecithins and their ratios of saturated to unsaturated fatty acids (S/U) in % (m/m).

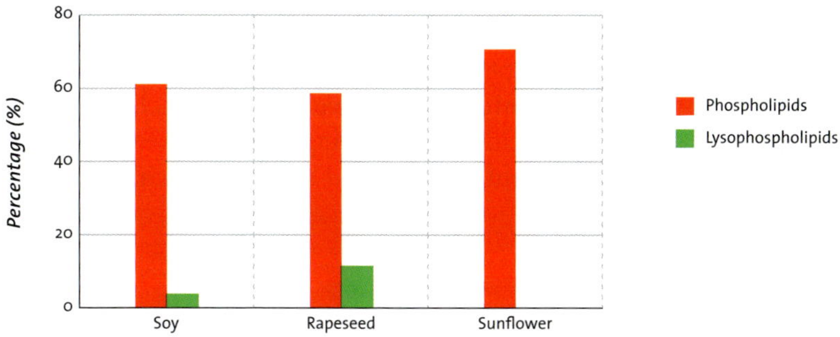

Figure 21.

Ratio of phospholipids to lysophospholipids in deoiled soy, rapeseed and sunflower lecithin.

However, a comparison of the volume yields achieved with the deoiled lecithins (see Fig. 19) with the percentages of monoacyl phospholipids in Fig. 21 does show a correlation between baking activity and the content of lyso compounds. Deoiled rapeseed and soy lecithin show comparatively good baking activity and contain 11.8 and 4.9% (m/m) lysophospholipids respectively, whereas the sunflower lecithin contains no lysophospholipids and results in poorer volume yields in comparison with loaves with no addition.

After the addition of PC with defined fatty acid radicals, micro-baking trials with subsequent determination of the loaf volume showed that the baking activity of this class of compounds depends on the chain length, the degree of saturation and the quantitative distribution of saturated and unsaturated fatty acids or the number of fatty acid radicals. Dilauryl PC proved to have extremely good baking activity (+54%), although the volume yield decreased considerably as the fatty acid chain length increased. Dioleyl PC showed practically no baking activity compared to distearoyl PC. LPC from egg lecithin (volume increase up to +34%) showed much better baking activity than PC from egg lecithin (no more than +11%). PC from deoiled soy lecithin resulted in volume yields of up to 10%, while PC from deoiled sunflower lecithin and egg lecithin showed practically no baking activity. In this connection the distribution of the fatty acids of the lecithins showed a much higher proportion of saturated fatty acids in PC from egg lecithin (S/U 4.9) than in PC from deoiled soy lecithin (S/U 2.1). It was therefore possible to demonstrate a dependency of baking volume yield on the composition of the fatty acid radicals, especially on the proportion of saturated and unsaturated fatty acids, in the case of the individual PC compounds, but for whole lecithins no such correlation could be established on the basis of the available analytical data. A connection was also found between lysophospholipid content and the volume yield achieved with the lecithins. Both of the deoiled lecithins containing LPE (rapeseed and soy) showed far better baking volume yield than deoiled sunflower lecithin, which contains no lyso compounds.

6.3.4.5 Verification of the results of the baking process in normal baking trials

In order to verify the changes in loaf volume achieved in the micro-baking trials by addition of the various lecithin and phospholipid classes, selected lecithins were used in normal baking trials on the basis of 300 g flour. The loaf volumes were compared with those of loaves to which no addition had been made. The changes in loaf volume in per cent are shown in Fig. 22.

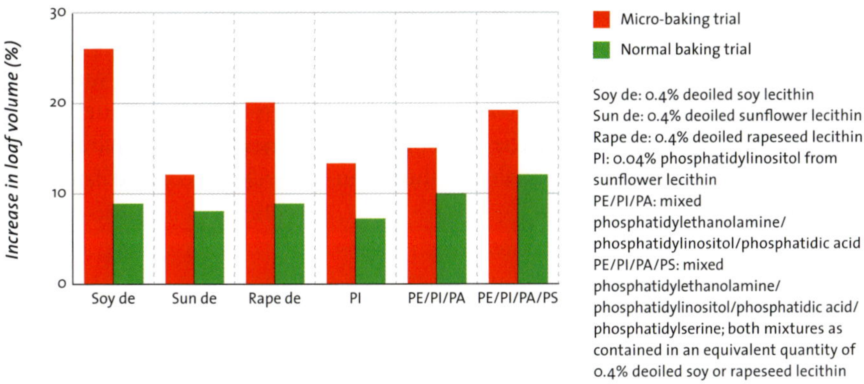

Figure 22.

Increase in the volume of rolled loaves in normal baking trials in comparison with the volume yields achieved in the micro-baking trials.

In the normal baking trials, the addition of the selected deoiled lecithins and the phospholipid classes resulted in smaller volume increases than that achieved in the micro-baking trials. These enlarged the loaf volume by an average 8–12%, as compared to loaves without addition. The commercial lecithins brought about a volume increase corresponding to 35–67% of the yield achieved in the micro-baking trials. The phospholipid classes caused volume increases 33–46% lower than in the micro-baking trials. However, the trends were similar, and the results of the micro-baking trials were therefore confirmed by the normal trials.

6.3.5 Influence of the lecithins and phospholipid classes on dough properties

6.3.5.1 Preliminary remarks

Micro-rheological measurements in the Farinograph and Extensograph make it possible to determine the influence of the raw lecithins and individual phospholipid classes on optimally developed dough. The purpose of these tests was to determine the effects of the lecithins and the phospholipid classes derived from deoiled soy and sunflower lecithin during paste formation (Farinograph) and their influence on the extensibility of the optimally kneaded dough (Extensograph).

6.3.5.2 Effect on the developing dough

Liquid and deoiled soy lecithin, deoiled rapeseed lecithin and deoiled sunflower lecithin were added at a concentration of 0.2 to 1.0% (m/10 g flour) and a paste formed. The changes in the most important measurement parameters as a percentage in the Farinograph are shown in Table 5, in comparison with doughs without added lecithin.

Table 5.

Changes in the measurement parameters in the Farinograph brought about by the addition of lecithins as a function of the amount added, in comparison with doughs without added lecithin. Coefficient of variation at n = 3 determinations ≤ 5%.

Quantity added %	Measurement parameters	Soy, deoiled %	Rapeseed, deoiled %	Sunflower, deoiled %
0.2	Dough development time (min)	−10	−22	−3
	Dough stability (min)	+5	+12	−21
	Dough softening (FU)	−13	+16	−4
	Dough resistance (min)	−1	0	−12
0.4	Dough development time (min)	+1	−6	−11
	Dough stability (min)	−2	−5	−21
	Dough softening (FU)	+2	+39	−4
	Dough resistance (min)	−1	0	−16
0.6	Dough development time (min)	+18	−1	−8
	Dough stability (min)	−10	−11	−14
	Dough softening (FU)	+17	+59	−4
	Dough resistance (min)	−1	−8	−17
0.8	Dough development time (min)	+25	+1	−12
	Dough stability (min)	−14	−19	−13
	Dough softening (FU)	+15	+63	−1
	Dough resistance (min)	−1	−11	−17
1.0	Dough development time (min)	+22	+1	−24
	Dough stability (min)	−13	−19	−15
	Dough softening (FU)	+15	+64	−1
	Dough resistance (min)	−1	−11	−19

The dough development time is the time required by the pasted flour up to the dough optimum (maximum of the curve). The dough optimum is the point at which the standardized dough has a consistency of 550 FU +/- 10. Dough stability is defined as the duration of the dough optimum (maximum of the curve), and dough softening (kneading tolerance) as the difference between dough consistency at the maximum of the curve and dough resistance after kneading for 20 minutes. Dough resistance is the sum of dough development time and dough stability.

At 0.2%, the addition of deoiled soy and rapeseed lecithin shortened the dough development time by as much as 10–22%, and with higher levels of addition the dough development time was prolonged by up to 25%. The addition of deoiled sunflower lecithin shortened the dough development time: the larger the amount added, the shorter was the dough development time. The stability of the dough increased by 5–12% through the addition of 0.2% deoiled soy and rapeseed lecithin, but decreased by up to 19% when a larger amount was added.

The addition of deoiled sunflower lecithin also reduced the stability of the dough by as much as 12–19%, in proportion to the amount added. On the other hand, the addition of 0.2% deoiled soy lecithin resulted in less dough softening, while a larger amount softened the dough by up to 17%. Deoiled rapeseed lecithin led to very considerable dough softening, ranging from 16% softening at an addition of 0.2% to 64% softening at an addition of 1.0%. The addition of deoiled sunflower lecithin had no effect on dough softening. Whereas the addition of deoiled soy lecithin had practically no effect on dough resistance, the addition of deoiled rapeseed and sunflower lecithin reduced dough resistance by as much as 19%, as compared to doughs with no added lecithin.

The Farinograph changes caused by the additions of phospholipid classes derived from deoiled soy and sunflower lecithin are shown in Table 6. These were added at 0.02 to 0.1% (m/10 g flour) and the results compared against doughs without such additions. The dough development time was prolonged by 40–46% by the addition of 0.02% PC, PE and PA. In the case of PE and PA this development time decreased as the amount added increased.

Table 6.

Changes in the measurement parameters in the Farinograph brought about by the addition of phospholipid classes derived from deoiled soy and sunflower lecithin as a function of the amount added, in comparison with doughs without such an addition. Coefficient of variation at *n* = 3 determinations ≤ 5%.

Quantity added %	Measurement parameters	PC sunflower %	PE sunflower %	PI soy %	PA soy %
0.02	Dough development time (min)	+44	+46	−22	+40
	Dough stability (min)	−20	−19	+5	−20
	Dough softening (FU)	+51	+22	−7	+22
	Dough resistance (min)	−1	0	+1	0
0.04	Dough development time (min)	+47	+44	−3	+37
	Dough stability (min)	−21	−14	+3	−17
	Dough softening (FU)	+48	+22	+1	+22
	Dough resistance (min)	0	+1	+1	+2
0.06	Dough development time (min)	+49	+26	−1	+37
	Dough stability (min)	−21	−14	+3	−14
	Dough softening (FU)	+49	+22	+2	+28
	Dough resistance (min)	−1	−1	+1	+2
0.08	Dough development time (min)	+42	+23	+4	+29
	Dough stability (min)	−21	−11	+1	−12
	Dough softening (FU)	+49	+22	+11	+24
	Dough resistance (min)	+1	−1	+1	+2
0.1	Dough development time (min)	+42	0	+7	+24
	Dough stability (min)	−19	−10	−1	−12
	Dough softening (FU)	+46	+22	+9	+15
	Dough resistance (min)	−1	−7	+1	+1

PC sunflower: phosphatidylcholine derived from deoiled sunflower lecithin
PE sunflower: phosphatidylethanolamine from deoiled sunflower lecithin
PI soy: phosphatidylinositol from deoiled soy lecithin
PA soy: phosphatidic acid from deoiled soy lecithin

In the case of PC there was hardly any change as the amount added increased. PI shortened the dough development time by 22% at an addition of 0.02%. However, as the added PI was increased the dough development time became prolonged, by up to 7 per cent. The addition of PC, PE and PA reduced the stability of the dough by 10–21%, while the addition of PI had practically no effect on these parameters (+5 to –1%). The addition of PC, PE and PA softened the dough by 15–51%; PI added at 0.02% shortened the dough softening, but at higher levels of addition actually prolonged the dough shortening by as much as 11%. These phospholipid classes had practically no effect on dough resistance over their addition ranges of 0.02 to 0.1% (m/10 g flour).

6.3.5.3 Effect on the fully developed dough

In order to determine the effects of the lecithins and phospholipid classes on the fully developed dough, deoiled soy, rapeseed and sunflower lecithin were added at concentrations between 0.2 and 1.0% (m/10 g flour) and the phospholipid classes PC, PE, PI and PA at concentrations of 0.02, 0.04, 0.1 and 1.0% (m/10 g flour). The optimally kneaded dough (FU 550) was subjected to a micro-tensile test. The measurements obtained for resistance (R) and extensibility (E) are shown in Table 7 and Table 8. The optimally kneaded dough without additions showed a resistance of 0.23 ± 0.01 N and an extensibility (E) of 5.8 ± 0.4 cm.

Table 7.

Micro-tensile tests on doughs made with the addition of lecithins. Resistance (R) and extensibility (E) of the doughs.

Quantity added %	Soy, deoiled R / E cm	Rapeseed, deoiled R / E cm	Sunflower, deoiled R / E cm
0.2	0.24 / 5.8	0.24 / 6.0	0.25 / 5.9
0.4	0.25 / 5.8	0.20 / 7.2	0.28 / 4.9
0.6	0.21 / 6.2	0.20 / 7.0	0.30 / 4.7
0.8	0.22 / 6.2	0.20 / 7.3	0.27 / 4.8
1.0	0.21 / 6.2	0.20 / 7.1	0.27 / 4.6

As Table 7 shows, the addition of deoiled soy or deoiled rapeseed lecithin resulted in reduced resistance and increased extensibility. The increase in extensibility brought about by the addition of deoiled rapeseed lecithin (17–22%) was greater than that achieved with deoiled soy lecithin (7%). On the other hand, the addition of deoiled sunflower lecithin increased the resistance, and reduced the extensibility by up to 20%.

As Table 8 shows, the addition of the four phospholipid classes tested – PC, PE, PI and PA – increased the resistance of the dough and at the same time reduced its extensibility. The addition of as little as 0.02% PI reduced the extensibility by 14%, as compared to doughs without added phospholipids. The greatest reduction in extensibility was brought about by the addition of 1.0% PE (71% reduction). The addition of 0.2% PC and PE initially increased the extensibility of the dough by 12 and 17% respectively, as compared to doughs without added phospholipids.

Table 8.

Micro-tensile tests on doughs made with the addition of phospholipids. Resistance (R) and extensibility (E) of the doughs.

Quantity added %	PC sunflower R / E cm	PE sunflower R / E cm	PI soy R / E cm	PA soy R / E cm
0.02	0.23 / 6.5	0.24 / 6.8	0.25 / 5.0	0.26 / 4.6
0.04	0.25 / 5.8	0.25 / 5.1	0.26 / 4.6	0.26 / 4.2
0.10	0.28 / 5.0	0.28 / 4.9	0.30 / 3.2	0.28 / 3.5
1.00	0.36 / 3.0	0.60 / 2.0	0.69 / 1.9	0.57 / 2.1

PC sunflower: phosphatidylcholine derived from deoiled sunflower lecithin
PE sunflower: phosphatidylethanolamine from deoiled sunflower lecithin
PI soy: phosphatidylinositol from deoiled soy lecithin
PA soy: phosphatidic acid from deoiled soy lecithin

6.3.5.4 Effects on fermentation behaviour and gas retention capacity

After the baking activity and rheological behaviour of the lecithins had been examined in micro-baking trials and paste formation and tensile tests, a final experiment was to show the effect of selected lecithins on the fermentation process, in particular on fermentation behaviour and gas retention capacity.

In order to test gas retention capacity, rounded and unrounded doughs either without additions or with the addition of 0.4% deoiled soy or sunflower lecithin were fermented for periods from 0 to 100 minutes at room temperature in a measuring cylinder filled with paraffin; the changes in volume were read off every 5 minutes. The objective was to determine the contribution made by the lecithins to the gas retention capacity of the dough. In the case of the unrounded doughs, the dependence of gas retention capacity on the addition of lecithin was expected to be particularly apparent, since the gluten network of such doughs is not yet developed and gas retention depends primarily on the lecithin used. The results for unrounded and rounded doughs are shown in Fig. 23.

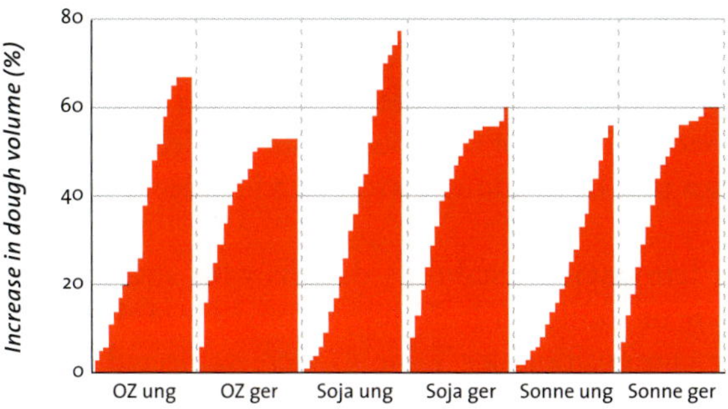

OZ: without addition
ung: unrounded
Soja: soy lecithin
ger: rounded
Sonne: sunflower lecithin.
Each bar represents the fermented volume of the dough as a function of time (readings every 5 minutes)

Figure 23.

Volume increase in % in rounded and unrounded doughs with and without the addition of 0.4% deoiled lecithin over a period of 0 to 100 minutes.

The volumes of the doughs after a fermentation period of 100 min were 37 to 56 ml. There was a clear difference in the course of fermentation between unrounded and rounded doughs. Whereas the volume of the unrounded doughs increased potentially over 100 minutes, the measurements for the rounded doughs resulted in a logarithmic curve. In the case of the doughs without an addition and with the addition of deoiled soy lecithin, the volume of the rounded doughs initially increased much more vigorously than that of the unrounded doughs over 0–65 min, but then the increase in volume over time was much less than with the unrounded doughs. In the case of the rounded dough with the addition of deoiled sunflower lecithin, the volume increase between 10 and 50 minutes fell from 85% to 15% per unit of time (5 min), whereas in the case of the unrounded dough the volume increase fell from 100% to 45% per unit of time.

Whereas the volume of the unrounded dough amounted to 23% of the volume increase of the rounded dough after a fermentation time of 10 minutes, it reached 68% of the volume increase of the rounded dough after only 50 minutes. In Fig. 23 this effect is seen in the form of a flattening curve for the rounded doughs. In the case of the rounded doughs without an addition and with the addition of deoiled soy lecithin, the volume at the end of the fermentation time was 21–22% less than with the corresponding rounded doughs. In the dough containing deoiled sunflower lecithin, the curves for the rounded and unrounded doughs corresponded to those of the doughs without an addition and with the addition of deoiled soy lecithin. On the other hand, the volume yield of the unrounded dough was much lower than that of the other unrounded doughs. The volume increase after 100 min was therefore 7% greater with the rounded dough than with the unrounded dough. After 100 minutes the volume increase of the unrounded dough with deoiled soy lecithin was 15% greater than with the dough without added lecithin, and exceeded that of the dough with deoiled sunflower lecithin by 28%. The volume increase of the rounded dough with deoiled soy lecithin after 100 min corresponded to that of the dough with deoiled sunflower lecithin. It was 13% greater than that of the dough without added lecithin.

To determine the effects on fermentation stability, doughs with and without the addition of 0.4% deoiled soy or sunflower lecithin or a mixture of the phospholipid classes PE/PI/PA (for composition see Table 4) were subjected to fermentation periods of different lengths. From the loaf volumes resulting from the micro-baking trials, the volume yields were then determined in comparison with loaves without additions. These results are shown in Fig. 24.

In line with the previous tests, the deoiled lecithins used in this case and the reconstituted sample PE/PI/PA increased the loaf volume as compared to loaves without such an addition. In the case of the loaves without addition, loaf volume was 23% lower after a fermentation time of 65 minutes than after a fermentation time of 55 minutes. On the other hand, the additions used stabilized the loaf volume. With the addition of deoiled soy lecithin, loaf volume increased again to 79 ml after a fermentation time of 65 minutes. In the case of added deoiled sunflower lecithin and PE/PI/PA, loaf volume declined slightly after 65 minutes but the volume yields were nevertheless 22–29% higher as compared to loaves without the addition.

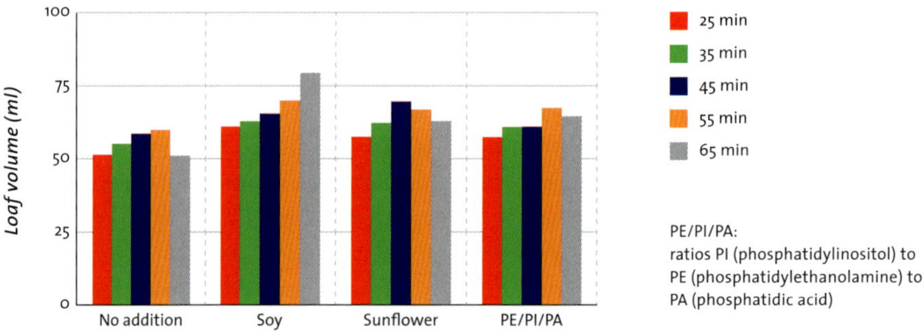

Figure 24.

Volume in millilitres (ml) of loaves with different fermentation times, with and without the addition of 0.4% deoiled soy and sunflower lecithin and PE/PI/PA.

6.3.5.5 Effect on frozen dough

In order to show the influence of the emulsifiers used on the baking efficiency of frozen doughs, the volume yields were determined with the addition of 0.6% deoiled soy, rapeseed and sunflower lecithin, 0.02% PI from deoiled sunflower lecithin, 0.3% DATEM RK 3084, and with a normal (0.7 g) and increased (1.0 g) amount of yeast in comparison with loaves with no addition. The volume yield was first determined with freshly made doughs that had not been frozen ('green' dough portions), and this was then compared with the volume yield achieved with frozen dough portions. The results of the micro-baking trials are shown in Table 9.

Table 9.

Loaf volumes achieved in millilitres (ml) and changes in volume in % of loaves made from 'green' and frozen dough portions with and without the addition of emulsifiers.

Addition	'Green' dough portions ml	Frozen dough portions %
Without the addition but with 0.7 g yeast	33.5	−5.2
Without the addition but with 1.0 g yeast	36.6	+5.0
Soy lecithin, deoiled, 0.6%	41.8	+5.0
Sunflower lecithin, deoiled, 0.6%	39.4	+8.0
Rapeseed lecithin, deoiled, 0.6%	41.0	+4.0
DATEM RK 3084 0.3%	44.8	+3.0
Phosphatidylinositol, sunflower, 0.02%	37.6	−8.0

A comparison of the volume yields between 'green' (fresh) and frozen dough portions without added emulsifiers and with 0.7 g yeast showed that freezing of the dough without such an addition impairs the baking properties. It was found that the reduced volume yield could be made up for by adding a larger amount of yeast or by the addition of deoiled lecithins or DATEM. Adding these increased the volume yield of the frozen dough portions by 3–8% as compared to the green dough portions. PI reduced the loaf volume of frozen dough portions by 8% as compared to the green dough.

6.3.5.6 Summary

The rheological tests on doughs with the addition of deoiled lecithins derived from soy, rapeseed or sunflower seed, as well as with the addition of isolated phospholipid classes from deoiled soy and sunflower lecithin, led to the following results: doughs with the addition of 0.2 to 1.0% deoiled soy or rapeseed lecithin showed the same trends in paste-forming and tensile tests. The dough development time was longer than with doughs without such an addition. Dough stability was reduced, and dough softening increased. Resistance decreased, while the extensibility of the dough increased. By contrast, the addition of 0.2 to 1.0% deoiled sunflower lecithin shortened the dough development time, increased resistance and thus reduced the extensibility of the dough. On the other hand, this addition reduced dough stability, as with the addition of deoiled soy and rapeseed lecithin. Deoiled sunflower lecithin had practically no influence on dough softening. The addition of 0.6% deoiled sunflower or rapeseed lecithin served to maintain the baking properties of frozen dough portions. The addition of 0.4% deoiled soy or sunflower lecithin improved the gas retention capacity of rounded doughs as compared against doughs without this addition.

In the paste formation tests PC, PE and PA showed a similar trend: when added at a concentration of 0.02 to 0.1% they prolonged the dough development time, reduced the stability of the dough and made it softer. By contrast, PI in this same concentration range shortened the dough development time and scarcely softened the dough. Nor did PI have any appreciable effect on dough stability. Dough resistance was not improved by any of the four phospholipid classes. They greatly reduced the extensibility of the dough and increased its resistance. PE and PI had the most pronounced effect. Unlike the deoiled lecithins, the addition of 0.02% PI impaired the baking properties of frozen doughs as compared to the 'green' dough portions. The addition of 0.4% of the PI/PE/PA mixture prolonged the gas retention capacity in comparison with dough without such an addition.

6.3.6 Discussion

6.3.6.1 Baking activity and composition of the lecithins

Lecithins and other emulsifiers have been used for improving the properties of dough and bakery products for 80 years. In the present study, liquid and deoiled soy, rapeseed and sunflower lecithins were tested in respect of their baking activity. In baking trials they brought about an increase in the volume of the loaves when added over the measured concentration range of 0.2–1.0% as compared to loaves without such an addition. Differences between lecithins in their abilities to enlarge the loaf volume at a particular concentration, which were very marked in some cases, led to the conclusion that the decisive factor is the composition of the lecithins. The objective then became to determine which components had the decisive influence on baking activity.

Lecithins consist of the following lipid fractions: phospholipids, lysophospholipids, glycolipids, sphingolipids, free fatty acids and non-polar lipids. A comparison of the compositions of the lecithins tested with their baking activity in the optimum range of 0.2 to 1.0% (m/10 g flour) showed that their baking activity did not depend on the amount of the main lecithin fractions present, or on their relative ratios. This had already been demonstrated experimentally, in a similar form, by other authors (29,30,31). For example, deoiled sunflower lecithin has the highest phospholipid content (70.2%) of all the lecithins tested, but its baking activity is similar to that of liquid sunflower and rapeseed lecithins, which contain only 41.6–44.9% phospholipids, and is less than that achieved by deoiled and liquid soy lecithins, which also have a lower phospholipid content (53.3–61.3%).

Micro-baking trials showed that deoiled sunflower lecithin has an unfavourable effect on baking activity. Unlike soy and rapeseed lecithin, sunflower lecithin does not contain LPE. But the poor baking activity of deoiled sunflower lecithin cannot be attributed merely to the absence of lysophospholipids. If that were the case, deoiled rapeseed lecithin, which contains 11.8% LPE, would have much better baking activity than deoiled soy lecithin with its 4.9% LPE. The same applies to the sphingolipids. If the lower baking activity of the sunflower lecithin were due to its low sphingolipid content of 6.4%, liquid rapeseed lecithin should create a higher volume yield because of its higher sphingolipid content (9.8%).

Further, the glycolipid content of the lecithins does not seem to correlate with greater baking activity. With its glycolipid content of 17.3% and a maximum increase in loaf volume of 25%, deoiled sunflower lecithin does not show better baking activity than liquid rapeseed lecithin with 8.9% glycolipids and maximum 26% increase in loaf volume. Moreover, liquid soy lecithin with its 13.2% glycolipid content would create much poorer volume increases than would deoiled soy or rapeseed lecithin with higher glycolipid contents of 20.1 and 22.5%, respectively. But that is not the case.

The baking trials additionally showed that the quantity of non-polar lipids also is not decisive for lecithin baking activity. Liquid soy lecithin containing 22% non-polar lipids had an effect comparable to that of deoiled rapeseed lecithin, which contains only 3% non-polar lipids. Adams et al. (16) found that deoiled soy lecithin showed 10% less baking activity than liquid soy lecithin used at the same concentration.

So it may be concluded from the above results that the baking activity of lecithins depends on the composition and ratios of their main classes of phospholipid compounds. The reason for the different volume yields is therefore to be found either in the level of a single compound contained in a particular fraction (e.g. in a single phospholipid class) or in the interaction between several compounds (e.g. synergistic effects of some phospholipid classes) or between the lecithin fractions themselves (e.g. synergisms between phospholipids and glycolipids). Since the phospholipid fraction, with a content of 41.6–70.2%, is the major constituent of lecithin and has the most classes of compounds in its composition, we might reasonably assume that this is the fraction chiefly responsible for baking activity.

A comparison of the volume yields achieved in the micro-baking trials and in normal baking trials with the data from the literature shows a certain consistency in the measurements. The volume yield achieved depends on a number of parameters such as the way the dough is prepared (mixer or spiral kneader), the type of bread (round or rolled loaves, pan loaves), the amount of water added, the fermentation time, the type of flour, the wheat variety and the type of lecithin or batch used. In other words, it is possible to ascertain a certain consistency of the data, but for these same reasons it is not possible to compare the absolute values for the volume yields achieved.

In the micro-baking trials, liquid and deoiled soy lecithin increased the loaf volume by 7–37% at an addition of 0.2–1.0% (m/10 g flour) as compared to loaves without such an addition. In the normal baking trials, the addition of 0.4% deoiled soy lecithin increased the volume by 9% as compared to rolled loaves without the addition. The volume yield achieved according to Doose (1) with the use of 0.8% deoiled soy lecithin in normal baking trials was 25%; Adams et al. (16) reported that liquid soy lecithin at a concentration of 0.3–1.0% increased the volume of rolled loaves by 13 to 27%. In other studies, deoiled soy lecithin added at concentrations between 0.3 and 0.5% to rolled loaves and pan loaves resulted in a 5 to 13% increase in volume (2,9,21). The volume yields achieved with the addition of the lecithins in this study therefore correspond to the data existing in the literature.

Compared with DATEM, the deoiled lecithins that we tested resulted in lower volume yields. Whereas the DATEM RK 3084 increased the loaf volume by 34% at a concentration of only 0.3%, the volume yields achieved with the deoiled soy, rapeseed and sunflower lecithins ranged between 18 and 25% at 0.6% concentration. This result is consistent with other authors (9,16). So although the volume yields achieved with the deoiled lecithins are greater than those resulting from SSL (sodium stearoyl-2-lactylate), which enlarged the loaf volume by only 13% at an addition of 0.5% (2,32), DATEM is undoubtedly the most effective emulsifier. At concentrations from 0.1 – 0.5% it can enlarge loaf volume by as much as 62% (33).

Measurement of the crumb firmness of stored pan bread with the addition of 0.4% deoiled lecithin showed clearly that the crumb stays softer over a period of three days in loaves with the lecithins than in loaves without them. According to Forssell et al. (5), hydrolysed lecithins intensify this effect because of the sterically induced facilitated formation of inclusion compounds of amylose. However, our tests indicate that the lysophospholipid content alone does not influence the softness of the crumb during storage. If that were so, this effect would be much more pronounced with rapeseed lecithin because of its high level of lyso compounds (11.8%) than with soy lecithin (4.9%) and sunflower lecithin (no lyso compounds). But this does not happen. Deoiled soy lecithin results in a much softer crumb than rapeseed and sunflower lecithin, which scarcely differ in their degree of crumb effect.

Micro-baking trials were carried out in order to determine the effect of lecithins on the baking properties of frozen doughs. The addition of 0.6% deoiled lecithin not only maintained the volume yield of loaves made from frozen dough portions as compared to freshly made doughs; it actually increased it by 4 – 8%. Lecithin can therefore be recommended as an additive for maintaining baking activity in the manufacture of bakery products from frozen doughs.

The yeast cells and the gluten of the dough are subjected to considerable physical and chemical stress in frozen doughs. Although manufacturers work with efficient flash-freezing methods, it is impossible to avoid the formation of ice crystals. This results in inactivation of the yeast (3), or ice crystals impair the gluten structure because of the carbonic acid that forms during the freezing process. The latter comes about partly as a result of the carbon dioxide bound in the gas bubbles in the dough and partly through the use of carbon dioxide pellets as a coolant for freezing the dough (34). In the worst case the pH falls to the isoelectric point of the gluten or the proteins of the yeast, greatly reducing the original effect of the yeast and gluten. Abd El-Hady et al. (35) attribute the reduced baking activity to continuous decomposition of the polar lipids of the flour during storage at –20°C over a period of four weeks. Supplementation with lecithin or DATEM compensates for the loss of polar lipids in the frozen dough portions, thus maintaining the gas retention capacity of the frozen dough. According to Ghyczy (3), the yeast is better suspended in the dough by the addition of lecithin, which results not only in more even distribution but in lessened susceptibility to the physical stress from the ice crystals. According to Lösche (34), the alkaline emulsifiers added to the dough act as a buffer against the forming carbon dioxide and thus maintain the properties of the dough.

The rheological measurements underline the information on the different baking activity of the three deoiled lecithins obtained in the micro-baking trials. Whereas the addition of deoiled rapeseed and soy lecithin at quantities of 0.2–1.0% (m/10 g flour) prolonged the dough development time, reduced the stability of the dough and made it softer and therefore more extensible, deoiled sunflower lecithin had the opposite effect at the same concentrations. Using hydrolysed soy lecithin added at 0.5%, Pomeranz (24) observed rheological effects similar to those of the deoiled soy and rapeseed lecithins.

The addition of deoiled soy lecithin also improved the gas retention capacity more noticeably than deoiled sunflower lecithin. The effect was especially marked in the case of unrounded doughs, since gluten aggregation is much lower in their case than in the rounded doughs and the gas retention capacity is therefore due primarily to the added lecithins. Because of the reduced extensibility of dough to which deoiled sunflower lecithin has been added, its gas retention capacity is far inferior to that of doughs containing additives that increase extensibility (e.g. deoiled soy lecithin).

Moreover, lecithins were shown to shorten the fermentation time. Doughs to which 0.4% deoiled soy lecithin had been added had the same volume after 25 minutes' fermentation time as doughs without such an addition after 45 minutes. The shortening of the final proof time by the addition of lecithins, described by Mettler et al. (8), therefore corresponds to the results described here. The use of lecithin seems to lead to better retention of the carbon dioxide formed during fermentation because of the increased extensibility of the dough. This is possibly due to adherence of the phospholipids to the interface between gluten and water, which facilitates sliding of the gluten strands and thus makes the dough more extensible.

Baking trials on the micro and normal scale showed deoiled rapeseed lecithin to be a viable alternative to controversial soy lecithin in respect of baking activity. In Germany, especially, the use of rapeseed as a raw material for lecithin production would seem a good choice, since it is grown in all parts of the country. But deoiled sunflower lecithin also produces a fine, even texture of the loaves and could be an ideal replacement for soy in toast loaves in spite of its lower volume yield. The newly acquired information on the structure – effect relationships of the phospholipids in bakery products opens up the possibility of testing the baking activity of lecithins derived from plants whose fats or oils contain a high percentage of lauric acid. Examples might be the lecithins obtained during the production of palm kernel, coconut or babassu oil.

6.3.6.2 Influence of the phosphatidyl and fatty acid radicals

In these studies, the baking activity of individual phospholipid classes from lecithins was determined for the first time with the aid of the phospholipid classes isolated from deoiled soy and sunflower lecithin. Using reconstituted samples, the significance of the phospholipid fraction for the technical and functional properties of the lecithins was also investigated for the first time. The investigation into the structure – effect relationships of phospholipids as a function of the phosphatidyl derivative and the fatty acid radicals was rounded off by the use of phosphatidylcholine with defined fatty acid radicals in micro-baking trials.

It was demonstrated for the first time that PI is the only phospholipid class to show a high level of baking activity in the concentration range of 0.02 to 0.08%. These are the concentrations of PI that occur when deoiled soy, rapeseed and sunflower lecithins are used in quantities of 0.2 to 1.0% in doughs. But a higher PI concentration of 0.1% (m/10 g flour) reduced the loaf volume. PC from soy lecithin also enlarged the baked volume (9–10%), whereas PC from deoiled sunflower lecithin showed practically no baking activity. PE and PA from soy lecithin reduced the loaf volume by as much as 15%; PA from sunflower lecithin enlarged the loaf volume by 7% at a usage level of 0.06% (m/10 g flour). The individual phospholipid classes therefore showed very different levels of baking activity when compared with each other, but they also resulted in very different volume yields according to the type of lecithin. This had already been demonstrated in a similar form by Pomeranz et al. (11) with synthesized and isolated phospholipids. The maximum volume increase we achieved, namely 13%, also makes it plain that no single phospholipid class is fully responsible for the baking activity of the lecithins.

Taking this as a starting point, reconstituted samples were made up from the phospholipid fractions of deoiled soy, rapeseed and sunflower lecithin to determine the baking activity of lecithin on the basis of the isolated phospholipid classes. In this way it was possible for the first time to determine the components responsible for the baking activity of a lecithin from a reconstituted sample consisting of isolated phospholipid classes. In the case of deoiled soy and rapeseed lecithin, the reconstitution of their components PE, PI, PA and PS resulted in baking activity similar to that of the commercial product. An increase in the concentration, the removal of one of these components or the addition of a further phospholipid impaired the volume yield. Moreover, for deoiled sunflower lecithin the higher PI content and the absence of LPE and PS were identified as reasons for its poorer baking activity as compared to the other deoiled lecithins. Though the main component of the phospholipid fraction, PC has scarcely any influence on baking activity. These findings can also be applied to the other lecithins tested. Accordingly, liquid soy lecithin has baking activity comparable to that of deoiled soy and rapeseed lecithin in spite of its lower overall phospholipid concentration because here, too, PE/PI/PA/PS are present in an optimal composition. Liquid rapeseed lecithin has a higher phospholipid content than liquid soy lecithin, but has much poorer baking activity since it lacks PS.

A more precise reason for the differences in baking activity between the same phospholipid classes, depending on their origin (lecithin variety), was explained by the baking trials with PC containing defined fatty acids. These trials showed impressively the extent to which the composition of the fatty acid groups influences the baking activity of a phospholipid class. However, it must be remembered that according to the gas chromatographic analysis of the fatty acid methyl ester, neither hexanoic (caproic) acid nor lauric acid was present in the commercial lecithins tested. In other words, the influence of the fatty acids in the individual phospholipid classes of the lecithins is not as clear in reality as in the model experiment with PC. At an addition of 0.1%, PC from soy lecithin, whose fatty acid fraction consists of 47% stearic acid, increased the loaf volume by 10%, i.e. to almost the same extent as distearoyl PC at an addition of 0.5% (+13%). A comparison of PCs from egg lecithin and soy lecithin in connection with their fatty acid distribution also suggested that the baking activity of a phospholipid class is partly determined by the degree of saturation of the fatty acids. For example, egg lecithin PC with its much higher level of saturated fatty acids (S/U factor 4.9) proved to have much less baking activity than PC from soy lecithin with its S/U ratio of 2.1.

The results shown make it clear that the baking activity of the phospholipid classes does not stem from a single structural element. On the contrary: it is likely to be an outcome of factors such as the nature of the phosphatidyl derivative and the length, degree of saturation and number of fatty acid radicals for each phospholipid class.

The reason for the differences in baking activity between PC from soy and sunflower lecithin becomes clear when one considers the overall fatty acid distributions of the two lecithins. Whereas the fatty acid pattern of deoiled soy lecithin consists of 69% saturated fatty acids (mainly palmitic and stearic acid), deoiled sunflower lecithin is made up of nearly 50% unsaturated fatty acids, the main constituent being oleic acid (44%). That leads us to assume that PC from deoiled sunflower lecithin also contains more unsaturated oleic acid radicals than PC from deoiled soy lecithin, making the volume yield of the loaves therefore correspondingly lower.

According to Greene (36), attachment of the fatty acid radicals of the phospholipid classes to the hydrophobic side chains of the gluten results in aggregation of the dough through a reduction of the net charge of the positively charged constituents of the gluten; this manifests itself in firmer dough and is especially desirable with low-gluten flours. In the case of PS and PA, the results of the baking trials suggest that this effect is especially pronounced from using these compounds. The carboxyl group of PS and the double-negatively charged phosphate ion of PA presumably allow more effective reduction of the net charge of the proteins than the other phospholipid classes.

Rheologically, aggregation of the dough can be measured by reducing its extensibility, and in baking trials it did not increase the volume of the loaves as a function of the concentration. The addition of PA from deoiled soy lecithin at a concentration of 0.02 to 0.10% (m/10 g flour) led to a considerable reduction of extensibility and reduced the loaf volume by 7% compared to dough or loaves without such an addition. According to Kieffer et al. (37), zwitterions such as PC and PE also have a gluten-hardening effect. Here, too, it was assumed that the effect comes about through intermolecular ionic bonds.

References

1. Doose, O., 1951. Untersuchungen über die Wirkung von Lecithin auf die Teig- und Gebäckeigenschaften. Brot und Gebäck 5, 133-136.
2. Indrani, D., Rao, G., 1992. Effect of Improvers on the Quality of Whole Wheat Flour Bread. J. Food Sci. Technol. 29, 357-359.
3. Ghyczy, M., 1995. Neue Lecithinfraktionen. Chemie. Eigenschaften. Verwendung. Getreide, Mehl und Brot 49, 352-358.
4. Haas, F.S., Haas, J., Tiefenbacher, K., 1997. Method of Manufacturing Edible Waffle Products by Using Monostearyl Citrate. Patent 9747205. Application WO 1997-AT128.
5. Forssell, P., Shamekh, S., Harkonen, H., Poutanen, K., 1998. Effects of Native and Enzymatically Hydrolyzed Soya and Oat Lecithin in Starch Phase Transitions and Bread Baking. J. Sci. Food Agric. 76, 31-38.
6. Working, E.B., 1924. Lipoids, a Factor Influencing Gluten Quality. Cereal Chem. 1, 153-158.
7. Mettler, E., Seibel, W., Brümmer, J.-M., Pfeilsticker, K., 1991a. Experimentelle Studien der Emulgator- und Hydrokolloidwirkung zur Optimierung der funktionellen Eigenschaften von Weizenbroten. 1. Mitt.: Rezepturauswahl – Einfluss der Emulgatoren und Hydrokolloide auf das rheologische Verhalten von Weizenmehlen. Getreide, Mehl und Brot 45, 145-152.
8. Mettler, E., Seibel, W., El Baya, A.W., Pfeilsticker, K., 1991b. Experimentelle Studien der Emulgator- und Hydrokolloidwirkung zur Optimierung der funktionellen Eigenschaften von Weizenbroten. 2. Mitt.: Vergleichende analytische und rheologische Untersuchungen der Emulgator- und Hydrokolloidwirkung während der Teigführung. Getreide, Mehl und Brot 45, 206-210.
9. Jodlbauer, H.D., Freund, W., Senneka, J., 1992. Effect of Lecithin Fractions in the Manufacture of Bread and Bakery Products. Getreide, Mehl und Brot 46, 174-177.
10. Mettler, E., Seibel, W., Brümmer, J.-M., Pfeilsticker, K., 1992. Experimentelle Studien der Emulgator- und Hydrokolloidwirkung zur Optimierung der funktionellen Eigenschaften von Weizenbroten. 5. Mitt.: Einfluss der Emulgatoren und Hydrokolloide auf die funktionellen Eigenschaften von Weizenbroten. Getreide, Mehl und Brot 46, 43-47.
11. Pomeranz, Y., Carvajal, M.J., Shogren, M.D., Hoseney, R.C., Finney, K.F., 1970. Wheat Germ in Breadmaking. II. Improving Breadmaking Properties by Physical and Chemical Methods. Cereal Chem. 47, 429-437.
12. Munz, E., Brabender, C.W., 1940. Prediction of Baking Value from Measurements of Plasticity and Extensibility of Dough. I. Influence of Mixing and Molding Treatments Upon Physical Dough Properties of Typical American Wheat Varieties. Cereal Chem. 17, 78-100.
13. Horubalowa, A., Jakubczyk, T., Pasturczak, A., 1975. Physical Transformations of Starch. Gluten and Yeast Induced by Some Emulsifiers. Acta Alimentaria Polonica 1, 259-268.
14. Merrit, P.P., Bailey, C.H., 1945. Preliminary Studies with the Extensograph. Cereal Chem. 22, 372-391.
15. Helmerich, G., Koehler, P., 2004. Struktur-Wirkungsbeziehungen von Phospholipiden bei Backwaren. Verlag Dr. Hut. ISBN 3899631196. 9783899631197.
16. Adams, W., Funke, A., Gölitz, H., Schuster, G., 1991. Wirksamkeit von Emulgatoren in Backwaren. Getreide, Mehl und Brot 12, 355-361.
17. Sullivan, B., 1940. The Function of the Lipids in Milling and Baking, Cereal Chem. 27, 661.
18. Pratt, D.B., 1946. Lecithin Increases the Effect of Shortening in Bread. Food Industry 18, 16-19.
19. Olcott, H.S., Mecham, D.K., 1947. Characterization of Wheat Gluten. I. Protein-Lipid Complex Formation During Doughing of Flours. Lipoprotein Nature of the Glutenin Fraction. Cereal Chem. 24, 407-414.
20. Coppock, J.B.M., Cookson, M.A., Laney, D.H., Axford, D.W.E., 1954. The Role of Glycerides in Baking. II. The Influence of Flour Oil on Behaviour of Glycerinated Fats in Baking, and the Effect of Natural Monoglycerides Present in Flour Oils and Baking Fats on the Pharmacological Desirability of Using Glycerinated Fats in Baked Products. J. Sci. Food Agr. 5, 19-26.
21. Adler, L., Pomeranz, Y., 1959. Use of Lecithin in Production of Bread Containing Defatted Soy Flour as a Protein Supplement. J. Sci. Food Agr. 10, 449-456.
22. Grosskreutz, J.C., 1961. A Lipoprotein Model of Wheat Gluten Structure. Cereal Chem. 38, 336-348.
23. Zentner, H., 1964. The Oxidation of Mechanically Developed Doughs. J. Sci. Food Agr. 15, 629-634.

24. Pomeranz, Y., 1966. Review of Recent Studies of Wheat Flour Lipids in Breadmaking. Baker's Digest 40, 44-77.
25. Wassermann, L., 1983. Lipide als Backmittel. Fette, Seifen, Anstrichmittel 85, 117-121.
26. Jodlbauer, H.D., 1987a. Backmittel für Hefeteige. Patent DE 3616597 A1.
27. Jodlbauer, H.D., 1987b. Backmittel für Brötchenteige. Patent DE 3616594 A1.
28. Schneider, M., 1997. Anwendung von Phospholipiden für die Backanwendung. Patent DE 19618439.
29. MacRitchie, F., 1984. Baking Quality of Wheat Flours. Food Res. 29, 201-277.
30. McCormack, G., Panozzo, J., MacRitchie, F., 1991. Contributions to Breadmaking of Inherent Variations in Lipid Content and Composition of Wheat Cultivars. II. Fractionation and Reconstitution Studies. J. Cereal Sci. 13, 263-274.
31. Stampfli, L., Nersten, B., Molteberg, E.L., 1996. Effects of Emulsifiers on Farinograph and Extensograph Measurements. Food Chem. 57, 523-530.
32. Garti, N., Lindner, C., Pinthus, E.J., 1980. Evaluation of Food Emulsifiers in the Bread Making Industry. Baker's Digest 54(5), 24-37.
33. Koehler, P., Grosch, W., 1999. Study on the Effect of DATEM. 1. Influence of Fatty Acid Chain Length on Rheology and Baking. J. Agric. Food Chem. 47, 1863-1869.
34. Loesche, K., 2003. Es war das Gas und nicht der Kristall. Brot und Backwaren 5, 29-30.
35. Abd El-Hady, E., Shouk, A., Aaseen, A., 2002. Effects of Lipids and Emulsifiers on Frozen Bread Dough Quality. Egypt. J. Food Sci. 30, 77-94.
36. Greene, F.C., 1975. On the Mechanism of the Functionality of Surfactant Dough Conditioners. Baker's Digest 49(3), 16-26.
37. Kieffer, R., Kim, J.J., Belitz, H.-D., 1983. Einfluß niedermolekularer inonischer Verbindungen auf die Löslichkeit und die rheologischen Eigenschaften von Weizenkleber. Z Lebensm Unters Forsch 176, 176-182.

6.4 Sunflower lecithin emulsion applications

Estefania N. Guiotto, Mabel C. Tomás

Abstract

Lecithin is a mixture of acetone-insoluble phospholipids. These compounds contain lipophilic fatty acyl groups and hydrophilic head groups; this amphiphilic structure makes lecithin a good surface tension reducing agent, and thus a good emulsifier. Modifications are essential for achieving and adjusting optimal ratios between lecithin's hydrophilic and lipophilic properties, and for ensuring good food processing ability. The most common processes for lecithin modification are deoiling, fractionation with absolute ethanol to produce phosphatidylcholine or phosphatidylinositol-enriched fractions, or enzymatic hydrolysis of these compounds. Emulsifiers and stabilizers play an important role in emulsion stability. This chapter describes some of the main characteristics of lecithin such as phospholipid composition, hydrophilic-lipophilic balance (HLB) values, types and functionalities. It also discusses the effects of different modified sunflower lecithins and the addition of a thickening agent (chia mucilage) on the physical stability of functional oil-in-water (O/W) emulsions as a function of storage time.

6.4.1 Introduction

Lecithin is a term used to refer to commercial mixtures of different phospholipids (PL). At some point the term began to be applied to phosphatidylcholine (PC), the main PL in lecithins, but this usage is now increasingly rare. Lecithins are now accepted to be complex PL mixtures and are staples of commerce.

Initially, the major source of lecithin was from animals, mainly from egg yolk. The advent of the process of refining and marketing of vegetable oils made possible large-scale lecithin production at low cost. Phosphatidylcholines (PC), phosphatidylethanolamines (PE), phosphatidylserines (PS), phosphatidylinositols (PI), phosphatidylglycerols (PG), phosphatidic acid (PA), glycophospholipids (GPL) and their lyso forms are the major phospholipids found in lecithin from vegetable sources (1,2). This is now a major by-product of the plant oil industry and represents a multifunctional additive, being used in a wide range of industrial applications: dietetic, pharmaceutical, food, cosmetics et cetera (3). Sunflower lecithin is a promising alternative to soybean lecithin because it is considered the product of a non-genetically modified organism (non-GMO). The distribution of the main phospholipid components of sunflower lecithin (16% PC, 14% PI, 8% PE) appears to be rather similar to that of soybean lecithin (15% PC, 10% PI, 11% PE) (4).

The functional properties of lecithin are mainly the result of the surface-active character of its constituent phospholipids, and have found numerous applications in the food industry, for products such as emulsifiers, stabilizers, controlled-crystallization agents, viscosity modifiers, antioxidants and reducers or replacers of fat (1,5). Nowadays, the ability to change the PL concentrations or even the actual PL composition of native lecithin using chemical modification processes can originate modified lecithins with different physicochemical and functional properties customized for diverse industrial purposes (5).

Lecithin modification under industrial conditions combined with adequate techniques of analysis may be useful for evaluating the potential applications of the sunflower by-products to the production of new emulsifiers (6,7). The lecithin modification processes such as deoiling, fractionation with ethanol or enzymatic hydrolysis all are important for achieving an optimal balance between hydrophilic and lipophilic compounds (the hydrophilic-lipophilic balance or HLB) and to ensure a good food processing ability (8).

Phospholipids, natural surface-active agents

Phospholipids (PL) are nature's principal surface-active agents, since they have both hydrophilic (water-soluble) and lipophilic (oil-soluble) portions in their molecular structure. Physically, they tend to concentrate at the interface between oil and water and subsequently reduce the interfacial tension. Thereby, they facilitate the formation of an emulsion during the homogenization process (9). Most PLs consist of a glycerol backbone esterified with two fatty acids and a phosphate group which, in turn, may be esterified with monovalent alcohols (e.g. choline or ethanolamine), or polyvalent alcohols (e.g. glycerol or inositol) (3). Because of this amphiphilic character, these compounds have a tendency to be arranged in ordered structures such as lipid bilayer or micelle, trying to align in hydrophilic/lipophilic interfaces, with their fatty acid chains oriented to the lipophilic phase and their polar head groups oriented to the aqueous phase (10). The different PL phase structures at an oil – water interface influence emulsion formation and stability.

In food processing, the electrostatic particles (e.g. fat crystals) and polymeric (e.g. protein) stabilization can have positive synergistic effects on emulsion stabilization. The PL composition and their relative abundances directly influence the characteristics of the emulsion; hence, various emulsions with specific characteristics for different applications can be formed by changing the phospholipid compositions of lecithins (1).

Phosphatidylcholine (PC) forms a lamellar phase at the interface between oil and water, different to the reversed hexagonal phase of PE or the hexagonal phase of lysophospholipids, which lack one or both esterified fatty acids (11). This knowledge is used for the successful application of lecithins with adapted, modified or fractionated phospholipid composition. It is a fact that different phospholipids have different emulsifying properties and that the phospholipid pattern of a lecithin determines the emulsion type that is formed. For example, the PC-enriched fraction, due to its high PC/PE ratio and the lamellar phase structure of the PC at the interface between oil and water, is recognized to be a good oil-in-water (O/W) emulsifier, while the PI-enriched fractions are more appropriate for water-in-oil emulsions (W/O) (4,12).

Hydrophilic-lipophilic balance (HLB)

The main application of lecithin in the food industry is associated with its role as emulsifier agent for dispersions or emulsions. Emulsifier properties are often characterized by their HLB value. This is an empirical value that varies between 1 and 20, where the value 1 is assigned to a surfactant that is completely oil-soluble and the value of 20 to emulsifiers that are completely soluble in water (13). The HLB number therefore gives an indication of the relative affinity of a surfactant molecule for the oil and aqueous phases (14). Each surfactant is assigned an HLB number according to its chemical structure. A molecule with a high HLB number has a high ratio of hydrophilic groups to lipophilic groups and vice versa (see Table 1).

Table 1.

Group numbers for calculating the HLB number (from Davis (14)).

Hydrophilic group	Number	Lipophilic group	Number
$-SO_4^-Na^+$	38.7	$-CH-$	−0.475
$-COO^-K^+$	21.1	$-CH_2-$	−0.475
N (tertiary amine)	9.4	$-CH_3-$	−0.475
Ester (sorbitan ring)	6.8		
−COOH	2.1		
−O−	1.3		

A widely used, semi-empirical method for calculating the HLB number of a surfactant is according to Eq. 1:

$$\text{HLB} = 7 + \Sigma\,(\text{hydrophilic group numbers}) - \Sigma\,(\text{lipophilic group number}) \quad \mathbf{(1)}$$

The HLB number of a surfactant gives a useful indication of its solubility in either the oil and/or water phase and can be used to predict the type of emulsion that will be formed (14). A surfactant with a low HLB number (3 to 6) is predominantly hydrophobic, dissolves preferentially in oil, stabilizes water-in-oil emulsions, and forms reverse micelles in oil. A surfactant with a high HLB number (8 to 18) is predominantly hydrophilic, dissolves preferentially in water, stabilizes oil-in-water emulsions, and forms micelles in water. A surfactant with an intermediate HLB number (6 to 8) has no particular preference for either oil or water. Molecules with HLB numbers below 3 and above 18 are not particularly surface active and tend to accumulate preferentially in bulk oil or aqueous phases, rather than at an oil – water interface (14).

One of the major drawbacks of the HLB concept is that it does not take into account the fact that the functional properties of a surfactant molecule are altered significantly by changes in temperature or solution conditions (14).

Fig. 1 illustrates the approximate hydrophilic-lipophilic balance (HLB) values of different lecithins (15). The emulsification properties of lecithins are often characterized by their HLB value. The different modifications on lecithins produce emulsifiers with specific hydrophilic-lipophilic properties. High hydrophilic HLB values increase the ability of lecithins to form O/W emulsions. A PC-enriched fraction obtained from an ethanol fractionation process (HLB ≈ 7.5) is recognized to be a good O/W emulsifier, while PI-enriched fractions, with lower HLB values, present more lipophilic characteristics due to relatively lower PC content, relatively higher PI content, and consequent reduction of their PC/PE ratios, being suited for W/O emulsions (13,15).

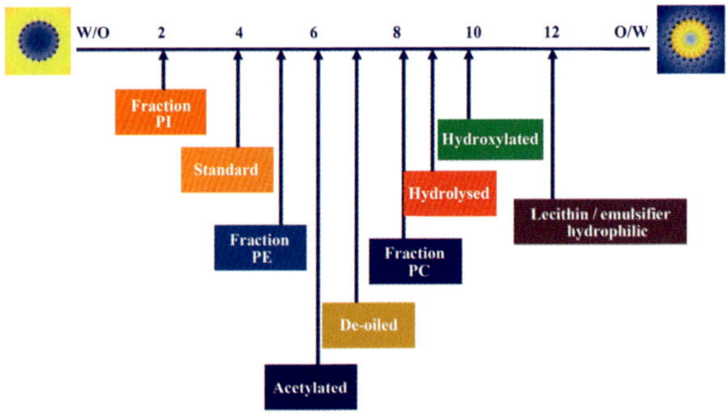

Figure 1.
Hydrophilic-lipophilic balance (HLB) of modified lecithins (from van Nieuwenhuyzen (15)).

On the other hand, in hydrolysed lecithins the relatively high concentration of hydrophilic phospholipids (mainly lysophospholipids) and relatively low levels of hydrophobic compounds (mainly phosphatidylinositols) increase the HLB value. Therefore, this type of lecithin presents the best properties as an O/W emulsifying agent (8,16).

Cabezas et al. (17) studied how to improve the O/W emulsifying properties of modified lecithins with low HLB values in mixtures with hydrolysed sunflower lecithins. The results suggest that the use of mixtures of modified sunflower lecithins with an optimal balance between hydrophilic and lipophilic compounds could revalue lecithin with a theoretical low capacity to form O/W emulsions.

Emulsions

Emulsions are heterogeneous systems composed of two immiscible liquids (usually oil and water), with one of the liquids dispersed as small spherical droplets in the other. Emulsions can be conveniently classified according to the distributions of their oil and aqueous phases. A system that consists of oil droplets dispersed in an aqueous phase is called an O/W emulsion (e.g. mayonnaise, milk, cream, soups and sauces). A system that consists of water droplets dispersed in an oil phase is called a W/O emulsion (e.g. margarine, butter and spreads) (18).

Both these emulsion systems are thermodynamically unstable from a physicochemical point of view, tending to eventually separate into two immiscible phases according to their kinetic stability. Physical destabilization mechanisms of emulsions include oil droplet size variation processes such as flocculation and coalescence, and particle migration phenomena like sedimentation and creaming (12,18).

Emulsions may become more kinetically stable by the addition of more emulsifiers and/or thickening agents prior to homogenization. Emulsifiers are surface-active molecules that adsorb to the surface of freshly formed droplets during homogenization, forming a protective membrane which prevents the droplets from coming close enough together to aggregate.

Thickening agents are ingredients added to O/W food emulsions with low oil content to obtain a stable system with good shelf life and modify their textural properties. They provide long-term emulsion stability by immobilizing the droplets of the dispersed phase, thereby increasing the viscosity of the emulsion's continuous phase (18). Hydrocolloids are widely used in different applications within the food industry, due to their thickening and gelling properties, syneresis control, emulsion stabilization et cetera (19). Among the hydrocolloids, it is worth mentioning mucilage, a complex carbohydrate of high molecular weight which is secreted when the seed is brought into contact with water and subsequently generates solutions with high viscosity. Many studies have examined the functional properties of gums from various plant species *(Salvia hispanica, Linux usitatissimum, Opuntia picus indica, Alyssum homolocarpum, Psyllium plantago)* (20–24).

In the case of chia, mucilage is an especially important component due to its physiological role. The mucilage extracted with water at room temperature from chia seeds yielded 38 g/kg with a composition of moisture 9.37%, protein 7.29%, crude fibre 11.42%, oil 3.83%, ash 10.27% and nitrogen-free extract (NFE) 56.24%. It is a new thickening agent that also can contribute to the health of consumers as a source of soluble dietary fibre (20).

6.4.2 Adding chia mucilage into O/W emulsions with sunflower lecithins. A case study

Stabilizers are crucial for emulsion stability. Optical characterization and droplet size distribution of oil-in-water emulsions formulated with different types and concentrations of modified sunflower lecithin, both a phosphatidylcholine-enriched fraction (PCF) and a deoiled sunflower lecithin (DL), with or without the addition of chia mucilage (CM: 0.75% wt/wt), have been evaluated as a function of storage time at 4±1°C (25).

The phospholipid compositions of the modified sunflower lecithins used in the study were as follows: for the PCF, PC 77.2%, PI 3.2%, PE 8.3%, PA 1.0%, other phospholipids 10.5%; for the DL, PC 36.4%, PI 35.8%, PE 11.6%, PA 4.5%, other phospholipids 11.7%. A marked difference between the two modified lecithins was observed. The PCF showed a PC/PE ratio of 9.3, which is a better level to use as an emulsifier agent in O/W emulsions compared to DL (3.1).

6.4.2.1 Rheological properties

The flow behaviour was analysed with a Haake RS600 controlled stress rheometer (Thermo Electron, Germany) using PP35-S serrated parallel plate measuring geometry (35 mm diameter), and measurements were performed at room temperature (25±1°C) in the 1–500 s^{-1} range. All measurements were performed at initial time (t=0) in triplicate. The experimental data (shear stress – shear rate) were fitted using the power law model according to Eq. 2:

$$\tau = k\dot{\gamma}^{n} \quad (2)$$

where τ is the shear stress (Pa); $\dot{\gamma}$ is the shear rate (s^{-1}); k is the consistency coefficient (Pa s^n) and n is the flow behaviour index (dimensionless).

O/W emulsions without chia mucilage (CM) exhibited typical Newtonian flow behaviour (n = 1), characterized by a linear relationship between the applied shear stress and the shear rate (i.e. constant viscosity) over the entire range of the parameters studied. The average viscosity (η) was 0.0014±0.0001 Pa s, with no significant differences ($p > 0.05$) for the different types and concentrations of modified sunflower lecithin (PCF, DL). The parameter k is sometimes referred to as the consistency index. For the case of a Newtonian fluid, the consistency index k is identically equal to the viscosity of the fluid, η.

Table 2 shows the power law parameters, consistency index (k) and flow behaviour (n) of the upward curve of the O/W emulsions with CM. The coefficients of determination (R^2) were ≥0.9875 for all samples, indicating that the power law model is adequate to describe the flow properties of emulsions with modified sunflower lecithins and CM. The k values for this type of emulsions varied between 0.680 – 0.720 and 0.549 – 0.565 Pa s^n for PCF and DL, respectively. A low value of this coefficient was observed for emulsion with CM alone (0.411 Pa s^n). Emulsions with chia mucilage and PCF presented a more consistent index with respect to the other ones; this finding could be related to a major PC – CM interaction.

Table 2.

Power law parameters for O/W emulsion with modified sunflower lecithins (PCF, DL) at different concentrations and with chia mucilage present.

Storage time (days)	PCF (% wt/wt) + CM 0.5	1.0	2.0	DL (% wt/wt) + CM 0.5	1.0	2.0
k (Pa s^n)	0.683±0.012^c	0.720±0.029^c	0.680±0.031^c	0.559±0.034^b	0.565±0.025^b	0.549±0.009^b
n	0.385±0.008^a	0.366±0.010^a	0.370±0.005^a	0.362±0.004^a	0.377±0.007^a	0.383±0.009^a
R^2	0.9875	0.9901	0.9889	0.9891	0.9921	0.9884

Values followed by different letters differ significantly (p <0.05) among rows, according to Tukey's test. Data are means ± SD (n=3). CM: chia mucilage (0.75% wt/wt).

The different O/W emulsions with CM analysed presented a pseudoplastic flow behaviour (n≤ 1). The addition of emulsifier provided a significantly higher consistency to the emulsion in comparison with those containing CM alone, mainly PCF.

6.4.2.2 Optical characterization of O/W emulsions

The physical stabilities of the O/W emulsions (30:70 wt/wt) immediately after the homogenization step were followed by measuring the evolution of backscattering (% b) with a QuickScan Vertical Scan Analyzer (Coulter Corp., Miami, FL). The backscattering of monochromatic light (λ = 850 nm) by the emulsions was determined as a function of the height of the sample tube (65 mm) for quantifying the rate of the different destabilization processes during the first 60 min. This methodology allowed a differentiation between particle migration (sedimentation, creaming) and particle size variation (flocculation, coalescence) processes (26). The basis of the vertical scan analyser profiles has been exhaustively studied by Mengual et al. (27).

The destabilization of the O/W emulsions was followed by monitoring of the sequential profiles obtained in the bottom zone I (10–15 mm), the area where a clarification of the emulsion is produced by the migration of the emulsified droplets from the bottom towards the top of the tube (i.e. creaming), and in the upper zone II (50–55 mm), this area being the location of the potential destabilization of the cream phase of the emulsions through coalescence as a function of time.

Fig. 2 shows the backscattering (% b) profiles corresponding to emulsions containing 1.0% of the modified sunflower lecithins (PCF, DL) (25).

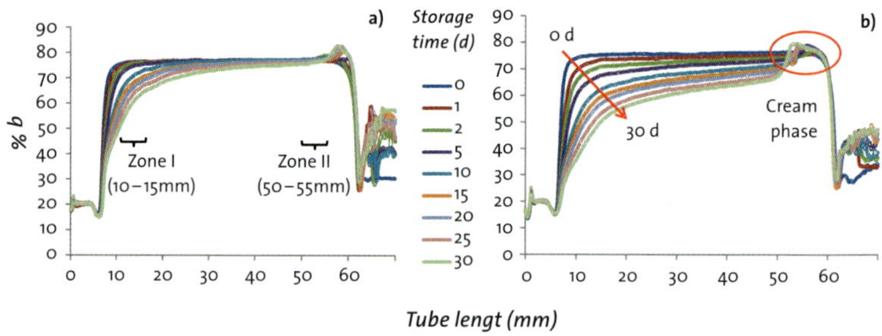

Figure 2.

Backscattering (% b) profiles of O/W emulsions with 1.0% wt/wt of modified sunflower lecithins: a) PCF, b) DL as a function of storage time (d) at 4±1°C. Zone I = 10–15 mm, zone II = 50–55 mm. The length 0 mm was located at the bottom of the tube (25).

The comparative evolution of % b values corresponding to O/W emulsions containing different concentrations of modified sunflower lecithins (PCF, DL) is visualized in Fig. 3. At the initial stage of storage (t = 0), the QuickScan profiles corresponding to zone I did not show significant variations (p <0.05) of % b values between the different emulsions studied. That is, the p values were >0.05. During storage at 4±1°C, O/W emulsions with 0.5 and 1.0% DL presented a faster destabilization than the other ones. On the other hand, an emulsion with 2.0% PCF was the most stable system. After 30 d, emulsions with 0.5 and 1.0% DL showed a sharp decrease in % b of 40 and 38%, respectively, while for 2.0% of this emulsifier the decrease represented about 21%. Emulsions with PCF showed a significant increase (p ≤0.05) of the emulsion stability as a function of increasing concentration. The influence of storage produced a decrease of the % b values (25, 18 and 12%) for emulsions with 0.5, 1.0 and 2.0%, respectively. It is interesting to note that the diminution of % b level at the bottom of the tube can be related to the creaming process during storage.

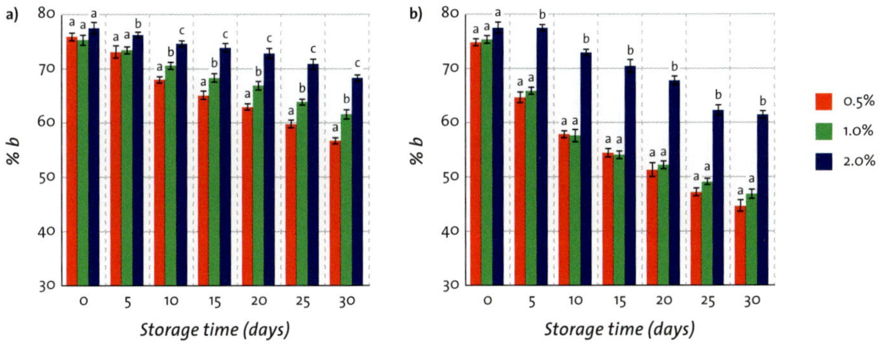

Figure 3.

Backscattering (% b) values of O/W emulsions with a) PCF, b) DL, in a range of concentrations 0.5 to 2.0% wt/wt (zone I 10–15 mm). Different letters differ significantly (*p* <0.05) at the various storage times, according to Tukey's test. Average values ± SD (n = 2).

Zone II (50–55 mm) is characterized by the accumulation of oil droplets after the creaming process (cream phase). The difference of the density between the oil and water phases produces the migration of the oil droplets to the upper zone of the tube, i.e. gravity separation. In this zone, it can generally be observed that the addition of both emulsifiers (PCF, DL) allowed the formation of a stable cream phase versus time during the studied storage period. Similar average % b for emulsions was recorded for the different emulsifiers (1.0 and 2.0%). However, emulsion with 0.5% DL presented a significant decrease ($p \leq 0.05$) during the storage time (data not shown). This observation could suggest that at this level of emulsifier the droplets were not protected by a sufficiently strong interfacial film. These results could be associated with the occurrence of the cream phase destabilization by coalescence. The differences of the emulsifying properties of different modified sunflower lecithins may be related to the PC/PE ratio, since the PCF is a better O/W emulsifier than DL (7).

In the cases of O/W emulsions with chia mucilage and sunflower lecithins, the % b values corresponding to zone I (10–15 mm) showed an increase as a function of increasing concentration of modified lecithins, mainly with PCF, while 1.0 and 2.0% wt/wt DL did not exhibit significant differences ($p \geq 0.05$) (Fig. 4). In particular, O/W emulsions with a PCF level of 2.0% presented higher % b values than those obtained with DL over the same concentration ($p \leq 0.05$), while for 0.5 and 1.0% wt/wt of these emulsifiers, no significant differences ($p \geq 0.05$) were observed.

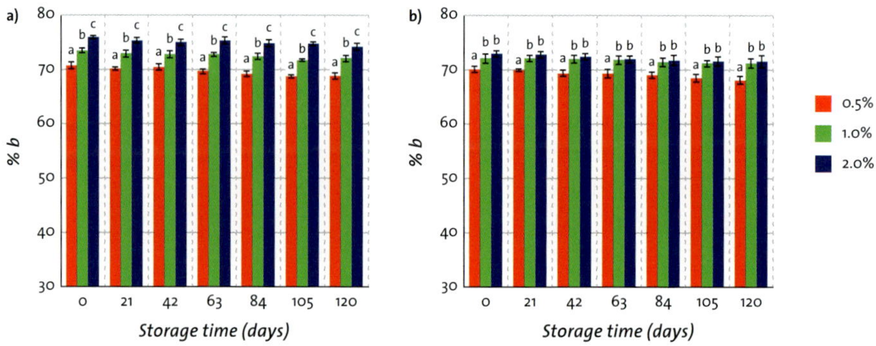

Figure 4.

Backscattering (% b) values of O/W emulsions with a) PCF, b) DL, in a range of concentrations 0.5 to 2.0% wt/wt and all with chia mucilage present (0.75% wt/wt) (zone I 10–15 mm). Different letters differ significantly (*p* <0.05) within each storage time, according to Tukey's test. Average values ± SD (n = 2).

The % b values of the cream phase (zone II) of these O/W emulsions showed a similar behaviour to that observed in zone I (data not shown). All emulsions remained stable during storage at 4±1°C, the average % b of PCF with 0.5, 1.0 and 2.0% being 69.9±0.6, 72.7±0.4 and 75.5±0.3, respectively, as compared to 69.2±0.5, 71.9±0.4 and 72.2±0.6, respectively, in the DL cases.

It is worth mentioning that the O/W emulsions formulated with different concentrations of modified sunflower lecithin (PCF, DL) and CM (0.75% wt/wt) exhibited similar profiles with higher stability against creaming during the period analysed (120 d) (Fig. 4) than the emulsions previously described (30 days, see Fig. 3). This fact could be mainly attributed to the addition of CM, which would prevent the individual droplets from moving and, therefore, would decrease the probability of the formation of aggregates/ fusion of droplets contributing to the emulsion stability. Emulsions with CM alone presented good stability during the storage period with % b levels of 70%. Avila-de la Rosa et al. (28) studied the viscoelastic response of chia seed mucilage in the vicinity of the oil – water interface and observed that its emulsification properties seemed to arise from the formation of spatial structures surrounding the oily phase.

6.4.2.3 Particle size distribution

Particle size distribution was determined with a Malvern Mastersizer 2000E analyser (Malvern Instruments Ltd., UK) with a Hydro 2000MU dispersion unit. Optical parameters were refractive indexes of sunflower oil and water 1.47 and 1.33, respectively. Samples of each emulsion were diluted in a water container, at a stirring rate of 2,000 rpm, in which a laser beam passed through a transparent internal cell where the diluted emulsion was recirculated. The light scattered at various angles from droplets of different sizes passed through a complex optical system, creating an angular light scattering pattern. The software of the equipment then translated this pattern into the corresponding particle size distribution (18). The De Brouckere mean diameter D (4,3) associated with volume particle distribution was determined during storage time.

Fig. 5 shows the comparative particle size distribution in volume for O/W emulsions obtained with different modified sunflower lecithins (PCF, DL) and concentrations (0.5 – 2.0% wt/wt) at initial time (t = 0) (25). These distributions presented a monomodal or bimodal character depending on the concentration of modified lecithin added. In this sense, 2.0% modified lecithin (PCF, DL) showed a bimodal character, but these small populations of large particles did not affect the stability of emulsions. Regarding emulsions with DL, a displacement to a smaller droplet size as a function of increasing concentrations was observed.

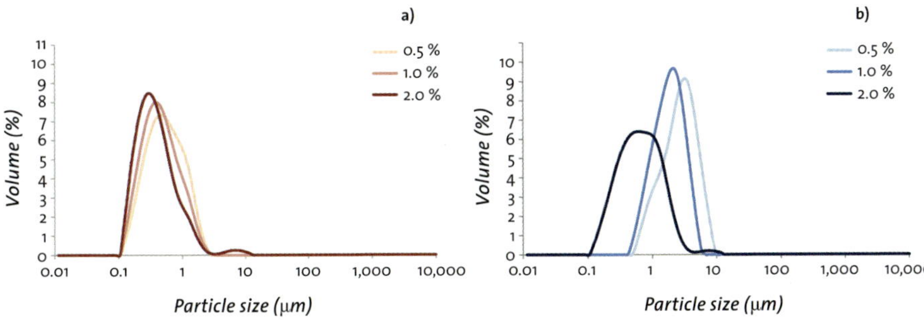

Figure 5.

Particle size distribution in volume of O/W emulsions with different concentrations (0.5 – 2.0% wt/wt) of modified sunflower lecithins: a) PCF, b) DL at initial time (t = 0) (from Guiotto et al. (25)).

O/W emulsions formulated with PCF presented similar mean diameter D (4,3) values and no significant differences (p >0.05) were recorded between the different concentrations (pa). O/W emulsions with DL showed a significant decrease (p ≤0.05) of mean diameter D (4,3) values as a function of increasing concentration of this emulsifier agent. Emulsions with 2.0% DL exhibited mean diameter D (4,3) values between 0.69 and 0.81 µm (micrometres), without significant differences (p >0.05) during refrigerated storage time. Furthermore, emulsions with 0.5 and 1.0% DL increased significantly (p ≤0.05) compared to the corresponding mean diameter D (4,3) values at the initial time (data not shown).

Table 3.

De Brouckere mean particle diameter D (4,3) (µm) of O/W emulsions with modified sunflower lecithins (PCF, DL) during storage at 4±1°C.

Storage time (days)	PCF % wt/wt			DL % wt/wt		
	0.5	1.0	2.0	0.5	1.0	2.0
0	0.61±0.02^{ab}	0.53±0.03^{a}	0.47±0.02^{a}	1.72±0.05^{d}	1.39±0.03^{c}	0.69±0.01^{b}
7	0.62±0.02^{a}	0.61±0.03^{a}	0.55±0.01^{a}	2.51±0.04^{d}	1.92±0.04^{c}	0.78±0.02^{b}
15	0.57±0.01^{a}	0.51±0.03^{a}	0.53±0.02^{a}	2.63±0.02^{d}	1.90±0.03^{c}	0.81±0.02^{b}
30	0.61±0.03^{a}	0.53±0.04^{a}	0.54±0.02^{a}	2.82±0.05^{d}	1.87±0.04^{c}	0.81±0.01^{b}

Mean values followed by different letters differ significantly (p <0.05) among rows, according to Tukey's test.
Average values ± SD (n = 2).

D (4,3) values correlate with the destabilization shown in Fig. 3, in which the decrease of % b values was mainly associated with a migration of particles (gravity separation) rather than a droplet size increase (coalescence). Furthermore, emulsions with PCF were more stable due to the formation of droplets of smaller diameter with respect to emulsions with DL.

The particle size distributions of O/W emulsions with modified sunflower lecithins (PCF, DL) and CM at the initial and final time of storage are shown in Fig. 6 (25). A displacement of the curves towards to the right (large droplets) can be seen as a function of a decrease of concentration of these emulsifier agents.

At initial time, the particle size distribution of the O/W emulsions with different concentrations of PCF and CM exhibited monomodal character. Moreover, emulsions with DL and CM showed a bimodal character for all concentrations studied.

During the storage time it was possible to observe the evolution of the particle size distribution from a monomodal to a bimodal profile for O/W emulsions with PCF and CM. In contrast, emulsions with DL and CM presented an increase in the large droplet population. These changes in particle size distribution did not significantly influence the physical stability of the O/W emulsions.

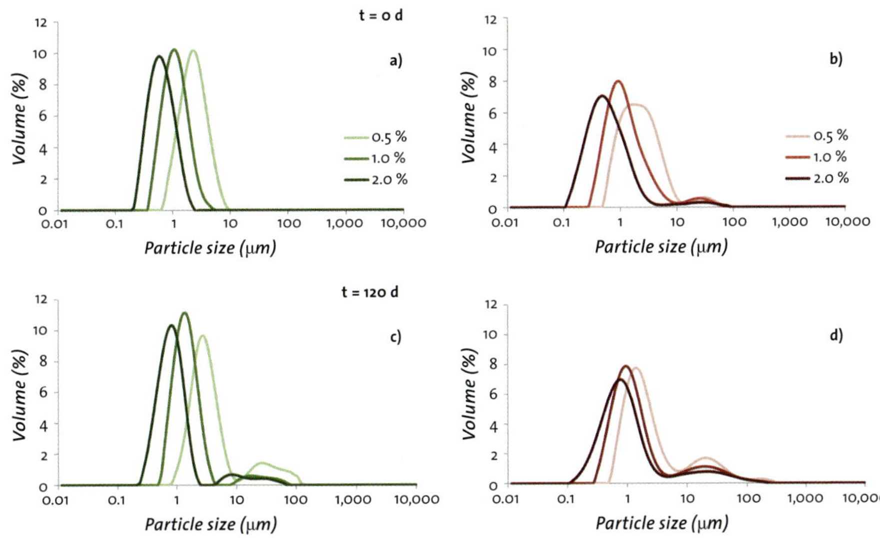

Figure 6.

Particle size distribution in volume of O/W emulsions with different concentrations (0.5–2.0% wt/wt) of modified sunflower lecithins, all with chia mucilage present (0.75% wt/wt): a), c) PCF; b), d) DL at initial (t = 0) and final storage time (t = 120 d) (from Guiotto et al. (25)).

Table 4.

De Brouckere mean particle diameter D (4,3) (μm) for O/W emulsions formulated with modified sunflower lecithin and chia mucilage, during storage time at 4±1°C.

Storage time (days)	CM % wt/wt	PCF % wt/wt			DL % wt/wt		
	0.75	0.5	1.0	2.0	0.5	1.0	2.0
0	13.51 ± 0.40^{d}	2.24 ± 0.29^{ab}	1.90 ± 0.06^{a}	1.52 ± 0.16^{a}	4.90 ± 0.30^{c}	2.97 ± 0.31^{b}	1.86 ± 0.19^{a}
30	13.88 ± 0.20^{d}	9.73 ± 0.33^{c}	5.60 ± 0.28^{b}	3.76 ± 0.22^{a}	9.15 ± 0.14^{c}	5.61 ± 0.13^{b}	4.35 ± 0.41^{a}
60	15.78 ± 0.02^{e}	9.75 ± 0.35^{d}	5.77 ± 0.27^{bc}	3.01 ± 0.40^{a}	9.43 ± 0.29^{d}	6.54 ± 0.18^{c}	4.77 ± 0.13^{b}
90	16.28 ± 0.15^{e}	9.46 ± 0.56^{d}	5.36 ± 0.39^{c}	3.02 ± 0.18^{a}	8.49 ± 0.13^{d}	6.05 ± 0.27^{c}	4.30 ± 0.04^{b}
120	16.53 ± 0.35^{e}	9.17 ± 0.11^{d}	4.64 ± 0.19^{b}	2.58 ± 0.10^{a}	8.82 ± 0.11^{d}	6.26 ± 0.02^{c}	4.91 ± 0.32^{b}

Values followed by different letters differ significantly ($p < 0.05$) among rows, according to Tukey's test. Data are means ± SD (n = 2). CM: chia mucilage.

Table 4 shows that D (4,3) values decreased significantly ($p \leq 0.05$) as a function of enhancing modified sunflower lecithin (PCF, DL) concentration with chia mucilage (CM). D (4,3) values of emulsions formulated with CM alone were higher than other ones and exhibited an increasing tendency as a function of storage time (data not shown). Regarding refrigerated storage time, this parameter showed significant increases ($p \leq 0.05$) over the first

30 days and then remained stable for different O/W emulsions (Table 4). This fact was evidenced in terms of the particle size distributions through the formation and/or increase of a second population of large droplets (see Fig. 6).

On the other hand, the D (4,3) values for O/W emulsions with 0.5% emulsifiers did not show significant differences (p >0.05) between PCF and DL, both with CM from 30 days of storage. O/W emulsion with levels of 1.0% PCF exhibited D (4,3) values significantly lower (p ≤0.05) with respect to those obtained with DL for 0 and 120 d. At levels of 2.0%, from 60 days of storage, D (4,3) values of the O/W emulsion with PCF were significantly lower (p ≤0.05) compared to the emulsions with DL (Table 4).

The comparative evaluation of O/W emulsions with CM alone versus those with modified sunflower lecithins (PCF, DL) and CM (0.75% wt/wt) suggests a significant effect of the presence of PCF or DL on the particle size. The CM-modified lecithin interaction could contribute to the production of particles of smaller diameter and with higher interfacial area than those emulsions without this chia by-product.

6.4.3 Conclusions

The main application of lecithin in the food industry is associated with its role as emulsifying agent. Sunflower lecithin may be used as a food additive in view of its high phosphatidylcholine and essential fatty acid content. Its use in the manufacture of food could be increased by application of modification processes (deoiling, fractionation, enzymatic modification), resulting in lecithins with different physicochemical and functional properties for various industrial purposes. Thus, a case analysis of the O/W emulsions with phosphatidylcholine-enriched fraction from sunflower lecithin (PCF) exhibited higher stability in comparison with those obtained with deoiled sunflower lecithin (DL). This fact could be associated with the highest PC/PE ratio found in the PCF, which would favour their use as an emulsifier agent in this type of emulsion.

On the other hand, the addition of chia mucilage produced a good performance as a novel functional ingredient to be used as stabilizer or thickening agent in emulsions. Chia mucilage contributed to obtaining stable emulsions against creaming during refrigerated storage, probably due to the reduction of the mobility of oil particles by the formation of a tridimensional network.

Modified sunflower lecithin and chia mucilage interaction produces stable emulsions, which present high stability and low particle sizes as a function of increasing concentration of emulsifier, mainly in the case of PCF.

Acknowledgements

This work was supported with grants from Agencia Nacional de Promoción Científica y Tecnológica (ANPCyT, PICT 2016 0323), Consejo Nacional de Investigaciones Científicas y Técnicas (CONICET) and Universidad Nacional de La Plata (UNLP), Argentina. B. W. K. Diehl Director of Spectral Service GmbH, Cologne, Germany is acknowledged for his technical assistance.

References

1. McClements, D.J., 1997. Lipid-based emulsions and emulsifiers. In Akon, C.C., Min, D.B. (Eds), Food Lipids: Chemistry, Nutrition and Biotechnology. Marcel Dekker.
2. Szuhaj, B.F., 2005. Lecithins. In Shahidi, F. (Ed.), Bailey's Industrial Oil and Fat Products. John Wiley & Sons, Sixth Edition, Vol. 2, Chapter 13, 361-456.
3. Schneider, M., 2008. Major sources, composition and processing. In Gunstone, F.D. (Ed.), Phospholipid Technology and Applications. The Oily Press, Bridgewater, 21-40.
4. Van Nieuwenhuyzen, W., Tomás, M.C., 2008. Update on vegetable lecithin and phospholipid technologies. Eur. J. Lipid Sci. Technol. 110, 472–486.
5. Dashiell, G.L., 1988. Chap. 1. Lecithin in food processing applications. In Szuhaj, B.F. (Ed.), Lecithins, Sources, Manufacture & Uses. AOCS Press, Champaign, IL.
6. Cabezas, D.M., Diehl, B.W.K., Tomás, M.C., 2009. Effect of processing parameters on sunflower PC enriched fractions extracted with aqueous-ethanol. Eur. J. Lipid Sci. Technol. 111, 993-1002.
7. Guiotto, E.N., Cabezas, D.M., Diehl, B.W., Tomás, M.C., 2013. Characterization and emulsifying properties of different sunflower phosphatidylcholine enriched fractions. Eur. J. Lipid Sci. Technol. 115(8), 865-873.
8. Cabezas, D.M., Diehl, B.W., Tomás, M.C., 2015. Emulsifying properties of hydrolysed and low HLB sunflower lecithin mixtures. Eur. J. Lipid Sci. Technol. (in press).
9. Van Nieuwenhuyzen, W., 2014. The changing world of lecithins. Inform 25(4), 254-259.
10. Wendel, A., 2000. Lecithin: the first 150 years. Part I: From discovery to early commercialization. Inform 11, 885-892.
11. Van Nieuwenhuyzen, W., Szuhaj, B.F., 1998. Effects of lecithins and proteins on the stability of emulsions. Eur. J. Lipid Sci. Technol. 100, 282-291.
12. Bueschelberger, H.G., 2004. Lecithins. In Whitehurst, R.J. (Ed.), Emulsifiers in Food Technology. Blackwell Publishing, West Sussex.
13. Hernández, E., Quezada, N., 2008. Uses of phospholipids as functional ingredients. In Gunstone, F.D. (Ed.), Phospholipid Technology and Applications. The Oily Press, Bridgwater.
14. Davis, H.T., 1994. Factors determining emulsion type: hydrophile-lipophile balance and beyond. Colloids Surf. A 91, 9-24.
15. Van Nieuwenhuyzen, W., 1998. Lecithin open doors to baked goods. IFI 2:32- 36.
16. Carlsson, A., 2008. In Gunstone, F.D. (Ed.), Phospholipid Technology and Applications. The Oily Press, Bridgwater.
17. Cabezas, D.M., Madoery, R., Diehl, B.W.K., Tomás, M.C., 2012. Emulsifying properties of different modified sunflower lecithins. J. Am. Oil Chem. Soc. 89, 355-361.
18. McClements, D.J., 1999. Food Emulsions: Principles, Practice and Techniques. CRC Press, NY.
19. Vázquez-Ovando, J., Rosado-Rubio, G., Chel-Guerrero, L., Betancur-Ancona, D., 2009.

Physicochemical properties of a fibrous fraction from chia (Salvia hispanica L.). LWT - J. Food Sci. Techol. 42, 168-173.

20. Capitani, M.I., Ixtaina, V.Y., Tomás, M.C., Nolasco, S.M., 2013. Microstructure, chemical composition and mucilage exudation of chia (Salvia hispanica L.) nutlets from Argentina. J. Sci. Food Agric. 93(15), 3856-3862.
21. Mazza, G., Biliaderis, C.G., 1989. Functional properties of flax seed mucilage. J. Food Sci. 54, 1302-1305.
22. Sáenz, C., Sepúlveda, E., Matsuhiro, B., 2004. Opuntia spp mucilages: a functional component with industrial perspectives. J. Arid. Environ. 57, 275-290.
23. Koocheki, A., Mortazavi, S.A., Shahidi, F., Razavi, S.M.A., Taherian, A.R., 2009. Rheological properties of mucilage extracted from Alyssum homolocarpum seed as a new source of thickening agent. J. Food Eng. 91, 490-496.
24. Cheng, Z., Blackford, J., Wang, Q., Yu, L., 2009. Acid treatment to improve psyllium functionality. J. Funct. Foods 1, 44-49.
25. Guiotto, E.N., Capitani, M.I., Nolasco, S.M., Tomás, M.C., 2016. Stability of Oil-in-Water Emulsions with Sunflower (Helianthus annuus L.) and Chia (Salvia hispanica L.) By-Products. J. Am. Oil Chem. Soc. 93(1), 133-143.
26. Pan, L.G., Tomás, M.C., Añón, M.C., 2002. Effect of sunflower lecithins on the stability of Water in Oil (W/O) and Oil in Water (O/W) emulsions. J. Surfactants Deterg. 5, 135-143.
27. Mengual, O., Meunier, G., Cayré, I., Puech, K., Snabre, P., 1999. Turbiscan MA 2000: multiple light scattering measurement for concentrated emulsion and suspension instability analysis. Talanta 50, 445-456.
28. Avila-de la Rosa, G., Alvarez-Ramírez, J., Vernon-Carter, E.J., Carrillo-Navas, H., Pérez-Alonso, C., 2015. Viscoelasticity of chia (Salvia hispanica L.) seed mucilage dispersion in the vicinity of an oil-water interface. Food Hydrocolloids 49, 200-207.

6.5 Use of lecithin in the production of instant dairy powder

Waldemar Buxmann

6.5.1 Requirements for instant dairy powder

Against the background of an annual increase in milk production worldwide, the significance of dairy powder is growing enormously. Milk is dried into powder in order to facilitate storage and transportation, and the powders are often converted into milk again. Furthermore, the powders are also used as ingredients of numerous foods such as confectionery or bakery products, and milk proteins are becoming more and more common as constituents of sport nutrition products and infant formulas.

If dairy powders are to be used for these purposes, they must have specific functionalities. Besides flavour, colour, flowability etc. an essential characteristic is the ability to dissolve in cold liquids such as water and milk. Powders with these properties are termed instant (Fig. 1).

Instant powders are foods in powder form that disperse or dissolve quickly and completely, without lump formation, with little energy input, when stirred into cold water (1,2).

Figure 1.

Example of the behaviour of a non-instantized milk powder (left) and an instantized milk powder (right) in cold water.

The process of mixing with cold water, also termed reconstitution, is rather complex and can be divided into several actions which take place either successively or – more often – simultaneously. If powder is added onto the surface of a liquid, the liquid penetrates the pore network of the particles due to capillary action and the particles become **wet**. Subsequently, the particles **sink** into the liquid by overcoming the surface tension, followed by **dispersion** and eventually by **dissolution** of the particles in the liquid (3–6).

The instant properties of dairy powders can therefore be characterised by following four parameters.

Wettability: The ability of a powder to absorb water and become wet. In the process of wetting, the gas phase in the cavities between the particles is replaced by water. Wettability depends on the size, density, porosity, surface charge, surface structure and surface activity of the particles (7–9).

Sinkability is a function of density, i.e. the mass and the proportion of enclosed air. Higher-density and less air cause the particles to sink below the surface of the water. Increasing the weight of the particles, e.g. by agglomeration, increases their ability to sink. After a time the density of the particles decreases, since certain substances contained in the powder, such as lactose or minerals, dissolve quickly in water. At the same time, the density of the solution increases. This may cause the particles that have already sunk down to start rising up again (10).

Dispersibility: The milk powder agglomerates wetted with water should break down into primary particles immediately, with little mechanical assistance or completely without, and disperse uniformly in the water. In this process the structure and adhesive properties of the individual particles play an important role (10,11).

The **solubility** of a powder in water depends chiefly on its chemical composition. Generally speaking, it has less significance for the instant properties of a powder (1).

It is important to emphasize that there is no clear dividing line between the various stages of reconstitution. So it is difficult to determine the individual capabilities independently of each other. This means that all these quality parameters shall be targeted if good instant properties are to be achieved (9,12).

In sum, we may say that the instant properties of a powder mainly depend on its composition and structure and the affinity of its constituents to water (5,13).

Because of their surface composition and structure and resulting characteristics, spray-dried dairy powders tend to have rather poor instant properties. Two different effects may occur:

1) In the case of a powder with a large proportion of polar, hydrophilic or 'hydrate-forming' components on the surface, e.g. proteins, the following can be observed: when a powder of this kind comes into contact with water, the surface quickly becomes wet. As a result, a gelatinous hydrate film forms on the surface (Fig. 2). This is impervious to water and thus prevents further penetration of the water and wetting of the powder. As a result, lumps form that are wet on the outside and dry inside and float on the surface of the water. Well-known examples of such powders are whey protein concentrate (WPC) and isolate (WPI). See Fig. 2 (4,9).

2) The second effect occurs if the powder is rich in hydrophobic components (e.g. fat). Because of its hydrophobic (water repelling) nature, this surface fat prevents wetting with water through its surface tension. The powder floats on top of the water, completely unwetted. Examples include whole milk powder and also skimmed milk powder to which vegetable fats have been added (fat filled milk powder) (14).

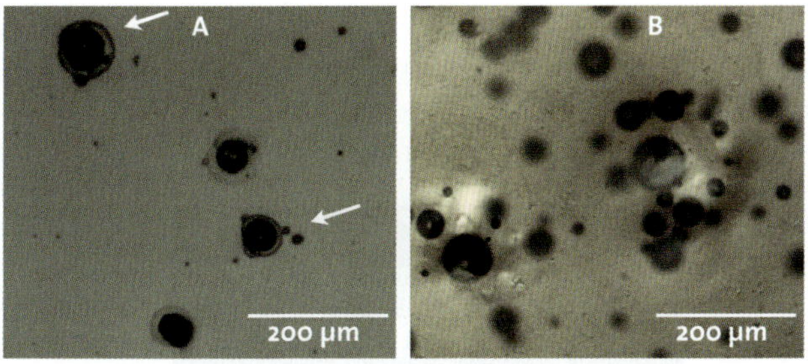

Figure 2.

Optical microscopy images of whey protein isolate (WPI) particles showing impermeable hydrate films at the interface between the particles and the water (marked with white arrows in A); thicker films can be seen in B when more WPI particles have been wetted with water. The particles were dispersed in water on slides before the analysis. White bar = 200 µm. From Ji, J. et al. (4) and (9), modified with the kind permission of the author.

It becomes clear that wetting of the powder with water is the most important step towards achieving instant properties (15). Ji et al. (4) describe wetting as the critical step that determines the speed at which dairy powders can be reconstituted.

The wettability of dairy powders depends primarily on the hydrophobicity and hydrophilicity of the surface of the powder. The smaller the angle of contact between the powder and the wetting agent (water), the greater is the wettability. Hydrophobicity and hydrophilicity can be influenced with surface active substances, and lecithin is by far the most important surfactant used. The phospholipids of the lecithin can reduce the speed of hydration and thus prevent or decelerate formation of the gelatinous structure of the proteins, so that the water has enough time to wet all the particles uniformly. Moreover, lecithin reduces the surface tension between the powder particles and the water so that the powder disperses more quickly in the water.

A further requirement for good wetting is that the powder particles are highly porous, with large cavities between them, and that they have a certain minimum size. Once the water has overcome the first barrier, i.e. either gelation or surface tension, it has to penetrate into the particles through the capillaries or cavities in order to wet the whole surface and the core of the particles. However, natural milk powders produced solely by spray drying have small, dense particles without capillaries. Large particles with a rough surface and a porous structure can be produced by agglomeration (Fig. 3).

Agglomeration is defined as a process in which single particles are brought together to form large, porous structures. In order to achieve good instant properties, the particles that have a size of 40 – 80 µm after spray drying must be enlarged to >100 or preferably 200 – 500 µm (16,17).

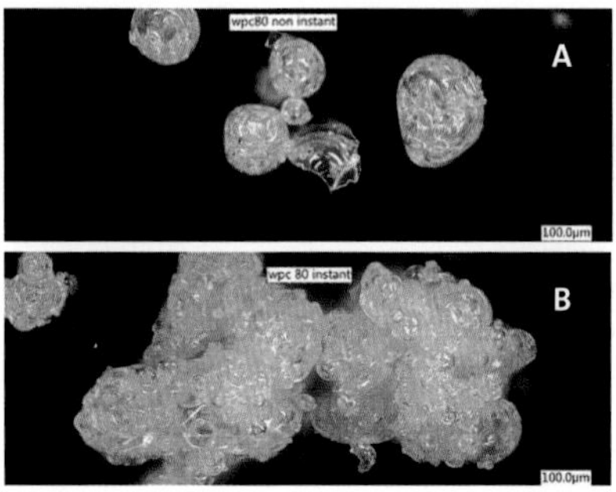

Figure 3.

Light microscope images of a non-agglomerated (A) and agglomerated (B) WPC 80 (whey protein concentrate) powder.

This leads to the conclusion that wettability, and therefore the instant properties of dairy powders, can be improved in the following way:

a) By modifying the polarity of the powder surface ➤ lecithination.
b) By modifying the structure of the powder ➤ agglomeration.
c) By combining the two ➤ agglomeration + lecithination.

6.5.2 Choice of the process

Powders can be agglomerated and lecithinated by different methods.

Büschelberger (14) describes the process from the point of view of lecithination, distinguishing between 'in-line' and 'post-process' lecithination. Schuck (10), on the other hand, distinguishes between 'straight-through' and 'rewet' processes, referring to the agglomeration process which may include lecithination.

The following sections describe how the instant properties of dairy powders can be adjusted specifically by optimizing the instantization process. This will be presented by integrating the respective points of view of both lecithination and agglomeration.

6.5.2.1 Straight-through agglomeration (in-line lecithination)

We speak of an in-line process when lecithination follows spray drying and agglomeration as part of a continuous process for the production of instant dairy powders. The main steps in the process are as follows:

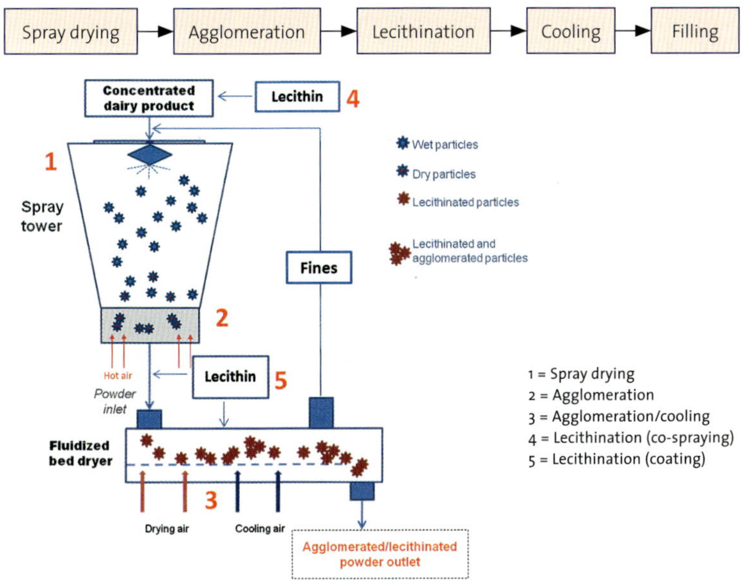

Figure 4.

Diagram of straight-through agglomeration including in-line lecithination.

Straight-through spray drying and agglomeration

The typical 'straight-through' spray drying and agglomeration process for the production of instant dairy powders is based on the two-stage or three-stage method. Both include the use of spray drying and fluidized bed drying (18).

The process starts with the production of a milk concentrate containing about 50% solids. In the case of milk powders containing fat, extra fat – e.g. palm or coconut – may be added at this stage by means of homogenization.

The milk concentrate is then atomized through one or more spray nozzles (at high pressure) or rotating discs (turning at high speed) in order to form small droplets that are dried in a countercurrent (80–220°C) as they fall through a drying chamber. This process is known as spray drying (Fig. 4, process step 1).

The powder can be agglomerated simultaneously with spray drying, for example through suitable arrangement of the nozzles and by returning the fines fractionated in the fluidized bed, which serve as nuclei for subsequent agglomeration (1).

In the **two-stage method**, the spray-dried and pre-agglomerated powder falls into an external vibrating fluidized bed dryer (Fig. 4, process step 3). This consists of one to four temperature sectors. In the first sector of the vibrating fluidized bed, the powder is dried with hot air at a temperature of 80–120°C to a residual moisture of 2–4%. In the next sectors of the fluidized bed, the powder is then cooled at different temperatures until it reaches room temperature (approx. 25°C). In this two-stage drying method, further agglomeration also takes place in the fluidized bed.

In the **three-stage method**, a static fluidized bed dryer is integrated at the bottom of the spray tower (Fig. 4, process step 2); this is in addition to the vibrating fluidized bed. In this process, the final and most important agglomeration step takes place in the static fluidized bed dryer. In this case the vibrating fluidized bed has the task of finally agglomerating the lecithinated powder (see below) or ensuring that the lecithin is distributed evenly over the surface of the powder and cools down slowly and evenly.

In-line lecithination

Powders can be treated with lecithin simultaneously with agglomeration by the straight-through method. This part of the instantization process is called lecithination. In this case, too, there are several ways of applying lecithin into or onto the powder.

1. In situ lecithination (= co-spraying).
In this method, lecithin is added to the milk concentrate, which is then spray dried (Fig. 4, process step 4). Here, the lecithin acts both as an emulsifier and as a wetting agent.

Since the active part of lecithin, the phospholipids may also form complexes with proteins (lipoproteins), they are no longer fully available as wetting agents. In the case of whole milk powder or fat filled skimmed milk powder, for example, the fat collects on the outside of the powder particles during spray drying. Since the lipoproteins accumulate in one place, the phospholipids may not cover the whole surface area (19).

Moreover, it has to be assumed that during subsequent agglomeration a large part of the lecithin will be found inside the particles instead of on the surface of the agglomerates. In other words, the effect of the lecithin will not be great enough to ensure good wetting.

The co-spraying method is often used, if the lecithin also needs to serve as an emulsifier in subsequent use of the powder, for example in calf milk replacers. Lecithin can also be added in combination with other co-emulsifiers in order to support the emulsifying effect.

2. In-line lecithination (= coating)
The method most widely used in the production of lecithinated instant dairy powders is lecithination by coating. In this process, lecithin is sprayed onto the powder as part of the straight-through method via a nozzle between the spray tower or static fluidized bed dryer and the inlet of the external vibrating fluidized bed dryer (Fig. 4, process step 5). Alternatively, it is also possible to spray the lecithin onto the milk powder in the vibrating fluidized bed dryer.

In order to ensure effective lecithination, the following parameters should be targeted

- The temperature of the lecithin should be 60 – 65°C in order to reduce the viscosity of the lecithin and improve its atomization and distribution over the surface of the powder.

- The temperature of the powder should be preferably about 50–55°C, also in order to improve the distribution of the lecithin over the surface of the powder (12).
- It is important to distribute the lecithin evenly over the surface of the particles in order to prevent it from accumulating in certain places. According to Millqvist-Fureby (20), the optimal amount of lecithin deposited on the surface of the particles is 8 mg/m^2. The optimal thickness of the phospholipid layer is 0.1– 0.3 µm (16). It is important for the phospholipids to take the form of a monolayer on the particle surfaces in order to achieve the desired wetting effect. If too much lecithin is applied per available surface area of a powder, double layers may form. In this case the non-polar parts of the phospholipid molecules protrude outwards, causing the required effect of reducing the surface tension to be lost (17).

In this procedure it is usually only the surface of the agglomerates that becomes coated with lecithin. Fig. 5 shows that the surface of the WPC 80 agglomerates, produced by the straight-through process, is almost completely and evently covered with lecithin, whereas scarcely any lecithin is to be found inside the agglomerates. In some cases new agglomerates are formed, when lecithin acts as a kind of adhesive (see Fig. 5, right). However, such agglomerates tend to be unstable and disintegrate easily.

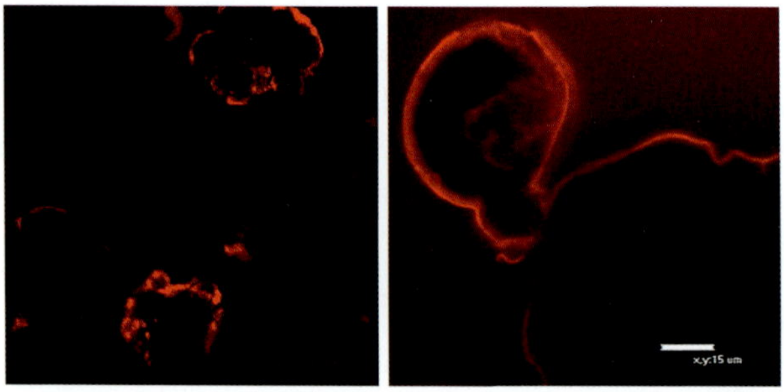

Figure 5.

CLSM images of instant WPC 80 (whey protein concentrate), produced by straight-through agglomeration combined with coating lecithination. The images show that only the surface of the agglomerates is covered with phospholipids; there are no phospholipids in the cores of the particles. Phospholipids are coloured red. Images made by the German Institute of Food Technologies (DIL) for Sternchemie.

6.5.2.2 Rewet agglomeration (post-lecithination)

An alternative process in which spray drying of the powder and its instantization by agglomeration and lecithination are separate processes is also widely used. This process is known as rewet agglomeration.

In certain cases it can be an advantage to separate spray drying from the agglomeration and lecithination processes, for example if it is not intended to instantize the whole amount of milk powder produced or if it is a long way from the place where the powder was initially produced to the area where the instantized powder is to be distributed, making long transportation necessary (21).

In this method, the already spray-dried powder is first rewetted (Fig. 6) and then agglomerated in a fluidized bed dryer. Rewetting is usually done with water or with an aqueous solution of milk solids (approx. 10%) (Fig. 6, process step 1) (1,22). The powder is agglomerated and lecithinated at the same time. The lecithin is added to the aqueous solution either as a powder or in liquid form.

In the fluidized bed dryer, the moist and sticky particles collide as a result of the turbulence and adhere together; agglomerates are formed (Fig. 6, process step 2). The main steps of the rewet agglomeration are as follows:

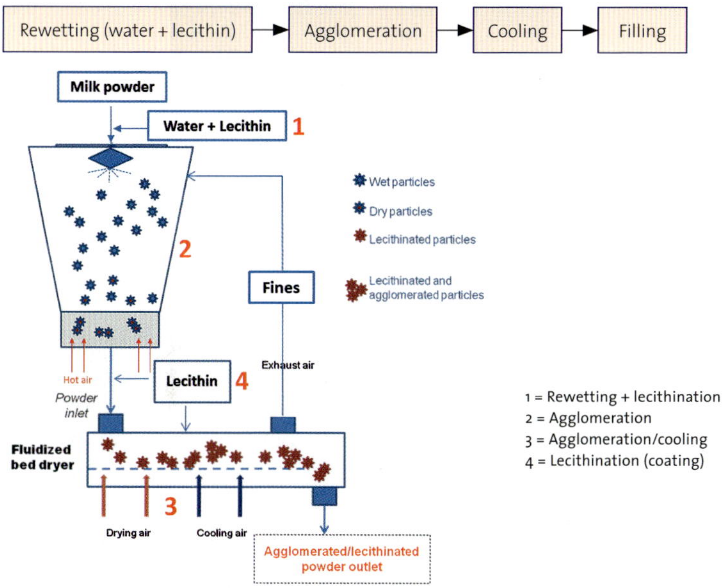

Figure 6.
Diagram of the rewet agglomeration including post-lecithination.

If the lecithin is sprayed onto the powder together with the wetting agent, especially water, it can also act as a bridge between the individual particles and thus facilitate agglomeration. Fig. 7 (A+B) shows the phospholipid coating of

WPI that has been agglomerated by rewetting with a dispersion of lecithin in water. The figure shows clearly that the phospholipids collect not only on the surface of the particles but increasingly in the spaces between the individual particles. In this way, large and porous agglomerates come about. Moreover, the phospholipids are also available as wetting agents when the agglomerates disintegrate into separate particles in water.

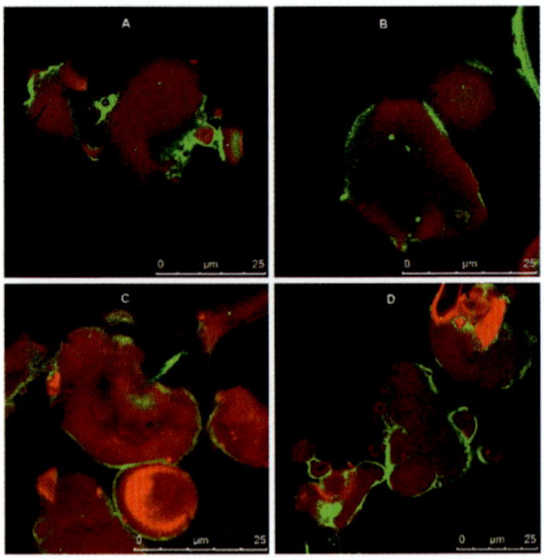

Figure 7.

CLSM images of WPI (whey protein isolate) rewetted and agglomerated (A+B) and coated (C+D) with a lecithin dispersion. The figure shows the coating of the WPI particles with lecithin (whey proteins red, phospholipids green). Whereas lecithin serves as a binding bridge in A+B, C+D show increased coating of the surface with lecithin. From Ji, J.et al. (9), with the kind permission of the author.

Alternatively, lecithin can also be sprayed onto the powder after agglomeration, i.e. between the rewetting tower and the fluidized bed (➤ coating) (Fig. 6, process step 4).

Fig. 7 C+D shows a WPI coated with a dispersion of lecithin and water by the Wurster process. During coating, the phospholipids are deposited almost solely on the surface of the particles. They form a fairly thick and homogeneous film. In this process the lecithin has less effect on the structure and size of the agglomerates (9).

6.5.3 Choice of lecithin

Since lecithin – or the active component of lecithin, the phospholipids – is a natural constituent of milk and therefore of dairy powders, it is largely accepted as a wetting agent and very widely used for the production of instant dairy powders (21).

In co-spraying lecithination, lecithin powder or standard liquid lecithin can be used.

For in-line lecithination or coating, standard liquid lecithin is too viscous to be sprayed directly, even at 65°C. For this reason it is usually liquefied further, i.e. made sprayable by mixing with oil, for example. In the past, up to now, butter oil (anhydrous milk fat, AMF) has often been used, since it was not permissible to add fats of different origin to milk powders. In the meantime, vegetable oils may be used too. This is a great advantage from the functional point of view, since oils with a low melting point (<18°C) are preferable in this application. So vegetable oils such as sunflower, soy or rapeseed oil are very widely used.

Blending lecithin and oil on-site is possible, but only with suitable equipment and shortly before spraying onto the powder. Such a set-up has several disadvantages. Besides logistical constraints due to the additional step in the process, there is a risk of separation of the blend into the a lecithin phase and the an oil phase. Such instability can occur due to ineffective mixing or extended standing times of the blend. If it goes unnoticed, this may cause non-uniform lecithination and the fluctuation of instant properties across production batches. Therefore, it is more common nowadays to choose ready-to-use, low-viscosity lecithin/oil blends that remain stable over a long period of time. Today, a number of lecithin producers offer ready-to-spray blends with the advantage that they can be connected to the spraying unit instantly and without further preparation or adjustment. A careful selection of specific lecithin types as well as of the oils used for adjustment makes it possible to adapt the blends to the specific processing, quality and functional requirements.

The concentration of the lecithin in the oil typically varies between 20–60%, depending on the necessary viscosity or required functionality. During spraying, the lecithin/oil blend is dosed in such a way that a final lecithin concentration of 0.2–0.5% on the powder is achieved.

It is also possible to mix standard liquid lecithin or lecithin powder with water. The resulting emulsion can then be sprayed onto the milk powder. This is the method used almost exclusively in the rewet instantizing process. The advantage of using lecithin powder lies in the fact that no additional fat is applied to the powder; that is especially important with high-protein powders. Once the water has been removed by spray drying, only phospholipids are left on the surface of the particles. In other words: more functionality and less impact on organoleptic properties of the powder. The disadvantages, however, are that suitable equipment, for example a homogenizer, is needed and that it involves an additional step in the production process.

The type of lecithin used will depend on the equipment and processes available (straight-through or rewet) and also on the type of dairy powder to be instantized. As already mentioned, dairy powders can be classified according to their degree of hydrophobicity. It may be said in general that hydrophobic powders, which are characterized by a higher fat content (whole milk powder, milk powder with added fat etc.), can best be lecithinated with higher-polarity lecithins. On the other hand, hydrophilic dairy powders, which have a greater protein content (whey protein concentrates and isolates), are prerferably instantized with low-polarity lecithins (17).

6.5.4 Maintenance of instant properties during storage

For the quality of dairy powder it is essential for the instant properties to remain unchanged for a long time after production, or at least they must not deteriorate significantly. In order to ensure this, the following factors must be taken into account during the production of instant dairy powders:

- Control of the process before, during and after lecithination.
- Type of lecithin.
- Storage conditions of the instantized powder.

Process control before, during and after lecithination:
It is important for the lecithin to be distributed evenly over the individual particles of the milk powder or agglomerates:

- To achieve this, the correct dosage must be chosen, e.g. set indirectly through adjustment of the flow rate during spraying of the lecithin.
- It must be ensured that the lecithin is added continuously, without interruptions.
- The lecithin/oil blend must not be separated.
- The lecithin must be properly atomized.
- Care must be taken to maintain the temperature of the powder surface and the retention time in the external vibrating fluidized bed dryer.
- After application of the lecithin, the powder should be cooled slowly to room temperature in the different temperature zones of the fluidized bed. This is especially important in the case of powders with a high fat content (12).

According to Pisecky (21) the effect of lecithin on powders with surface fat develops when phospholipids mix with the surface fat or dissolve there. In the case of whole milk powder, the milk fat consists of high-melting and low-melting triglycerides.

During lecithination, the lecithin/oil blend is sprayed onto whole milk powder with a temperature of about 60°C. At these temperatures, the entire milk fat fraction is in a liquid state, which means that the phospholipids disperse well in it. The subsequent treatment of the powder at 45°C in the fluidized bed facilitates the distribution of the phospholipids in the fat phase. During the slow cooling that follows in the fluidized bed, the higher-melting triglyceride fractions crystallize. Since phospholipids themselves do not form crystals, they remain in the still liquid, low-melting triglyceride fractions and accumulate there.

After slow cooling to 25°C in the respective temperature zones of the fluidized bed, the individual whole milk powder particles or areas of the particle surface are partially filled with large crystals of the high-melting fractions. At the same time, the whole particle surface is coated with the low-melting fractions containing the dissolved phospholipids. These ensure that the particle surface nevertheless remains hydrophilic and thus capable of wetting.

If whole milk powder is cooled down too quickly after application of the lecithin, the lecithin/oil blend become viscous too soon; the lecithin is then unable to spread adequately over the surface of the particles. Moreover, too rapid cooling causes a multitude of small fat crystals to form, which then cover the whole surface of the powder. Since lecithin only dissolves in liquid fractions, this means that some areas of the whole milk powder are not coated with lecithin. As a result, some areas of the powder remain hydrophobic and cannot be wetted. The instant properties of the powder are then considerably impaired (21).

This is a further reason why vegetable oils are preferable to milk fat for liquefying the lecithin.

Lecithin type and storage of the powder:
It is known that the instant properties of milk powders may change during storage. The main reason for this is migration of the phospholipids from the surface of the agglomerates into the interior of the particles. How soon migration takes place depends on the type of lecithin and the storage conditions. Since the various phospholipid fractions differ in their ability to form strong or stable complexes with the proteins of the milk powder, the composition of the

individual phospholipids may play a crucial role in determining how long the phospholipids remain on the surface of the milk powder.

McKenna (23) showed that during storage of a lecithinated whole milk powder for 6 months at different temperatures (between 20 and 35°C), the phospholipids accumulate near the contact surfaces of individual agglomerates. Moreover, some phospholipids diffuse into the interior of the particles (Fig. 8). This largely takes place through cracks in the particle surfaces; these may be caused by storage of the powder at high temperatures. It emerged that the oil has greater mobility at temperatures at which the oil phase of the lecithin/oil blend is liquid. This oil takes the dissolved lecithin with it. Consequently, the phospholipids are no longer fully available as wetting agents, since the surface is no longer evenly covered with phospholipids.

Accordingly, McKenna (23) showed that the wettability of the whole milk powder was significantly decreased after storage.

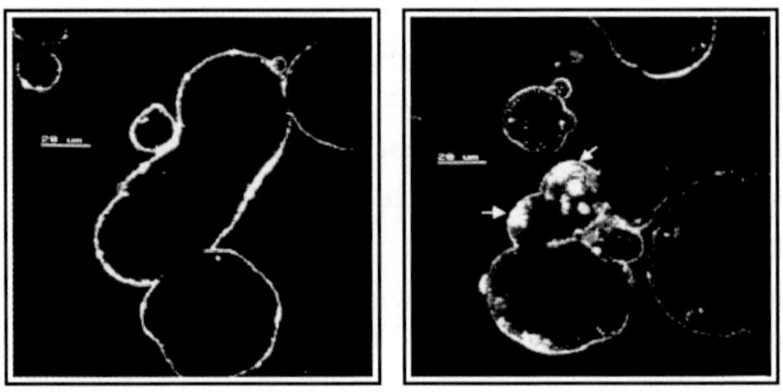

Figure 8.

Whole milk powder particles with fluorescent phospholipid layers: after production (left) and after 6 months of storage at different temperatures between 20 and 35°C (right). The arrow on the image (right) indicates the region where the phospholipids have diffused into the interior of the particles. From McKenna (23), with the kind permission of the author.

It becomes clear from the above that storage of the milk powder at low temperatures reduces the mobility of the lecithin/oil mixtures and thus makes it possible to maintain the wettability of the milk powder over a longer period.

6.5.5 Assessment of the instant properties of milk powders

The milk powder produced must have specific quality attributes. In the case of instantized dairy powders it is chiefly the instant properties that have to be evaluated. Appropriate measurements are essential for product development and also, as part of in-line control during the production process and also in follow-up testing for shelf-life stability. For this reason it should be possible to conduct the analyses quickly, easily and reproducible.

This section describes the standard analytical methods for assessing the instant properties of dairy powders, namely wettability, dispersibility and solubility.

Wettability:
The wettability of instantized dried dairy products is defined by ISO/TS 17758:2014 (24) as the time required for all the particles of a sample of the product to become wet when placed on the surface of water.

The International Dairy Federation (IDF) prescribes the following procedure for determining the wettability of skimmed and whole milk powder: add 10 g skimmed or whole milk powder to 250 ml water at a temperature of 25°C ± 2°C. Measure the time required for wetting the powder with a stopwatch.

In detail: Measure 250 ml of distilled water adjusted to 25°C into a dry 600 ml glass beaker. On the beaker of distilled water place a glass plate, and on this a stainless-steel ring. Place a 10 g sample of the powder on the glass plate, inside the stainless-steel ring. Then withdraw the glass plate quickly (within 2 seconds) so that the powder falls onto the surface of the water. Start recording the time as soon as the powder reaches the surface of the water.

This procedure is simple and very easy to carry out, so it is often the sole method used for determining the instant properties of numerous instant dairy powders.

Dispersibility:
Dispersibility is the percentage of the dry substance of the sample that can be dispersed in water (3).

A 10 g sample of the powder to be analysed is added to 100 g water at 25°C and stirred intensively with a spatula for 20 s. The mixture in the beaker is filtered through a filter with a mesh size of 200 µm, and the filtrate collected in a glass beaker. The total solids content of the filtrate is then determined by drying. Dispersibility is calculated as the percentage ratio of the total solids content of the filtrate to the mass of the test sample weighed (taking the water content and the total solids content of the test sample into account).

The IDF classifies the powder as instantly soluble if its dispersibility is at least 85% (whole milk) or 90% (skimmed milk).

Solubility:
Solubility is determined by stirring 10 g of milk powder with 100 ml water at a temperature of approx. 24°C for 90 sec. in a mixer running at high speed.

50 ml of the suspension is then removed and centrifuged for 5 min. After centrifuging, the supernatant is drawn off and the centrifuge tube filled with distilled water again. After centrifuging for a further 5 min, the amount of sediment in the centrifuge tube is read off. The sediment, expressed in ml, is the solubility index.

For rating the instant properties, Chever et al. (6) suggests an Instant Index. This states that a milk powder may be termed 'instant' if the following criteria are met simultaneously:

IC (instant criterion) for good instant milk powder:
WI (wettability index = wetting time) < 20 s*
DI (dispersibility index) > 95%
SI (solubility index) > 99%

*Well instantized powders may also have a WI < 10 sec.

References

1. Schubert, H., 1989. Instantisieren pulverformiger Lebensmittel (Instantisation of powdered food). Presentation at the annual meeting of process engineers, Berlin 1989. 27-29.
2. Carić, M., Milanović, S., 2002. Milk Powders. Physical and Functional Properties of Milk Powders. Encyclopedia of Dairy Sciences. 1874-1880, New York: Elsevier.
3. Schuck, P., 2011. Dehydrated Dairy Products. Milk Powder: Physical and Functional Properties of Milk Powders. Encyclopedia of Dairy Sciences (Second Edition). Academic Press, 117-124.
4. Ji, J., Fitzpatrick, J., Cronin, K., Maguire, P., Zhang, H., Song, M., 2016. Rehydration behaviours of high protein dairy powders: The influence of agglomeration on wettability, dispersibility and solubility. Food Hydrocolloids (58) 194-203.
5. Sharma, A., Jana, A. H., Chavan, R. S., 2012. Functionality of Milk Powders and Milk-Based Powders for End Use Applications - A Review. Compr. Rev. Food Sci. F. 11 (5), 518-528.
6. Hammes, M. V., Englert, A.H., Noreña, C.P., Cardozo, N.S., 2015. Study of the influence of soy lecithin addition on the wettability of buffalo milk powder obtained by spray drying. Powder Technology 277, 237-243.
7. Kim, E.H.-J., Chen, X.D., Pearce, D., 2009. Surface composition of industrial spray-dried milk powders 3. Changes in the surface composition during long-term storage. J. Food Eng. 94, 182–191.
8. Fang, Y., Selomulya, C., Chen, X., 2007. On measurement of food powder reconstitution properties. Dry. Technol., 26 (1), 3-14.
9. Ji, J., Cronin, K., Fitzpatrick, J., Song, M., 2017. Enhanced wetting behaviours of whey protein isolate powder: The different effects of lecithin addition by fluidised bed agglomeration and coating processes. Food Hydrocolloids 71, 94-101.
10. Schuck, P., 2011. Dehydrated Dairy Products. Milk Powder: Types and Manufacture. Encyclopedia of Dairy Sciences (Second Edition). Academic Press 108-116.
11. Goalard, C., Samimi, A., Galet, L., Dodds, J. A., Ghadiri, M., 2006. Characterization of the dispersion behaviour of powders in liquids. Part. Part. Syst. Char. 23 (2), 154-158.
12. Westergaard, V., 2004. Milk Powder Technology: Evaporation and Spray Drying, Niro A/S, 2004, Copenhagen, Denmark.
13. Toma, S. J. and Nakai, S., 1972. Effects of Lecithin on Reconstitutability of Dry Whole Milk. Can. Inst. Food. Sci. Technol. J. 5 (4), 223-225.
14. Bueschelberger, H.-G., 2004. Lecithins, in Emulsifiers in Food Technology. Oxford, UK Blackwell Publishing Ltd. 29-33.
15. Chever, S., Mejean, S., Dolivet, A., Mei, F., Den Boer, C.M., Le Barzic, G., Jeantet, R., Schuck, P., 2017. Agglomeration during spray drying: Physical and rehydration properties of whole milk/sugar mixture powders. LWT - Food Sci. Technol. 83, 33-41.
16. Pardun, H., 1988. Die Pflanzenlecithine, Gewinnung, Eigenschaften, Verarbeitung und Anwendung pflanzlicher Phospholipidpräparate. Verlag für chemische Industrie H. Ziolkowsky KG, Augusburg.
17. Müller, K., 2014. Lecithins in Agglomerated Foods. Presentation on the ILPS Short Course 2014, Gent.
18. Kim, E.H.-J., Chen, X.D., Pearce, D., 2009. Surface composition of industrial spray-dried milk powders. 1. Development of surface composition during manufacture. J. Food Eng. 94,163–168.
19. Lallbeeharry, P., Tian, Y., Fu, N., Wu, W.D., Woo, M.W., Selomulya, C., Chen, X.D., 2014. Effects of ionic and nonionic surfactants on milk shell wettability during co-spray-drying of whole milk particles. J. Dairy Sci. 97 (9), 5303-5314.
20. Millqvist-Fureby, A., Smith, P., 2007. In-situ lecithination of dairy powders in spray-drying for confectionery applications. Food Hydrocolloids. 21 (5–6), 920-927.
21. Pisecky, J., 2012. Handbook of Milk Powder Manufacture. GEA Process Engineering A/S, Copenhagen, Denmark.
22. Kalyankar, S.D., Deshmukh, M.A., Chopde, S.S., Khedkar, C.D., Lule, V.K., Deosarkar, S.S., 2016. Milk Powder. Encyclopedia of Food and Health, New York: Elsevier. 724-728.
23. McKenna, A., 2000. Effect of processing and storage on the reconstitution properties of whole milk and ultrafiltered skim milk powders. PhD Thesis, Massey University, Palmerston North.
24. ISO/TS 17758:2014 (en) Instant dried milk — Determination of the dispersibility and wettability

6.6 Lecithins as dietary supplements

Parris M. Kidd

6.6.1 Introduction

Dietary supplements are substances added to the daily diet with the aim of improving health. This review covers lecithin dietary supplements proven to improve human health. The commercial sources for lecithin dietary supplements have expanded considerably in recent years, and so has the clinical research on their health benefits. The body of peer-reviewed clinical and other scientific literature is huge, with thousands of relevant published peer-reviewed articles, so this review focuses on the clinical trial evidence.

For the purpose of this review lecithins are defined as concentrates of phospholipids and related glyceride substances prepared from natural sources. As this author has described in Chapter 2.2, phospholipids are essential for all life forms on this planet. Though phospholipids are often confused with fats (triglyceride lipids), their importance for life is dramatically more fundamental than other lipids because they build cell membranes. Without membranes, human cells and those of all known life forms could not exist.

The word 'phospholipids' (PL) appeared in English for the first time in Thudicum's 1884 classic *Treatise on the Chemical Constitution of the Brain* (1). Thudicum made a prophetic statement: "Phosphatides [an earlier term for phospholipids] are the centre, life, and chemical soul of all bioplasm whatsoever, that of plants as well as animals." His insight is now validated by the modern understanding of PLs as major molecular building blocks for the cell membranes that occur in all cells and underlie practically all life processes (2). The basic human physiology of PLs is reviewed by this author in Chapter 2.2 of this volume.

Lecithins have been commercially available as dietary supplements for over a century, with German inventors leading the way. In Germany, between 1900 and 1913 more than 100 lecithin or related PL preparations were commercially introduced. The very first lecithin dietary supplement was likely Dr H.C. Buer's Neura Lecithin Pearls, launched in 1907 in Cologne (3). In 1908, an English ad for Ovaltine claimed it contained "active lecithin, the best brain and nerve tonic known" (3). The first commercial lecithins were made from chicken eggs, but by the mid-1920s less expensive lecithins from soybeans began to displace egg lecithin in Germany. By 1934 soy lecithin was also being manufactured in the USA (3).

Clinical research on the health benefits of lecithins began in Germany in the 1920s, mainly on 'nerve tonic' applications but also to help counter alcohol intoxication (4). By 1941 there were numerous publications, primarily in German, on lecithins' relevance to the blood, brain and nervous system, heart and circulation, liver, and digestive system (4,5). By 1943 American clinicians were researching soy lecithin for lowering high blood cholesterol, liver problems and psoriasis, and by 1946 soy lecithin was being aggressively marketed in the U.S. as a 'health food' (4).

Egg lecithin has remained important, despite its higher production costs. In 1932 it was employed in China to help addicted individuals withdraw from opium, and in 1937 in India with a claimed 70 per cent degree of success (5). By 1938 in India, in a "unique campaign unparalleled in the history of drug addiction anywhere in the world", egg lecithin had been employed to help almost 8,000 people overcome opium addiction (6). Since that time, hundreds of clinical trials have been conducted on lecithins from a variety of natural sources.

6.6.1.1 Lecithin dietary supplement categories

The commercially available lecithins that have shown human benefits in clinical trials include (7,8,9):

- **Egg lecithins**, from chicken egg yolks, which contain PC (phosphatidylcholine), PE (phosphatidylethanolamine), PI (phosphatidylinositol) and SM (sphingomyelin), and often a low but substantial content of the omega-3 fatty acid DHA (docosahexaenoic acid) (7,10).
- **Soy lecithins**, which contain mostly PC, along with PE, PI, PA (phosphatidic acid) and traces of PS (phosphatidylserine). Numerous brands of soy lecithins have existed, with wide-ranging phospholipid and fatty acid profiles, as listed and discussed by Kuellenberg and associates (7) and Gundermann and associates (8). **Sunflower and canola lecithins** have PL profiles similar to soy lecithins, and though they have little specific clinical research support are becoming increasingly attractive for consumers with soy allergies.
- **Soy PS lecithins**, typically prepared from soy lecithin by enzymatically converting most of the PC into PS (11). Typically contains a minimum 20% PS and low amounts of PC while roughly retaining the PE, PI and PA profile of soy lecithin. Soy PS can also be enzymatically processed to insert omega-3 DHA and/or EPA on its tail groups, and the resultant **soy PS omega-3 lecithins** have been tested in clinical trials.

- **Krill lecithins,** typically prepared from the marine crustacean *Euphausia superba* (Antarctic krill) also contain omega-3 DHA and EPA, with EPA predominant. These lecithins take the form of a red 'krill oil' coloured by small amounts of the marine carotenoid astaxanthin. They contain from 18% to as much as 81% PL by weight, along with triglycerides and free fatty acids. Of the PL, most is PC and the remainder mainly PE and PS (12).
- **Fish lecithins**, typically prepared from the roe of herring or salmon, contain mostly PC along with PE, PI, PS and SM. These lecithins also have substantial omega-3 fatty acid content, mainly of DHA and EPA (13).
- **Algal lecithin**, produced from sustainably cultured *Nannochloropsis oculata*, has about 15% polar (charged) lipid content, which includes 6% PLs and 9% glycolipids; 25% EPA, bound to the polar lipids; along with triglycerides. Currently these preparations lack DHA (9).

The sections that follow review each of these lecithin categories from the perspective of their clinical advantages (and disadvantages) as dietary supplements.

6.6.2 Egg lecithins

Egg lecithins are generally prepared from powdered chicken egg yolk, as reviewed by Schneider in Chapter 3.2. According to Schneider, the egg lecithin phospholipid profile is approximately 73% PC, 18% PE, 2% PI and 3% SM, with practically no PS or PA. The fatty acid profile is about 41% saturated, 35% monounsaturated and 24% polyunsaturated with omega-6 arachidonic acid (AA) and omega-3 docosahexaenoic acid (DHA) included.

Though egg lecithins were the first to be commercialized, as mentioned above (3), due to their unfavourable production economics their current importance is restricted mainly to infant feed formulas, as sources of omega-6 AA and omega-3 DHA. Both these nutrients are essential for full-term infants (10) and can be life-saving for prematurely born infants (14). The rationale and other details of this application are comprehensively covered by Schneider.

6.6.3 Soy lecithins

This category of lecithins has an almost bewildering array of brand names, which were detailed by Kuellenberg and colleagues in 2012 (7) and Gundermann in 2011 (8) and therefore will not be listed here. Certain of these brands have been clinically investigated, mainly for circulatory, brain, liver and gastro-intestinal benefits.

6.6.3.1 Trials on blood lipids and other circulatory benefits

After decades of clinical trials the long hoped-for benefits of soy lecithins for cholesterol and triglyceride control remain unconfirmed. In a 1989 meta-analysis of 20 trials, Knuiman and colleagues concluded that soy lecithins did not consistently lower serum cholesterol (15). They noted that most of the trials were small and lacked suitable control groups. Four of the 20 trials did have controls in the form of vegetable oils, and soy lecithins did not perform better than any of them. However, this meta-analysis can be criticized for lumping together trials that employed very different soy lecithins.

One important variable between various soy lecithin preparations is the concentration of PC in relation to the other PLs in the soy lecithin preparation. The Knuiman meta-analysis included lecithins with PC content ranging from 10% to greater than 72% of their total PL profile (15). Another variable was the doses of soy lecithin used in these 20 trials, which varied from less than 1 gram per day to more than 50 grams per day. These two wide-ranging variables limit the power of this meta-analysis to draw clear conclusions.

One double-blind (DB) trial conducted subsequent to the Knuiman meta-analysis (15) did find a positive effect on blood lipids (16). It involved PPC (polyenephosphatidylcholine), a concentrated lecithin with a minimum 73% PC content, also classified as EPL (essential phospholipids) (8). In this small trial on 30 diabetic patients with hyperlipidaemia, EPL markedly improved low-density lipoprotein (LDL) cholesterol, total cholesterol and triglycerides (16).

Unfortunately, two subsequent DB trials with concentrated lecithins failed to demonstrate positive blood lipid benefits (17,18). But even with these negative clinical findings some researchers continue to believe soy lecithin has potential for effective blood cholesterol control. One suggestion is that by achieving relatively high concentrations of soy lecithins in the intestinal contents, it is possible to substantially impede intestinal cholesterol absorption (7).

The absorption of cholesterol requires bile fluid, which provides PC along with non-phospholipid substances that emulsify cholesterol and other dietary lipid molecules into microscopic micelles, which can then be absorbed by the intestinal lining cells (enterocytes). For further details of bile emulsification, see Chapter 2.2 in this volume. There is substantial experimental evidence that consuming multiple grams of lecithin can somehow block the enterocytes' ability to absorb cholesterol (17,19,20). Some estimates suggest 10 grams/day of a PPC lecithin with its very high PC content may be sufficient to impair intestinal cholesterol absorption (20).

Another rationale supporting soy lecithin's potential for cholesterol benefits is stimulation of reverse cholesterol transport, that is, removal of cholesterol from the tissues into the blood and followed by its transport to the liver for subsequent excretion (21). Reverse cholesterol transport is known to be mediated by high-density lipoprotein (HDL) cholesterol particles, for which the apolipoprotein-A1 (apo-A1) is a major building block. Interventions that elevate apo-A1 also tend to elevate high-density lipoprotein (HDL) cholesterol (22). From a small placebo-controlled trial, Polichetti and colleagues reported 3 grams/day of a soy lecithin very high in PC increased apo-A1 levels versus placebo (23). They suggested this soy lecithin might be beneficially upregulating reverse cholesterol transport.

Though the suggested mechanisms for soy lecithins to lower cholesterol have some evidence in favour, to date the clinical trial findings indicate soy lecithins remain unproven as dietary supplements for blood lipid management. The clinical trials have been criticized for various methodological flaws, one of which is lack of standardization of the fatty acid profiles of the various soy lecithin brands. It is necessary to consider that soy lecithins' circulatory benefits may well be limited by their lack of omega-3 fatty acid content.

As early as 1989, the Knuiman meta-analysis suggested that soy lecithins' fatty acid profiles might correlate with the presence or absence of cholesterol benefits (15). In the 3 decades since then, it has become clear that dietary supplementation with omega-3 fatty acids promotes cardiovascular health – at least in certain populations such as diabetics (24). Soy lecithins have negligible omega-3 content. Lecithins that can deliver substantial dosages of omega-3 fatty acids are rapidly becoming more attractive dietary supplement ingredients, a reality that will be further explored as this review progresses.

Turning away from blood lipids, there are hints from human and animal research that soy lecithin supplementation can improve blood vessel wall health. In their comprehensive 2012 review, Kuellenberg and colleagues reviewed Russian-language and other hard to find publications that seemed to suggest dietary Lipostabil® PC can remove cholesterol from atherosclerotic plaques as well as lower platelet aggregation tendency (7). They asserted that lecithins with their phospholipids were more effective than fish oils with their triglycerides, but also introduced evidence that lecithins with omega-3 fatty acids could be even more effective than those without omega-3s.

6.6.3.2 Memory and other brain trials

Soy lecithins have had many brain clinical trials, dating back to at least 1979 (25-31). The initial focus was on Alzheimer's disease, based mainly on the ability of PC and other lecithin PLs to provide choline, which the brain and other nervous system tissues can convert into acetylcholine, a major neurotransmitter. The strong evidence that the brain's acetylcholine neurotransmitter pathways decline prematurely in Alzheimer's as well as vascular dementia (see Little and colleagues (26) for references) led to the cholinergic hypothesis of geriatric memory dysfunction (27), which stimulated intensive clinical research with soy lecithins.

The fundamental basis for the many Alzheimer's/dementia trials with soy lecithins was the expectation that their PC content would provide choline that the brain could effectively use to replace acetylcholine in the affected cholinergic pathways, thereby achieving "cholinergic replacement therapy" (27). But the trial findings were inconsistent from the start, and a 2003 meta-analysis of 12 randomized controlled trials (RCTs) concluded the evidence did not support the use of soy lecithin in the treatment of dementia (29). To date (early 2019) there have been no further clinical trial findings that warrant changing this conclusion.

The failure of soy lecithins against the catastrophic memory loss of Alzheimer's may not rule out some usefulness for less severe memory problems. In 1993 Ladd and colleagues reported on a DB trial with college students, in which a single high dose of concentrated soy lecithin improved learning better than a placebo (30).

Soy lecithins have also been tested for stroke management. In 2001 Chinese researchers reported on a large trial that included 542 patients – the full report is in Chinese with the abstract in English (31). Patients were assigned within 48 hours after stroke onset to receive either basic stroke treatment (n=60), basic treatment plus the PC precursor citicoline (n=122 patients), or basic treatment plus a soy lecithin at 30 grams per day, for 28 days. The researchers reported that the soy lecithin group showed significantly better lowering of the infarct volume versus the basic and the citicoline groups. However a subsequent meta-analysis rated this trial as having high risk of bias (32), so additional trials are needed to replicate these results.

A soy lecithin also was tested against Parkinson's disease, in a small (16 patients) trial (33). Each patient received either 32 grams per day of the soy lecithin containing 25% PC, or a placebo (powdered skim milk). The lecithin group did not significantly improve versus the placebo group, but the researchers reported "some indications of a positive treatment effect" on memory, cognition and motility tests.

Various trials with soy lecithins for Friedreich's ataxia and Gilles de Tourette's syndrome failed to detect benefit. For specific references see Kuellenberg and colleagues (7).

As happened with the circulation – blood lipids trials, for brain applications the fatty acid profile of soy lecithins are also relevant to success or failure. One clue comes from an animal experiment. Zhou and colleagues used scopolamine, which impairs the brain's cholinergic pathways, to block cognitive (water maze performance) in mice, then tested three lecithins for reversing this 'scopolamine amnesia' effect (34). They found that two of the lecithins with PC of higher omega-3 content, namely a high-DHA squid lecithin and a high-EPA sea cucumber lecithin, blocked the scopolamine amnesia better than did an egg lecithin with practically zero omega-3 content.

6.6.3.3 Trials on alcoholic and other liver diseases

The liver is the body's "workhorse" organ, tasked with conducting a wide range of essential life functions (21). The liver's parenchymal cells (also called hepatocytes) do most of this work, and those that are most active have the most dense cell membrane networks (35). As is likely for all human cells, most of the hepatocytes' functions are performed by membrane systems, which rely on phospholipids as major building blocks for their growth, maintenance and repair (for details, refer to Chapter 2.2). Soy lecithin dietary supplements help meet this need by augmenting the daily intakes of PL from foods – for data on PL content of specific foods, see Cohn and others (19).

The liver is the main organ for processing absorbed triglycerides (fats), cholesterol and other dietary lipids, and for distributing them to other organs (21). When fats accumulate in the liver faster than the liver can clear them, fatty liver disease (FLD, often termed just 'fatty liver') develops (36,37). Fatty liver can become complicated by inflammation, then progress further into scarring (fibrosis) which ushers in cirrhosis that is life-threatening. Fatty liver even can progress to liver cancer, which has practically a zero chance of 5-year survival (37). The liver damage that progresses to fatty liver can be initiated by practically any kind of toxin, by viruses, medications, even by poor diet. The risk of developing fatty liver is also often higher with obesity, diabetes, high cholesterol, high blood pressure or high alcohol intake. In clinical trials, lecithins rich in PC, the most abundant PL in cell membranes, have shown substantial benefit for fatty liver and other liver damage.

Concentrated soy lecithins can improve alcoholic liver disease

Alcohol overuse can initiate fatty liver and promote its progression into more severe liver disease (21). For liver disease linked to excessive alcohol intake, the term alcoholic liver disease (ALD) is often used; otherwise NAFLD is used, for non-alcoholic fatty liver disease. The concentrated soy lecithins called essential phospholipids (EPL) can especially benefit ALD patients.

The extensive clinical and preclinical research on EPL for fatty liver was pulled together in a 2016 English-language review by Gundermann that included findings from publications in German, Italian, Polish, Russian, Chinese and Japanese (36). Gundermann clarified the definition of EPL: highly purified soy lecithins, standardized to a high content of PC (72–96%), and with as much as 52% of the PC molecules containing omega-6 linoleic acid, which is an 18-carbon, doubly unsaturated fatty acid. For clarification refer to Figure 1 in Gundermann (36). EPL preparations are better than 90% absorbed when taken by mouth (38), and radioactive labelling studies prove some of their PC molecules do become incorporated into the membranes of hepatocytes (8).

EPL lecithins have improved ALD in several DB trials (39-42), as previously reviewed by Gundermann (8,36), Kuellenberg and colleagues (7) and Kidd (35). Two of these trials found EPL (1,350 mg/day, with B vitamins) improved blood levels of liver enzymes and bilirubin better than did a placebo (39,40). A large trial by Lieber and colleagues found that after 2 years, patients who drank six or more drinks a day showed benefits from taking EPL (4 g/day) (42).

The findings from these DB trials and from many other human and animal studies on alcoholic liver disease indicate that EPL can improve biochemical variables such as transaminases, bilirubin and immunoglobulins as well as slow progression at the stage of fibrosis. Findings from a landmark study with baboon primates suggest EPL could be capable of even more clinical benefit.

In a long-term study, Lieber and colleagues fed baboons alcohol every day for periods of years, sufficient to cause extensive liver tissue damage and initiate fatty liver disease (43). Baboons fed EPL while receiving alcohol developed less liver damage. The EPL slowed liver tissue and cell damage, lowered the depletion of PC from the liver cell membranes, and partially protected the baboons against progression from fatty liver to fibrosis and cirrhosis.

Concentrated soy lecithins can improve other toxic liver damage

As with ALD, in clinical trials EPL preparations have shown benefit against NAFLD (non-alcoholic fatty liver disease) (8,35,36,44). In small DB trials, EPL when compared to placebo was found to improve NAFLD linked to diabetes and obesity (refer to Table 1 in Gundermann (36)).

Drugs used against tuberculosis can be liver-toxic. In a DB trial, patients being treated with isonicotinic acid hydrazide and ethambutol for active tuberculosis received either EPL (1,350 mg/day) or a placebo and continued to be treated with the drugs (45). After 3 months, the EPL patients showed significantly lower levels of liver damage enzymes versus those patients on placebo. This trial's outcome indicated EPL could protect against liver damage from these drugs.

The liver's hepatocyte cells are uniquely exposed to toxins, especially since the liver is the body's main tool for detoxifying "foreign" substances (xenobiotics). The liver's detoxication functions are performed mainly by P450 enzyme systems located in the internal membranes (endoplasmic reticulum) of the hepatocytes (2). The P450 enzymes generate a variety of by-products that can (and do) damage these cells' membrane systems. Providing the embattled liver with soy lecithin equips its hepatocytes and other cells with PC and other PLs that are literally building blocks for making new membrane mass and/or new cells (35).

Numerous case histories have been published that suggest EPLs and other concentrated soy lecithins can have healing effects against liver damage, including from radiation treatment; phenytoin, an antiepileptic drug; oral trichloroethylene poisoning; and numerous other toxic agents (35). EPLs have shown versatile antitoxic effects in animal studies that encompass some 43 different experimental models in eight different animal species, including primates (8).

These extensive animal data suggest that in humans EPLs are worth trying as dietary supplements to prevent or treat liver damage, whether from radiation therapy; chemicals, including mercuric chloride, carbon tetrachloride, organic solvents, galactosamine; drugs, including acetaminophen (the main active ingredient in Tylenol®), tetracycline, alloxan, rifampicin, cyclosporine A, anaesthetics; food additives including monosodium glutamate (MSG); choline deficiency; or a high-fat diet (tested in seven animal models).

EPL has a broad range of beneficial action mechanisms (8,35,36,43,44):

- Increasing hepatocyte membrane PL content, PL:cholesterol ratio, membrane fluidity (health significance discussed in Chapter 2.2).
- Lowering excessive liver fat content, blocking abnormal LDL cholesterol oxidation.
- Improving hepatocyte mitochondrial performance and energy generation.
- Improving antioxidant defence, lowering oxidative "free radical" activity.
- Working in synergy with S-adenosylmethionine (SAMe) for cell renewal and repair.
- Blocking collagen build-up linked to liver fibrosis (scarring).

Concentrated soy lecithins fight viral liver diseases

Liver cells and their membrane systems are vulnerable to destruction not only from alcohol or other toxins but also from viral infections. EPL preparations have been clinically tested against viral liver infections, with considerable success.

EPL was tested against hepatitis ("hep") B or C viral infections in at least nine DB trials that involved at least 3,860 patients (8). EPL consistently lowered blood liver enzyme levels that indicated liver damage, improved liver cell structure, lessened cell death and tissue necrosis (as judged from biopsies), and improved clinical well-being (8,35). In one small DB trial, EPL (2.3 g/day, for 1 year) may have completely "turned off" hepatitis B virus activity in three of the 15 patients who received it (46). In another DB trial, EPL 1,350 mg/day (fortified with B vitamins) cleared hepatitis B antigens from the blood significantly better than placebo, after just 30 days (47).

EPL also can improve severe liver damage in hepatitis B patients. In a DB trial, 20 of the 25 patients who received EPL (1,350 mg/day, for 1 year) showed improved blood marker and biopsy results, versus six of 25 who received a placebo (48). The EPL group continued to improve well past the end of the trial, which suggested that EPL likely set in motion long-term healing processes that continued after it was no longer being taken daily.

For chronic viral infections, other trials indicated EPL may work better against hepatitis C than hepatitis B. In a large European trial, patients with chronic hepatitis B or C received alpha interferon (a proven antiviral therapy), and either EPL (1.8 g/day) or placebo, for 24 weeks (49). The patients that received interferon + EPL showed significantly better lowering of liver damage enzyme levels than those who received only interferon. This positive response to EPL came from the patients with hepatitis C (71% of those receiving EPL improved, versus 51% of the placebo group). Observation for another 24 weeks without interferon confirmed the hepatitis B group did not effectively respond to EPL.

Gundermann discussed a later DB trial in which hepatitis C patients received pegylated interferon and ribavirin, along with high-dose EPL (4.5 g/day) or a placebo, for 18 months (8). EPL reportedly caused significant lowering of liver enzymes, and when continued for 3 years also significantly improved the fibrosis (scarring) score.

Overall, in cases of chronic viral hepatitis infection EPL consistently improved clinical symptoms such as tiredness and liver enlargement, the liver's tissue structure, blood bilirubin and liver enzyme levels, and other liver health indicators (8). The minimum effective EPL dose is probably 1,350 mg/day, for at least 6 months.

These clinical findings with the EPL soy lecithin concentrates, containing more than 72% PC, strongly suggest they are safe and effective dietary supplements for liver health. EPLs are quite popular in some countries, most notably Russia, and further clinical exploration via large, well-controlled trials could lead to their expanded use in more countries.

Alcoholic and non-alcoholic fatty liver as well as viral liver cases are on the rise worldwide, being already epidemic in some parts of the world, yet the existing treatment options for these deadly diseases are very limited (50,51). EPL is practically free of adverse effects and is likely to have emulsifying effects that would increase the absorption of substances mixed with it. Combinations of EPL with B vitamins are already fairly common; other combinations that could have exciting liver results include with S-adenosylmethionine (SAMe) (51),

N-acetylcysteine (NAC) (52), and silybin or other milk thistle flavonolignans (53). Omega-3s are another obvious option for combining with EPL, especially since a fish lecithin rich in DHA and EPA was previously found beneficial for six patients with chronic liver disease (54).

6.6.3.4 Soy lecithin protects the gastrointestinal (GI) tract

As introduced in Chapter 2.2, PC and other phospholipids are important protective nutrients for the gastrointestinal (GI) tract. PC and other PL molecules form a hydrophobic (water-repellent), continuous barrier located in the mucus layer of the GI tract. This barrier likely has the thickness of just one PL molecule – see Fig. 4 in Chapter 2.2. But it effectively protects the one-cell-thick GI epithelium against self-destruction by 'auto-digestion', as epithelial cells release strong digestive acids into the intestinal contents (55,56). It also helps exclude waterborne acids and other chemicals, bacteria such as *Helicobacter pylori,* and other agents that could penetrate the mucus layer and breach the epithelial lining with major adverse consequences (57).

Ulcerative colitis (UC) is an aggressive inflammatory colon disease that can hit children or adults. In UC patients the mucus PC content of the small and large intestine can be depleted to just 30% of normal (58). In a DB trial with UC patients, Stremmel and collaborators (55) used Sterpur P-30, a granulated soy lecithin containing 30% PC, in a delayed-release formulation that ensured delivery of PL into the distal intestine. Taken at 6 g/day for 3 months, Sterpur led to clinical remission in 53% of the patients, with 90% experiencing measurable clinical improvement, versus 10% and 10% in the placebo group.

In a later DB, placebo-controlled trial with patients suffering from chronic active UC and dependent on steroid medication for managing the disease, Stremmel and collaborators extended these findings using a patented delayed-release lecithin with 33% PC, at 2 g/day of PC for three months (59). Half the PC patients (15 of 30) achieved greater than 50% improvement and were able to discontinue steroids, versus 10% (three in 30) of the placebo recipients. Fully 80% of the PC group were able to discontinue steroids without having their symptoms worsen. PC improved a number of other GI measures significantly better than the placebo.

Karner, Stremmel and colleagues performed a later DB, placebo-controlled trial on UC patients with bloody diarrhoea, using a patented delayed-release soy lecithin named LT-02, which was almost pure (94%) PC (60). They tried several doses and the highest dose (3.2 g/day of PC) worked best. This dose group improved 51.7% in disease activity score, versus 33.3% for the placebo. Tissue structure observed by histology improved twice as much in the PC group, by 40% vs 20% for the placebo.

The only adverse effects of the soy lecithin LT-2 on these UC patients was bloating, not sufficiently severe to cause patients to drop out of the trials (60). The excellent tolerability of PC lipid replacement therapy stands in contrast to the high incidence of adverse effects from pharmaceutical UC medications (59).

Soy lecithins also have been used to lessen the adverse effects of NSAIDs (non-steroidal anti-inflammatory drugs) on the GI mucus barrier. Some 20–40% of long-term NSAID users experience ulcerative bleeding of the stomach or small intestine, responsible for thousands of deaths annually and linked to depletion of the mucus PC by the drug (61). In trials that used the NSAIDs aspirin or ibuprofen pre-combined with soy lecithin, the GI damage was markedly lowered yet the drugs' desired anti-inflammatory actions were fully preserved (61,62). This begs the question: why aren't all aspirin and ibuprofen drugs pre-combined with a lecithin prior to sale?

Schneider, Stremmel and colleagues discussed the central importance of PC and other lecithin PLs for lipid-based therapy of the GI tract (63). Their reasoning parallels earlier concepts of lipid replacement therapy (64) and membrane-lipid therapy (8,65,66).

6.6.3.5 Soy lecithins and lipid/membrane replacement therapy

In 2003 Nicolson presented the concept of lipid replacement therapy (LRT) as a specific strategy for "actual replacement of damaged cellular lipids with undamaged lipids to ensure proper structure and function of cellular structures, mainly cellular and organelle membranes" (64). As early as 1993 Gundermann had suggested lecithin PLs could be "membrane therapeutics", based mainly on their benefits for liver diseases known to involve damage to cell membranes (see Table 1 and Reference 48 in Gundermann (8)). In 2008 and 2017 Escriba and colleagues published a comprehensive review of cell membranes that included membrane pathologies which might be targeted by membrane-lipid therapy (65,66).

Nicolson and colleagues conducted several trials using soy lecithins in combination with a great diversity of other foods, so that their reported successes cannot be credited solely to LRT, or membrane lipid replacement as they now call it (67). In 2018 Hirose and colleagues reported on a DB, placebo-controlled trial against fatigue and menopausal symptoms in middle-aged women, which was motivated by their interest in LRT (68). They used a soy lecithin from Kikkoman (PC 24%, PE 20%, PI 12%, remainder not disclosed), at 600 mg/day or 1,200 mg/day, versus a placebo for 8 weeks. The high dose significantly increased vigour on the Profile of Mood States (POMS) questionnaire as well as cardiovascular measures – diastolic blood pressure and the cardio-ankle vascular index (CAVI). Further work is necessary to establish whether LRT contributed to these benefits.

As summarized by Escriba (66), cell membrane lipid abnormalities have been documented in many diseases, notably Alzheimer's, cancers, fatigue, hypertension, infectious diseases, kidney diseases, metabolic disorders, respiratory diseases, schizophrenia, skin diseases and stroke. Clinical trials are urgently needed to explore just how well soy (or other) lecithins can "feed" our cell membranes with phospholipids and accompanying fatty acids to repair or optimize their functional performance.

6.6.3.6 Soy lecithins as delivery vehicles for enhancing bioavailability

Milk thistle *(Silybum marianum)* has been used as a medicinal plant since at least the time of ancient Egypt (53). A standardized extract from milk thistle called silymarin contains silybin and other flavonol bioflavonoids that have powerful antioxidant, anti-inflammatory and other protective effects but are very poorly absorbed when taken by mouth (68). Soy lecithin's extremely high oral absorption (90%) (38) plus its low cost made it an obvious choice for combining with standardized milk thistle extracts to dramatically increase their absorption efficiency. These molecular matrices are known as phytosomes (69). Silybin and other silymarin flavonols used in phytosome form exhibit superior liver benefits over their non-phytosome formulations. This phytosome dietary supplement has substantial promise for NAFLD, which is an increasingly common health problem worldwide (53).

Soy lecithins are increasingly being used as delivery vehicles for poorly absorbed nutrients and herbal extracts, and the variety of preparations extends beyond the scope of this review. For additional in-depth information on these topics consult the references of Ostlund (70), Gananath (71) and Cox (72).

6.6.4 Soy PS lecithins

The phospholipid phosphatidylserine (PS) is universally present in all human cells and in all other known cells. In humans it is particularly abundant in our brain cells. As introduced by this author in Chapter 2.2 on the physiology of lecithin phospholipids, and further discussed by Schneider in Chapter 3.2 on animal phospholipids, PS is not only one of the core cell membrane building blocks but also has been found to be crucial for memory and other brain functions. PS now has a long and impressive record of clinical efficacy and safety.

6.6.4.1 PS as dietary supplement: Background

Clinical trials with PS for brain function began in the mid-1980s, using PS complexed with PC and other phospholipids. The first commercial PS lecithin was prepared from bovine (cow) brain cortex and was often called BC-PS. After an initial series of clinical trials, by 1993 it was clear that BC-PS was beneficial against memory loss in ageing subjects (73–82), and that it could improve quality of life in Alzheimer's patients (82–83). But by that time bovine brain was no longer a safe source of PS (84).

The advent in Europe of "mad cow disease", technically bovine spongiform encephalopathy (BSE), spelled the end of bovine brain PS as a dietary supplement. Humans consuming products from BSE animals were at risk of acquiring abnormal prion proteins that could trigger brain pathology and lead to dementia. As BSE spread around the world it became too risky to continue to make PS from bovine brain. Enterprising lecithin manufacturers developed a method for generating a PS lecithin from a plant source – soy lecithin.

Soy lecithins rarely contain measurable PS. But an enzyme can be used to remove the choline group from PC and replace it with serine (enzymatic head group exchange). This soy PS lecithin was soon tested in a number of DB clinical trials and found to work as well as the bovine PS. To date (early 2019), soy PS lecithin has more published DB trials than bovine PS lecithin (see reference (75) for access to a list) (85-95).

In the United States, after reviewing the body of clinical research on PS, and its exemplary safety record, the FDA (Food and Drug Administration) in 2003 granted two qualified health claims (QHC) for PS, announcing that they would "exercise enforcement discretion" where such claims were made for PS dietary supplements (96). Their conditions for granting this prestigious QHC status included that the product must supply an effective dose of PS and that the PS ingredient raw material must be competently manufactured.

The U.S. FDA's willingness to grant PS not just one but two QHCs is still unique compared to other nutrients, and represents their clear recognition that this particular lecithin offers tangible benefit to the human brain. To date the evidence from numerous randomized, controlled, DB trials makes PS the best-documented dietary supplement for memory enhancement. PS is also the best-tolerated brain supplement overall, with a practically flawless record of safety after decades of commercial use (97).

6.6.4.2 Premier dietary supplement for memory

PS stands out above all other dietary ingredients for its memory benefits. The most definitive trial with PS for memory is that of Crook and collaborators (80). This was a multicentre, DB, randomized, placebo-controlled trial, and included 149 subjects of ages 50–75 years, diagnosed with AAMI (age-associated memory impairment). AAMI was defined as a non-disease condition in people over age 50, with (a) absence of dementia – score of 27 or higher on the Folstein Mini – Mental State Examination, (b) adequate intellectual function – raw score of 32 or higher on the Vocabulary subtest of the Wechsler Adult Intelligence Scale, and (c) poor memory performance, at least 1 standard deviation below the mean for young adults on either the Benton or Wechsler memory scales. The clinical endpoints were (a) a clinical global rating scale, (b) non-computerized neuropsychological tests and (c) a computerized psychometric battery to simulate cognitive tasks of daily life. Its primary measures were Name Face Association (Acquisition, Delayed Recall), Facial Recognition, Tel. No. Recall with Interference, and Misplaced Objects Recall. Its secondary measures were Names Acquisition, Divided Attention, and Selective Reminding (Acquisition, Delayed Recall). This was arguably the most sophisticated memory trial conducted up to that time.

The trial went for 12 weeks (80). The PS group (total 74 subjects) received 300 mg PS per day, while 75 subjects received the placebo. The PS group performed significantly better vs placebo on Facial Recognition. In a subgroup of subjects with relatively poorer memory performance, PS performed significantly better benefit than placebo on all five primary psychometric measures: Name Face Acquisition, Name Face Delayed Recall, Facial Recognition, Tel. No. Recall, and Misplaced Objects Recall (all $p<0.05$ at week 12). The PS subjects also improved significantly better on secondary measures: Names Acquisition,

Divided Attention, global cognitive improvement, and a visual analogue scale of global improvement (all $p<0.05$ at week 12). The authors commented, "People most likely to respond (to PS treatment) appear to be those who score above the range of cognitive performance associated with dementing disorders ... but who perform in the low range of normality for persons of the same age."

This trial has been the lynchpin for the clinical and commercial success of PS lecithin. Its data are supported by other DB trials that tested PS for memory, most notably Crook and others (83), Villardita and others (76), Amaducci and others (77), Cenacchi and others (82), Gindin and others (85) and Kato-Kataoka's group (88).

The Kato-Kataoka trial established that a 100 mg daily dose of soy PS lecithin can be very helpful for memory (88). This was also a DB, randomized, placebo-controlled trial. It recruited subjects of average age 59-plus years, with mild memory impairment but without dementia. They received either 100 mg of PS per day (the "PS100" group), 300 mg of PS per day (the "PS300" group) or a placebo, for 6 months. After being tested at baseline and at 6 months they were tested again at 9 months, that is, 3 months after the PS dosing had ceased.

The resulting data from this trial clearly indicated that the PS100 group showed a significantly better improvement versus the placebo on the overall MMSE. On the Delayed Word Recall subtest of the MMSE, both the PS100 and PS300 groups scored significantly better than the placebo group. On the Delayed Word Recall subtest of the HDS-R (Hasegawa Dementia Scale), the PS300 group was significantly superior to placebo; the PS100 group had a higher variance that likely robbed it of significance (refer to Fig. 3A in the publication) (88).

This Kato-Kataoka trial produced other interesting data (80). When the groups were divided into "low-score" and "high-score" subgroups based on their starting (baseline) memory performance, the low-score subgroup of PS300 scored significantly superior to placebo at 9 months, whereas the high-score subgroups did not perform better with PS than with placebo. This pattern is reminiscent of the Crook (80) trial: among subjects with memory impairment but not dementia, those who were worse off seemed to derive the most benefit from PS.

The benefits of PS can take weeks or months to appear – though there are anecdotal reports of noticeable benefit after just a few days. The iconic Crook (80) trial found significant benefit at 3 weeks, the first measuring point past baseline (80). The Kato-Kataoka (88) trial findings suggested that 6–9 months may be required for full benefit from doses of either 100 mg or 300 mg per day (88). But in this trial, benefits continued to build for 3 months after the PS daily dosing was discontinued, suggesting that PS initiates regenerative-type brain changes that continue for months after the PS supplementation has ended. This is an exciting aspect of PS that is consistent with its capacity to boost nerve growth factor (NGF) actions in experimental animals (98).

NGF was the first growth factor discovered, and is known to be crucial for brain cholinergic pathways fundamental to memory (99). In rats, PS given by mouth was found to increase the density of receptors for NGF in the hippocampus zone of the brain (98). This effect of PS could help account for the continued build-up of memory benefits even months after it ceased being taken daily as a supplement.

The molecular receptors for NGF are localized in cell membranes, and PS has a variety of cell membrane actions that likely facilitate NGF receptor action. As reviewed in Chapter 2.2, PS facilitates such membrane actions as electrical signal conduction, the generation of energy by the mitochondrial double membrane system, neurotransmitter release via vesicles at the synapses, and the sensitivity of neurotransmitter receptors on post-synaptic membranes.

PS is more concentrated in the brain than in the other organs, and there it is most concentrated in the hippocampus, which is the zone most involved in generating new memories (100). Signalling proteins that control nerve cell survival and differentiation are influenced by PS in brain cell membranes, so that PS is very likely important for neurogenesis – the production and differentiation of brain stem cells into functioning neurons and/or support cells (101).

Although the most studied daily intake of PS for memory enhancement has been 300 mg per day, higher doses are safe and may be more effective. For example, in a small human study using PET (positron emission tomography), 500 mg/day of PS significantly improved energetics (measured as glucose metabolism) in the hippocampus, basal ganglia/thalamus, auditory cortex and the brain as a whole (102).

6.6.4.3 Attention and behaviour

PS can improve attention, learning and behaviour in children with attention deficit hyperactivity disorder (ADHD) (95,103). ADHD is characterized by difficulties with sustaining attention, behavioural problems, and sometimes also excessive motor activity. This author participated with two physicians – Richard Kunin,

MD and Carol Ann Ryser, MD – in the conduct of three small case history series that recruited ADHD children and administered them soy PS under medical supervision. All three series suggested soy PS was beneficial for most of the children.

The largest of these case series was Dr Ryser's consecutive in-office study of 21 children aged 4–19 years (103). PS at 200–300 mg/day for up to 4 months benefited 19 of them (90%), with attention and learning most consistently improved. PS also consistently benefited the depression and anxiety commonly seen in ADHD children. Behaviour and academic performance often improved, though oppositional conduct proved resistant to PS treatment. These children showed no adverse effects or drug interactions with PS, consistent with its 30-plus years of clinical use. PS also seemed to extend the benefits experienced by children who were also receiving omega-3-enriched fish oil or gamma-linolenic acid (GLA) (omega-6)-enriched evening primrose oil.

A 2014 DB RCT by Hirayama and colleagues confirmed the promise of these earlier case histories (95). ADHD children aged 4–14 years (total 36) received either a placebo or soy PS 200 mg per day, for 2 months. PS improved overall ADHD symptoms (attention deficits) and impulsivity significantly better than the placebo. Children in the PS group also developed improved self-control both in their classrooms and at home, whereas children on placebo did not.

6.6.4.4 Stress and mood

PS has been widely represented in the marketplace as a supplement effective for coping with stress by lowering abnormally high levels of stress hormones such as cortisol and ACTH (adrenocorticotropic hormone). However in clinical trials PS supplementation has not consistently lowered either of these hormones (104-106), even at doses as high as 750 mg per day for 10 days (105).

PS was reported to lower blood cortisol levels in two small trials on physical stress (intensive cycling) (107,108). Unfortunately both these trials had a "crossover" design that would have required PS practically saturate the tissues within 10 days, and conversely to "wash out" from the tissues within 10 or 14 days. Because PS is a lipid-phase substance this timeline for build-up and washout is so highly unlikely that these crossover trials have very limited value.

In another crossover trial involving physical stress, healthy men (11 total) were subjected to "overtraining" by an intense weight workout programme (109). They received PS (800 mg per day) or a placebo, for 2 weeks. PS reportedly lessened muscle soreness and may have lowered cortisol, but in this trial again a short washout period (maximum 3 weeks) creates doubt about its findings. Another crossover trial that involved resistance exercise had no washout period, and failed to find changes in cortisol or testosterone from 400 mg per day of PS over 14 days (106). A trial with men running downhill used a longer washout period (4 weeks), yet found that PS at 750 mg per day did not significantly lessen muscle soreness (110).

PS has shown better results against mental stress and for improving mood. In a DB trial, Benton and colleagues randomized 48 healthy male undergraduate students to receive either PS (300 mg per day) or a placebo, for 30 days (111). After that period they were subjected to a standard psychological stressor, namely having to take a time-limited mental arithmetic test. Of those who received PS, a subgroup that had scored high for neuroticism reported a significantly lower feeling of stress from the test, as compared against others with high neuroticism who received the placebo. After the test, the mood of these individuals on placebo was significantly worse than for those who received PS.

The Benton trial's finding of PS improving mood in young healthy men complements the finding from an earlier small DB trial, which indicated bovine PS improved mood in elderly depressed women (112).

PS may help stabilize mental focus under stressful conditions. In a DB RCT with 20 male golfers aged 20–55, those who received 200 mg per day of PS for 6 weeks (as a nutritional bar, with protein, carbohydrates, fats and vitamins) showed significantly better targeting of their shots than those who received placebo bars (113). Another DB RCT employed EEG (electroencephalography) to demonstrate that the same PS bar may have improved the relaxation state in young men subjected to auditory stress (114).

The widely purported anti-stress benefits of soy PS lecithin really rest on limited clinical trial findings, often with very high – and therefore very costly – PS daily doses. Results from trials using soy PS lecithins in combination with phosphatidic acid – the "PAS" ingredient – seem to offer more promise against stress.

PAS lecithin complex for stress and premenstrual syndrome (PMS)

PAS is a lecithin complex that includes phosphatidic acid (PA), PS and other lecithin phospholipids (115,116,117). PA is a phospholipid that lacks a polar head group like serine, choline, ethanolamine or inositol (for the basic layout of a PL molecule refer to Fig. 1 in Chapter 2.2). The molecular layout of PA is illustrated in Fig. 1, below. PAS lecithin has a substantial PA content and has shown promise for stress and PMS (premenstrual syndrome).

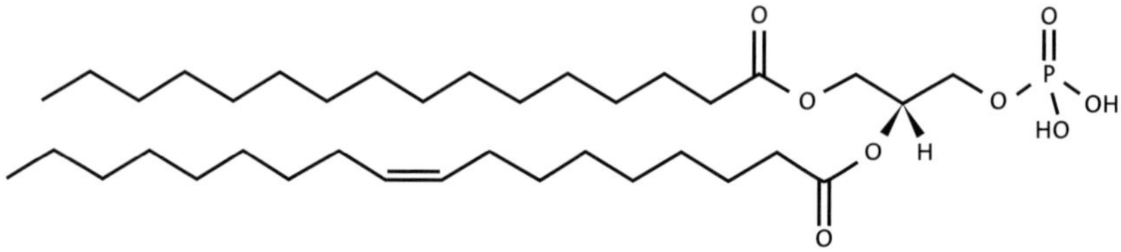

Figure 1.

A phosphatidic acid (PA) molecule, with the carbon-3 having the usual phosphate group attached but without choline, ethanolamine, inositol or serine. In this example the carbon-1 has palmitic acid (C16:0) attached as the acyl group, and carbon-2 has oleic acid (C18:1) attached. This molecule is PA (16:0/18:1) or more technically 1-palmitoyl-2-oleoyl-sn-glycero-3-phosphate. From ChemSpider (118).

Physiologically, PA is a major cell membrane constituent that likely facilitates synaptic vesicle release and recycling (119). PA also may be released from the membrane milieu and influence activity in the cell interior (120). Further, metabolites of PA, especially lyso-PAs (LPAs), which have just one fatty acid tail, also are messenger substances that may promote nerve cell generation and maturation as well as regulate synaptic functions (121,122). PAS is a patented complex that contains (1) PA, (2) PS and (3) other lecithin PLs (PC, PE, PI and lyso-PLs), in ratios that vary by weight.

The first clinical trial of PAS was a DB RCT published in 2004 (115). This PAS consisted of a 1.25 to 1 to 2.7 ratio of PA to PS to the other PLs, by weight. Groups of healthy male and female subjects aged 20–45 years had their salivary cortisol measured, then for 21 days received either 400 mg/day PAS, 600 mg/day PAS, 800 mg/day PAS, or a placebo. They then reported to a laboratory where they gave saliva and blood samples, filled out psychological questionnaires, and took the Trier social stress test (TSST). This test included having to give a 5-minute job application speech with just 3 minutes to prepare, followed by counting backwards from 2,083 to zero in steps of 17 as fast as possible. Then they had to give more saliva and blood samples.

As expected, in this trial the stress of taking the TSST test caused cortisol to increase in all the treatment groups (115). The group that received 400 mg/day PAS had significantly less cortisol elevation (in both saliva and serum) than the placebo group, but the two groups receiving higher PAS dosages did not differ significantly from placebo.

In a later DB RCT trial, the researchers used different PAS compositions, which also varied between the two PAS subject groups (116). For the "PAS 400" group the profile was (1) PA+lyso-PA, (2) PS+lyso-PS and (3) "other phospholipids and glycerides", in a ratio of 1 to 1 to 2.35 by weight. The "PAS 200" group received a ratio of 1 to 1 to 6.7 by weight. Healthy men (no women) aged 20–45 years were initially tested on the Trier Chronic Stress Inventory, then classified High Chronic Stress or Low Chronic Stress and randomly allocated to receive either the PAS 400 lecithin (providing 400 mg PA and 400 mg PS per day), the PAS 200 (providing 200 mg PA and 200 mg PS per day) or a placebo. They received the treatment for 42 days, then were subjected to saliva and blood sampling for stress hormones, psychometric tests and the TSST.

In agreement with the previous trial, the TSST caused saliva and serum cortisol to increase in all three groups of men (116). The only men to show a significantly lower cortisol response were those classified High Chronic Stress and receiving the high PAS dose (PAS 400 lecithin, providing 400 mg PA + 400 mg PS + 940 mg phospholipids and other glycerides per day). This was also the only subgroup to show significantly lower elevation of serum ACTH (adrenocorticotropic hormone, another stress hormone) versus the placebo.

It seems, then, that a combination of PA with PS and a variety of other phospholipids can dampen the effects of an acute stressor, especially in men tested and found to be experiencing relatively high stress.

Premenstrual syndrome has both mental and physical symptoms. The mental "affective" symptoms include angry outbursts, anxiety, confusion, depression, irritability and social withdrawal. Physical "somatic" symptoms include abdominal bloating, breast tenderness or swelling, headache, joint or muscle pain, swelling of extremities and weight gain (123). In a DB RCT, women aged 18–45 years with a PMS diagnosis received PAS or a placebo for three menstrual cycles (117).

This trial employed PAS 400 lecithin, again defined as 400 mg PA (including lyso-PA) + 400 mg PS (including lyso-PS) + 940 mg other PLs and glycerides, per day, versus a placebo (117). The women filled in symptom questionnaires daily, over a baseline cycle then for three subsequent cycles on PAS or placebo. At the end, those on PAS were found to have significantly lower scores for overall symptom severity, versus those on placebo. PAS also worked significantly better for reducing physical symptoms. PAS slightly lowered cortisol levels (2.7%), performing significantly better than the placebo group, which increased 19.8 per cent. PAS clearly deserves further, enthusiastic investigation for PMS.

PAS Lecithin for memory and other cognition

Results of two trials with PAS for memory problems were reported in 2014 (124). One DB RCT assessed PAS on elderly subjects with memory problems. Another DB RCT assessed PAS on patients with Alzheimer's disease. A few positive benefits were seen.

The first trial included men and women aged 60 – 80 years who reported some memory problems but scored 25 – 30 on the Folstein Mini – Mental State Examination and therefore likely did not have dementia (124). They received either a PAS mix (300 mg PS + 240 mg PA per day) or a placebo, for 3 months. Their memory was tested using the Wechsler Memory Scale (WMS). Both groups scored improved memory that was not significantly different between them. A subgroup analysis of those beginning with less memory loss (WMS>68) found better improvement with PAS than with placebo.

The second trial included men and women aged 50 – 90 years and diagnosed definite or probable Alzheimer's (124). They received the same PAS formula and dose, or a placebo, for 2 months. The PAS group showed no functional decline on seven Activities of Daily Living (ADL) measures. The placebo group significantly declined, and the difference between the two groups was statistically significant. During the still double-blind phase of the trial, patients were given a choice whether or not to continue on the test medication after the trial ended; six of the 14 PAS patients agreed to continue versus zero of 11 receiving the placebo, and this difference was statistically significant. This finding was consistent with earlier data from non-controlled administration of PAS to elderly dementia patients.

Considering that PAS is reported to improve brain function in older men and women either with mild memory loss or with probable Alzheimer's, this specialized lecithin formulation clearly deserves further investigation for its cognitive benefits. With the caveat that its full phospholipid profile should be well characterized and standardized.

6.6.4.5 Soy PS omega-3 lecithins

In the late 1990s the PS research community began debating an "omega-3 PS" – a molecular species of PS that would carry omega-3 EPA and/or EPA as its acyl tail groups, esterified at the sn-1 and/or sn-2 positions. Initial animal research (rodents, baboons) had suggested that fatty acids might be more efficiently absorbed when in PL forms, rather than in the triglyceride form as occurs in commercial fish oils. As the reasoning went, with DHA and EPA showing clinical promise of brain benefit, and with PS already proven to have brain benefit, linking DHA and/or EPA into PS molecules might be conducive to a synergy of brain benefit. The first-generation PS omega-3 lecithin did improve memory in elderly subjects (89,90,91).

This PS omega-3 was called PS-DHA, though it also contained EPA. In a DB RCT, 122 elderly subjects not demented but complaining of memory loss received either a placebo or 300 mg/day of this PS omega-3, which also supplied 79 mg/day of DHA+EPA in a 3:1 ratio, for 15 weeks (124). The PS group improved better than placebo on verbal immediate recall. When this trial ended the investigators donated additional PS omega-3 at 100 mg/day to the subjects for another 15 weeks (91). They reported that the group formerly on placebo and switched to PS improved significantly in sustained attention and memory, while those continuing on PS maintained their improved cognitive status. The researchers also tracked a variety of physiological measures in their subjects, and concluded their omega-3 PS-DHA was both effective and safe to use at 100 mg/day for at least 30 weeks (90).

Another obvious potential application for PS omega-3 was to improve on the early success of plain soy PS lecithin for ADHD (103). Vaisman and collaborators recruited children aged 8–13 years with impaired visual sustained attention, as assessed using the computerized Test of Variables of Attention (TOVA) (125). The children were randomly allocated to receive either (a) a PS omega-3 lecithin (580 mg/day), providing about 300 mg/day of the PS omega-3, 168 mg/day of other PLs, 95 mg/day DHA and 156 mg/day EPA (these mg amounts add up to more than 500 mg because the omega-3 molecules are built into the PS and PL molecules); (b) a fish oil preparation (799 mg/day), consisting of triglycerides providing 96 mg/day DHA and 153 mg/day EPA; or (c) a placebo of rapeseed oil (742 mg/day), also triglyceride in nature but providing mostly omega-9 fatty acids and no DHA or EPA. All three preparations were emulsified in a dairy chocolate-flavoured spread prior to consumption.

The trial went for 3 months (125). Both the PS omega-3 lecithin and fish oil groups showed significantly improved attention, via improved total TOVA scores, faster response times, and lowered response time variabilities. The PS omega-3 lecithin also significantly improved errors of commission, and was statistically superior to the fish oil on all the attention scores. The placebo group did not show any significant improvements. The PS omega-3 lecithin significantly benefited about 60% of the children in that group, whereas only about 35% of those on fish oil showed significant benefit. However, biochemical testing for omega-3 status raised doubt on whether the PS omega-3 lecithin worked by improving the children's omega-3 status.

This trial monitored DHA and EPA in both the plasma PL and the PL of the red blood cell membranes, as percentages of the total fatty acids present (refer to its Table 4) (125). The percentage of DHA+EPA in the red blood cells typically is low in ADHD (126) individuals, and indeed was low at the beginning of this trial. At the trial's end EPA was up slightly but DHA was lower, so the total DHA+EPA was not increased above baseline. Thus the PS omega-3 lecithin had not substantially raised DHA and EPA in these children's tissues. One plausible explanation is that the total phospholipid actions of the PS and other PL molecules were responsible for the superior clinical benefit of the PS omega-3 lecithin, rather than just the modest amounts of DHA and EPA provided by the preparation.

Another DB RCT was conducted on ADHD children with a PS omega-3 lecithin preparation (127,128). Children aged 6–13 years (total 147) received either a placebo or 300 mg/day of a PS that also provided 40 mg/day of DHA and 80 mg/day of EPA, for 15 weeks. The PS group improved significantly more than the placebo group on the Restless/Impulsive subscale of the Conners Rating Scale – Parent as well as the Emotional Impact on Parent subscale of the Child Health Questioner. The boys subgroup improved better than placebo on the Conners teacher DSM-Inattentive subscale. Additional subgroup analysis suggested that children who were more hyperactive/inattentive experienced more complete benefits.

The researchers then did an open-label extension of this trial, in which 102 of the children continued on the PS omega-3 and 48 were switched from placebo to half-dose PS omega-3 (150 mg/day), for another 15 weeks (127). The "switched" children significantly improved on oppositional and hyperactivity subscales, and the continuers significantly improved on family activities and psychosocial subscales.

This trial also monitored physical, biochemical and other parameters related to safety (128). Its overall conclusions were that the PS omega-3 at 300 mg/day for 15 weeks or 150 mg/day for 30 weeks was well tolerated and safe, with no negative effect on body weight and growth, and that this PS improved ADHD, most specifically in children who were the most hyperactive-impulsive and more emotionally and behaviourally "dysregulated" (127,128).

The fascinating saga of PS lecithins clearly will continue. They have revolutionized the marketability, value and overall commercial relevance of lecithins as dietary supplements.

6.6.5 Krill lecithins

Krill are small, shrimp-like oceanic crustaceans with a huge biomass that provides essential food for numerous marine animals. Antarctic krill *(Euphausia superba)* live in the most frigid seas, and survive because their cell membranes can remain fluid – and therefore functional – at these very low ocean temperatures (the relationship of membrane fluidity to function is discussed in Chapter 2.2). This fluidity is mainly attributable to a high density of unsaturated double bonds contributed by omega-3 DHA and EPA. Antarctic krill are the commercial source of krill oil, a red oily fluid that has substantial phospholipid content and meets the definition of a lecithin preparation.

Krill lecithin (KL) preparations from Antarctic krill have lipid profiles that include 18–81% phospholipids by weight, along with triglycerides, other glycerides, free fatty acids and cholesterol (12). Krill lecithins usually have about 35% of their fatty acid molecules as omega-3s, with a DHA:EPA ratio of roughly 1:2. The PL fraction usually has 42–50% of its fatty acid molecules as omega-3, and the triglyceride fraction about 25% omega-3, which is comparable to the triglycerides of fish oils.

As with any product from nature, the lipid profiles of krill (and therefore of their oil) vary substantially with diet, season, geographic origin and processing techniques. One crucial factor is that krill begin to degrade immediately once caught, and this can result in the production of oil lower in PL and higher in free fatty acids. Independent lab analyses sometimes reveal marked deviations of KL brands from their labelled content (12).

The phospholipids of krill oil carry most of the omega-3 fatty acids, and are dominated by PC with much smaller amounts of PE and PS (12). The red colour of KL derives from its low content of astaxanthin, a bright red carotenoid membrane antioxidant (129).

Krill lecithin emerged as a dietary supplement in 2003, after a DB RCT indicated it could relieve both the mental and inflammatory symptoms of premenstrual syndrome (PMS), a potentially life-threatening condition (130). In this trial (131), women with PMS were randomized to receive either a Neptune Krill Oil (NKO) KL (composition not disclosed) or a fish oil (FO) for 3 months. For the first 30 days, the krill group received 2 g/day of KL (said to provide about 600 mg/day of DHA+EPA, with about 800 mg/day of PL) while the FO group received 2 g/day of an "18:12" fish oil (12% DHA and 18% EPA), which would have provided about 600 mg/day of DHA+EPA. Then, over the subsequent 60-day period the women in each group received 2 g/day of their assigned supplements for just 10 days: 8 days leading up to their menstrual period and 2 days during menstruation. The KL group recorded significantly greater improvements in a majority of the PMS symptoms than the FO group, and their use of painkillers (mainly ibuprofen, acetaminophen or aspirin) was also significantly lowered versus the FO group. Subsequent RCTs have examined various KL brands for a variety of health applications, as covered below.

6.6.5.1 Krill lecithin blood lipids trials

Various brands of KL have been investigated for the management of blood lipids (cholesterol, triglycerides) in several RCTs. In the first such trial (132) the KL composition was not disclosed but was stated to be Neptune Krill Oil (NKO) containing PL carrying DHA and EPA, with "various potent antioxidants, including vitamins A and E, astaxanthin, and a novel flavonoid." A total of 120 subjects with hyperlipidaemia (total cholesterol above 194 mg/dL; triglycerides above 204 mg/dL) were divided into four groups. The "low-dose" KL group received either 1 or 1.5 g/day, depending on body mass index (BMI); the "high-dose" group received KL at 2 or 3 g/day. The fish oil group received 3 g/day of a standard 18:12 fish oil that supplied 900 mg/day of omega-3s (180 mg EPA and 120 mg DHA per gram). The fourth group received a placebo (microcrystalline cellulose). The trial lasted 12 weeks.

In this trial both the KL and the FO performed significantly better than the placebo for lowering total cholesterol and triglycerides, and for elevating HDL cholesterol (132). Due to a surprising lowering of LDL cholesterol by the placebo, neither the KL nor the FO achieved statistical significance for lowering LDL vs the placebo. The high-dose (but not low-dose) KL outperformed the FO for lowering triglycerides though not for elevating HDL cholesterol. This trial seemed to show that a relatively high intake of a KL (2–3 g/day) could improve blood lipid status better than 3 g/day of a basic fish oil, in hyperlipidaemic subjects.

The next trial was on the Aker brand of KL (133). It recruited healthy subjects (113 total) with slightly elevated total blood cholesterol and/or triglyceride levels, and randomly allocated them into three groups. The subjects received either (a) the KL at 3 g/day, providing 543 mg/day of DHA+EPA, (b) a FO at 1.8 g/day, providing 864 mg/day of DHA+EPA, or (c) neither KL nor FO, as a control group. The trial lasted 7 weeks, and found no statistical superiority of the KL group versus the FO or the control group.

The apparent failure of the KL in this trial was probably not due to any problem with compliance in taking the daily doses, since the krill subjects' plasma DHA+EPA levels did show marked increases (133). The degree of increase was comparable with the FO group, and since the daily KL dose provided just 64% of the DHA+EPA provided by the daily FO dose, the researchers suggested KL was just as efficient or perhaps more efficient than FO for delivering omega-3s into the blood. However, this reasoning cannot account for the clinical finding that the KL group of subjects did not experience superior improvements of blood lipids, oxidative stress or inflammation as compared to the subjects who received fish oil.

The Aker KL preparation used by Ulven and colleagues for this 7-week trial (133) was subsequently subjected to a 12-week trial, this time not on healthy subjects but on subjects with borderline high or high fasting triglyceride (TG) levels (150–499 mg/dL) (134). A total 267 subjects were randomized into five groups, to receive eight capsules/day of either an olive oil placebo or the KL at 0.5 g/day 1.0 g/day, 2.0 g/day, or 4.0 g/day, for 12 weeks. These doses were stated as supplying approximately 0, 100, 200, 400 and 800 mg/day of DHA+EPA.

The data from this trial had such high variability that the researchers decided to pool all the data from the four krill groups into one larger group for comparison to the placebo group (134). This data treatment allowed them to conclude that the KL lowered triglyceride levels significantly better than the placebo, but meant that no dose-related pattern could be identified. The data also indicated the KL did not improve cholesterol measures better than the placebo, even when just the highest KL dose (4 g/day) was considered.

This trial by Berge and colleagues also tracked the subjects omega-3 index (O3I) (134). This index is a measure of the DHA+EPA content of the tissues. It is obtained by measuring DHA and EPA in the cell membranes of circulating erythrocytes (red blood cells) and calculating their combined percentage as a molecular (mol/mol) proportion of all the fatty acids present in the membrane. A minimum 8% O3I, equivalent to at least eight DHA or EPA molecules out of every 100 fatty acid molecules, has been correlated with lowered risk for adverse cardiovascular events (135,136). In this Berge trial (136), the O3I increased from the baseline 3.65% level to 6.3% after 12 weeks in the 4 g/day KL group. Notably, this increased O3I level was still substantially lower than the desirable 8% minimum O3I (135,136).

An Aker KL preparation was also used in a DB, crossover RCT on endothelial function and other cardiovascular factors in subjects with type 2 diabetes (137). A total of 47 subjects received either the KL (providing 1,000 mg/day of omega-3 fatty acids) or a placebo (olive oil, amount not disclosed) for 4 weeks. Then after a 2-week washout each subject was switched to the opposite treatment for four more weeks. The data analysis concluded the KL group showed significantly better improvement over the olive oil group, on endothelial function and on insulin resistance. But the very short washout period complicates these findings, since it is highly unlikely these cell membrane fatty acids could have been "washed out" from the membranes within such a short period.

In this trial by Lobraico and colleagues (137), most of the subjects from both groups were then continued for another 17 weeks on the KL alone. Their HDL cholesterol significantly increased, but since there was no control group at this stage of the trial the clinical usefulness of this finding is limited.

A KL from Enzymotec Ltd was tried on healthy subjects aged 18–49 years with triglyceride levels lower than 200 mg/dL, total cholesterol lower than 240 mg/dL, and LDL cholesterol lower than 160 mg/dL (138). This was also a crossover trial: each treatment phase lasted 4 weeks, separated by a washout phase of 8 weeks. The daily dosages of KL and FO were matched to both provide approximately 600 mg/day of total omega-3 fatty acids, plus the KL provided 1.8 mg/day of astaxanthin. Corn oil was used as a placebo. The clinical outcome was that both the KL and the FO failed to lower triglyceride levels. Surprisingly, both raised LDL and total cholesterol, significantly more than the placebo and comparably with each other.

The reason(s) for these unhealthy cholesterol effects are not clear (138). The FO had a very high omega-6 content, different from other fish oils, which the researchers stated was due to blending with corn oil and which helps account for its cholesterol elevation effect (139,140). But why this particular KL would have this negative health effect is unclear. The KL did improve the O3I (from a mean of 4.94%, up to 5.97%), better than did the FO (from 4.96% to 5.43%). But again, neither the 3 g/day of the KL nor the 3 g/day of FO raised the O3I sufficiently to cross the 8% threshold for lowering cardiovascular risk. Earlier research strongly suggested that a 4-week period of dosing with each supplement was much less time than required for full delivery of the KL or FO fatty acids into cell membranes (135,141).

A Rimfrost KL preparation, stated to be relatively high in phospholipids (47% PLs by weight), was tested by Berge and colleagues in a small, non-blinded, short-term pilot trial without a comparison group (142). A total of 17 young, healthy participants received 4.5 g/day of this preparation, stated as providing 832.5 mg/day of DHA+EPA, for 28 days. Triglycerides significantly decreased and HDL cholesterol significantly increased, but LDL and total cholesterol also showed significant (though small) increases. The LDL particle profile, measured as percentage of highly atherogenic small particles, did not change in the direction of increasing cardiovascular risk.

This Berge pilot trial (142) did have some encouraging positive outcomes: several antioxidant and anti-inflammatory biomarkers increased. Plasma choline and betaine also increased, probably linked to the increased plasma levels of PC and other PLs. Both choline and betaine are important methyl group donators, being required in numerous metabolic pathways including the homeostatic control of homocysteine, a major cardiovascular risk factor (143-146).

From the beginning of this Berge trial, the subjects' mean O3I was higher than the 8% desirable threshold (8.2%) (142). By the trial's end (the 28th day), the O3I had jumped to 9.9%, a highly significant improvement compared to baseline ($p<0.0001$). This trial has limited meaning due to its small size, short term of dosing, and not being an RCT. But this highly concentrated Rimfrost KL preparation seemed to have important positive effects even though it had the negative effect of elevating cholesterol. It deserves a larger, randomized controlled trial.

A KL from Erbozeta S.r.l. (composition not disclosed) was tested in mildly overweight, high-triglyceride subjects in a small DB RCT (147). The 25 subjects received either 1,000 mg/day of the Erbozeta KL (providing 240 mg of DHA+EPA) or 2,000 mg/day of FO ethyl esters (providing 1,700 mg of DHA+EPA) for just 4 weeks. Then after a 4-week washout period each was "crossed over" to the other treatment for another 4 weeks. The results led the researchers to conclude this KL lowered triglycerides less effectively than the FO but surpassed the FO for elevating HDL cholesterol and apo-A1 while lowering CRP (C-Reactive Protein), a biomarker of inflammation. But this trial's design was fatally flawed.

As with crossover trials conducted with other KL brands, the initial 4-week dosing period was probably too short to saturate the subjects' cell membranes with DHA and EPA – there is ample evidence the minimum time period is 5–6 months (141,148). This KL material, whatever its composition, could be more fairly evaluated via a larger non-crossover trial.

6.6.5.2 Potential for brain benefits

The 2003 trial of the NKO brand established that a KL could significantly improve emotional symptoms of PMS (feeling overwhelmed, stressed, irritable, depressed), and worked better than fish oil (131). In 2007 the NKO manufacturer claimed a successful outcome from a non-randomized trial on adults with ADHD (149), though the trial has not been published in a peer-reviewed journal.

In 2013, a DB RCT was published that compared a KL from Nippon Suisan Kaisha Ltd versus a fish oil, for cognitive function in elderly subjects (150). This KL contained about 45% PL, and about 14% DHA and EPA in a 1:2.1 ratio. Healthy retired men who were not cognitively impaired (total 42) received 2 g/day of either the KL, sardine oil as a typical fish oil (FO), or medium-chain triglycerides (MCT) as a placebo, for 12 weeks. At weeks 6 and 12 the subjects were tested on a working memory task and a calculation task. Their brain blood oxygenation was measured via oxyhaemoglobin (OHb) concentration, and electrical activity by EEG. The benefits from this trial were apparent after 12 weeks though not after 6 weeks, with the KL group showing significantly more benefit than the FO or MCT groups on some of the measures.

When this trial ended at 12 weeks (150), on the working memory task both the KL and the FO group showed significantly greater increases in brain oxygenation versus the placebo group. The KL group's P300 latencies (per EEG) became significantly faster than placebo, whereas the FO group's latencies did not. Faster P300 latencies suggest faster information processing, and are known to slow with ageing. On the calculation task, the KL group's oxygenation increased significantly over placebo but the FO's did not. Interestingly, the KL did not increase plasma DHA and EPA levels significantly better than either the FO or the placebo, which suggests its unique phospholipid and astaxanthin content likely contributed more to its superior brain effects than its fatty acid profile.

The body's endocannabinoid system (ECS) is an intricate network of messenger substances generated from fatty acids attached into cell membrane phospholipids and acting via cell membrane receptors to regulate a range of physiological processes (150). The 'endocannabinoid' term means the system also can be influenced by plant cannabinoids. One such possible influence is on body weight regulation. Banni and colleagues did a DB RCT on healthy men and women who were overweight or obese, conditions associated with higher levels of endocannabinoids (152). They compared 2 g/day of either an Aker KL, a fish oil (FO) or an olive oil control group (OO), for 4 weeks, for increasing ECS cannabinoids in overweight and obese subjects. The Aker KL, providing 309 mg/day of DHA and EPA in a 1:2 ratio, lowered plasma levels of the major cannabinoid (2-AG) 2-arachidonoylglycerol in the obese subjects. The KL worked better than the FO, which provided 390 mg/day of DHA and EPA in a 1:1 ratio; and better than the olive oil. Though the Aker KL did not significantly change body weight, considering the current intense interest in the human benefits of oral cannabinoids this finding deserves replication and follow-up.

6.6.5.3 Arthritis and other inflammation trials

For a 2007 DB RCT (153), Deutsch recruited 90 patients diagnosed with osteoarthritis and/or rheumatoid arthritis and/or cardiovascular disease, and with C-reactive protein (CRP) values 1.0 mg/dl or higher. CRP is a well-documented indicator of systemic inflammation (154). The patients received either 300 mg/day of NKO KL or a placebo (microcrystalline cellulose), for 30 days. Standard questionnaires were used to assess joint stiffness, functional impairment and pain. Patients were allowed to take acetaminophen for pain as needed.

By 7 days the KL had improved joint pain, stiffness and functional impairment, along with CRP, significantly better than the placebo (153). These benefits were sustained at days 14 and 30. Patients taking the KL were able to significantly reduce their acetaminophen consumption by 31.6% at day 30, superior to the placebo (5.9%). No adverse effects were reported.

A 2016 DB RCT tested a KL from Sunsho Pharmaceutical Co., Ltd for mild knee pain judged as probably osteoarthritic but not requiring analgesic medication (155). A total 47 subjects attending a Japanese orthopaedic clinic, mostly female, received 2 g/day of either the KL preparation (delivering 110 mg/day DHA and 240 mg/day EPA), or a safflower oil placebo lacking DHA or EPA. At the trial's end (after 30 days), the KL had improved pain, stiffness and range of motion, all significantly better than the placebo.

6.6.5.4 Immune system activity following exercise

Based on their observation that supplementation with fish oil could increase immune killer T-cell activity in the recovery period following exercise, Da Boit and colleagues did a similar trial with an Aker KL (156). They also wanted to determine whether this KL could alter exercise performance.

For this DB RCT 37 young, healthy subjects were recruited. First blood samples were taken, then they did a maximal exercise cycling test which increased their workload minute by minute until they became exhausted (156). Their gas exchange and heart rate were monitored. For 6 weeks they received 2 g/day of either the Aker KL (providing 120 mg DHA + 240 mg EPA + 244 micrograms astaxanthin), or a placebo that "reflected the fatty acid composition of the average European diet". At the trial's end the maximal exercise test regimen was repeated.

The KL group showed significant increases in red cell DHA (by 21%), EPA (by 75%) and omega-3 index (by 28%, from 5.3% to 6.8%), while the control group did not (156). Though the KL group did not cross the 8% O3I threshold, the rate of increase in the index suggested the 2 g/day intake may well be sufficient to achieve this goal within a reasonable period, perhaps another 6 weeks. For the fish oil, there is good evidence that its mere 360 mg/day of DHA+EPA is unlikely to cross the 8% O3I threshold for at least 40 weeks (10 months) (141).

In this trial the Aker KL group surpassed the placebo in raising subjects' natural killer (NK) cell activity (156). These immune cells are very important for eliminating cancer cells. The KL also surpassed the placebo in releasing IL-2 cytokine from subjects' monocyte cells. This cytokine can stimulate T-cell activation and proliferation (157).

Despite its success at elevating immune cell activity, the Aker KL did not improve exercise performance better than did the placebo (156).

6.6.5.5 Antioxidant status following exercise

An NKO KL preparation was tested on a small number of trained rowers subjected to exhaustive exercise (158). A total of nine rowers received 1 g/day of KL, while eight received a placebo, for 6 weeks. At the beginning and at the end of the trial they had to row at full pace for 2,000 metres on a simulator. Rowing elevated TBARS (thiobarbituric acid reactive substances), a marker of oxidative damage, as well as the pro-inflammatory cytokine TNF-alpha and the antioxidant enzyme superoxide dismutase (SOD) in both groups. However the KL showed significantly less elevation of TBARS versus the placebo, indicating it may have lowered post-exercise oxidative damage. The KL did not improve exercise performance.

6.6.5.6 Resistance training exercise

In a 2018 DB RCT, experienced resistance training subjects received 3 g/day of either a Rimfrost KL or an olive oil placebo, and followed an intensive resistance training programme for 8 weeks (159). The KL group did not improve over the placebo on performance, body composition, muscle strength, self-reported recovery or "mental vitality".

6.6.5.7 Additional clinical benefits of krill lecithins

Krill lecithin can improve eye health. Dry eye disease (DED) is an inflammatory disease that affects 5–34% of people worldwide depending on population and has many features in common with autoimmune disease (160). DED is characterized by increased osmolarity of the tear film and eye inflammation, mediated by autoreactive T helper cells that infiltrate the lacrimal (tear-producing) glands and the surface of the eye. Deinema and colleagues conducted a DB RCT on subjects with mild to moderate DED (161). Subjects received either a KL (Nutra-Life OceanClean brand) that provided 1,455 mg/day of DHA+EPA, a fish oil that provided 1,500 mg/day, or 1,500 mg/day of an olive oil placebo, for 90 days. The KL surpassed both the placebo and the FO in lowering pro-inflammatory cytokine 17A in the tear film, and improving the Ocular Surface Disease Index score.

Two KL brands have improved blood sugar management (132,162). The 2004 trial on subjects with high blood lipids reported that 1–3 g/day of the NKO brand for 3 months lowered fasting blood glucose levels by 5–6 per cent (132). In a 2018 small DB RCT with high-triglyceride healthy subjects, a high daily dose of the Rimfrost brand (4 g/day, providing about 665 mg/day of DHA+EPA) was found to lower fasting blood glucose by about 5 per cent, significantly better than the control group, which received high-oleic sunflower oil (162).

In this same Rimfrost trial (162,163), monocyte cells drawn from the KL subjects showed altered gene expression consistent with the observed lowering of blood sugar. These changes were not seen in the FO group, but did have some similarity to the sunflower oil control group, which actually received a substantial dosage of monounsaturated (omega-9) fatty acids from the sunflower oil. High-oleic sunflower oil and olive oil, both relatively high in omega-9s, may not make good placebos for krill trials.

6.6.5.8 Effective source of choline and methyl groups

Choline supports numerous life processes, and is a conditionally essential nutrient since some people cannot make enough of it to maintain good health (143). For most people dietary choline is fully essential (164), and recommended daily intake levels have been set in the U.S. and EU (144). Human choline deficiency is associated with structural alterations in DNA and chromosomes, and with changes in gene expression linked to increased risk of cancer (143). In the U.S., only 11% of adults may achieve the recommended intakes (144). The main dietary source of choline is the phospholipid PC (phosphatidylcholine), and krill lecithins have relatively high PC content.

Choline coming from PC in KL dietary supplements can readily be used to make acetylcholine, a key neurotransmitter both within and outside the brain. Choline also is a rich source of methyl groups (–CH3), able to donate three methyls per choline molecule. Methyl is required for a great many metabolic pathways including brain production of neurotransmitters and the sleep hormone melatonin, myelin insulation of nerve cells, and the production, repair and epigenetic regulation of DNA (143,144). In a non-randomized pilot study with 17 healthy volunteers, 4.5 g/day of a Rimfrost KL providing 1,750 mg of PC increased plasma choline levels by 28%, along with the choline derivatives betaine, dimethylglycine and sarcosine, all of which contribute to methyl metabolism (165).

6.6.5.9 The bioavailability controversy

Immediately krill lecithin emerged as a commercial dietary supplement in 2003, it was claimed to be superior to fish oil. The fact that KL continues to be called 'krill oil' when it is actually a lecithin may be related to this comparison, which fuels ongoing and sometimes contentious market competition between KL and FO products. Obviously claims of superiority for KL over FO (or vice versa) can only be resolved by well-designed human research studies.

Though there are a number of compositional differences between KL and FO, during the 15 years that have passed since the first KL clinical trial a great deal of research effort has been expended on efforts to prove it is better absorbed than fish oil and therefore must also work better. Advocates of KL have consistently suggested that the phospholipids in KLs make their DHA and EPA content more bioavailable than do the triglycerides that make up FOs (which have zero phospholipids).

Indeed, animal studies do suggest that when taken by mouth, omega-3 DHA and EPA attached to phospholipid molecules can deliver more DHA and EPA to the tissues, milligram for milligram, compared to the omega-3 triglycerides of fish oils (166,167). Multiple human studies have compared KL omega-3 bioavailability against triglyceride or ethyl ester FO, but the findings have been inconclusive. Crossover design – switching the subject between KL and FO, generally after a short washout period – has certainly hindered meaningful results (168). In 2014 Salem and Kuratko published an in-depth critique of the bioavailability studies published to that point (169).

Salem and Kuratko rightfully pointed out that studying absorption after giving a single dose, or after daily dosing over a few days, is not representative of long-term supplementation (169). They criticized the design of several of the studies, including those that provided less DHA+EPA from the KL relative to the FO, then had the blood levels come out not significantly different and concluded this meant the KL was more bioavailable. They also pointed out that often the EPA to DHA ratios were not matched between the KL and FO preparations. In one trial a substantial amount of omega-6 linoleic acid was added into the FO, sufficient to bias the trial outcome in favour of the KL (138).

The Salem and Kuratko critique (169) also pointed out that sometimes the double-blinding and the dosing compliance in comparison trials were questionable because the KL and FO groups were not matched for numbers of capsules to take each day, or the capsules differed in colour (recall that krill oil is bright red whereas fish oil has no red colour). In some trials the subjects were not matched for blood DHA and EPA levels at the beginning of the study. Some subjects showed a decrease in blood DHA and EPA that indicated they were not taking the capsules. At times the subjects' prior dietary intake of omega-3s was not considered before they were "randomly" assigned to KL or FO groups. All these criticisms weaken the contention that KLs deliver DHA and EPA to human tissues more efficiently than FOs.

Salem and other colleagues then published their own bioavailability comparison (170). In a DB RCT they randomized 66 healthy subjects of average age 35 years into three groups. All received 1.3 g/day of DHA+EPA (approximately 522 mg/day DHA and 816 mg/day EPA), in the form of either a Neptune KL (44% PL), a triglyceride fish oil or an ethyl ester fish oil, as six red capsules per day for 4 weeks. They found that the subjects' omega-3 index values were not significantly different between the three groups at the end of the trial period. The subjects' dosing compliance was estimated excellent – over 90 per cent. The findings from this well-designed study suggest that 4 weeks of dosing with high levels of a KL are unlikely to deliver more DHA and EPA into human tissues than dosing with the equivalent amount of fish oil.

Still, there is persuasive evidence that KLs may be superior to FOs for delivering DHA and EPA, to the brain at least. In 2018 Patrick published a very well reasoned analysis that strongly suggests phospholipids with DHA or EPA attached at their sn-1 carbons are more successful than triglycerides at delivering them into the brains of rats, mice or humans (171). Patrick discussed persuasive evidence from radiolabelling and other studies that indicates DHA (or EPA) at this sn-1 position is not detached from the PL molecule by the gut's digestive lipase enzymes, and reaches the blood circulation as lyso-PL molecules (PL with one tail group remaining, which from krill can often be DHA or EPA). For the DHA-lyso-PC at least, Patrick discusses specialized transport proteins that can move these molecules across the blood – brain barrier and into the brain tissue.

Patrick also presented evidence that DHA molecules not attached to PL are not eligible for such facilitated transport and must enter the brain by simple diffusion, which is less efficient than active transfer by transport proteins (171). DHA is far more abundant than EPA in the human brain; in the iconic Framingham Heart Study, elderly subjects with relatively high plasma and DHA-PC level (the precursor of DHA-lyso-PC) had significantly lower risk of progression to dementia in the subsequent 9-year period (172). The omega-3 index has been reported to positively correlate with blood flow to the human brain: the higher the index, the greater the brain's blood flow (173).

Whatever the extent that krill lecithins deliver DHA and/or EPA to the human brain and other tissues, continuing efforts to prove superior bioavailability to fish oils – or other omega-3 sources – will likely continue to breed confusion and frustration in the marketplace. The real issue – and test – for krill lecithins is: when a certain brand of well-standardized KL is tested in a double-blind, randomized controlled trial, as a dietary supplement at a specified daily dose over a specified time period, for a specified set of health benefits...Does it work?

6.6.6 Fish lecithins

Like krill, fish that live in relatively cold waters use omega-3 fatty acids to keep their cell membranes sufficiently fluid at low temperatures. Certain cold-water fish are currently sustainably harvested in amounts sufficient to yield lecithins that are relatively high in omega-3s. One such is the Norwegian herring fishery.

The roe (egg masses) that might otherwise be discarded as a by-product have become a convenient source of fish lecithins (FL) that are high in PL and provide DHA and EPA in a ratio that approximates healthy human breast milk (around 3:1 DHA:EPA). Two such branded products from Arctic Nutrition have been tested in clinical trials.

6.6.6.1 Herring roe fish lecithin trials

The first trial was with an FL named MOPL30, described as 30% PL by weight, of which about three-quarters was PC and the rest other PLs, and about 45% omega-3s by weight, with a DHA:EPA ratio of about 3:1 (174). In this non-blinded trial, male and female subjects (total 21) received 1,738 mg DHA and 616 mg EPA per day, for 2 weeks. Over this relatively short period their average omega-3 index increased from 3.5 to 5.8%. Their blood triglycerides fell markedly and HDL cholesterol increased by 5.5 per cent.

As discussed in Section 6.6.5.8, phosphatidylcholine (PC) is a major dietary source of choline, a nutrient essential for human health (144,164). In this trial MOPL30 significantly increased the blood levels of both choline and its functionally important metabolite betaine (174). There was also a small but significant improvement in the subjects' blood sugar response to a classic glucose tolerance test. Their plasma acylcarnitine levels fell, suggesting MOPL30 may have improved their energy generation efficiency. Overall, this trial suggested the MOPL30 herring roe FL had substantial potential for human benefit. Randomized controlled trials could extend these promising findings.

The second herring roe FL trial was with Romega 30 from Arctic Nutrition, which contained 305 PL (175). In a randomized, single-blind, crossover trial, Romega was compared against a high-triglyceride fish oil for delivering DHA and EPA into PC in the blood plasma. The uptake periods studied were both "acute" (the first 12 hours) and "chronic" (2 weeks). The daily Romega dosage provided 1,810 mg of DHA, 628 mg of EPA and 137 mg of DPA (docosapentaenoic acid, omega-3); the fish oil provided 1,843 mg DHA, 575 mg EPA and 259 mg DPA. After adjusting the data to compensate for the differing doses from the two preparations, the researchers concluded Romega 30 delivered significantly more DHA and EPA into the subjects' plasma PC within the first 12 hours, and that for the 2-week period the preparations seemed evenly matched. This trial deserves replication over a longer period and without the crossover feature, for the same reasons discussed above, in Section 6.6.5.9. on krill lecithin bioavailability.

These herring roe FL have other potential applications, such as for psoriasis as indicated by positive anecdotal results (176). Meanwhile, the Norwegian Atlantic herring *(Clupea harengus)* fishery continues to be sustainable, certified by the influential Marine Stewardship Council, and classified as a "species of least concern" by the International Union for Conservation of Nature (IUCN) (177).

6.6.6.2 Salmon roe fish lecithins against cancer

A fish lecithin called MPL (for 'marine phospholipids') from Bio-Sea was stated to come from salmon roe and to contain 30% PL by weight, with more than half of its PL fatty acids being omega-3 – see Taylor and colleagues for its detailed composition (178). This group did a small non-blinded trial on 17 cancer patients with various types of tumours and suffering weight loss ('cancer cachexia'). The patients received 1.5 g/day of MPL for 6 weeks, providing 300 mg/day of DHA+EPA. On average their weight stabilized, with ten of the 17 showing weight gain, and their appetite and quality of life improved. Their HDL increased. The omega-3 index did not appear to increase, though the red cell arachidonic acid (omega-6) did significantly decrease, suggesting a possible anti-inflammatory effect of MPL.

In a subsequent trial some of the same researchers tested MPL, branded Vitalipin from Membramed Health Food GmbH, against prostate cancer (179). A total of 93 prostate cancer patients with tumours in various stages of development, some with metastases, received 1.5 g per day of MPL, providing 256 mg/day DHA and 223 mg/day EPA, for 3 months. Another 31 men aged over 70, some with enlarged prostates but none with cancer, served as controls and received the same daily MPL dosage. A number of changes occurred in both groups, but their clinical significance is hard to assess because most of the men had surgery shortly before the trial. The omega-3 index was not measured, but in both groups the plasma EPA and DHA increased and AA (arachidonic acid) decreased, also CRP (C-Reactive Protein) fell, all consistent with progression to a less inflammatory state. The researchers speculated these and other changes they observed could retard the progression of tumour metastases.

A subsequent DB RCT compared MPL to a fish oil against cachectic weight loss in patients with pancreatic cancer (180). A total of 33 patients received either 1.5 g/day of MPL providing 525 mg/day PL and 300 mg/day DHA+EPA or a fish oil that provided 300 mg/day DHA+EPA, for 6 weeks. About half the patients in both groups gained weight during the trial, without increase in appetite, which the researchers interpreted was related to lowering of inflammation by both treatments. The MPL was better tolerated than the fish oil, however, and the MPL patients showed better dosing compliance, that is, better acceptance of this supplement compared to the fish oil.

Clearly the fish lecithins prepared from sustainably harvested herring roe and from salmon roe have great promise for future application as omega-3 lecithins. It could be rewarding to further assess them as dietary supplements against memory decline or the numerous other brain problems that involve chronic inflammation, as well as for inflammation-related problems of other organs, including the liver, for which non-omega-3 lecithins have shown promise but not breakthroughs (see Section 6.6.3.3 above). A 1999 trial with a salmon roe lecithin on six patients with serious liver disease and diabetes suggested liver benefits, such as elevation of HDL and two apolipoproteins known to be produced by the liver (181). Patients received 1,600 mg/day of PL, providing about 502 mg/day DHA and 181 mg/day EPA, for 6 months.

6.6.7 Algal lecithins

Algae are the planet's primary producers of long-chain omega-3s such as DHA and EPA. The DHA and EPA in fish and krill come from algae they consume, and since fish and krill stocks are declining worldwide it makes sense to pursue algae as omega-3 sources. Technologies have been developed for large-scale culturing of algae and extraction of their omega-3s, some of which are attached to phospholipids. One commercial algal lecithin source is *Nannochloropsis oculata,* cultured in freshwater tanks and yielding an oil with EPA (about 25% by weight) attached to phospholipids (6%) and glycolipids (9%). This algal omega-3 lecithin was compared against krill oil for bioavailability, in both humans and rats.

In an acute human bioavailability study (182), either this algal lecithin (AL) or NKO krill lecithin (KL) were fed to ten healthy men right after a high-fat breakfast, then blood samples were taken at regular intervals from one or two hours to ten hours after the meal. The design was a crossover one, with each subject receiving both lecithins one week apart. Both preparations supplied about 1.5 grams of total DHA+EPA: the AL supplied only EPA and no DHA, the KL supplied 1.02 g EPA and 0.54 g DHA.

By the five-hour mark of this trial the AL had raised plasma EPA significantly higher than the KL (182). The increase was proportionally greater than their relative concentrations: though the AL provided about 50% more EPA, it raised

total PL was PS. All the PL classes contained DHA and/or EPA. Certain clam fisheries are already being sustainably harvested, and on this basis clam lecithins could become worthwhile for further research into their health benefits.

6.6.9 Future directions

Lecithin dietary supplements sustain every one of our more than 30 trillion cells, by providing a variety of different phospholipids that build cell membranes (196). These phospholipid molecules also contribute those fatty acids built into their molecular structure (2). Certain of these fatty acids, most notably omega-3 DHA and EPA, are essential for good health via their multiple actions that include optimizing cell membrane structure and function. Now we can measure our DHA and EPA nutritional status, and improve it as needed simply by taking omega-3 lecithin supplements.

The advent of the omega-3 index ushered in an exciting new phase for disease risk management. This index represents the omega-3 status of practically all of our cells AND predicts our risk for cardiovascular, brain, joint and other diseases and disorders (135,136,173,197,198). And anyone can learn their nutritional omega-3 status by having their index measured (199). It is an easily accessible, inexpensive and convenient test that does not require a doctor's prescription, yet it is a better predictor of cardiovascular health than the typical measures of cholesterol, triglycerides, homocysteine or CRP (C-reactive protein) (136). Omega-3 lecithins with substantial content of DHA and EPA have tremendous potential for use as dietary supplements to improve this index.

Bioavailability enhancement continues to be an exciting application for lecithins. As introduced in section 6.6.3.6, due to their very high absorption rate (38,167), emulsifying properties and safety for long-term use, lecithins are excellent for combining with nutrients and herbals to enhance their absorption. Following the milk thistle phytosome breakthrough (69), considerable success has been achieved by combining lecithins with curcuminoids from turmeric root (72,200,201).

Curcuminoids are known to be extremely beneficial when directly supplied to cells, but when taken by mouth are very poorly absorbed so that even multi-gram doses may show no benefit (201,202). However, when combined with soy (or recently sunflower) lecithins, curcuminoids' absorption can be increased by at least 30 times (201,203). After numerous clinical trials certain of these combinations are proven to benefit the circulation (especially capillary systems), brain, retina (actually improving vision), nervous system, joints, prostate, muscles and skin, lessen damage from cancer chemotherapy and radiation therapy, and reduce pain (72,201,202).

The lecithins have a long and storied history as dietary supplements. With the continuing diversification of their sources and biochemical versatility, their future applications promise to be ever more beneficial for human health and welfare. Humanity owes a great debt to lecithin's pioneers.

Acknowledgements

Lecithins have been a major focus for this author since 1987. The considerable intellectual stimulation, encouragement and funding opportunities received from Dr Michael Schneider and Mr Peter Rohde are gratefully acknowledged.

References

1. Thudicum, J.L.W., 1884. A Treatise on the Chemical Constitution of the Brain. Hamdon, CT: Archon Books, 1962 (with a new historical introduction by Drabkin).
2. Alberts, B., Johnson, A., Lewis, J., et al., 2015. Molecular Biology of the Cell (6th Edition). New York, NY: Garland Science.
3. Shurtleff, W., Aoyagi, A., 2016. History of Lecithin and Phospholipids (1850-2016): Extensively Annotated Bibliography and Sources. Lafayette, CA: Soyinfo Center. Retrieved from www.soyinfocenter.com.
4. Shurtleff, W., Aoyagi, A., 2004. History of Soy Lecithin - Page 1. Lafayette, CA: Soyinfo Center. Retrieved from www.soyinfocenter.com
5. Chopra, R.N., Chopra, G.S., 1937. The treatment of opium habit with lecithin and glucose. The Indian Medical Gazette May 1937, 265-269.
6. Chopra, R.N., Chopra, I.C., 1957. Treatment of drug addiction. Experience in India. United Nations Office on Drugs and Crime. Retrieved from www.unodc.org
7. Kuellenberg, D., Taylor, L.A., Schneider, M., Massing, U., 2012. Health effects of dietary phospholipids. Lipids Health Dis 11:3.
8. Gundermann, K.-J., Kuenker, A., Kuntz, E., Drozdzik, M., 2011. Activity of essential phospholipids (EPL) from soybean in liver diseases. Pharmacol Rep. 63, 643-659.
9. Ma, X.-N., Chen, T.-P., Yang, B., et al., 2016. Lipid production from Nannochloropsis. Mar. Drugs 14, 61.
10. Carlson, S.E., Ford, A.J., Werkman, S.H., et al., 1996. Visual acuity and fatty acid status of term infants fed human milk and formulas with and without docosahexaenoate and arachidonate from egg yolk lecithin. Pediatr. Res. 39(5), 882-888.
11. U.S. Food and Drug Administration, 2006. GRAS Notices: GRN No. 186. Soy lecithin enzymatically modified to have increased phosphatidylserine. Retrieved from www.accessdata.fda.gov
12. Araujo, P., Zhu, H., Breivik, J.F., et al., 2014. Determinations and structural elucidation of triacylglycerols in krill oil by chromatographic techniques. Lipids 49, 163-172.
13. Dupont, P., 2010. Use of lecithin as a medication for the treatment of psoriasis. United States Patent 7,759,325.
14. Carlson, S.E., Montalto, M.B., Ponder, D.L., et al., 1998. Lower incidence of necrotizing enterocolitis in infants fed a preterm formula with egg phospholipids. Pediatr. Res. 44(4), 491-498.

15. Knuiman, J.T., Beynen, A.C., Katan, M.B., 1989. Lecithin intake and serum cholesterol. Am. J. Clin. Nutr. 49, 266-268.
16. Kirsten, R., Heintz, B., Nelson, K., et al., 1994. Polyenephosphatidylcholine improves the lipoprotein profile in diabetic patients. Int. J. Pharmacol. Ther. 32(2), 53-56 (abstract only).
17. Oosthuizen, W., Vorster, H.H., Vermaak, W.J.H., et al., 1998. Lecithin has no effect on serum lipoprotein, plasma fibrinogen and macro molecular protein complex levels in hyperlipidemic men in a double-blind controlled study. Eur. J. Clin. Nutr. 52, 419-424.
18. Weiland, A., Bub, A., Barth, S.W., et al., 2016. Effects of dietary milk- and soya-phospholipids on lipid-parameters and other risk indicators for cardiovascular diseases in overweight or obese men - two double-blind, randomised, controlled, clinical trials. J. Nutr. Sci. 5, e21.
19. Cohn, J.S., Kamili, A., Wat, E., et al., 2010. Dietary phospholipids and intestinal cholesterol absorption. Nutrients 2, 116-127.
20. Kesaniemi, Y.A., Grundy, S.M., 1986. Effects of dietary polyenephosphatidylcholine on metabolism of cholesterol and triglycerides in hypertriglyceridemic patients. Am. J. Clin. Nutr. 43(1), 98-107 (abstract only).
21. Hall, J.E., 2016. Guyton and Hall Textbook of Medical Physiology (Thirteenth Edition, International). Philadelphia, PA: Elsevier.
22. Schaefer, E.J., Anthanont, P., Asztalos, B.F., 2014. HDL metabolism, composition, function, and deficiency. Curr Opin Lipidol. 25(3), 194-199.
23. Polichetti, E., Janisson, A., Iovanna, C., et al., 1998. Stimulation of the apo A1-high density lipoprotein system by dietary soyabean lecithin in humans. J. Nutr. Biochem. 9(11), 659-664 (abstract only).
24. O'Mahoney, L.L., Matu, J., Price, O.J., et al., 2018. Omega-3 polyunsaturated fatty acids favourably modulate cardiometabolic markers in type 2 diabetes: a meta-analysis and meta-regression of randomized controlled trials. Cardiovasc. Diabetol. 17:98.
25. Peters, B.H., Levin, H.S., 1979. Effects of physostigmine and lecithin on memory in Alzheimer disease. Ann. Neurol. 6(3), 219-221 (abstract only).
26. Little, A., Levy, R., Chuaqui-Kidd, P., Hand, D., 1985. A double-blind, placebo-controlled trial of high-dose lecithin in Alzheimer's disease. J Neurol. Neurosurg. Psychiatr. 48, 736-742.
27. Bartus, R.T., Dean, R.L., Pontecorvo, M.J., Flicker, C., 1985. The cholinergic hypothesis: a historical overview, current perspective, and future directions. Ann. NY Acad. Sci. 444, 332-358.
28. Zeisel, S.H., 2006. Choline: critical role during fetal development and dietary requirements in adults. Annu. Rev. Nutr. 26, 229-250.
29. Higgins, J.P., Flicker, L., 2000. Lecithin for dementia and cognitive impairment. Cochrane Database Syst. Rev. 2000 Issue 4. Art. No. CD001015.
30. Ladd, S.L., Sommer, S.A., LaBerge, S., Toscano, M., 1993. Effect of phosphatidylcholine on explicit memory. Clin Neuropharmacol. 16(6), 540-549.
31. Shi, F., Zhou, J., Meng, D., 2001. Curative effect of soybean lecithin on cerebral infarction (Article in Chinese). Zhonghua Yi Xue Za Zhi 81(21), 1301-1303.
32. Secades, J.J., Alvarez-Sabin, J., Castillo, J., et al., 2016. Citicoline for acute ischemic stroke: a systematic review and formal meta-analysis of randomized, double-blind, and placebo-controlled trials. J. Stroke Cerebrovasc. Dis. 25(8): 1984-1996.
33. Tweedy, R.A., Garcia, C.A., 1982. Lecithin treatment of cognitively impaired Parkinson's patients. Eur. J. Clin. Invest. 12(1), 87-90 (abstract only).
34. Zhou, M., Xue, Y., Sun, H., et al., 2016. Effects of different fatty acids composition of phosphatidylcholine on brain function of dementia mice induced by scopolamine. Lipids Health Dis. 15:135.
35. Kidd, P.M., 1996. Phosphatidylcholine: a superior protectant against liver damage. Altern Med Rev. 1(4), 258-274.
36. Gundermann, K.-J., Gundermann, S., Drozdzik, M., Prasad, V.G.M., 2016. Essential phospholipids in fatty liver: a scientific update. Clin. Exp. Gastroenterol. 9, 105-117.
37. Poole, L.G., Dolin, C.E., Arteel, G.E., 2017. Organ-organ crosstalk and alcoholic liver disease. Biomolecules 7, 62 (16 pages).
38. Zierenberg, O., Grundy, S.M., 1982. Intestinal absorption of polyenephosphatidylcholine in man. J. Lipid Res. 23, 1136-1142.
39. Knuechel, F., 1979. Double blind study in patients with alcohol-toxic fatty liver. Med. Welt 30, 411-416.
40. Schuller-Perez, A., San Martin, F.G., 1985. Controlled study using multiply-unsaturated phosphatidylcholine in comparison with placebo in the case of alcoholic liver steatosis. Med. Welt 72, 517-521.

41. Panos, M.Z., Polson, R., Johnson, R., et al., 1990. Polyunsaturated phosphatidyl choline for acute alcoholic hepatitis: a double-blind, randomized, placebo-controlled trial. Eur. J. Gastroenterol. Hepatol. 2(5), 351-355.
42. Lieber, C.S., Weiss, D.G., Grozmann, R., et al., 2003. Veterans Affairs cooperative study of polyenylphosphatidylcholine in alcoholic liver disease. Alcohol. Clin. Exp. Res. 27(11), 1765-1772.
43. Lieber, C.S., Robins, S.J., Li, J., et al., 1994. Phosphatidylcholine protects against fibrosis and cirrhosis in the baboon. Gastroenterol. 106(1), 152-9. See also the editorial: Mezey, E., 1994. Prevention of alcohol-induced hepatic fibrosis by phosphatidylcholine. Gastroenterol. 106(1), 257-259.
44. Lieber, C.S., Leo, M.A., Cao, Q., et al., 2007. The combination of S-adenosylmethionine and dilinoleoylphosphatidylcholine attenuates non-alcoholic steatohepatitis produced in rats by a high-fat diet. Nutr. Res. 27(9), 565-573.
45. Marpaung, B., Tarigan, P., Zein, L.H., et al., 1988. Tuberkulostatische Kombinations - therapie aus INH, RMP und EMB (Article in German). Therapiewoche 38, 734-740.
46. Jenkins, P.J., Portmann, B.P., Eddleston, A.L., Williams, F., et al., 1982. Use of polyunsaturated phosphatidylcholine in HBsAg negative chronic active hepatitis, results of prospective double-blind controlled trial. Liver 2, 77-81.
47. Visco, G., 1985. Polyunsaturated phosphatidylcholine associated with vitamin B complex in the treatment of acute viral hepatitis B. Clin. Ther. 114, 183-188.
48. Ilic, V., Begic-Janev, A., 1991. Therapy for HBsAg-positive active hepatitis. Effect of "essential" phospholipids. Med. Welt 85, 523-525.
49. Niederau, C., Strohmeyer, G., Heingtes, T., et al., 1998. Polyunsaturated phosphatidyl-choline and interferon alpha for treatment of chronic hepatitis B and C. A multicenter, randomized, double-blind, placebo-controlled trial. Hepato-Gastroenterology 45, 797-804.
50. Dajani, A., Abu Hammour, A., 2016. Treatment of nonalcoholic fatty liver disease: where do we stand? An overview. Saudi. J. Gastroenterol. 22(2), 91-105.
51. Noureddin, M., Mato, J.M., Lu, S.C., 2015. Nonalcoholic fatty liver disease: Update on pathogenesis, diagnosis, treatment, and the role of S-adenosylmethionine. Exp. Biol. Med. (Maywood) 240(6), 809-820.
52. Atkuri, K.R., Mantovani, J.J., Herzenberg, L.A., Herzenberg, L.A., 2007. N-acetylcysteine – a safe antidote for cysteine/glutathione deficiency. Curr Opin Pharmacol. 7, 355-359.
53. Federico, A., Dallio, M., Loguercio, C., 2017. Silymarin/silybin and chronic liver disease: a marriage of many years. Molecules 22, 191.
54. Hayashi, H., Tanaka, Y., Hibino, H., et al., 1999. Beneficial effect of salmon roe phosphatidylcholine in chronic liver disease. Curr Med. Res. Opin. 15(3), 177-184.
55. Stremmel, W., Merle, U., Zahn, A., et al., 2005. Retarded release phosphatidylcholine benefits patients with chronic active ulcerative colitis. Gut 54, 966-971. See also accompanying editorial: Gibson, P.R., Muir, J.G., 2005. Reinforcing the mucus: a new therapeutic approach for ulcerative colitis? Gut 54, 900-903.
56. Stremmel, W., Gauss, A., 2013. Lecithin as a therapeutic agent in ulcerative colitis. J. Digest. Dis. 31(3-4), 388-390.
57. Stremmel, W., Hanemann, A., Ehehalt, R., et al., 2010. Phosphatidylcholine (lecithin) and the mucus layer: evidence of therapeutic efficacy in ulcerative colitis? J. Digest. Dis. 28, 490-496.
58. Braun, A., Treede, I., Gotthardt, D., et al., 2009. Alterations of phospholipid concentration and species composition of the intestinal mucus barrier in ulcerative colitis: a clue to pathogenesis. Inflamm Bowel Dis. 15(11), 1705-1720.
59. Stremmel, W., Ehehalt, R., Autschbach, F., Karner, M., 2007. Phosphatidylcholine for steroid-refractory chronic ulcerative colitis. A randomized trial. Ann. Intern. Med. 147, 603-610.
60. Karner, M., Kocjan, A., Stein, J., et al., 2014. First multicenter study of modified release phosphatidylcholine "LT-02" in ulcerative colitis: a randomized, placebo-controlled trial in mesalazine-refractory courses. Am. J. Gastroenterol. 109, 1041-1051.
61. Lanza, F.L., Marathi, U.K., Anand, B.S., Lichtenberger, L.M., 2008. Clinical trial: comparison of ibuprofen-phosphatidylcholine and ibuprofen on the gastrointestinal safety and analgesic efficacy in osteoarthritic patients. Aliment. Pharmacol. Ther. 28, 431-442.
62. Cryer, B., Bhatt, D.L., Lanza, F.L., et al., 2011. Low-dose aspirin-induced ulceration is attenuated by aspirin-phosphatidylcholine: a randomized clinical trial. Am. J. Gastroenterol. 106(2), 272-277.
63. Schneider, H., Braun, A., Fuellekrug, J., et al., 2010. Lipid based therapy for ulcerative colitis-modulation of intestinal mucus membrane phospholipids as a tool to influence inflammation. Int. J. Mol. Sci. 11(10), 4149-4164.

64. Nicolson, G.L., 2003. Lipid replacement as an adjunct to therapy for chronic fatigue, anti-aging and restoration of mitochondrial function. J. Am. Nutraceutical. Assoc. 6(3), 22-28.
65. Escriba, P.V., Gonzalez-Ros, J.M., Goni, F.M., et al., 2008. Membranes: a meeting point for lipids, proteins and other therapies. J. Cell Mol. Med. 12(3), 829-875.
66. Escriba, P.V., 2017. Membrane-lipid therapy: A historical perspective of membrane-targeted therapies - from lipid bilayer structure to the pathophysiological regulation of cells. Biochim. Biophys. Acta. 1859, 1493-1506.
67. Nicolson, G.L., Ferreira, G., Settineri, R., et al., 2018. Mitochondrial dysfunction and chronic disease: treatment with membrane lipid replacement and other natural supplements. In Oliveira, P.J. (Ed.), Mitochondrial Biology and Experimental Therapeutics. Cham: Springer International Publishing AG, part of Springer Nature.
68. Hirose, A., Terauchi, M., Osaka, Y., et al., 2018. Effect of soy lecithin on fatigue and menopausal symptoms in middle-aged women: a randomized, double-blind, placebo-controlled study. J. Nutr. 17:4.
69. Kidd, P., Head, K., 2005. A review of the bioavailability and clinical efficacy of milk thistle phytosome: a silybin-phosphatidylcholine complex (Siliphos®). Altern. Med. Rev. 10(3), 193-203.
70. Ostlund, R.E., Spilburg, C.A., Stenson, W.F., 1999. Sitostanol administered in lecithin micelles potently reduces cholesterol absorption in humans. Am. J. Clin. Nutr. 70, 826-831.
71. Gananath, K., Sri Nataraj, K., Ganga Rao, B., 2017. Phospholipid complex technique for superior bioavailability of phytoconstituents. Adv. Pharm. Bull. 7(1), 35-42.
72. Cox, K.H.M., Pipingas, A., Scholey, A.B., 2015. Investigation of the effects of solid lipid curcumin on cognition and mood in a healthy older population. J. Psychopharmacol. 29, 642-651.
73. Delwaide, P.J., Gyselynck-Mambourg, A.M., Hurlet, A., Ylieff, M., 1986. Double-blind randomized controlled study of phosphatidylserine in demented patients. Acta. Neurol. Scand. 73, 136-140.
74. Palmieri, G., Palmieri, R., Inzoli, M.R., et al., 1987. Double-blind controlled trial of phosphatidylserine in patients with senile mental deterioration. J. Clin. Trials. 24, 73-83.
75. Ransmayr, G., Ploerer, S., Gerstenbrand, F., Bauer, G., 1987. Double-blind placebo-controlled trial of phosphatidylserine in elderly patients with arteriosclerotic encephalopathy. J. Clin. Trials. 24, 62-72.
76. Villardita, C., Grioli, S., Salmeri, G., et al., 1987. Multicentre clinical trial of brain phosphatidylserine in elderly patients with intellectual deterioration. J. Clin. Trials. 24, 84-93.
77. Amaducci, L., SMID Group, 1988. Phosphatidylserine in the treatment of Alzheimer's disease: results of a multicenter study. Psychopharmacol. Bull. 24, 130-134.
78. Feunfgeld, E.W., 1989. Double-blind study with phosphatidylserine (PS) in Parkinsonian patients with senile dementia of Alzheimer's type (SDAT). Prog. Clin. Biol. Res. 317, 1235-1246.
79. Hershkowitz, M., Fisher, M., Bobrov, D., Rabinowitz, M., 1989. Long-term treatment of dementia Alzheimer type with phosphatidylserine: effect on cognitive functioning and performance in daily life. In Bazan, N.G., Horrocks, L.A., Toffano, G. (Eds), Phospholipids in the Nervous System: Biochemical and Molecular Pathology. Padova: Liviana Press, 279-288.
80. Crook, T.H., Tinklenberg, J., Yesavage, J., et al., 1991. Effects of phosphatidylserine in age-associated memory impairment. J. Neurol. 41, 644-649.
81. Engel, R.R., Satzger, W., Gunther, W., et al., 1992. Double-blind cross-over study of phosphatidylserine vs. placebo in patients with early dementia of the Alzheimer type. Eur. Neuropsychopharmacol. 2, 149-155.
82. Cenacchi, T., Bertoldin, T., Farina, C., et al., 1993. Cognitive decline in the elderly: a double-blind, placebo-controlled multicenter study on efficacy of phosphatidylserine administration. Aging (Milano) 5, 123-133.
83. Crook, T., Petrie, W., Wells, C., Massari, D.C., 1992. Effects of phosphatidylserine in Alzheimer's disease. Psychopharmacol. Bull. 28, 61-66.
84. Kidd, P.M., 1996. Phosphatidylserine, membrane nutrient for memory. A clinical and mechanistic assessment. Altern. Med. Rev. 1, 70-84.
85. Gindin, J., Noviko, M., Kedar, D., et al., 1995. The effect of plant phosphatidylserine on age-associated memory and mood in the functioning elderly. Rehovot: Geriatric Institute for Education and Research and Department of Geriatrics, Kaplan Hospital. Internal Report of Lipogen Ltd.
86. Jaeger, R., Purpura, M., Geiss, K.-R., et al., 2007. The effect of phosphatidylserine on golf performance. J. Int. Soc. Sports Nutr. 4, 23-27.
87. Vaisman, N., Kaysar, N., Zaruk-Adasha, Y., et al., 2008. Correlation between changes in blood fatty acid composition and visual sustained attention performance in children with inattention: effect of dietary n-3 fatty acids containing phospholipids. Am. J. Clin. Nutr. 87, 1170-1180.

88. Kato-Kataoka, A., Sakai, M., Ebina, R., et al., 2010. Soybean-derived phosphatidylserine improves memory function of the elderly Japanese subjects with memory complaints. J. Clin. Biochem. Nutr. 47, 246-255.
89. Vakhapova, V., Cohen, T., Richter, Y., et al., 2010. Phosphatidylserine containing omega-3 fatty acids may improve memory abilities in non-demented elderly with memory complaints: a double-blind placebo-controlled trial. Dement. Geriatr. Cogn. Disord. 29, 467-474.
90. Vakhapova, V., Richter, Y., Cohen, T., et al., 2011. Safety of phosphatidylserine containing omega-3 fatty acids in non-demented elderly: a double-blind placebo-controlled trial followed by an open-label extension. BMC Neurol. 11, 79.
91. Vakhapova, V., Richter, Y., Cohen, T., et al., 2014. Phosphatidylserine containing omega-3 fatty acids may improve memory abilities in non-demented elderly individuals with memory complaints: results from an open-label extension study. Dement. Geriatr. Cogn. Disord. 38, 39-45.
92. Hellhammer, J., Hero, T., Franz, N., et al., 2012. Omega-3 fatty acids administered in phosphatidylserine improved certain aspects of high chronic stress in men. Nutr. Res. 32, 241-250.
93. Manor, I., Magen, A., Keidar, D., et al., 2012. The effect of phosphatidylserine containing omega-3 fatty acids on attention deficit-hyperactivity disorder symptoms in children: a double-blind placebo-controlled trial, followed by an open-label extension. Eur. Psychiatry 27, 335-342.
94. Manor, I., Magen, A., Keidar, D., et al., 2013. Safety of phosphatidylserine containing omega-3 fatty acids in ADHD children. Eur. Psychiatry 28, 386-391.
95. Hirayama, S., Terasawa, K., Rabeler, R., et al., 2014. The effect of phosphatidylserine administration on memory and symptoms of attention-deficit hyperactivity disorder: a randomised, double-blind, placebo-controlled clinical trial. J. Hum. Nutr. Diet. 27 (Suppl 2), 284-291.
96. Taylor, C.L., 2003. Phosphatidylserine and cognitive dysfunction and dementia (qualified health claim: final decision letter). FDA (U.S. Food and Drug Administration) Center for Food Safety and Applied Nutrition, Office of Nutritional Products, Labeling and Dietary Supplements, May 13.
97. U.S. Food and Drug Administration, 2007. GRAS Notices: GRN No. 223. Phosphatidylserine. Retrieved from accessdata.fda.gov
98. Nunzi, M.G., Guidolin, D., Petrelli, L., et al., 1992. Behavioral and morpho-functional correlates of brain aging: a preclinical study with phosphatidylserine. In Bazan, N.G., Murphy, M.G., Toffano, G. (Eds), Neurobiology of Essential Fatty Acids. Adv. Exp. Med. Biol. 318, 393-398. Boston, MA: Springer.
99. Hall, J.M., Gomez-Pinilla, F., Savage, L.M., 2018. Nerve growth factor is responsible for exercise-induced recovery of septohippocampal cholinergic structure and function. Frontiers Neurosci. 12, Article 773.
100. Monti, J.M., Baym, C.L., Cohen, N.J., 2014. Identifying and characterizing the effects of nutrition on hippocampal memory. Adv. Nutr. 5, 337S-343S.
101. Kim, H.-Y., Huang, B.X., Spector, A.A., 2014. Phosphatidylserine in the brain: metabolism and function. Prog. Lipid Res. October, 1-18.
102. Klinkhammer, P., Szelies, B., Heiss, W.-D., 1990. Effect of phosphatidylserine on cerebral glucose metabolism in Alzheimer's disease. Dement. Geriatr. Cogn. Dis. 1, 197-201.
103. Kidd, P.M., 2000. Attention deficit/hyperactivity disorder (ADHD) in children: rationale for its integrative management. Altern. Med. Rev. 5(5), 402-428.
104. Nerozzi, D., Aceti, F., Mella, E., et al., 1987. Phosphatidylserine and impaired memory in the elderly. Clin. Ther. 120, 399-404.
105. Kingsley, M.I., Wadsworth, D., Kilduff, L.P., et al., 2006. Effects of phosphatidylserine on exercise capacity during cycling in active males. Med. Sci. Sports Exerc. 38, 64-71.
106. Parker, A.G., Gordon, J., Thornton, A., et al., 2011. The effects of IQPLUS Focus on cognitive function, mood, and endocrine response before and following acute exercise. J. Int. Soc. Sports Nutr. 8:16.
107. Monteleone, P., Maj, M., Beinat, L., et al., 1992. Blunting by chronic phosphatidylserine administration of the stress-induced activation of the hypothalamo-pituitary-adrenal axis in healthy men. Eur. J. Clin. Pharmacol. 41, 385-388.
108. Starks, M.A., Starks, S.L., Kingsley, M., et al., 2008. The effects of phosphatidylserine on endocrine response to moderate intensity exercise. J. Int. Soc, Sports Nutr. 5, 11-16.
109. Fahey, T.D., Pearl, M., 1998. The hormonal and perceptive effects of phosphatidylserine administration during two weeks of resistive exercise-induced overtraining. Biol. Sport 15, 135-144.
110. Kingsley, M.I., Kilduff, L.P., Mceneny, J., et al., 2006. Phosphatidylserine supplementation and recovery following downhill running. Med. Sci. Sports Exerc. 38(9), 1617-1625.
111. Benton, D., Donohoe, R.T., Sillance, B., et al., 2001. The influence of phosphatidylserine on mood and heart rate when faced with an acute stressor. Nutr. Neurosci. 4(3), 169-178.

112. Maggioni, M., Picotti, G.B., Bondiolotti, G.P., et al., 1990. Effects of phosphatidylserine therapy in geriatric patients with depressive disorders. Acta. Psychiatr. Scand. 81, 265-270.
113. Jaeger, R., Purpura, M., Geiss, K.-R., et al., 2007. The effect of phosphatidylserine on golf performance. J. Int. Soc. Sports Nutr. 4, 23-27.
114. Baumeister, J., Barthel, T., Geiss, K.R., 2008. Influence of phosphatidylserine on cognitive performance and cortical activity after induced stress. Nutr. Neurosci. 11, 103-110.
115. Hellhammer, J., Fries, E., Buss, C., et al., 2004. Effects of soy lecithin phosphatidic acid and phosphatidylserine complex (PAS) on the endocrine and psychological responses to mental stress. Stress 7, 119-126.
116. Hellhammer, J., Vogt, D., Franz, N., et al., 2014. A soy-based phosphatidylserine/phosphatidic acid complex (PAS) normalizes the stress reactivity of the hypothalamus-pituitary-adrenal-axis in chronically stressed male subjects: a randomized, placebo-controlled study. Lipids Health Dis. 13:121.
117. Schmidt, K., Weber, N., Steiner, M., et al., 2018. A lecithin phosphatidylserine and phosphatidic acid complex (PAS) reduces symptoms of the premenstrual syndrome (PMS): results of a randomized, placebo-controlled, double-blind clinical trial. Clin. Nutr. ESPEN 24, 22-30.
118. ChemSpider, 2018. Phosphatidic Acid. ID 4446637. Retrieved from www.chemspider.com
119. Raben, D.M., Barber, C.N., 2017. Phosphatidic acid and neurotransmission. Adv. Biol. Regul. 63, 15-21.
120. Sergeant, S., Waite, K.A., Heravit, J., McPhail, L.C., 2001. Phosphatidic acid regulates tyrosine phosphorylating activity in human neutrophils. J. Biol. Chem. 276(7), 4737-4746.
121. Garcia-Morales, V., Montero, F., Gonzalez-Forero, D., et al., 2015. Membrane-derived phospholipids control synaptic neurotransmission and plasticity. PLoS Biol. 13(5), e1002153.
122. Ramesh, S., Govindarajulu, M., Suppiramaniam, V., et al., 2018. Autotaxin-lysophosphatidic signaling in Alzheimer's disease. J. Int. Mol. Sci. 19(7), 1827.
123. Hofmeister, S., Bodden, S., 2016. Premenstrual syndrome and premenstrual dysphoric disorder. Am. Fam. Physician 94(3), 236-240.
124. More, M.I., Freitas, U., Rutenberg, D., 2014. Positive effects of soy-derived phosphatidylserine plus phosphatidic acid on memory, cognition, daily functioning, and mood in elderly patients with Alzheimer's disease and dementia. Adv. Ther. 31, 1247-1262.
125. Vaisman, N., Kaysar, N., Zaruk-Adasha, Y., et al., 2008. Correlation between changes in blood fatty acid composition and visual sustained attention performance in children with inattention: effect of dietary n-3 fatty acids containing phospholipids. Am. J. Clin. Nutr. 87, 1170-1180.
126. Parletta, N., Niyonsenga, T., Duff, J., 2016. Omega-3 and omega-6 polyunsaturated fatty acid levels and correlations with symptoms in children with attention deficit hyperactivity disorder, autistic spectrum disorder and typically developing controls. PLoS ONE 11(5), e0156432.
127. Manor, I., Magen, A., Keidar, D., et al., 2012. The effect of phosphatidylserine containing omega-3 fatty acids on attention deficit-hyperactivity disorder symptoms in children: a double-blind placebo-controlled trial, followed by an open-label extension. Eur. Psychiatry 27, 335-342.
128. Manor, I., Magen, A., Keidar, D., et al., 2013. Safety of phosphatidylserine containing omega-3 fatty acids in ADHD children. Eur. Psychiatry 28, 386-391.
129. Kidd, P.M., 2011. Astaxanthin, cell membrane nutrient with diverse clinical benefits and anti-aging potential. Altern. Med. Rev. 16(4), 355-364.
130. Ducasse, D., Jaussent, I., Olie, E., et al., 2016. Personality traits of suicidality are associated with premenstrual syndrome and premenstrual dysphoric disorder in a suicidal women sample. PLoS ONE 11(2), e0148653.
131. Sampalis, F., Bunea, R., Pelland, M.F., et al., 2003. Evaluation of the effects of Neptune Krill Oil on the management of premenstrual syndrome and dysmenorrhea. Altern. Med. Rev. 8(2),171-179.
132. Bunea, R., El Farrah, K., Deutsch, L., 2004. Evaluation of the effects of Neptune Krill Oil on the clinical course of hyperlipidemia. Altern. Med. Rev. 9(4), 420-428.
133. Ulven, S.M., Kirkhus, B., Lamglait, A., et al., 2011. Metabolic effects of krill oil are essentially similar to those of fish oil but at lower doses of EPA and DHA, in healthy volunteers. Lipids 46, 37-46.
134. Berge, K., Musa-Veloso, K., Harwood, M., et al., 2014. Krill oil supplementation lowers triglycerides without increasing low-density lipoprotein cholesterol in adults with borderline high or high triglyceride levels. Nutr. Res. 34, 126-133.
135. Harris, W.S., 2007. Omega-3 fatty acids and cardiovascular disease: a case for omega-3 index as a new risk factor. Pharmacol. Res. 55, 217-223.

136. Harris, W.S., Tintle, N.L., Etherton, M.R., Vasan, R.S., 2018. Erythrocyte long-chain omega-3 fatty acid levels are inversely associated with mortality and with incident cardiovascular disease: the Framingham Heart Study. J. Clin. Lipidol. 12, 718-727.
137. Lobraico, J.M., DiLello, L.C., Butler, A.D., et al., 2015. Effects of krill oil on endothelial functions and other cardiovascular risk factors in participants with type 2 diabetes, a randomized controlled trial. BMJ Open Diabetes Research & Care 3, e000107.
138. Ramprasath, V.R., Eyal, I., Zchut, S., Jones, P.J.H., 2013. Enhanced increase of omega-3 index in healthy individuals with response to 4-week n-3 fatty acid supplementation from krill oil versus fish oil. Lipids Health Dis. 12, 178.
139. Nichols, P.D., Kitessa, S.M., Abeywardena, M., 2014. Commentary on a trial comparing krill oil versus fish oil. Lipids Health Dis. 13, 2.
140. Ramprasath, V.R., Eyal, I., Zchut, S., Jones, P.J.H., 2014. Response to commentary on a trial comparing krill oil versus fish oil. Lipids Health Dis. 13, 17.
141. Flock, M.R., Skulas-Ray, A.C., Harris, W.S., et al., 2013. Determinants of erythrocyte omega-3 fatty acid content in response to fish oil supplementation: a dose-response randomized controlled trial. J. Am. Heart Assoc. 2(6), e000513.
142. Berge, R.K., Ramsvik, M.S., Bohov, P., et al., 2015. Krill oil reduces triacylglycerol level and improves related lipoprotein particle concentration, fatty acid composition and redox status in healthy young adults - a pilot study. Lipids Health Dis. 14, 163.
143. Zeisel, S.H., 2012. Dietary choline deficiency causes DNA strand breaks and alters epigenetic marks on DNA and histones. Mutat. Res. 733(1-2), 34-38.
144. Zeisel, S.H., 2017. Choline, other methyl-donors and epigenetics. Nutrients 9, 445.
145. Imbard, A., Benoist, J.-F., Esse, R., et al., 2015. High homocysteine induces betaine depletion. Biosci. Rep. 25, Art. e00222.
146. Zhao, G., He, F., Wu, C., et al., 2018. Betaine in inflammation: mechanistic aspects and applications. Frontiers Immunol. 9, Art. 1070.
147. Cicero, A.F.G., Rosticci, M., Morbini, M., et al., 2016. Lipid-lowering and anti-inflammatory effects of omega 3 ethyl esters and krill oil: a randomized, cross-over, clinical trial. Arch. Med. Sci. 12(3), 507-512.
148. Katan, M.B., Deslypere, J.P., van Birgelen, A.P.J.M, et al., 1997. Kinetics of the incorporation of dietary fatty acids into serum cholesteryl esters, erythrocyte membranes, and adipose tissue: an 18-month controlled study. J. Lipid Res. 38, 2012-2022.
149. Neptune Technologies & Bioresources, 2007. Neptune releases final ADHD study results showing significant beneficial effects of NKOTM and welcomes new cognitive expert to its team. Retrieved from www.nutraceuticalsnow.com/articles/2007/01/18
150. Konagai, C., Yanagimoto, K., Hayamizu, K., et al., 2013. Effects of krill oil containing n-3 polyunsaturated fatty acids in phospholipid form on human brain function: a randomized controlled trial in healthy elderly volunteers. Clin. Interv. Aging 8, 1247-1257.
151. Galleli, C.A., Calcagnini, S., Romano, A., et al., 2018. Modulation of the oxidative stress and lipid peroxidation by endocannabinoids and their lipid analogues. Antioxidants 7, Art. 93.
152. Banni, S., Carta, G., Murru, E., et al., 2011. Krill oil significantly decreases 2-arachidonoylglycerol plasma levels in obese subjects. Nutr. Metab. 8, Art. 7.
153. Deutsch, L., 2007. Evaluation of the effect of Neptune Krill Oil on chronic inflammation and arthritic symptoms. J. Am. Coll Nutr. 26(1), 39-48.
154. Ridker, P.M., 2016. A test in context. High-sensitivity C-reactive protein. J. Am. Coll Cardiol. 67(6), 712-723.
155. Suzuki, Y., Fukushima, M., Sakuraba, K., et al., 2016. Krill oil improves mild knee joint pain: a randomized control trial. PLoS ONE 11(10), Art. e0162769.
156. Da Boit, M., Mastalurova, I., Brazaite, G., et al., 2015. The effect of krill oil supplementation on exercise performance and markers of immune function. PLoS ONE 10(9), Art. e0139714.
157. Walsh, N.P., Gleeson, M., Shephard, R.J., et al., 2011. Position Statement. Part One: Immune function and exercise. Exerc. Immunol. Rev. 17, 6-63.
158. Skarpanska-Stejnborn, A., Pilaczynska-Szczesniak, L., Basta, P., et al., 2010. Effects of supplementation with Neptune Krill Oil (Euphausia superba) on selected redox parameters and pro-inflammatory markers in athletes during exhaustive exercise. J. Hum. Kinet. 25, 49-57.
159. Georges, J., Sharp, M.H., Lowery, R.P., et al., 2018. The effects of krill oil on mTOR signaling and resistance exercise: a pilot study. J. Nutr. Metab. 2018.

160. Messmer, E.M., 2015. The pathophysiology, diagnosis, and treatment of dry eye disease. Dtsch. Arztebl. Int. 112, 71-82.
161. Deinema, L.A., Vingrys, A.J., Wong, C.Y., et al., 2017. A randomized, double-masked, placebo-controlled clinical trial of two forms of omega-3 supplements for treating dry eye disease. Ophthalmology 124(1), 43-52.
162. Rundblad, A., Holven, K.B., Bruheim, I., et al., 2018. Effects of krill oil and lean and fatty fish on cardiovascular risk markers: a randomised controlled trial. J. Nutr. Sci. 7.
163. Rundblad, A., Holven, K.B., Bruheim, I., et al., 2018. Effects of fish and krill oil on gene expression in peripheral blood mononuclear cells and circulating markers of inflammation: a randomised controlled trial. J. Nutr. Sci. 7.
164. Institute of Medicine, National Academy of Sciences USA, 1998. Dietary reference intakes for folate, thiamin, riboflavin, niacin, vitamin B12, panthothenic acid, biotin, and choline. National Academy Press, Washington D.C., 390-422.
165. Bjorndal, B., Bruheim, I., Lysne, V., et al., 2018. Plasma choline, homocysteine and vitamin status in healthy adults supplemented with krill oil: a pilot study. Scand. J. Clin. Lab. Invest. Sep 27, 1-6.
166. Wijendran, V., Huang, M.-C., Diau, G.-Y., et al., 2002. Efficacy of dietary arachidonic acid provided as triglyceride or phospholipid as substrates for brain arachidonic acid accretion in baboon neonates. Pediatr. Res. 51, 265-272.
167. Destaillats, F., Oliveira, M., Schmid, V.B., et al., 2018. Comparison of the incorporation of DHA in circulatory and neural tissue when provided as triacylglycerol (TAG), monoacylglycerol (MAG), or phospholipid (PL) provides new insights into fatty acid bioavailability. Nutrients 10, Art. 620.
168. Schuchardt, J.P., Schneider, I., Meyer, H., et al., 2011. Incorporation of EPA and DHA into plasma phospholipids in response to different omega-3 fatty acid formulations - a comparative bioavailability study of fish oil vs. krill oil. Lipids Health Dis. 10, Art. 145.
169. Salem, N., Kuratko, C.N., 2014. A reexamination of krill oil bioavailability studies. Lipids Health Dis. 13, Art. 137.
170. Yurko-Mauro, K., Kralovec, J., Bailey-Hall, E., et al., 2015. Similar eicosapentaenoic and docosahexaenoic acid plasma levels achieved with fish oil or krill oil in a randomized double-blind four-week bioavailability study. Lipids Health Dis. 14:99.
171. Patrick, R.P., 2018. Role of phosphatidylcholine-DHA in preventing APOE4-associated Alzheimer's disease. J. Am. Soc. Exp. Biol. Oct 5:fj201801412R.
172. Schaefer, E.J., Bongard, V., Beiser, A.S., et al., 2006. Plasma phosphatidylcholine docosahexaenoic acid content and risk of dementia and Alzheimer disease. Arch. Neurol. 63, 1545-1550.
173. Amen, D.G., Harris, W.S., Kidd, P.M., et al., 2017. Quantitative erythrocyte omega-3 EPA plus DHA levels are related to higher regional cerebral blood flow on brain SPECT. J. Alzheimer's Dis. 58, 1189-1199.
174. Bjorndal, B., Strand, E., Gjerde, J., et al., 2014. Phospholipids from herring roe improve plasma lipids and glucose tolerance in young adults. Lipids Health Dis. 13:82.
175. Cook, C.M., Hallaraker, H., Saebo, P.C., et al., 2016. Bioavailability of long chain omega-3 polyunsaturated fatty acids from phospholipid-rich herring roe oil in men and women with mildly elevated triacylglycerols. Prostagland. Leukot. Essent. Fatty Acids. 111, 17-24.
176. Arctic Nutrition, 2018. Havard (72) got rid of his psoriasis with herring roe. Retrieved from www.arcticnutrition.no
177. International Union for Conservation of Nature (IUCN). Red List of Threatened Species. Retrieved from www.iucn.org
178. Taylor, L.A., Pletschen, L., Arends, J., et al., 2010. Marine phospholipids - a promising new dietary approach to tumor-associated weight loss. Support. Care Cancer 18(2), 159-170.
179. Kuellenberg de Gaudry, D., Taylor, L.A., Kluth, J., et al., 2014. Effects of marine phospholipids extract on the lipid levels of metastatic and nonmetastatic prostate cancer patients. Int. Sch. Res. Notices 2014, Article ID 249204.
180. Werner, K., Kuellenbergy de Gaudry, D., Taylor, L.A., et al., 2017. Dietary supplementation with n-3-fatty acids in patients with pancreatic cancer and cachexia: marine phospholipids versus fish oil - a randomized controlled double-blind trial. Lipids Health Dis. 16:104.
181. Hayashi, H., Tanaka, Y., Hibino, H., et al., 1999. Beneficial effect of salmon roe phosphatidylcholine in chronic liver disease. Curr Med. Res. Opin. 15(3), 177-184.

182. Kagan, M.L., West, A.L., Zante, C., Calder, P.C., 2013. Acute appearance of fatty acids in human plasma - a comparative study between polar-lipid rich oil from the microalgae Nannochloropsis oculata and krill oil in healthy young males. Lipid Health Dis. 12:102.
183. Kagan, M.L., Levy, A., Leikin-Frenkel, A., 2015. Comparative study of tissue deposition of omega-3 fatty acids from polar-lipid rich oil of the microalgae Nannochloropsis oculata with krill oil in rats. Food Funct. 6, 186-192.
184. Robertson, R.R., Guiheneuf, F., Bahar, B., et al., 2015. The anti-inflammatory effect of algae-derived lipid extracts on lipopolysaccharide (LPS)-stimulated human THP-1 macrophages. Mar. Drugs 13, 5402-5424.
185. Miller, M.R., Pearce, L., Bettjeman, B.I., 2014. Detailed distribution of lipids in Greenshell™ mussel (Perna canaliculus). Nutrients 6, 1454-1474.
186. Ulbricht, C., Chao, W., Costa, D., et al., 2009. An evidence-based systematic review of green-lipped mussel (Perna canaliculus) by the Natural Standard Research Collaboration. J. Diet. Suppl. 6(1), 54-90.
187. Mickleborough, T.D., Vaughn, C.L., Shei, R.-J., et al., 2013. Marine lipid fraction PCSO-524TM (lyprinol®/omega XL®) of the New Zealand green lipped mussel attenuates hyperpnea-induced bronchoconstriction in asthma. Resp. Med. 107, 1152-1163.
188. Coulson, S., Vecchio, P., Gramotnev, H., Vitetta, L., 2012. Green-lipped mussel (Perna canaliculus) extract efficacy in knee osteoarthritis and improvement in gastrointestinal dysfunction: a pilot study. Inflammopharmacol. 20(2), 71-76.
189. Stebbings, S., Gray, A., Schneiders, A.G., Sansom, A., 2017. A randomized double-blind placebo-controlled trial to investigate the effectiveness and safety of a novel green-lipped mussel extract - Biolex® - for managing pain in moderate to severe osteoarthritis of the knee. BMC Compl. Altern. Med. 17:416.
190. Kean, J.D., Sarris, J., Scholey, A., et al., 2017. Reduced inattention and hyperactivity and improved cognition after marine oil extract (PCSO-524) supplementation in children and adolescents with clinical and subclinical symptoms of attention-deficit hyperactivity disorder (ADHD): a randomised, double-blind, placebo-controlled trial. Psychopharmacol. 234, 403-420.
191. Sukumaran, S., Pittman, K.B., Patterson, W.K., et al., 2010. A phase 1 study to determine the safety, tolerability and maximum tolerated dose of green-lipped mussel (Perna canaliculus) lipid extract, in patients with prostate and breast cancer. Ann. Oncol. 21, 1089-1093.
192. Hosokawa, M., Shimatani, T., Kanada, T., et al., 2000. Conversion to docosahexaenoic acid-containing phosphatidylserine from squid skin lecithin by phospholipase D-mediated transphosphatidylation. J. Agric. Food Chem. 48(10), 4550-4554.
193. Lee, B., Sur, B.-J., Han, J.-J., et al., 2015. Oral administration of squid lecithin-transphosphatidylated phosphatidylserine improves memory impairment in aged rats. Prog. Neuro-Psychopharmacol. Biol. Psychiatry 56, 1-10.
194. Friend of the Sea, 2019. Squid & Cuttlefish. Retrieved from www.friendofthesea.org
195. Liu, Z.-Y., Zhou, D.-Y., Wu, Z.-X., et al., 2018. Extraction and detailed characterization of phospholipid-enriched oils from six species of edible clams. Food Chem. 239, 1175-1181.
196. Roy, A.L., Conroy, R.S., 2018. Toward mapping the human body at a cellular resolution. Mol. Biol. Cell 29, 1779-1785.
197. Messamore, E., McNamara, R.K., 2016. Detection and treatment of omega-3 fatty acid deficiency in psychiatric practice: rationale and implementation. Lipids Health Dis. 15:25.
198. Lupton, J.R., Blumberg, J.B., L'Abbe, M., et al., 2016. Nutrient reference value: non-communicable disease endpoints – a conference report. Eur. J. Nutr. 55(Suppl 1), 1-10.
199. www.omegaquant.com.
200. Kidd, P.M., 2009. Bioavailability and activity of phytosome complexes from botanical polyphenols: the silymarin, curcumin, green tea, and grape seed extracts. Altern. Med. Rev. 14(3), 226-246.
201. Kidd, P.M., 2015. Unlocking the Power of Curcumins with Meriva®. St. George, UT: TWIP, The Wellness Imperative People.
202. Gupta, S.C., Patchva, S., Aggarwal, B.B., 2013. Therapeutic roles of curcumin: lessons learned from clinical trials. The AAPS Journal 15, 105-218.
203. Gota, V.S., Maru, G.B., Soni, T.G., et al., 2010. Safety and pharmacokinetics of a solid lipid curcumin particle formulation in osteosarcoma patients and healthy volunteers. J. Agric. Food Chem. 58(4), 2095-2099.

7 Lecithin in the Feed Industry

7.1 Feed lecithins for ruminants, poultry and pig fattening

Roland Adelmann

In the food industry lecithin is mainly used as a technical component, but in animal feed, over and above the technical aspect, it also supplies functional nutrients. Lecithin has emulsifying properties and is an aid to digestion, particularly of fats. But its individual components are also important. Lecithin is mostly a mix of phospholipids, of which the most biologically active are phosphatidylcholine (PC), phosphatidylethanolamine (PE), phosphatidylserine (PS) and phosphatidylinositol (PI). All these are basic building blocks for the body's cell membranes and as such play an active role in metabolism.

The PC in lecithin is a bioavailable source of choline. Being necessary for the synthesis of the neurotransmitter acetylcholine, choline plays a special role in the nervous system. Via its metabolite betaine, choline is a methyl group donor in a way similar to methionine. However, there is also a mutual dependency between the methyl group donors methionine and choline and the coenzyme involved, 5-methyltetrahydrofolate (MTHF). If, for example, methionine and MTHF fail to supply sufficient quantities of methyl groups, choline has to fill the gap, at which point it could be classed as an essential nutrient (1).

Choline is normally classed as vitamin-like only, because although the body needs it, the substance can also be synthesized, at least in tiny amounts. Since choline is involved, as a PC building block, in transporting lipids from the liver via lipoproteins, there is also a connection between choline deficiency and fatty degeneration of the liver (2), a condition that can occur in a variety of animal species. For high-yielding dairy cows in particular, this can be a problem during early lactation (3).

7.1.1 Ruminants

Ruminants' feed basis consists of plants. In relation to their dry mass, plants contain 3–10% lipids, of which some are fats (triglycerides) but a large proportion are phospholipids (4). Due to evolutionary selection for this type of feed, ruminants are capable of digesting phospholipids from lecithins in their rumen. Various working groups such as Jenkins et al. (5) or Baldi and Pinotti (6) have confirmed this. Wettstein and collaborators do in fact see a greater benefit in using lecithins as a substitute for vegetable oils, because lecithins could help to prevent depressed digestibility of fibrous material (7). This is especially true of lecithins containing low levels of polyunsaturated fatty acids (8). Grummer et al. described a similar effect (9), and proved that after infusion into the abomasum (fourth stomach compartment), raw soy lecithin increased the volume of milk and fat more than did soy oil (triglycerides).

Phospholipids can be broken down very quickly in the rumen. This is demonstrated by Neill et al. in the case of both choline chloride and the choline that is naturally present in the PC of lecithin (10). However, choline chloride is digested more rapidly in the rumen than natural choline compounds (11).

Free choline does not accumulate in the rumen, which indicates it is absorbed through the wall of the rumen (12). The phospholipids that have not been broken down are taken up by the microbes in the rumen and used for their own growth. This is backed up by findings from the U.S. National Research Council (NRC) showing that some 85 to 90% of the fatty acids leave the rumen as free fatty acids, whereas the remaining 10 to 15% of the fatty acids are bound to microbial phospholipids (13). Neill et al. cite the figure of 10% for choline bound to protozoa (10). In a later paper, Dawson et al. demonstrated similarly low values, since lambs derive no more than 30% of the PC directly from the feed (14). According to the NRC, the microbial lecithin produced in the rumen is processed into a fat emulsifier by pancreatic phospholipase (13).

If lecithins are fed in larger quantities, some will pass through the rumen and act as emulsifiers to improve digestibility in the colon (15), as may be seen from better digestion of fat in the colon. Another option is to use rumen-protected choline, e.g. in the form of coated choline chloride. Various publications note a positive effect on milk yields, feed conversion and profitability through the use of rumen-protected choline (16,17).

No choline requirements have in fact been specified for dairy cows, but Donkin (18) has taken data from human and calf needs to derive a maintenance requirement of 4–6 g choline per animal per day. A cow in calf needs additional energy for the growing foetus and thereafter, in full lactation, for the production of milk. An assessment of various studies points to requirements of between 10 and 50 g of rumen-protected choline per cow per day for the period around calving (19). A point to note is that during this time a cow excretes large quantities of methionine in the milk, while at the same time insufficient methyl precursors are supplied by the metabolism of MTHF (6). Thus it could be said that, in the case of high-yielding dairy cows, choline is a limiting nutrient for milk production, at least at the onset of lactation (20). One can try to regain a balance by supplementing with rumen-protected choline (21). Cooke et al. report that protected choline can also lower the incidence of fatty liver triggered by feed restriction (22) in dairy cows.

In dairy cows the post-partum phase is comparable with restricted feeding, as the cow loses more energy through the milk than can be replaced by feeding due to a limited uptake capacity. In order to redress the balance the cow mobilizes energy stored in body fat; to do so however, fat components must be temporarily stored in the liver. The result can be a fatty liver, a problem that afflicts many newly lactating cows. Choline is helpful here in aiding the transport of fat from the liver (22).

Additional choline does not necessarily need to be supplied as choline chloride; the natural source of choline, PC as present in lecithin, can be used too. De Nardi et al. (23) see an advantage in using lecithin because, compared with protected choline chloride, comparable milk yields are obtained at half the dosage if the choline comes from lecithin. The statement that in the rumen natural choline compounds are broken down more slowly than choline chloride is borne out by the work of Atkins et al. (11).

Choline supplementation could be even more effective if lecithin were sluiced through the rumen into the small intestine. This would make a natural choline compound directly available for further metabolism. Not enough research has been carried out in this field of animal nutrition to date.

7.1.2 Poultry and pig fattening

In poultry and pig fattening, the cost of feed accounts for a substantial proportion of the overall costs and thus has an appreciable effect on the proceeds that can be achieved. In order to maximize profits, use is made of price fluctuations in the various feedstuffs by choosing the cheapest of the feed components with a comparable effect for use in the feed mixture. This is especially true of oils as high-energy components. Oils can be replaced with lecithin, with a corresponding saving in price (24,25). Further advantages are a saving in time and reduced costs for mixing and pelletizing the feed, since the emulsifying properties of the lecithin increase the lubricity of the feed components by "mediating" between water and oil, and thus permit higher throughput of the production plant.

Besides the "oil replacement" argument, lecithin offers a number of other advantages. Lecithin improves the digestibility of fat (26,27), and thus feed conversion and growth. This applies to both pigs (28,29) and poultry (30). The improved digestibility of fat in the animal's body is due primarily to the effect of lecithin as an emulsifier (31,32). By forming micelles, lecithin performs a function similar to that of bile (33). Bile also contains lecithins (4); these ensure that the feed fats are converted into an emulsion with a much larger lipid surface, so that the lipases in the gut are better able to break the fats down and prepare them for faster and more thorough absorption. This assistance of fat digestion is especially interesting in the case of young animals such as weaner pigs (35,36), whose digestive systems are not yet fully developed. But such additional help may be necessary with older animals, too, if their liver is already damaged. The liver then secretes too little bile, which reduces the emulsification of fat in the mash and may result in constipation and bloating, including the propagation of harmful microorganisms in the intestinal tract (37).

The effect of emulsifiers is so far-reaching that the addition of an emulsifier can make up for a reduced fat content of the feed (38). Jones et al. demonstrate this with weaner pigs, in which they balance the digestibility of tallow by adding lecithin (39). This means that either a fat with a lower energy content might be used, or the amount of fat or oil in the feed could be deliberately reduced. In both cases the addition of lecithin leads to better absorption of the feed fat and thus to faster growth through more complete utilization of the nutrients (40).

Besides fat digestion, lecithin assists animal metabolism in other ways. In laying hens, fatty liver disease is a constantly recurring problem. Excess lipids are normally removed from the liver with the aid of lipoproteins. But if there are not enough of these present, the addition of lecithin helps to reduce the lipid level in the liver of laying hens (41). In broilers, too, it is possible to reduce fat deposition in the liver by adding lecithin (42).

Not only does lecithin act as an emulsifier or an aid to preventing fatty liver disease. It also supplies basic components of the cell membranes of the body. For instance, the phospholipids contribute to the health of the gut. If, for example, piglets are not provided with enough energy and protein during the suckling phase, their small intestine is unable to adjust to its future function. The addition of lecithin helps the intestine to develop properly nevertheless (43).

But lecithins also play an active part in metabolism. This applies especially to phosphatidylcholine, phosphatidylethanolamine, phosphatidylserine and phosphatidylinositol, which are present in lecithin as a natural phospholipid mixture. Choline, in particular, is necessary for synthesis of the neurotransmitter acetylcholine (44), i.e. for conduction of stimuli from the parasympathetic nervous system, a part of the autonomic nervous system. As the opponent of the sympathetic nervous system, the parasympathetic system is responsible for reducing stress. Consequently, through synthesis of acetylcholine, the lecithin component phosphatidylcholine can help to alleviate stress in pigs and poultry (45). It can be deduced from this that the animals then behave more calmly (46), which could be expected to result in fewer losses and better growth. The problem of splay leg piglets is also being discussed in connection with acetylcholine. These piglets have a low acetylcholine level from birth and therefore difficulty in standing properly. As a preventive measure, the sows are given an increased dose of choline through their feed before farrowing. This can be done in the form of choline chloride or with lecithin as a source of phosphatidylcholine (36,47).

Lecithins are said to have other health-promoting qualities, too. They have antioxidant properties, for example in that they inactivate oxidation-inducing metals or strengthen the effect of phenolic antioxidants (48). But they also seem to promote the healing of ulcers in the gastrointestinal tract (49). In the feeding of horses, use has been made of this function for some long time (50). And it is likely to be a wise measure with fattening pigs too, since these frequently suffer from gastric lesions as a result of too finely ground feed.

Not only does lecithin have a favourable effect on the development and growth of the animals. Besides leading to better feed conversion and growth in broilers (42), it also has an effect on the foods produced from the animals. For example, its use in fattening feed enhances the sensory quality of pork (51). The reason behind this may be that lecithins supply essential fatty acids, since the middle hydroxyl group of the glycerol skeleton of most lecithins contains polyunsaturated fatty acids (52). This is an interesting argument in favour of the use of lecithin with laying hens, since linoleic acid is regarded as beneficial for egg production.

To summarize: lecithin may be regarded as a useful feed component in the fattening of pigs and poultry. It offers technical advantages, but it also has a positive influence on the animals' performance. Moreover, the effect of the lecithin reaches into the foods produced from the animals – a potential gain in quality of life for the consumers.

References

1. Zeisel, S., Da Costa, K.-A., 2009. Choline: An Essential Nutrient for Public Health. Nutr. Rev. 615-623.
2. Schugar, R., Huang, X., Moll, A., Brunt, E., Crawford, P., o.D. 2013. Role of Choline Deficiency in the Fatty Liver Phenotype of Mice Fed a Low Protein, Very Low Carbohydrate Ketogenic Diet. PLoS ONE 8(8): e74806. Retrieved from https://doi.org/10.1371/journal.pone.0074806
3. Grummer, R., 2011. Managing the Transition Cow-Emphasis on Ketosis and Fatty Liver Syndrome. Retrieved from https://www.vtdairy.dasc.vt.edu/pub-pres/docs/cow-colleges/2011-cowcollege/18-grummer-paper.pdf
4. Husveth, F., 2011. Physiological and reproductional aspects of animal production. Debreceni Egyetem, Nyugat-Magyarországi Egyetem, Pannon Egyetem.
5. Jenkins, T., Gimenez, T., Cross, D., 1989. Influence of phospholipids on ruminal fermentation in vitro and on nutrient digestion and serum lipids in sheep. J. Anim. Sci., 529-537.
6. Baldi, A., Pinotti, L., 2006. Choline metabolism in high-producing dairy cows: Metabolic and nutritional basis. Can. J. Anim. Sci., 207-212.
7. Wettstein, H.-R., Sutter, F., Machmüller, A., Scheeder, M., Kreuzer, M., 1999. Effect of Partial Replacement of Rumen-Protected Fat by Various Plant Lecithins on Nutrient Digestion and Milk Composition in Dairy Cows. In Stone, G., Forbes, T., Stuth, J., Byers, F., Nutritional Ecology of Herbivores.
8. Wettstein, H.-R., Quarella Forni, M., Kreuzer, M., Sutter, F., 2000. Influence of plant lecithin partly replacing rumen-protected fat on digestion, metabolic traits and performance of dairy cows. J. Anim. Physiol. a. Anim. Nutr., 165-177.
9. Grummer, R., Armentano, L., Marcus, M., 1987. Lactation response to short-term abomasal infusion of choline, inositol, and soy lecithin. J. Dairy Sci., 2518-2524.
10. Neill, A., Grime, D., Snoswell, A., Northtrop, A., Lindsay, D., Dawson, R., 1979. The low availability of dietary choline for the nutrition of the sheep. Biochem. J., 559-565.
11. Atkins, K., Erdman, R., Vandersall, J., 1988. Dietary choline effects on milk yield and duodenal choline flow in dairy cattle. J. Dairy Sci., 109-116.
12. Broad, T., Dawson, R., 1976. Role of choline in the nutrition of the rumen protozoon Entodinium caudatum. J. Gen. Microbiol., 391-397.
13. NRC (National Research Council), 2001. Nutrient requirements of dairy cattle. Washington: National Research Council.
14. Dawson, R.M., Grime, D., Lindsay, D.B., 1981. On the intensitivity of sheep to the almost complete microbial destruction of dietary choline before alimentary-tract absorption. Biochem. J., 499-504.
15. Jenkins, T., Fotouhi, N., 1990. Effects of lecithin and corn oil on site of digestion, ruminal fermentation and microbial protein synthesis in sheep. J. Anim. Sci., 460-466.
16. Erdman, R., Sharma, B., 1991. Effect of Dietary Rumen-Protected Choline in Lactating Dairy Cows. J. Dairy Sci., 1641-1647.
17. Mohsen, M., Gaafar, H., Khalafalla, M., Shita, A., Yousif, A., 2011. Effect of rumen protected choline supplementation on digestibility, rumen activity and milk yield in lactating Friesian cows. Slovak. J. Anim. Sci., 44(1), 13-20.
18. Donkin, S., 2002. Rumen-protected choline: Potential for improving health and production in dairy cows. Tri-State Dairy Nutrition Conference.
19. Pinotti, L., 2012. Vitamin-Like Supplementation in Dairy Ruminants: The Case of Choline. Retrieved from INTECH open science: http://dx.doi.org/10.5772/50770
20. Pinotti, L., Baldi, A., Dell'Orto, V., 2002. Comparative mammalian choline metabolism with emphasis on the high-yielding dairy cow. Nutrition Research Reviews, 315-331.
21. Brüsemeister, F., Südekum, K.-H., 2006. Rumen-protected choline for dairy cows: the in situ evaluation of a commercial source and literature evaluation of effects on performance and interactions between methionine and choline metabolism. Anim. Res., 93-104.
22. Cooke, R., Silva del Rio, N., Caraviello, D., Bertics, S., Ramos, M., Grummer, R., 2007. Supplemental Choline for Prevention and Alleviation of Fatty Liver in Dairy Cattle. J. Dairy Sci., 2413-2418.
23. De Nardi, R., Marchesini, G., Tenti, S., Contiero, B., Andrighetto, I., Segato, S., 2012. Lecithin as a supplement for mid-lactating dairy cows. 20th Int. Symp. "Animal Science Days". Kranjska gora.

24. Azman, M.A., Ciftci, M., 2004. Effects of replacing dietary fat with lecithin on broiler chicken zootechnical performance. Revue Med. Vet. 155(8-9), 445-448.
25. Cox, W.R., Richie, S.J., Sifri, M., Bennett, B., Kitts, D.D., 2000. The impact of replacing dietary fat with lecithin on broiler chicken performance. Poultry Sci. 79, 67.
26. Huang, J., Yang, D., Wang, T., 2007. Effects of replacing soy-oil with soy-lecithin on growth performance, nutrient utilization and serum parameters of broilers fed corn-based diets. Asian-Aust. J. Anim. Sci. 20(12), 1880-1886.
27. Jin, C.F., Kim, J.H., Han, I.K., Jung, H.J., Kwon, C.H., 1998. Effects of various fat sources and lecithin on the growth performance and nutrient utilization in pigs weaned at 21 days of age. AJAS. 11(2), 176-184.
28. Schulze, U., 2009. Powerfutter gegen das Sommerloch. Top Agrar. 7, 22-24.
29. Danek, P., Rozkot, M., 2008. Influence of lecithin emulsifier on piglets after weaning. Research in Pig Breeding 2, 42-44.
30. Xing, J.J., Van Heugten, E., Li, D.F., Touchette, K.J., Coalson, J.A., Odgaard, R.L., Odle, J., 2004. Effects of emulsification, fat encapsulation, and pelleting on weanling pig performance and nutrient digestibility. J. Anim. Sci. 82, 2601-2609.
31. Emmert, J.L., Garrow, T.A., Baker, D.H., 1996. Development of an experimental diet for determining bioavailable choline concentration and its application in studies with soybean lecithin. J. Anim. Sci. 74, 2738-2744.
32. Danek, P., 2009. Evaluation of lecithin as an emulsifier for weanling piglets. Research in Pig Breeding 3(1), 11-14.
33. Kammlott, E., 2003. Untersuchungen an pankreasgangligierten Schweinen zur Nährstoffverdaulichkeit (praecaecal/in toto) unter dem Einfluss einer Enzymsubstitution sowie zur Wirksamkeit ergänzender Maßnahmen (Einsatz von Omeprazol oder Lecithin bzw. fraktionierte Futter- und/oder Enzymgabe). Dissertation. Hannover.
34. O'Doherty, P.J., Kakis, G., Kuksis, A., 1973. Role of luminal lecithin in intestinal fat absorption. Lipids. 8(5), 249-255.
35. Danek, P., Paseka, A., Smola, J. F., Ondracek, J., Beckova, R., Rozkot, M., 2005. Influence of lecithin emulsifier on the utilisation of nutrients and growth performance in weanling pigs. Czech J. Anim. Sci. 50(10), 459-465.
36. Hertrampf, J.W., 1994. Fett und Lecithin in der Ferkelfütterung. Kraftfutter 1, 12-18.
37. Kleine-Klausing, H. 2003. Gesunde Sauen - ohne die Leber geht nichts. Nutztierpraxis Aktuell, 7.
38. Bohr, B., Rovers, M., 2014. Mit Emulgatoren Futterkosten sparen. DGS Magazin. 44, 16-18.
39. Jones, D.B., Hancock, J.D., Harmon, D.L., Walker, CE., 1992. Effects of exogenous emulsifiers and fat sources on nutrient digestibility, serum lipids, and growth. J. Anim. Sci. 70, 3473-3482.
40. Todorova, M., Ignatova, M., Petkova, M., 2011. Effect of lecithin supplementation in standard diet for weaned pigs on growth performance and blood cholesterol level. Archiva Zootechnica 14(4), 45-50.
41. An, B.K., Nishiyama, H., Tanaka, K., Ohtani, S., Iwata, T., Tsutsumi, K., Kasai, M., 1997. Dietary safflower phospholipid reduces liver lipids in laying hens. Poultry Science 76, 689-695.
42. Khosravinia, H., Chethen, P.S., Umakantha, B., Nourmohammadi, R., 2015. Effects of lipotropic products on productive performance, liver lipid and enzymes activity in broiler chickens. Poultry Sci. J. 3(2), 113-120.
43. Lopez-Pedrosa, J.M., Ramírez, M., Torres, M.I., Gil, A., 1999. Dietary phospholipids rich in long-chain polyunsaturated fatty acids improve the repair of small intestine in previously malnourished piglets. J. Nutr. 129, 1149-1155.
44. Billah, M., Anthes, JC., 1990. The regulation and cellular functions of phosphatidylcholine hydrolysis. Biochem. J. 269, 281-291.
45. Colares, J.R., Schemitt, E.G., Hartmann, R.M., Moura R.M., Morgan-Martins, M.I., Fillmann, H.S., Fillmann, L.S., Marroni, N.P., 2016. Effect of lecithin on oxidative stress in an experimental model of rat colitis induced by acetic acid. J. Coloproctology 36 (2), 97-103.
46. Holland, J.L., Kronfeld, D.S., Meacham, T.N., 1996. Behaviour of horses is affected by soy lecithin and corn oil in the diet. J. Anim. Sci. 74, 1252-1255.
47. Kuhn, M., Frohmann, B., Petersen, A., Rübesam, K., Jatsch, C., 1998. Utilization of crude soybean lecithin as a native choline source in feed rations of fattening pigs. Eur. J. Lipid Sci. Technol. 100(3), 78-84.

48. Doert, M., 2010. Zur Wirkung von Phospholipiden in der thermisch induzierten Lipidperoxidation. Dissertation. Berlin.
49. Schneider, H., Braun, A., Füllekrug, J., Stremmel, W., Ehehalt, R., 2010. Lipid based therapy for ulcerative colitis - modulation of intestinal mucus membrane phospholipids as a tool to influence inflammation. Int. J. Mol. Sci. 11, 4149-4164.
50. Venner, M., Lauffs, S., Deegen, E., 1999. Treatment of gastric lesions in horses with pectin-lecithin complex. Equine Vet J. Suppl. Apr, (29), 91-96.
51. D'Souza, D.N., Blake, B.L., Williams, I.H., Mullan, B.P., Pethick, D.W., Dunshea, F.R., 2015. Dietary lecithin supplementation can improve the quality of the M. longissimus thoracis. Animals, Basel, 5(4), 1180-1191.
52. Melisch, P., 1999. Untersuchungen zum Stoffwechselweg des Phosphatidylcholins beim Geflügel. Dissertation, Göttingen.

7.2 Phospholipids in aquaculture

Irina Ostroumova, Anatoliy Lyutikov, Alla Shumilina

Phospholipids fulfil important structural and functional roles in animal organisms including fish, by forming the basis of cell membranes. Natural fish feed is rich in phospholipids, composed predominantly of lecithin and with a great number of omega-3 polyunsaturated fatty acids, while lipids of artificial nutrition commonly used in aquaculture are represented by neutral lipids-triacylglycerols which contain far less omega-3 fatty acids. A major source of lipids in fish feeds is fish oil, which during manufacturing and refining loses practically all its phospholipids and finally consists almost completely of triacylglycerols (triglycerides) (1). In aquaculture the fish being kept entirely on artificial feeds may experience a shortage of these essential phospholipid compounds, although they are able to synthesize phospholipids when biochemical precursor are available to them.

Over the past few decades active studies were carried out, the purpose of which is a search for sources of phospholipids for use in feeding fish of various species and ages, including developing specification of the needs for them and of ingredients for their synthesis. In addition further new properties were discovered of phospholipids and specific fractions that play important roles in the adaptation of fish to environmental conditions, and which influence fish growth rate, development and physiological state.

7.2.1 Phospholipids introduced into feed of rainbow trout (Oncorhynchus mykiss) protect against fatty degeneration of the liver

Interest in phospholipids for feeding fish arose long ago. In the middle of the last century a disease of rainbow trout featuring impaired liver triglyceride metabolism caused extensive damage to trout farms in many countries. Fatty liver, often progressing to ceroid liver degeneration and concomitant anaemia, developed in fish being bred in cages and tanks and maintained on artificial feed mixtures. The trout's skin became black; often the eyeballs became abnormally protruded. The liver coloration took on a sandy hue, the hepatocytes (liver parenchymal cells) became overfilled with fat, which over time turned into ceroid including fatty oxidation breakdown products. The liver cell structure became disrupted and they lost their ability to function. The liver degeneration was usually accompanied by blood pathology: the haemoglobin content and red blood cell (erythrocyte) count became reduced, the gills turned pale, and large numbers of small, immature, non-haemoglobinated erythrocytes appeared in the blood. The blood white cell ratio rapidly deteriorated. The fish stopped taking food and died in large quantities. The disease often occurred due to feeding errors, especially when the trout were maintained on frozen fish.

Efforts to protect the trout from this disease became successful by adding sunflower phospholipids into their feeds (2). It was known from veterinary health and medicine that choline phospholipids have a lipotropic function – they remove fat from the liver by contributing to the formation of lipoprotein particles that distribute lipids to the body's tissues and organs. In cases of phospholipid deficiency fats are delayed in the liver, overfill it and cause 'fatty liver' progressing to ceroid degeneration. The addition of sunflower phospholipids into the trout diet resulted in the absence of pathological characteristics in rainbow trout of various ages, by their receiving frozen fish with the addition of 10% phospholipids for four months. The fish grew well, had practically 100% survival rate, and normal liver and blood clinical picture (Table 1).

Table 1.

Results of trout farming on feeds with phospholipids.

Content of phospholipids in feed, %	Two-year-old fish				Three-year-old fish				Four-year-old fish			
	Individual fish weight, g	Mortality, %	Haemoglobin, g/l	Red blood cell count, mln/mm³	Individual fish weight, g	Mortality, %	Haemoglobin, g/l	Red blood cell count, mln/mm³	Individual fish weight, g	Mortality, %	Haemoglobin, g/l	Red blood cell count, mln/mm³
0	70	75	23.0	0.467	189	19	27.5	0.611	849	0	44.8	0.810
5	72	23	55.5	1.074	228	13	57.3	1.103	908	0	93.9	1.322
10	86	1	70.7	1.170	263	0.3	82.9	1.266	1012	0	99.5	1.445

Phospholipids not only protected the trout from ceroid liver degeneration – they also could reverse the disease. If the disease had not progressed too far and the trout could still respond to their feed, then the addition of phospholipids to the diet cured the fish of the anaemia and degenerative processes that had already begun.

When included in feeds that don't cause fatty liver degeneration, phospholipids promote growth of fish. The final weight of rainbow trout was higher by

25–65%, of Atlantic salmon *(Salmo solar)* underyearlings by 27–36% and yearlings by 15–32%, compared to control fish not fed phospholipids. Phospholipids seem to promote high intensity of proteolysis and protein assimilation in the trout body that favours active growth (3).

7.2.2 Phospholipids as an essential factor in starter feed for fish larvae

The most vulnerable point in the early post-embryogenesis of a fish is its transition to external feeding. If in the wild its high mortality at this stage is often related to a deficiency of necessary food, then in conditions of aquaculture high mortality is related to a non-conformity of nutrient composition in feeds including lipids to larva-specific requirements.

The larvae of most fish species begin to feed in the first days after hatching at early developmental stages, when maternal resources in the form of yolk sac (Fig. 1) are not yet completely expended and their alimentary system is in a formative stage: stomach not yet formed, pancreas inactive, abdominal digestion practically doesn't function, intracellular digestion prevails. In spite of their alimentary system being incomplete, at this stage of development larvae possess the ability for very rapid growth. Daily gains, particularly in fishes living in warm waters, may reach 100% per day, in cold waters 10–20% per day.

In the wild a fry feeds on phytoplankton and small forms of zooplankton, lipids of which consist of about 50–70% phospholipids. In artificial starting feeds triacylglycerols usually prevail and are one of the causes of the low growth rate and high mortality of fry.

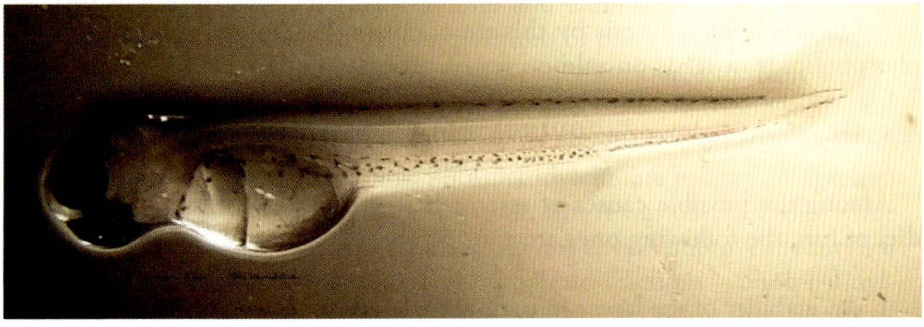

Figure 1.

Muksun (*Coregonus muksun*) larva with yolk sac a day before beginning of feeding. Weight 6.9 mg, length 8.0 mm.

In experiments on sea bass (*Dicentrarchus labrax*) (8), the incorporation of 11.7% soybean phospholipids in the feed, of which PC amounted to 3.5% and PI to 1.6%, ensured increased growth and survival and also reduced the number of larvae with vertebral developmental disorder and jaw deformation (Table 2).

According to the data from authors who worked with common carp larvae (12), PI in the starting diet prevented skeletal deformation: the percentage of juvenile carp with deformities was very low – not more than 2%. With only PC in the starting diet, there were many more fish with skeletal anomalies – 32%.

Table 2.

Biological parameters for sea bass larvae fed on diets with different quantities of soybean phospholipids containing phosphatidylcholine and phosphatidylinositol. This table was prepared according to the data of Cahu et al. (8).

Content of soybean phospholipids in diet %	Mass mg	Survival rate %	Number of larvae with skeletal deformation %
2.7	2	22	35
11.6	32	73	2

The impact of PI on the development of larvae of various fish species is still insufficiently studied. It is known that PI fulfils diverse functions in the cell, including having membrane actions that regulate calcium entry into the cell.

Good results are obtained with a ratio of PC to PI in the starting feeds of nearly 2:1, with PI accounting for 1.3–1.6% according to data of various authors.

7.2.4 Impact of highly unsaturated fatty acids of phospholipids and triacylglycerols on fish being reared in aquaculture

The basic feature of lipids of water-living organisms is their high degree of unsaturation. As components of phospholipids forming the basis of cell membranes, essential polyunsaturated fatty acids ensure lipid fluidity, increase the permeability of membranes under conditions of relatively low habitat temperatures, and promote adaptation of cold-blooded animals to changes of temperature conditions and some other environmental factors, for example change of hydrostatic pressure.

In experiments on sea bass (*Dicentrarchus labrax*) (8), the incorporation of 11.7% soybean phospholipids in the feed, of which PC amounted to 3.5% and PI to 1.6%, ensured increased growth and survival and also reduced the number of larvae with vertebral developmental disorder and jaw deformation (Table 2).

According to the data from authors who worked with common carp larvae (12), PI in the starting diet prevented skeletal deformation: the percentage of juvenile carp with deformities was very low – not more than 2%. With only PC in the starting diet, there were many more fish with skeletal anomalies – 32%.

Table 2.

Biological parameters for sea bass larvae fed on diets with different quantities of soybean phospholipids containing phosphatidylcholine and phosphatidylinositol. This table was prepared according to the data of Cahu et al. (8).

Content of soybean phospholipids in diet %	Mass mg	Survival rate %	Number of larvae with skeletal deformation %
2.7	2	22	35
11.6	32	73	2

The impact of PI on the development of larvae of various fish species is still insufficiently studied. It is known that PI fulfils diverse functions in the cell, including having membrane actions that regulate calcium entry into the cell.

Good results are obtained with a ratio of PC to PI in the starting feeds of nearly 2:1, with PI accounting for 1.3 – 1.6% according to data of various authors.

7.2.4 Impact of highly unsaturated fatty acids of phospholipids and triacylglycerols on fish being reared in aquaculture

The basic feature of lipids of water-living organisms is their high degree of unsaturation. As components of phospholipids forming the basis of cell membranes, essential polyunsaturated fatty acids ensure lipid fluidity, increase the permeability of membranes under conditions of relatively low habitat temperatures, and promote adaptation of cold-blooded animals to changes of temperature conditions and some other environmental factors, for example change of hydrostatic pressure.

2. Lack of energy, since phospholipids fulfil not only a structural and functional role in the early fry organism but also are used as its energy source, since they are digested better than neutral fats.
3. Enhanced food attractiveness to the larvae. Phospholipids incorporated into feeds, especially phosphatidylcholine, have attractant properties that increase fry appetite and growth.
4. As natural antioxidants, phospholipids help protect lipids of the feed from oxidation.
5. By limiting lipoprotein particle formation, the deficiency of phospholipids in the larvae inhibits the spread of fatty acids and neutral lipids throughout the body. In the larvae of many fish, within the first days of feeding large fat vacuoles appear in the intestinal epithelium, which indicates a delay in the transport of fat. This very limitation of transfer of exogenous fats from the intestinal tract to other tissues is currently considered the prime factor that makes phospholipids indispensable in the feeding of fish in early post-embryogenesis (1). The enrichment of diets by phospholipids, in particular by lecithin of egg yolk or soybean, as well as feeding on living *Artemia,* reduce the number of vacuoles and restore normal growth and viability of the fish. With age, the fatty vacuoles in enterocytes (intestinal lining cells) disappear, suggesting that endogenous synthesis of phospholipids by the fish organism has become more intensive.

7.2.3 Special aspects of phosphatidylcholine and phosphatidylinositol influence on fish fry

From the very first experiments on the incorporation of phospholipids in the diets of fry, it was noted that they not only increase growth rates but also reduce the occurrence of individuals with vertebral and cranial bone deformities. In studying the effects of the fractional compositions of phospholipids, a stimulating effect of phosphatidylcholine on growth was demonstrated in many studies on various fish species. One mechanism of this effect is a contribution to providing for the energy needs of the rapidly growing organism. Of the various fractional compositions of phospholipids, phosphatidylcholine principally specifies the lipotropic and growth-stimulating function. But it was observed that while increasing growth rate, phosphatidylcholine did not always improve the survival rate of fry. When different fractions of phospholipid fractions – phosphatidylcholine (PC) and phosphatidylinositol (PI) – were compared on the early fry of common carp (*Cyprinus carpio)* and gilt-head bream (*Sparus aurata)* (12,14), PC increased their growth while PI reduced skeletal deformation.

Investigations by Japanese authors in the 1980s and 1990s (4) established that in the period of intensive early growth of fish larvae, phospholipids are an essential nutritional factor and should be supplied with food in the required amount. The requirement is reduced with age, since the fish develop the ability to synthesize phospholipids. This was affirmed by many authors in the aquaculture of various fish species with the incorporation of sunflower, soybean or egg phosphatides (phospholipids) into the starting feeds. The addition of 3–5% sunflower phospholipids into the diet of larvae of whitefishes (Coregonidae) – peled (*Coregonus peled),* broad whitefish (*C. nasus)* and muksun (*C. muksun),* – increased their food intake index, promoted their growth and enhanced the viability (5). Especially high positive effects from the addition of phospholipids were obtained at low temperatures, around 11.6°C (6–16°C). At the age of 44 days, the final weight of peled on a feed with phospholipids was almost twice as high as control peled without phospholipids. At higher temperatures, around 14.4°C (11–19°C), these differences were less significant – individual weight was higher by 16%.

The indispensability of phospholipids in starting feeds for fish larvae was proven by investigations conducted both on marine fish species (6–8) and on freshwater ones – trout and salmon (9,10), cyprinids (Cyprinidae) (11,12) and pikeperch (*Sander lucioperca)* (13). The best growth rates of pikeperch larvae with the addition of 1.4, 4.7 or 9.5% soybean lecithin to the diet were obtained at 9.5%, but there an influence of phospholipids upon the fish survival rate was not observed. It was the same for pikeperch larvae that received a diet with different lecithin contents or without it. In our experiments the best growth and survival rates with the use of 7, 10 or 14% soybean lecithin in feeds for muksun larvae from the first days of feeding were obtained at 7 and 10%.

With the addition of phospholipids into the diets of fry raised to a few grams' weight, a positive effect was often absent, due to activation of their innate phospholipid synthesis by this time. Poston (9) studied the effect of 4% phospholipids in feeds on salmon fry of various weights – 0.18, 1.0, 1.7 and 7.5 g. An increase in the growth rate was achieved in the first three groups, especially in the lightest one (0.18 g), which achieved 50%.

Among the possible causes of essentiality of phospholipids for the starting diet of fish, the following ones are considered.

1. An inability of the larvae in periods of intensive augmentation of cell mass to synthesize sufficient phospholipids to keep up with the need for vital cell membranes.

25–65%, of Atlantic salmon *(Salmo solar)* underyearlings by 27–36% and yearlings by 15–32%, compared to control fish not fed phospholipids. Phospholipids seem to promote high intensity of proteolysis and protein assimilation in the trout body that favours active growth (3).

7.2.2 Phospholipids as an essential factor in starter feed for fish larvae

The most vulnerable point in the early post-embryogenesis of a fish is its transition to external feeding. If in the wild its high mortality at this stage is often related to a deficiency of necessary food, then in conditions of aquaculture high mortality is related to a non-conformity of nutrient composition in feeds including lipids to larva-specific requirements.

The larvae of most fish species begin to feed in the first days after hatching at early developmental stages, when maternal resources in the form of yolk sac (Fig. 1) are not yet completely expended and their alimentary system is in a formative stage: stomach not yet formed, pancreas inactive, abdominal digestion practically doesn't function, intracellular digestion prevails. In spite of their alimentary system being incomplete, at this stage of development larvae possess the ability for very rapid growth. Daily gains, particularly in fishes living in warm waters, may reach 100% per day, in cold waters 10–20% per day.

In the wild a fry feeds on phytoplankton and small forms of zooplankton, lipids of which consist of about 50–70% phospholipids. In artificial starting feeds triacylglycerols usually prevail and are one of the causes of the low growth rate and high mortality of fry.

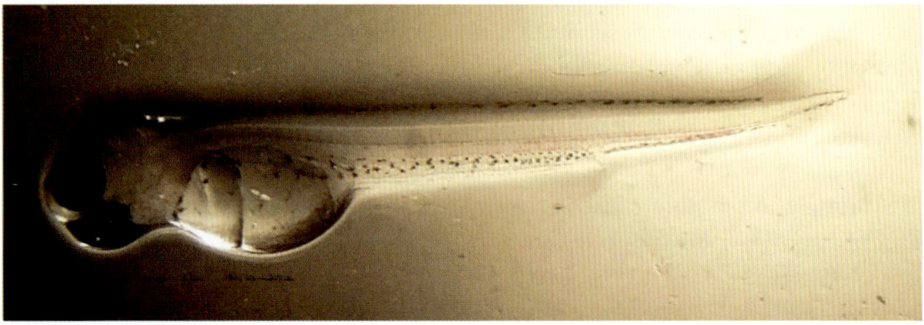

Figure 1.

Muksun (*Coregonus muksun*) larva with yolk sac a day before beginning of feeding. Weight 6.9 mg, length 8.0 mm.

Fish are able to synthesize only saturated fatty acids and monounsaturated fatty acids (having just one carbon – carbon double bond). Parent polyunsaturated acids with three double bonds (linolenic acid 18:3ω3) or two (linoleic acid 18:2ω6) are synthesized by just a few microalgae (15). They are transferred to animals and up the food chain by feeding, and are regarded as indispensable substances for life. Water ecosystems are the main source of these important biochemical components, especially the omega-3 (ω3) family. Linolenic acid (18:3ω3) serves as the precursor of the omega-3s most important for the physiology of all animals, including fish: the long-chain eicosapentaenoic acid (EPA, 20:5ω3) and docosahexaenoic acid (DHA, 22:6ω3). Linoleic acid (18:2ω6) is the essential omega-6 and precursor of arachidonic (AA, 20:4ω6). EPA, DHA and AA play critical roles in the maintenance of the normal physiological state of fish, owing mainly to their participation in the development and functioning of the nervous, endocrine and immune systems. The possibilities of their synthesis from parent short-chain forms differs between various fish species.

Marine organisms that are food to fish, whether zooplankton or benthic (deeper-dwelling) organisms, have relatively high content of the long-chain fatty acids omega-3 EPA and DHA. Perhaps for this reason, marine fish consuming these daily in sufficient quantity lack an ability to transform linolenic into EPA and DHA. Therefore their diet under aquaculture must include feed components comprising these omega-3 long-chain fatty acids.

As compared to marine waters, natural fish food in fresh waters have far less EPA and DHA but more linolenic acid. Freshwater fish are able to transform it into long-chain EPA, DHA and other long-chain omega-3 fatty acids. At the same time, the addition of EPA and DHA to a freshwater fish diet, in particular for fry, in a quantity of 1 – 2% quite often has a favourable effect.

Investigation of these problems has been conducted over many years in different directions. For example, whether polyunsaturated fatty acids from food enter the body of fish as components of phospholipids or as components of triacylglycerols turned out to be important. It was established that an omega-3 excess in the diet may adversely affect the growth of fish, since this causes the fatty acids to cease to enter into phospholipids and rather accumulate in triacylglycerols (16). Total omega-3 fatty acid content of lipids is not an indicator of their sufficiency in an organism, if they are not components of phospholipids.

In recent studies, EPA and DHA have been introduced into muksun larva feed. In the larval body they become included with neutral fats as well as phospholipids, but more so in the neutral lipids (17). This did not affect the growth and survival of the fish. At the time of the addition of the omega-3 fatty acids to their diet – 17 days – the muksun larvae (weight 15.6 mg) probably already had the ability to synthesize them from maternal omega-3 linolenic acid.

The higher effectiveness on fish physiology of omega-3 long-chain fatty acids as part of phospholipids rather than as part of neutral fats was acknowledged by a number of authors. Cod larvae used EPA and DHA more efficiently when incorporated into phospholipids than in neutral fats, as evidenced by the larger size of the larvae and the more developed digestive system (18). Larvae of cod that were fed neutral lipids with omega-3 long-chain fatty acids had less mass and much more fat in the liver than larvae fed phospholipids with these same fatty acids. Furthermore, other signs suggest that the feeds with more omega-3s incorporated into phospholipids than into neutral fats are more useful to the physiology and health of cod fry.

After examining ratios of phospholipids to triacylglycerols in feeds, Sargent et al. (19) concluded that an optimal diet for larvae of marine fish should contain 10% phospholipids rich in highly unsaturated omega-3 fatty acids, and less than 5% triacylglycerols. This ratio is typical for yolk eggs of marine fish, larvae by themselves, and their natural food – marine zooplankton.

The provision for fish in aquaculture, especially fry, of feeds with sufficient content of highly unsaturated long-chain fatty acids is a serious problem. It may be noted that vegetable oils are principally enriched in fatty acids of the omega-6 type. Some of them contain a certain quantity of omega-3 fatty acids in the form of linolenic acid (soybean oil, rapeseed oil and especially flaxseed oil), but all are absolutely lacking in the long-chain omega-3 EPA and DHA. Fish oil consists mainly of neutral fats (triacylglycerols) poor in highly unsaturated fatty acids. Probably, a solution to this problem lies in the use of marine phospholipids, which are examined in depth in Chapter 3.4 of this book (20).

7.2.5 The effect of the storage time of concentrated fish feeds with phospholipids on the degree of oxidation and hydrolysis of lipids

Phospholipids are natural antioxidants. However, lipids of concentrated fish feeds, in which there are a lot of high-protein components and mineral salts, in spite of their content of phospholipids are vulnerable to oxidation during their storage. According to our data (21), the change of oxidation parameters for lipids in whitefish starting feeds with soybean and sunflower phospholipids during the first two months after their manufacture was inconsiderable, but in the course of the third and fourth storage months at temperatures of 22–26°C processes of lipid oxidation and hydrolysis markedly intensified, to exceed the maximum admissible concentrations for fish fry. Feeds with liquid phospholipids were more vulnerable to such oxidation than dry ones.

Notably, the same dry phospholipids (such as soy lecithin) when not included in the composition of the feed and stored separately under the same conditions in an opened package had practically no change in quality and remained normal. Since the testing of starter diets on whitefish larvae was carried out in the first month and a half, when the quality of lipids was still good, the larvae had high rates of growth and survival.

In recent years, the attention of a number of authors was drawn to the antioxidant activity of lipids in feeds with different levels of omega-3 polyunsaturated fatty acids. Feeds with content of linolenic acid in quantities of 0, 0.5, 1.0, 1.5 and 2.0% had various impacts on the growth and antioxidant properties of sturgeon fry (22). The best growth rate and convertibility were demonstrated by feeds with 0.5 and 1.0% linolenic acid. These fry had the highest level of superoxide dismutase (SOD), an enzyme that protects an organism from the reactive oxygen radical species called superoxide. The lowest level of malondialdehyde – a lipid peroxidation marker – was from the feed with 0.5% linolenic acid.

Table 3.

Fractional and fatty acid compositions of liver lipids in peled underyearlings receiving feeds with different shelf lives.

Parameters	Storage life of feeds	
	1.5 months %	3.5 months %
Phospholipids	37.5–50.9	13.7–19.1
Neutral fat	42.9–53.0	53.1–67.5
Saturated fatty acids	33.5–37.1	17.0–20.1
Monounsaturated fatty acids	22.5–23.8	41.7–43.8
Eicosapentaenoic acid	3.6–4.0	2.7–3.1
Docosahexaenoic acid	21.8–23.2	10.1–12.6
Total ω6 (omega-6)	8.0–8.5	8.8–10.8
Total ω3 (omega-3)	27.8–29.9	15.6–21.8
ω3/ω6 (ratio omega-3/omega-6)	3.3–3.7	1.8–2.3

Negative phenomena in the body of fish consuming expired feed start to appear before they affect the growth and survival of the fish. In peled fry being kept on feeds with longer storage life, liver phospholipids sharply decreased (23), liver neutral fats increased, the content of monounsaturated fatty acids was augmented while the content of DHA was reduced by half (Table 3). Since the liver is indicative of processes occurring in the fish organism as a whole, it may be concluded that the use of feeds stored for a long time which do not yet have an adverse impact on fish-breeding and biological parameters (growth, survival rate) still deteriorate the quality of fish production, as nutritional substances indispensable to human health are being lost from the fish. The long-chain fatty acids omega-3 EPA and DHA play important roles in the functioning of the human nervous and cardiovascular systems, and are required for the normal work of the brain, heart and eyes. And fish is a main source of these elements in our nutrition. In connection with the results detailed here, we should pay attention to the shelf lives of feeds. They are usually established on the basis of analysis of fish-breeding and biological data.

In conclusion, it should be noted that materials presented in this article suggest a necessity of further research on the expansion of use of phospholipids in aquaculture. A range of dosing of phospholipids in the composition of feeds for fishes of various ages, species and ecologies is required. For this purpose consideration must be given not only to the improvement of the efficiency of fish-breeding processes, but also the enhancement of the nutritional value of fish production by means of its enrichment with nutrients essential to human health, including phospholipids with a high content of docosahexaenoic and eicosapentaenoic omega-3 fatty acids.

References

1. Tocher, D.R., Bendiksen, E.A., Campbell, E.A., Bell, J.G., 2008. The role of phospholipids in nutrition and metabolism of teleost fish. Aquaculture 280, 21-34.
2. Privolnev, T.I., Streltsova, S.V., Brizinova, P.N., Ostroumova, I.N., Korolyova, N.V., 1964. Protection of rainbow trout from a liver lipoid degeneration by inclusion of phosphatides in the diet. DAN SSSR 156, No.5, 1241-1243.
3. Brizinova, P.N., Streltsova, S.V., 1969. Intensity of digestion of feed protein component by Rainbow Trout. Izvestiya GosNIORKH, 68, 46-59.
4. Kanazawa, A., 1993. Essential phospholipids of fish and crustaceans. In INRA (Ed.), Fish Nutrition in Practice, IV International Symposium on Fish Nutrition and Feeding. Biarritz, June 1991. Paris, 519-530.
5. Knyazeva, L.M., 1988. Results and prospects for breeding and feeding of whitefishes in conditions of industrial rearing. Sbornik nauchnykh trudov GosNIORKH, Iss. 275, 26-37.
6. Coutteau, P., Geurden, I., Camara, M.R., Bergot, P., Sorgeloos, P., 1997. Review on the dietary effects of phospholipids in fish and crustacean larviculture. Aquaculture 155, No.1-4, 149-164.

7. Koven, W.M., Passa, G., Kolkovski, S., Tandler, A., 1998. The effect of dietary phosphatidyl-choline and its constituent fatty acid on microdiet ingestion and fatty acid absorption rate in gilt-head sea bream, *Sparus auratus,* larvae. Aquacult. Nutr. 4, No.1, 39-45.
8. Cahu, C., Zambonino, I.J., Takeuchi, T., 2003. Nutritional components affecting skeletal development in fish larvae. Aquaculture 227, No.1-4, 254-258.
9. Poston, H.A., 1990. Performance of rainbow trout fry fed supplemental soy lecithin and choline. Prog. Fish-Cult. 52, No.4, 218-225.
10. Poston, H.A., 1990. Effect of body size on growth, survival and chemical composition of Atlantic salmon fed soy lecithin and choline. Prog. Fish-Cult. 52, No.4, 226-230.
11. Szlaminska, M., Escaffre, A.M., Charlon, N., Bergot, P., 1993. Preliminary data on semisynthetic diets for goldfish (*Carassius auratus* L.) larvae. In INRA (Ed.), Fish Nutrition in Practice, IV International Symposium on Fish Nutrition and Feeding. Biarritz, June 1991. Paris, 607-612.
12. Geurden, I., Marion, D., Charlon, N., Coutteau, P., Bergot, P., 1998. Comparison of different soybean phospholipidic fractions as dietary supplements for common carp, *Cyprinus carpio*, larvae. Aquaculture 161, No.1-4, 225-235.
13. Hamza, N., Mhetli, M., Khemis, I.B., Cahu, C., Kestemont, P., 2008. Effect of dietary phospholipid levels on performance, enzyme activities and fatty acid composition of pikeperch (*Sander lucioperca*) larvae. Aquaculture 275, 274-282.
14. Sandel, E., Nixon, O., Lutzky, S., Ginsbourg, B., Tandler, A., Uni, Z., Koven, W., 2010. The effect of dietary phosphatidylcholine/phosphatidylinositol ratio on malformation in larvae and juvenile gilthead sea bream (*Sparus aurata*). Aquaculture 304, 42-48.
15. Sushchik, N.N., 2008. Role of indispensable fatty acids in trophometabolic interactions in freshwater ecosystems (review). Zhurnal Obshchey Biologii 69, No.4, 299-316.
16. Watanabe, T.,1982. Lipid nutrition in fish. Comp. Biochem. Physiol. 73B, No.1, 3-15.
17. Lyutikov, A.A., 2019. Incorporation of fatty acids of ω 3 - family into feeds for muksun larva. Voprosy rybolovstva 20, No.1, 123-127.
18. Wold, P.-A., Hoehne-Reitan, K., Cahu, C.L, Infante, J.Z., Rainuzzo, J., Kjørsvik, E., 2009. Comparison of dietary phospholipids and neutral lipids: effects on gut, liver and pancreas histology in Atlantic cod (*Gadus morha* L.) larvae. Aquacult. Nutr. 15, Iss.1, 73-84.
19. Sargent, J., McEvoy, L., Estevez, A., Bell, G., Bell, M., Henderson, J., Tocher, D., 1990. Lipid nutrition of marine fish during early development: current status and future directions. Aquaculture 179, 217-229.
20. Lovaas, E. Marine phospholipids. In Fascination of omega-3 phospholipids, chapter 3.4 in this volume.
21. Ostroumova, I.N., Shumilina, A.K., Lyutikov, A.A., 2019. To enhance an attention to a quality of feed raw materials and feeds that have an influence on efficiency of fish farming and fish production nutritive value. Astrakhan: Izdatelstvo AGTU. Retrieved from http://astu.org/Content/Page/5833
22. Wang, X., Li, Q., Wang, J., Li, E., Qin, J.G, Chen, L., 2019. Effects of dietary alpha - linolenic acids on growth performance, lipid metabolism and antioxidant responses of juvenile Russian sturgeon *Acipenser gueldenstaedtii.* Aquacult. Nutr. 25, 184-193.
23. Ostroumova, I.N., Shumilina, A.K., Lyutikov, A.A., 2018. Influence of shelf-life of feeds on fractional and fatty acid compositions of liver lipids in whitefishes. Vestnik Rybokhozyaystvennoy Nauki 5, No.3 (19), 60-67.

8 Other Fields of Application of Lecithin

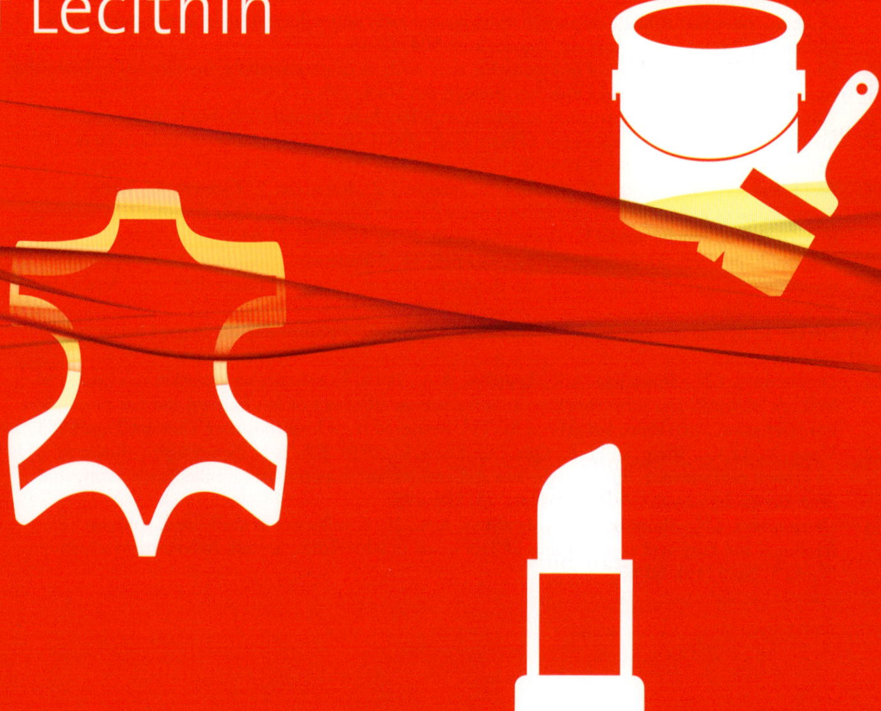

8.1 Lecithins in pharmaceutical products

Gerhard Helmerich

Lecithins are commonly used excipients in the pharmaceutical industry. Their unique lipid molecular structure makes them particularly useful for pharmaceutical applications. Lecithins when utilized as excipients are mainly used as wetting, dispersing, emulsifying and stabilizing agents.

Both natural (from plant or animal sources) and synthetic phospholipids are used as pharmaceutical excipients. Soybeans are the main source from which the lecithins are isolated and further processed for the pharmaceutical industry.

In this chapter different points are described that are of importance for the use of lecithins in the pharmaceutical industry, such as health benefits and claims, and from a chemistry, manufacturing and control (CMC) perspective its use as excipient, processing aspects, stability, safety and regulatory aspects.

8.1.1 Lecithin excipient applications

Lecithins play important roles in formulations containing liposomes or micelles, including suspensions, emulsions, solid dispersions and mixed micelles. They also can be valuable as excipients in intramuscular and intravenous injection preparations, parenteral nutrition formulations, and topical products such as creams and ointments. Products used for parenteral administration and containing lecithins include liposomes, oil-in-water emulsions, mixed micelles for intravenous use, slow-release vaccine vehicles, and drug suspensions for intramuscular and subcutaneous administration. Lecithins are also commonly used as a component of enteral and parenteral nutrition formulations. Often egg lecithins are used, at a concentration of 1.2% in 10–30% oil-in-water emulsions for parenteral nutrition.

Oil-in-water emulsions can also be used as carriers for oil-soluble drug substances. In such cases, egg phospholipids are the dominant lecithin form used as emulsifiers. In mixed micellar formulations, comprising phospholipids and cholate salts, exclusively soybean phospholipids are being used. These formulations are either used as solubilizers for poorly water-soluble compounds, or the soybean phosphatidylcholine is used as an active principle, often through its content of mainly polyunsaturated fatty acids for the treatment of liver disorders.

Lecithins are also used as suppository bases, to reduce the brittleness of the suppository but also to enhance absorption properties in intranasal insulin formulations.

Liposomes also include lecithins as a major component. Their molecular layers have been used to encapsulate drug substances. In this respect, their potential as novel delivery systems has been extensively investigated. This application generally requires purified lecithins combined in specific proportions.

Lecithins and derivatives have also been used as pulmonary surfactants in the treatment of neonatal respiratory distress syndrome. Respiratory distress syndrome is a disease in infants, characterized by an immature lung epithelium. For its treatment, mixtures of phospholipids, other lipids and proteins can be inhaled. Thus, in the product TOBI® Podhaler®, distearoylphosphatidylcholine is used as emulsifier during the manufacturing process to generate porous particles. This inhalable product is used to treat *Pseudomonas aeruginosa* lung infections.

Poloxamer lecithin organogels have been used in topical formulations for the delivery of non-steroidal anti-inflammatory drugs.

Natural phospholipids are generally more preferred than synthetic phospholipids for pharmaceutical formulations, as they are available in large scale, with reproducible quality, and at the same time with lower total production costs. Natural lecithins are well accepted by regulatory authorities and are produced using less chemicals and solvents and at higher yields than synthetic phospholipids. While natural lecithins can be used for practically any route of pharmaceutical administration, synthetic phospholipids so far have only been used in pharmaceutical formulations for parenteral administration.

Possible reasons for selecting synthetic phospholipids over natural phospholipids are manifold. First of all, there might not be a natural source for certain phospholipids. For example, PEGylated phospholipids such as mPEG-DSPE can be used to extend the circulation time of intravenously administered liposomes, but do not occur in nature.

In addition, the selection of synthetic phospholipids might be rooted in fear of unknown impurities and possible allergens that might occur in lecithins from natural sources.

Also, for liposomal dosage forms used in active or passive drug targeting, the specific liposomal phospholipids are considered to make essential contributions to the targeting properties of the liposomes. Therefore, the phospholipids should meet quality criteria comparable with new drug substances, and here synthetic phospholipids might be more manageable. Unfortunately, such products are vulnerable to difficulties with scale-up at a late stage of development, due to limited availability of synthetic phospholipids in larger volumes (multi-kilogram or metric ton scale).

8.1.2 Industrial processing aspects

As reviewed in depth elsewhere in this book, lecithins are prepared by extracting and concentrating phospholipids from naturally occurring products such as soybeans, eggs, sunflower and canola seeds, with options to purify individual phospholipids from these lecithins. The manufacturing process usually involves extraction, fractionation and purification steps. Non-toxic solvents and chromatographic procedures with low consumption of energy and minimum waste are most valuable for lecithin manufacture.

Application of validated processes and sourcing of raw materials with reproducible composition can result in very high batch quality of lecithins, with low variability in composition and in other quality key indicators, such as for example water content, viscosity and peroxide value. Their usually low production costs make lecithins invaluable in a broad range of manufacturing processes.

Natural-source phospholipids purified from lecithins can be further processed to saturated phospholipids by hydrogenation, or further treated with enzymes to partially remove fatty acids (phospholipase A) or to convert the polar group (phospholipase D), topics also covered in other chapters.

Stability and storage conditions

Lecithins decompose at extreme pH and at high temperature, i.e. when heated. Fluid or waxy lecithins should be stored at room temperature. At temperatures below 10°C phase separation may occur.

Phospholipids are stable in their solid state if protected from oxygen, heat and light. Liposomal phospholipids tend to more readily degrade via oxidation and hydrolysis.

8.1.3 Safety

Generally, lecithin phospholipids have little or no acute toxicity. They are usually well tolerated even when administered at doses in the range of grams per kilogram body weight, as with the use of oral doses up to 80 grams daily to treat tardive dyskinesia. The metabolic pathways for the body's utilization of phospholipids are well understood.

The safety of phospholipids delivered by the intravenous route is complicated by their tendency to form particles that are recognized by macrophages of the reticuloendothelial system (RES). Uptake by the RES is dependent on particle size and composition. However the well-established product Essentiale/ Lipostabil can be safely used to intravenously deliver at least 4 g of soybean phosphatidylcholine per day.

When used in topical formulations, lecithins are usually considered non-irritant and non-sensitizing ingredients. The concentration of lecithin applied in creams and ointments is usually between 2 and 15%, for both rinse-off and leave-in products.

Lecithins invariably contain a variety of poorly chemically defined materials. Therefore, care should always be taken in the use of unpurified lecithins in topical or injectable dosage forms, as interactions with the active substance or other excipients might be possible. A specification for soybean lecithin is contained in the Japanese Pharmaceutical Excipients publication (JPE). Literature from reputable suppliers is a rich source of information on the different lecithin grades available and their myriad applications in pharma formulations.

8.1.4 Regulatory aspects

Lecithins as natural phospholipid complexes are described in pharmacopoeias and relevant regulatory guidance documentation of the U.S. Food and Drug Administration (FDA) and European Medicines Agency (EMA). They are very well known to health authorities and have an impressive track record as excipients with very high tolerability and biocompatibility.

In the U.S., most soy lecithins are classed GRAS (Generally Recognized as Safe) by the Food and Drug Administration under the Code of Federal Regulations (21CFR Sec 582.1400) and specifications of the Food Chemicals Codex. The FDA designation is lecithin. The CAS No. 8002-43-5 for lecithin covers a broad range of natural lecithin preparations, including soybean lecithins, sunflower seed lecithins and egg lecithins.

Lecithins are included in the FDA Inactive Ingredient Database for a variety of applications, such as oral, otic, buccal, vaginal, topical, epidural, intravenous, intramuscular, and as inhalation aerosol. A number of phospholipids such as DPPG and DOPC are present in approved products in Europe and in the USA.

In the European Union, most lecithin products conform to EC Directive 96/77 EC Lecithin (No. E322). Most are also approved by the World Health Organization as a food additive class under Codex Alimentarius Standard INS 322. The WHO does not place any limit on the oral intake of lecithin.

Monographs for egg lecithin have been established in the US Pharmacopeia NF 31, the Chinese Pharmacopoeia (2010) and in the Japanese Pharmaceutical Excipients (2004).

Chemically modified lecithins sometimes require special labelling. For those enzymatically modified, for example, labels read 'Enzymatically Modified Lecithin'.

Some lecithins available on the market are kosher and halal approved.

Synthetic phospholipids are not mentioned in pharmacopoeias. As with hydrogenated natural phospholipids, they can be found in the Inactive Ingredient Guide of the FDA Center for Drug Evaluation and in the relevant reference documents of FDA and EMA.

References/Further reading

1. Anonymous, 1991. Intranasal insulin formulation reported to be promising. J. Pharm. 247, 17.
2. Cada, D.J., Levien, T., Baker, D.E., 2004. Morphine sulfate extended-release liposome injection. Thomas Land, St. Louis, MO.
3. Cada, D.J., Levien, T., Baker, D.E., 2008. Formulary drug reviews – clevidipine butyrate injectable emulsions. Hosp. Pharm. 43, 903-912.
4. Driscoll, D., 2006. Lipid injectable emulsions: pharmacopeial and safety issues. J. Pharm. Res. 23, 1959-1969.
5. Franckum, J., Ramsay, D., Das, N.G., Das, S.K., 2004. Pluronic lecithin organogel for local delivery of anti-inflammatory drugs. Int. J. Pharm. Compd. 8(2), 101-105.
6. Fricker, G., Kromp, T., Wendel, A., Blume, A., Zirkel, J., Rebmann, H., Setzer, C., Quinkert, R.-O., Martin, F., Müller-Goymann, C., 2010. Phospholipids and lipid-based formulations in oral drug delivery. J. Pharm. Res. 27, 1469-1486.
7. Gambling, D., Hughes, T., Martin, G., Horton, W., Manvelian, G., 2005. A comparison of depodur, a novel, single-dose extended-release epidural morphine, with standard epidural morphine for pain relief after lower abdominal surgery. Anesth. Analg. 100, 1065-1074.
8. Gregory, T.J., Steinberg, K.P., Spragg, R., Gadek, J.E., Hyers, T.M., Longmore, W.J., Moxley, M.A., Cai, G.Z., Hite, R.D., Smith, R.M., Hudson, L.D., Crim, C., Newton, P., Mitchell, B.R., Gold, A.J., 1997. Bovine surfactant therapy for patients with acute respiratory distress syndrome. Am. J. Respir. Crit. Care Med. 155, 1309-1315.
9. Grit, M., Zuidam, N.J., Underberg, W.J., Crommelin, D.J., 1993. Hydrolysis of partially saturated egg phosphatidylcholine in aqueous liposome dispersions and the effect of cholesterol incorporation on hydrolysis kinetics. J. Pharm. Pharmacol. 45, 490-495.
10. Growdon, J.H., Gelenberg, A.J., Doller, J., Hirsch, M.J., Wurtman, R.J., 1978. Lecithin can suppress tardive dyskinesia. N. Engl. J. Med. 298, 1029-1030.
11. Hammad, M.A., Mueller, B.W., 1998. Increasing drug solubility by means of bile salt-phosphatidylcholine-based mixed micelles. Eur. J. Pharm. Biopharm. 46, 361-367.
12. Hanin, I., Pepeu, G., 1990. Phospholipids: Biochemical, Pharmaceutical, and Analytical Considerations. Plenum Press, New York.
13. Hart, I.R., Fogler, W.E., Poste, G., Fidler, I.J., 1981. Toxicity studies of liposome-encapsulated immunomodulators administered intravenously to dogs and mice. Cancer Immunol Immunother 10, 157-166.
14. Japan Pharmaceutical Excipients Council, 2004. Japanese Pharmaceutical Excipients. Tokyo: Yakuji Nippo, 834-844.
15. Schneider, M., 1992. Achieving purer lecithin. Drug Cosmet. Ind. 150(2), 54, 56, 62, 64, 66, 101-103.
16. Rowe, R.C., Sheskey, P.J., Quinn, M.E., 2009. Handbook of pharmaceutical excipients. Pharmaceutical Press, 6th edition, London.
17. Scherphof, G.L., 1993. Phospholipid metabolism in animal cells. Phospholipids Handbook. New York: Marcel Dekker, 777-800.
18. Shah, P., Bhalodia, D., Shelat, P., 2010. Nanoemulsions: a pharmaceutical review. Syst. Rev. Pharm. 1, 24-32.
19. Vandevanter, D.R., Geller, D.E., 2011. Tobramycin administered by the TOBI® Podhaler® for persons with cystic fibrosis: a review. J. Med. Devices 4, 179-188.
20. Van Hoogesvest, P., Liu, X., Fahr, A., Leigh, M.L.S., 2011. Role of phospholipids in the oral and parenteral delivery of poorly water soluble drugs. J. Drug Delivery Sci. Technol. 21, 5-133.
21. Van Hoogevest, P., Wendel, A., 2014. The use of natural and synthetic phospholipids as pharmaceutical excipients. Eur. J. Lipid Technol. 116, 1088-1107.
22. Vertzoni, M., Markopoulos, C., Symillides, M., Goumas, C., Imanidis, G., Reppas, C., 2012. Luminal lipid phases after administration of a triglyceride solution of danazol in the fed state and their contribution to the flux of danazol across caco-2 cell monolayers. Mol. Pharm. 9, 1189-1198.

8.2 Lecithin: Emulsifying properties and cosmetic applications

Jean-Pierre Arnaud

8.2.1 Introduction

Every chocolate lover will have noticed this component in the list of ingredients of his/her favourite chocolate bar. In pure form, capsules or granules, the range of food supplements that contain lecithin demonstrate its anti-cholesterol and protective benefits for the cardiovascular system. Lecithin therefore has many functional and nutritional properties. Cosmeticians have long valued its properties in cosmetic applications such as dispersion of pigments, hydration in personal care products, and in drug targeting using liposomes etc. This substance, which nature exploits to create emulsions and is in the composition of cell membranes of all living organisms, has countless potential applications and the advantages of a natural image, complete non-toxicity and optimal physiological properties. In 1984, there were over 700 cosmetic products registered in the USA containing lecithin, mainly in make-up (54%), personal care products (25%) and hair products (19%) (1). Since then, its use has increased considerably thanks to more neutral organoleptically refined qualities, the discovery of new applications and a high demand by consumers for natural ingredients. Its properties rely on its amphiphilic nature and therefore its affinity with polar and non-polar phases. It is used in cosmetics for its dispersing properties to increase the intensity of the colours by improved homogeneity of pigments and colourants (2), reducing both viscosity and the flow threshold during moulding of lipsticks, preventing crystallization of fatty materials and hydrating the skin. In hair products, lecithin is used for its conditioning effect, to increase colouration and the resistance of hair dyes to washing while preventing irritation of the scalp.

In this chapter, we will describe the emulsifying properties of this multipurpose ingredient enabling the production of highly sophisticated formulations, fluid emulsions, lamellar phases and organogels for innovative galenic preparations. Thanks to their biomimetic structures, these emulsions possess a high degree of affinity with the skin and can be used to transport active ingredients and increase their efficacy.

8.2.2 Lecithin

This complex natural substance was highlighted by N.T. Gobley (3) in 1874 as a fatty phosphorus substance in egg yolk and brain tissues. Lecithin as well as partially hydrolysed lecithins have been used in food products since the start of the 20th century and meet the legal definition of a food ingredient, under the denomination of the international system SIN 322 (i) and 322 (ii), mentioning as a determining parameter the content in substances insoluble in acetone higher than 60% (or 56% for hydrolysed lecithins) (4). This fraction, essentially composed of phospholipids and glycolipids, represents the "active" part of lecithin, which gives it its various properties. Lecithin is therefore a polar fatty material. This definition includes a variety of lecithins the content and polar lipid composition of which determine its properties. Phospholipids and glycolipids are essential components of the cell membranes of all living organisms. Phospholipids are arranged there in bilayers to form a flexible and fluid liquid crystal structure, the selective permeability of which regulates exchanges with its environment. Homeostasis of the cells depends on them and demonstrates the importance of these components for optimum functioning of our cutaneous and other cells. For each of these classes of membrane lipids, two different structures exist depending on the nature of the skeleton of the molecule: glycerol and sphingosine are each composed of glycerolipids or sphingolipids. In the plant kingdom, which is of interest to the cosmetics industry, glycerophospholipids and glyceroglycolipids are widely predominant.

8.2.2.1 Origin

Although historically extracted from egg yolk, egg lecithin is almost exclusively reserved for pharmaceutical applications to disperse large quantities of oils into injectable solutions in complete safety. These emulsions constitute a technical feat and show the relevance of these lipids since they resist sterilization and their stability is retained for several years. Cerebral sources of egg lecithin were abandoned for reasons we can only guess at, whereas marine origins such as fish eggs and krill are marketed for food supplements and their richness in omega-3-type fatty acids. However, their odour does not enable cosmetic applications except in very low doses as an active agent, but not as a functional ingredient.

Since the beginning of the 20th century, lecithin has been extracted from soy as a sub-product of oil refining and this is still the main source of lecithin, with more than 100,000 tons per year extracted, used principally in food processing (chocolate-making, margarine, mayonnaise, milk powders etc.). Depending on the extraction protocol, soy lecithin can then be transformed into numerous fractions to obtain different functionalities. Sunflower and rapeseed lecithins exist but in much lower volumes and their colours are darker, which limit their use in cosmetic applications. However, these alternative sources to soy are taking on increasing importance, driving improvement in the extraction and discoloration processes for wider use in food and cosmetics.

8.2.2.2 Composition

Polar lipids of plant origin are comprised mostly of glycerophospholipids and glycolipids in lower quantities. In the first category, phosphatidylcholine (PC), phosphatidylethanolamine (PE), phosphatidylinositol (PI) and phosphatidic acid (PA) are generally considered responsible for the physicochemical behaviour of lecithin (5) (Fig. 1 and 2). Other classes are also present such as sphingolipids but their very low content in the plant sources used in cosmetics and the prohibitive cost of their purification limit their use to that of an active substance, not as an emulsifier.

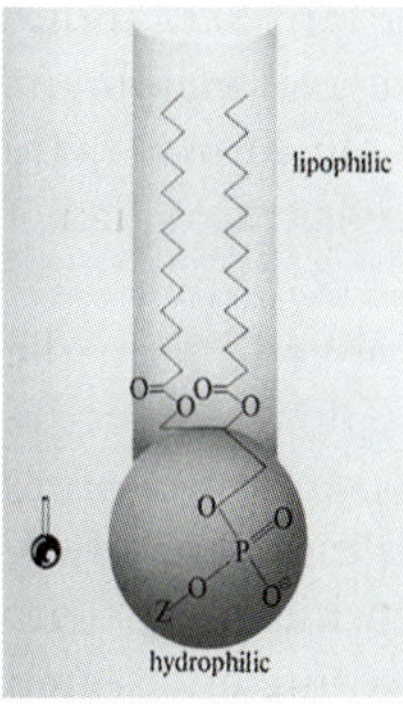

Figure 1.
General structure of a glycerophospholipid.

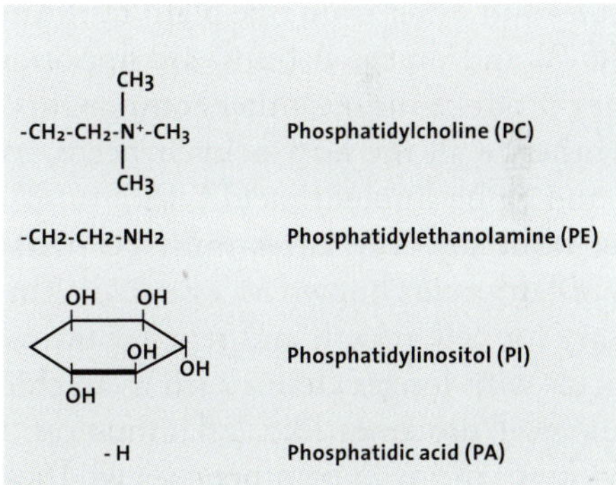

Figure 2.
Nature of the different Z groupings and type of corresponding glycerophospholipid.

Among all the possibilities offered by the choice of source and the process of obtainment, the compositions most frequently used in cosmetics are described in Table 1. Liquid lecithins (1,2) are directly extracted during oil refining. Deoiling enables the concentration of polar lipids (3,4), then fractionation with ethanol enables the concentration of PC (5), which is then isolated by chromatography (6,7).

Table 1.

Typical compositions of different fractions of lecithins currently used in cosmetics.

Description	1 Lecithin %	2 Lecithin %	3 Deoiled lecithin %	4 Deoiled hydro-genated lecithin %	5 PC-enriched lecithin %	6 PC %	7 Hydro-genated PC %
Origin	Soy	Sunflower	Soy	Soy	Soy	Soy	Soy
State	Liquid	Liquid	Powder	Powder	Wax	Wax	Powder
Insolubility in acetone	>60	>60	>97	>97	>97	>97	>97
PC	16	15	24	24	51	>93	>93
PE	13	5	19	19	14	–	–
PI	8	14	14	14	1	–	–
PA	2	1.5	4	4	1	–	–

More stable references to oxidation exist (4,7), obtained after hydrogenation of the fatty acids. Fatty acids are comprised of on average 50% of these molecules and differ from one plant to the next (Table 2). The fatty acids omega-3 (ω-3) and omega-6 (ω-6) are important elements of the cell membrane and they produce various other components in the organism. The human body can synthesize all the fatty acids it needs, except for two: linoleic acid C18:2 (ω-6) and alpha-linolenic acid C18:3 (ω-3), both present in polar lipids originating from soy. The latter must be transported exogenously and is in the group of fatty acids known as 'essential'. These two categories of fatty acids are necessary for cell growth and repair, but may also be used to synthesize other fatty acids with longer chains such as arachidonic acid, which is present in key physiological processes like inflammation. Contrarily to vegetable oils, lecithins possess antioxidant properties which enable them to be easily preserved and protects other materials sensitive to oxidation (6). They are therefore used as an antioxidant in numerous synergic preparations as Fig. 3 shows. Here, lecithin increases the induction time of oxidation of evening primrose oil, therefore delaying its appearance by more than 50%, and has a synergy with α-tocopherol (156%), whereas α-tocopherol alone only increases the induction time by 26%.

Table 2.
Typical composition in fatty acids of different lecithins.

	Deoiled sunflower lecithin %	Deoiled soy lecithin %	Deoiled and hydrogenated soy lecithin %
Fatty acids	50	50	50
Including:			
C16:0	15	20	20
C18:0	3	4	80
C18:1	13	12	
C18:2 (ω-6)	69	57	
C18:3 (ω-3)		7	

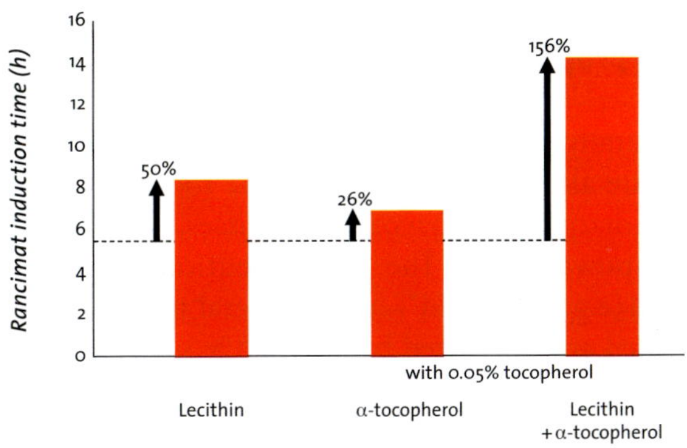

Figure 3.
Reduction of oxidation of evening primrose oil by sunflower lecithin.

To increase the field of application of lecithins, hydrogenation is used to saturate the double bonds of the polyunsaturated fatty acids of plant lecithins (linoleic and linolenic acids) to eliminate the phenomena of oxidation and to obtain a white colour and a neutral odour. Hydrogenated lecithins, by definition, therefore contain mainly saturated fatty acids in C16:0 and C18:0 (Table 2). The saturated fatty chains are aligned, enabling more rigid structures (interfaces, membranes, liposomes etc.) presenting with higher viscosity and lower permeability, which affects the characteristics of the finished product.

8.2.2.3 Safety

The safety of 17 different lecithins of natural origin was studied by the CIR (Cosmetic Ingredient Review) Expert Panel in 2001, then again in March 2015 (1). For a level of use of finished products of up to 15%, they did not show any cutaneous irritation, sensitization, phototoxicity or photosensitization in rinsed and non-rinsed products. Lecithins and hydrogenated lecithins are slightly irritating to the eyes, but they are not mutagenic or toxic for reproduction. Present in the natural state in numerous foods and used as a food ingredient, lecithins present no risk orally. Although protection of the environment is an increasing concern of consumers, lecithins also have the benefit of being very easily biodegraded since more than 98% are biochemically degraded after 28 days according to the test by the OCDE 301D.

8.2.3 Emulsifying properties

Polar lipids are naturally existing emulsifiers and constitute the functional part of lecithins. They are already widely used in the food-processing industry for chocolate-making, margarines, mayonnaise, sauces, reconstituted milks, milk and instant cocoa powders etc. The pharmaceutical industry has developed injectable solutions (e.g. Intralipid®) for parenteral nutrition where up to 30% oil may be emulsified and stabilized by only 0.5% soy lecithin or egg lecithin. These emulsions prepared under high pressure resist sterilization and have a stability of several months. The principal benefits of lecithin-based emulsions are:

- Natural composition obtained by extraction from field crops, without chemical transformation or only slightly transformed (hydrogenated lecithin), in comparison with conventional emulsifiers obtained by chemical synthesis.
- They are very safe, which is confirmed by numerous studies.
- They possess physiological benefits such as an increase in hydration, prevention of dehydration and reduction in inflammation.
- They have a unique fresh feel, very soft and non-greasy.
- Increased bioavailability of the active substances present in cosmetic emulsions.

Phosphoric esters of the phospholipids or the carbohydrate grouping of the glycolipids constitute the hydrophilic part of these molecules and will be focused on the more polar medium, and in particular water, whereas their chains of fatty acids will be dissolved in the non-polar phase generally comprised of fatty bodies. Fig. 4 gives details of some of the phases formed by phospholipids (polymorphism) encountered with commercially available phospholipids (7).

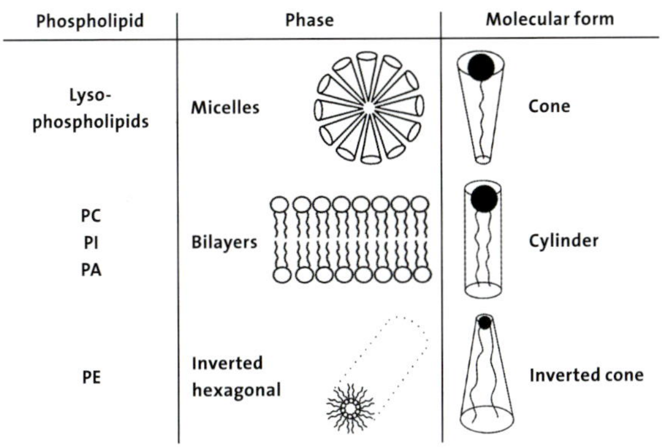

Figure 4.
Structures formed by the different fractions of phospholipids.

Lysophospholipids have a lower non-polar part as they only possess one chain of fatty acids. Their representation is that of a cone and they preferentially form micelles in the presence of water. Phases in bilayers are generally formed by fractions of lipids such as PC, PI or PA above the transition temperature (< 0°C for unsaturated and > 50°C for saturated). Moreover, at neutral pH, as PA and PI are partially dissociated, the introduction of these charged groups is shown by a greater quantity of water between the lamellar phases. The distance between the bilayers of a mixture of phospholipids (124 Å) is higher than the one obtained with pure PC (64 Å), which is expressed by a higher inter-planar distance and a greater incorporation of water: 64% and 50%, respectively (8). PE, with a smaller polar part than that of PC, possesses an inverted conical form organized into inverted hexagonal structures where water is located inside the cylindrical micelles (7). These pure forms are rarely marketed except for PC, which is available industrially. In the native state, the fractions present are equalized in PC, PE and PI. Even in the presence of PE and above 20 to 50% (molar) of lipids such as PC, PI or PA, phospholipids associate in the form of bilayers. This is the case during the use of deoiled commercial lecithin.

Lysophospholipids and PE intercalate to form bilayers (7).

In the case of deoiled commercial lecithin, containing PC, PE, PI and PA (among others), the different forms associated in the presence of water and at ambient temperature are the following:

- Low water content: inverted hexagonal structures.
- More than 20% water: lamellar phases.
- More than 64% water: lamellar phases and very diluted aqueous dispersion of phospholipids.

The formation of these phases varies equally depending on temperature, degree of unsaturation of the fatty chains, ionic strength, pH and the presence of divalent cations (Ca, Mg etc.). The global electrostatic charge of the system also plays an important role in the stability of the systems formed. Although PC and PE are globally neutral at pH 6–7, as PA and PI are partially dissociated and therefore negatively loaded, each molecule supports a load. So, a commercial deoiled lecithin possesses a negatively loaded fraction of 25 to 35% (molar) depending on its context in these loaded lipids (8). Electrostatic repulsions play a more important role in limiting coalescence than the presence of interfacial layers. An interfacial film of at least two monolayers of phospholipids is necessary for stabilization of the droplets as well as the presence of lipids loaded negatively like PI and PA to ensure repulsion of the droplets (8). In the case of commercial deoiled lecithins, the stability of the emulsions is obtained when this film exceeds 80 Å and the size of the globules varies between 3 and 5 μm (8).

Although there is a limited number of studies on the organization of non-hydrogenated phospholipids in the oil – water interface, it seems that they are organized in the form of monomolecular films and that the thermodynamic equilibrium is reached after a relatively long period of 12 to 15 h. The presence of a three-dimensional network has been demonstrated. In the case of PC, the polar heads will be linked to each other and the adsorption layer will be at least one thickness of three molecules. So, hydrated PC will form liquid crystal forms inside the interfacial frontier. This shows that the adsorption of lecithin around the droplets gives the interface a viscoelastic nature. The emulsions will therefore be more stable thanks to the liquid crystal structures (9).

In the case of hydrogenated lecithins, the phospholipidic composition is similar to deoiled unsaturated lecithins but the fatty acids are saturated and their fatty chains are not in a fluid state below 53–55°C. The phases formed are thus in a "gel" state and not liquid crystal as previously described. The effect of this jelled state on the structure and stability of the emulsions is not described but certainly plays an important role via a contribution of rigidity and viscosity.

Phospholipids are mild surfactants which after hydration do not form micelles like strong detergents but rather lamellar structures. They do not therefore form spontaneously oil – water emulsions and require an energy supply in the form of intense agitation or high-pressure homogenization to obtain stable emulsions (10).

Lecithins can be used alone but more generally in combination with other emulsifiers. From an operating mode point of view, powder (deoiled) and wax (fractionated) lecithins are dispersed in the aqueous phase under agitation. Liquid lecithins, containing a proportion of oil, are simply dissolved in the fatty phase. Certain scientific works suggest that an incorporation in oil will facilitate migration of the phospholipids to the interface of the emulsions and will minimize the necessary quantity of emulsifier (8).

8.2.3.1 Fluid emulsions

At very low concentrations, certain phospholipids like pure PC, under high agitation and in combination with other emulsifiers, form micelles which can disperse small quantities of oils or slightly polar components. Pure PC alone does not enable the formation of stable emulsions and requires the addition of other loaded phospholipids or other emulsifiers. These suspensions are not completely transparent but are opalescent and present a very soft feel (Table 3). Fluid emulsions are generally realized by combinations of lecithins and of co-emulsifiers presenting a higher polarity similar to that of the aqueous phase (Table 4). By increasing the fatty phase level, these systems become milky and this type of association enables the stabilization of up to 10% oil.

Table 3.
Opalescent micellar purified PC-based solution.

Ingredients	%
Water	Qsp 100
Sodium phytate*	0.1
Preservatives	0.3
Glycerin	2.0
Polyglyceryl-10 laurate	2.0
PC	0.25
Tocopherol	0.02

* sequestering agent

Table 4.
Perfume-based and hydrogenated PC-based fluid emulsion.

Ingredients	%
Water	Up to 100
Sodium phytate	0.1
Preservatives	0.3
Glycerin	2.0
Polyglyceryl-10 laurate	2.0
Hydrogenated PC	2.0
Perfume	5.0

8.2.3.2 Lamellar emulsions

Most cosmetic creams are produced thanks to a single layer of surfactants at high hydrophilic-lipophilic balance (HLB) in the oil – water interface which may destructure the *stratum corneum.* An alternative approach is the combination of phospholipids, acids and fatty alcohols in long chains which enable firstly the creation of a stabilizing layer separating droplets of fatty materials, then forming a very resistant and high-viscosity three-dimensional network stabilizing the droplets of oil and therefore the formulations produced. The systems thus formed are lamellar emulsions, the layers of which are highlighted in electron microscopy (Fig. 5).

Figure 5.
Electron microscopy with a lamellar gel (Cryo-TEM).

Liquid crystal forms may also be visible in optic microscopy with polarized light forming a Maltese cross pattern. These emulsions are stabilized by the establishment of lamellar phases in continuous phases and around the droplets. This gives them high viscosity and a gel-type texture without jellifying/thickening or co-emulsifying. A rheological study of these demonstrates the presence of a marked flow threshold and predominant elastic behaviour (Fig. 6). These lamellar emulsions have a very similar structure to that of the lipids of the *stratum corneum* and cover it with a kind of second skin which progressively transports the water and the active substances that it contains (Fig. 7).

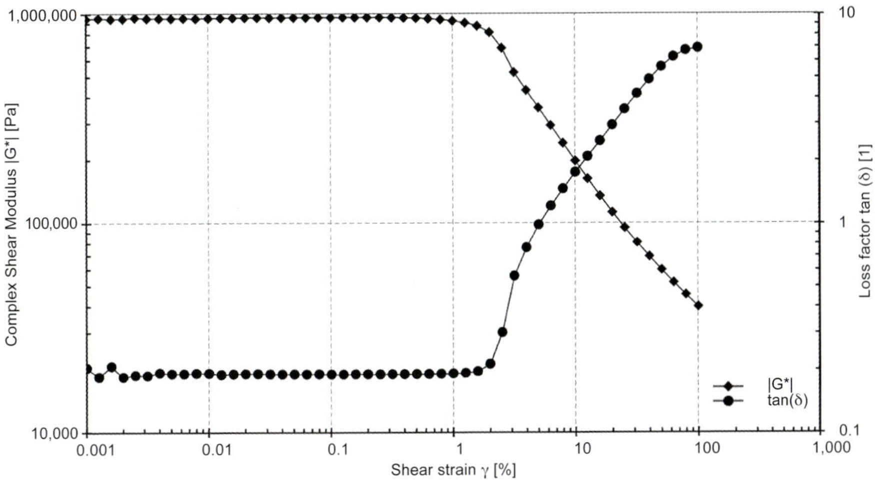

Figure 6.

Amplitude scanning with shear deformation at 25°C, on an emulsion with 5% emulsifier, 10% fatty phase (C8/C10 triglycerides) and 1% preservative.

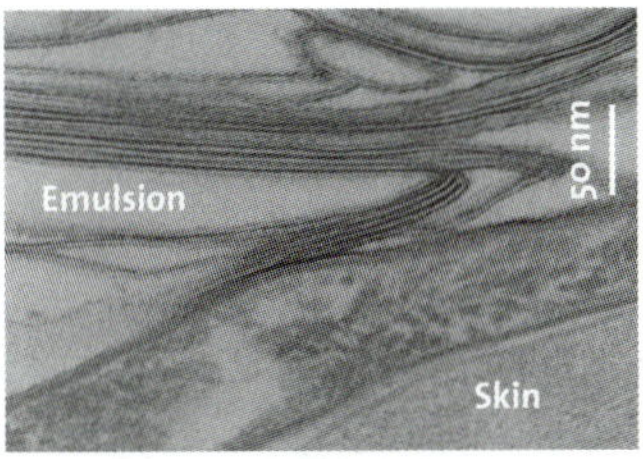

Figure 7.

Electron microscopy with a lamellar emulsion on the surface of an explant of skin.

Thanks to this network, 'emulsions' with no fatty phase other than phospholipids can be obtained with a very gentle feel linked to an increase in the cohesion of the corneocytes. Conversely, very rich formulas containing up to 20% vegetable butters enable very rapid absorption of the fatty phase and give the product a very pleasant feel. The sensory characteristics of these lamellar emulsions have been demonstrated with panels of 24 experts. The principal sensory descriptors of these emulsions are their fresh effect, their smoothness, the absence of stickiness and their ultra-sophisticated soft feel on application (11).

The hydrating properties of lecithin are often reported (8) and are due both to the essential fatty acids which comprise it (12) as well as its ability to bind water on these polar groupings or in the form of the different structures that it may form, such as lamellar phases or liposomes. These lamellar structures form a barrier to dehydration, reducing transepidermal water loss (TEWL) by 23%, 8 hours after application (11) (Fig. 8).

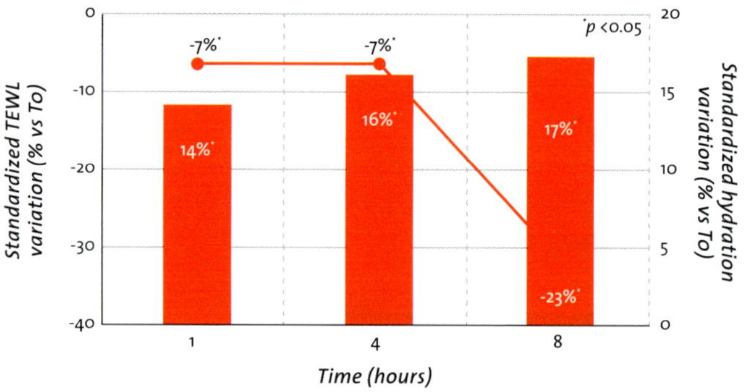

Figure 8.

Transepidermal water loss and hydration of the skin using a lamellar gel of phospholipids and alcohols/fatty acids (ten volunteers).

These emulsions also constitute a form of targeting system enabling the transportation of active molecules through the *stratum corneum* to reach the viable cells of the epidermis. It has been demonstrated using Franz's cells and explants of human skin that such emulsions containing model molecules of different molar masses and distribution coefficients, caffeine or a hexapeptide, merge with the *stratum corneum* and penetrate the skin gradually while respecting its integrity. In these experiments, percutaneous absorption of caffeine, an amphiphilic molecule, was 2.3 times greater in the control aqueous solution after 24 h exposure while being absorbed 1.9 times more rapidly. The same caffeine concentration in the receptor medium was therefore obtained after only 12 h with the phospholipidic emulsion, whereas 24 h was necessary for the control solution (13). The bioavailability of hexapeptide, a molecule of high molar mass (870 Da) and low distribution coefficient (-1.13), was also increased by around 8.5 times. It was therefore concluded that improved penetration thanks to phospholipids is the result of two mechanisms. First, a modification in the solubility of the "vectorized" molecules which increases the distribution coefficient between the skin and the formula. The second mechanism is linked to modification of the lipidic intercellular mechanism increasing diffusivity of the active agents into the skin (13).

It should also be noted that an emulsion applied to the skin will gradually lose its water. During this dehydration, the lecithin content in the water increases, gradually forming lamellar structures halting evaporation (14). Finally, only reversed micelles are present in the oil when all the water is evaporated. With an initial emulsion containing 5% natural emulsifier (lecithin and alcohols/fatty acids), 20% oil and 75% water, the residual product on the skin after evaporation will contain 20% natural emulsifier in oil. This composition plays a determining role in hydration of the skin as it can capture the water contained in the ambient atmosphere and forms once more a liquid crystal structure re-establishing hydration and the shiny appearance of the skin, whereas molecules like glycerol cannot produce the same effect (14).

8.2.3.3 Lecithin organogels

The organogels were observed by the addition of small quantities of water into native lecithin resulting in a drastic increase in viscosity. They were then studied for the transdermal transport of lipophilic molecules (15). These oily gels are transparent isotropic gels comprised of an non-polar phase, a surfactant (lecithin) and a polar phase such as water, glycerin and some glycols. Typically, the addition of very low quantities of water transforms dispersion of lecithin in an non-polar medium such as oil in a viscoelastic gel. Initially, the molecules of phospholipids are dispersed at random in the polar phase and incorporation of the polar phase causes the formation of inverse micelles of lecithin. The molecules of water bind to the polar heads of the phospholipids and thus form bridges between adjacent phospholipids. Reverse spherical micelles are formed for low water content. The addition of water causes the formation of inverse cylindrical micelles; these tubular flexible micelles will combine and form a three-dimensional viscoelastic and thermoreversible network. Transparency and stability depend widely on the water content. The gradual addition of water causes a very rapid increase in viscosity which goes up to a maximum then decreases when excess water is added, finally leading to separation in two phases (16).

Lipophilic, hydrophilic and amphiphilic active substances can be incorporated into them. In the near-absence of water, certain unstable active substances present improved stability (ascorbic acid). In particular, they represent a new oily jellified galenic form which immediately loses its viscosity by the addition of excess water.

All the difficulty resides in the formulation of an acceptable product since the possibilities of modifications of the following formulas are quite slim. By using a large proportion of oil and purified PC, jellification is obtained with an extremely low water content. The water must be added drop by drop so as not to exceed the maximum viscosity, which would cause near-immediate fluidification (Table 5). Very different formulas can be obtained using hydrogenated PC. In this case, the aqueous fraction must be replaced by a mixture of glycerol and butylene glycol, which are less polar. This type of formula may contain active substances unstable in the presence of water such as vitamin C, without causing their deterioration. As the quantity of polar phase is much greater than in the previous case in Table 5, this type of gel is easier to produce and retains its viscosity/transparency on a wider range of polar phases (Table 6). These formulations are cold-formed in the case of unsaturated lecithins or hot-formed for hydrogenated lecithins and under agitation in both cases.

Table 5.

Organogel obtained with soy PC (unsaturated).

Ingredients	%
PC	10.00
Dicaprylyl ether	89.90
Tocopherol	0.05
Water	0.05

Table 6.

Organogel obtained with PC (hydrogenated).

Ingredients	%
Hydrogenated PC	0.50
Glycerin	35.00
Butylene glycol	34.50
Dimethicone	30.00

8.2.4 Conclusion

Phospholipids are multipurpose ingredients and each formulator will choose them for their benefits depending on the nature of its products or their claims. Naturalness, feel, tolerance, efficacy and hydration are some of their benefits. New plant origins are appearing and original fractions are tested to increase functionality and efficacy. Very cutting-edge works of characterization and formulation have been conducted in the last few years, resulting in innovative and stable formulas. The increased demand each year for more active cosmetic products and the capacity of phospholipid-based emulsifiers to transport active agents into the epidermis is a major innovation in the domain of emulsions. This should result in the generalized use of these ingredients, which are highly beneficial for the cosmetics industry.

References

1. Cosmetic Ingredient Review, 2015. Safety assessment of lecithin and other phosphoglycerides as used in cosmetics, January 9.
2. Lefur, A., Bezivin, C., Lepais, M., Cottin, P., 2008. Phospholipids and pigment dispersions. Parfums, cosmétiques, actualités 202, 75-81.
3. Gobley, N.T., 1874. Sur la lécithine et la cérébrine. J. Pharm. Chem. 20, 98-103, 161-166.
4. Codex Alimentarius, 1989. Class names and the international numbering system for food additives, CAC/GL 36, Adopted in 1989, Revision 2008, Amendment 2016.
5. Schneider, M., 2008. Major sources, composition and processing. In: Gunstone, F.D. (Ed.), Phospholipid Technology and Applications. The Oily Press, Bridgwater, 21-40.
6. Belhaj, N., Arab-Tehrany, E., Linder, M., 2010. Oxidative kinetics of salmon oil in bulk and in nanoemulsion stabilized by marine lecithin. Process. Biochem. 45, 187-195.
7. Li, J., Wang, X., Zhang, T., Wang, C., Huang, Z., Luo, X., Deng, Y., 2015. A review on phospholipids and their main applications in drug delivery systems. Asian J. Pharm. Sci. 10, 81-98.
8. Rydhag, L., Witton, I., 1981. The function of phospholipids of soybean lecithin in emulsions. J. Am. Oil Chem. Soc. 58, 830-837.
9. Pichot, R., Watson, R.L., Norton, I.T., 2013. Phospholipids at the interface: Current trends and challenges. Int. J. Mol. Sci. 14, 11767-11794.
10. Van Hoogevest, P., Prusseit, B., Wajda, R., 2013. Phospholipids: Natural functional ingredients and actives for cosmetic products. SÖFW J. 139(8), 9-14.
11. Belhaj, N., Borel, M., Bezivin, C., 2014. Phospholipid-based emulsifiers give much more. Personal Care, September, 95-98.
12. Prottey, C., 1977. Investigations of function of essential fatty acids in the skin. Brit. J. Dermatol. 97, 29-38.
13. Belhaj, N., Ollier, V., Bezivin, C., Arnaud, J.-P., Loing, E., 2014. Skin delivery: Phospholipid lamellar system for the improvement of bioavailability of active molecules. IFSCC Magazine, 28th Congress, 3431-3439.
14. Friberg, S.E., 1990. Micelles, microemulsions, liquid crystals, and the structure of stratum corneum lipids. J. Soc. Cosmet. Chem. 41, 155-171.
15. Luisi, P.L., Scartazzini, R., Haering, G., Schurtenberger, P., 1990. Organogels from water-in-oil microemulsions. J. Colloid Polym. Sci. 268, 356-374.
16. Raut, S., Bhadoriya, S.S., Uplanchiwar, V., Mishra, V., Gahane, A., Jain, S.K., 2012. Lecithin organogel: A unique micellar system for the delivery of bioactive agents in the treatment of skin aging. Acta Pharm. Sinica B 2(1), 8-15.

8.3 Lecithins in the technical industries

Rüdiger Ziegelitz

8.3.1 Introduction

Besides describing the basic functional properties of lecithin, this chapter of the book deals with the prospects for its use in the different sectors of the chemical and engineering industries.

Because of the breadth of the topic, much of the relevant information is scattered over the various specialist fields. There are a great many sources in the form of patents and the literature. It was this abundance that gave rise to the idea of a streamlined contribution limited to explaining the basics.

Lecithins unfold their numerous attributes for the benefit of process engineering, production and cost reduction. Moreover, their natural origin makes the readily biodegradable lecithins ideal auxiliary agents permitting compliance with ecological requirements. For decades, the active constituents of lecithin have been used by various industries in the manufacture of technical products. Lecithins come onto the market in different forms: liquid, solid, powdered, fractionated, modified or compounded.

8.3.2 Properties and functions

Because of their chemical structure, lecithins can be classified among the amphiphilic molecules, i.e. they have a hydrophilic, polar head group and a non-polar, lipophilic part made up of hydrocarbon groups. They are therefore interface- and surface-active compounds. Thanks to their molecular structure they assemble themselves at the boundary between two phases and reduce the interfacial tension. In doing so they arrange themselves at the interfaces in such a way that the hydrophilic head group faces the phase with the stronger forces of attraction.

Surface-active substances function as emulsifiers, dispersants and wetting agents, foaming and defoaming agents, anti-static additives, flotation auxiliaries and solubilizers.

Emulsions

The characteristics of an emulsifier depend on whether the hydrophilic or the lipophilic part of the molecule is dominant. If the lipophilic part predominates, the emulsifier forms water-in-oil (W/O) emulsions. Water is the dispersed phase, and the oil is outside. If the lipophilic part predominates, the emulsifier produces oil-in-water (O/W) emulsions, the oil being the dispersed phase and the water outside.

What matters in practice is which emulsifier is most suitable for producing the desired emulsion. In the HLB system after Griffin, the surfactant substances are allocated numeric values from 1 to 20. The abbreviation 'HLB' stands for "hydrophilic-lipophilic balance" and gives information on the size and strength of the lipophilic and hydrophilic groups in an amphiphilic substance. Lipophilic compounds are given a low number, and hydrophilic emulsifiers a higher number.

Substances with HLB values up to 4 are typical W/O emulsifiers. Those with HLB values greater than 6 form O/W emulsions. Native lecithin is a W/O emulsifier, deoiled lecithin an O/W emulsifier. Lecithins modified by acetylation or hydroxylation have HLB values from 9–10, enzymatically hydrolysed lecithins (lysolecithins) 8–9.

Hydrolysed fractions like lysophosphatidylcholine, lysophosphatidylethanolamine and lysophosphatidylinositol have HLB values from 9–10, 8–8.5 and 10–11, respectively (see also Chapter 4.1).

Dispersions

The term dispersion describes the distribution of solid substances in liquids. The dispersant must have an effect very similar to that of an emulsifier. It has to wet the substance to be dispersed and must itself be soluble or at least readily dispersible in the liquid.

As a rule, the more hydrophobic lecithins lend themselves better to dispersing solids in oily liquids, while the more hydrophilic lecithins are more suitable for dispersing solids in aqueous or alcoholic liquids.

Dispersed substances with a hydrophobic surface interact with the oleophilic long-chain fatty acids of the lecithin, while dispersed substances with a hydrophilic surface interact with the hydrophilic head groups.

Lecithins are widely used as dispersants, especially for dispersing and stabilizing pigments in paints (whether water-based or solvent-based), and also for carbon papers, typewriter ribbons, putty, rubbers and plastics.

pholipids form complexes with proteins. The solubility and emulsifying and crystallizing behaviour of the proteins is modified, and this is especially interesting for cosmetics. Lecithin is also known to form complexes with carbohydrates and polysaccharides (glucose, dextrins, starch and other hydrocolloids). This complex formation obviously takes place via hydrogen bridges. Inclusion into the phospholipid micelles may play a role, too. Such complex formations or inclusions also help to explain the numerous influences of lecithin observed on the viscosity of genuine or colloidal solutions (e.g. glucose, starch or carboxymethyl cellulose). Lecithin can reduce the viscosity of W/O emulsions made, for example, with simple polyethylene glycol (PEG), sugar or glycerol esters as emulsifiers.

Antioxidant properties

The ability of lecithin to form complexes with metals, metal oxides and metal salts is closely connected with its ability to act as an antioxidant. With phospholipids, complexes and antioxidant properties go hand in hand.

Of course its original tocopherol content already gives lecithin antioxidant qualities. But in the deoiling process the tocopherol is removed, thus weakening the antioxidant effect.

The greatest practical significance of the antioxidant properties of lecithin lies in its synergistic effect as a co-antioxidant, for example with phenolic antioxidants, ascorbic acid, tocopherol and quercetin. Lecithin can be used as an antioxidant in plastics and rubber to delay ageing, in foods (fats, oils) and in animal feed to prevent rancidification, in petrol and other fuels to prevent gumming, in transformer and turbine oil to counteract sludge formation, in paints and coating materials, for car-body underfloor protection, and in lubricating oils, cosmetic formulations and vitamin preparations.

Properties as a biocide

For a long time the bactericidal and fungicidal properties of lecithin were discussed and relativized as a secondary effect of the wetting function. It is now certain that lecithin itself has such properties. They can be explained by the configuration phosphoric acid – phosphoric acid ester – amine in the lecithin molecule. Phosphoric acid and its esters and amines are known to be active fungicides. Lecithins are therefore widely used in agrochemicals.

Secondary functions

Lecithins are also used as moisturizers and viscosity regulators, not least because of their pronounced ability to bond with carbohydrates, polysaccharides, starch or CMC (inclusion compounds, complexes) or due to their hydrophilic/lipophilic relationships, these being ambivalent and varying in degree. Or simply as wetting agents. The softening properties of lecithins can also influence viscosities.

8.3.3 Applications

8.3.3.1 Soaps

Foam stabilization, refatting/superfatting, strengthening of the dispersing and emulsifying effect even in hard water, and antioxidation are the attributes that can be achieved in soaps by adding lecithin. As a replacement for synthetic wetting or refatting agents or antioxidants, lecithin enables soap manufacturers to follow the trend towards natural substances.

In the 20th century it was already discovered that phospholipids lower the surface tension of soap solutions. Although this reduces the height of the foam, it greatly improves its stability. The foam has finer bubbles and is denser – a valuable attribute for shaving soaps and also shaving creams, which are much more significant in terms of quantity. These are saponified vegetable or animal oils which have a strong defatting effect on the skin if used without additives. As a natural superfatting agent, lecithin closes a gap in the market for the usual synthetic refatting agents like carboxylic acid alkylolamides. Being a natural phospholipid, lecithin is also an important constituent of the epidermis and makes the skin feel supple and smooth. In soaps containing a large proportion of fat, lecithin is also a useful antioxidant, especially when autoxidation is catalysed by an active metal. In this case, certain complex-forming properties of the lecithin become apparent. The addition of 2–5% lecithin, preferably acetylated and hydroxylated lecithins, is very effective.

8.3.3.2 Paints, varnishes and printing inks

A large proportion of the lecithin used in the non-food sector finds its way into the paint and varnish industry. The reason is that because of its numerous functions, lecithin, as an inexpensive and effective additive, has kept pace with the meteoric developments in this industry from the classic solvent-based paints and varnishes to dispersion and emulsion paints, water and powder paints to radiation-hardened paints and lacquers.

Since the introduction of water-based paint systems, refined, fractionated, modified and compounded lecithin types have become more and more important. The early 1930s saw the start of basic investigations into the attributes of lecithin as a grinding aid and for facilitating the dispersion and wetting of pigments. These investigations were continued extensively in the 1940s and 1950s.

The development of lecithin as a valuable additive for paints and varnishes is inseparably linked with the name Max Kronstein (American Lecithin). Kronstein demonstrated the benefits of lecithin in numerous scientific studies, publications and patents and paved the way for its use in the industry. He also proved that processed lecithins (refined, modified, fractionated) are among the best wetting and dispersing agents.

Solvent-based paints and varnishes, pigmented and clear
These include oil paints, chlorinated rubber paints, epoxy paints and coatings, alkyd resin varnishes, nitrocellulose and compound nitrocellulose lacquers, and polyester and polyurethane paints.

In pigmented paints, especially, lecithin has practical significance as a wetting agent and for dispersing and stabilizing the pigments (1).

- Lecithin wets and coats the pigments and stabilizes them against precipitation.
- Lecithin facilitates the stirring up of heavily precipitated pigments.
- As a grinding aid, lecithin shortens grinding and mixing times and reduces the energy input.
- The pigments can be dispersed faster and more thoroughly.
- Lecithin influences the viscosity, and
- Permits an increased pigment content.
- Lecithin improves the spreading properties.
- The better dispersion of the pigment brought about by lecithin gives the paints enhanced gloss.
- Lecithin helps to prevent corrosion.

All these benefits are especially obvious in metal oxide pigments, particularly iron oxide, zinc oxide, tin oxide, molybdenum oxide and lead oxide pigments. It is also possible to treat the metal oxides with lecithin separately, remove the excess lecithin and pulverize them. Pigments treated in this way are equally suitable for both solvent-based and water-based paints, and in dispersion paints they offer protection against corrosion as well as good dispersibility (2).

The influence of lecithin on viscosity is easily demonstrated. When as little as 0.5 part spoon of lecithin was added to a mixture of:

- 45 parts lithopone
- 5 parts zinc oxide
- 38.5 parts barium sulphate
- 11 parts linseed oil,

the mixture was at once wetted and became homogeneous and liquid.

Similar results were achieved with carbon and ochre. The amount of lecithin required depends on the total surface area of the pigments and should be enough to form a monomolecular layer. The amounts added lie between 0.01 and 6%. As a rule, 2–3%, based on the pigment, is enough. Heavy pigments may necessitate larger amounts of lecithin. In the case of non-metallic pigments, e.g. carbon black, the usage level is 1–2%. The lecithin is best added at the grinding stage.

Paints containing more than one pigment have a tendency to precipitate unevenly, according to the density and size of the pigments. This does not only apply during storage; it continues during the drying process, as long as the paint layer is still fluid.

Lecithin counteracts this undesirable phenomenon and promotes the homogeneity of the whole system by enveloping the pigments uniformly. The speed at which the different pigments settle is largely harmonized.

- Floating is suppressed.
- Colour nuances are maintained.

Bleached lecithins are usually a good choice, but hydroxylated and acetylated lecithin types can be used, too.

The lecithin should preferably be added before grinding. The properties of lecithin are especially valuable in quality tints containing a white basic pigment and the colour pigment. Treating both with lecithin helps to prevent floating. These properties can also be used for thick-layer paints.

Lecithin is particularly effective and useful in *anti-corrosive formulations.*

It is capable of forming complexes with metal oxides, especially zinc oxide, but also with iron, lead and tin oxides. The reactions take place with the nitrogen-containing bases bound to phosphoric acid.

Complexes of this kind are highly active anti-corrosion agents. Deoiled lecithin types are better complex formers than those containing oil.

An interesting phenomenon was discovered by Joseph Eichberg, namely that the anti-corrosive properties of lecithin can be enhanced greatly by preheating for one hour to temperatures between 130 and 165°C (3). After 42 days in the humidity chamber, steel strips coated with lecithin pretreated in this way showed only 1% corrosion.

The wetting and spreading effect of lecithins also leads us to expect promising results in *clear varnishes,* for example for furniture wood with nitrocellulose sealers that has to be capable of quick stacking and sanding. Very small additions of lecithin (0.1%) can be very helpful.

The increasing need to comply with environmental regulations has led to the development of 'high-solid' systems, i.e. paints (and varnishes) with a high solid content and reduced amounts of solvents. Formulations suitable for a high dry matter content (50–80%) are difficult to achieve. Lecithins help to increase the proportion of pigment and reduce the problems of precipitation, separation, mixing times and flowability. Usage levels are correspondingly higher – between 1 and 10%, based on the pigment. Inexpensive native lecithins achieve the desired effect, although deoiled, chemically modified or even compounded lecithin products may be more suitable in special cases.

Polymers dispersed in water or emulsified (dispersion paints) have their market in coating absorbent surfaces such as masonry, concrete, aerated concrete blocks, bricks, and also wood.

Polymers typically used for this purpose are natural and synthetic latices (styrene-butadiene), polyvinyl acetate, polyvinyl propionate, and acrylate resins with particle sizes from 0.05 to 5 µm. These are also known as latex and dispersion paints.

Although water as a solvent is very much in the "away from organic solvents" trend, it raises new technical problems that demand new solutions for its applications.

Pigment dispersion, emulsion stability and spreading make greater demands on wetting agents and emulsifiers in an aqueous phase. Moreover, there are problems like corrosion resistance, foaming, behaviour when applied with a roller, and viscosity. In this context lecithin acts as a wetting agent and co-dispersant, stabilizes the pigment dispersion and facilitates spreading. It improves the evenness of the coating and prevents a milky appearance. Lecithin also gives the paint anti-corrosive properties, facilitates roller or brush application by adjusting the viscosity, counteracts wrinkling when a fresh layer of paint is applied, and minimizes foam.

Finally, in *water-based paints and varnishes,* lecithin offers all the advantages it already has in solvent-containing paints, these being:

- Better dispersion of the pigments.
- Reduction of grinding times by about 40% (less energy input).
- Reliable uniformity and stability of colour (also when several pigments are used).
- Greater covering power, especially with darker colours.
- Lecithin reduces drop formation and cracking. There is less tendency for the paint to run off.
- Pigments like zinc oxide, iron oxides and red lead that cannot be dispersed in water, or only with difficulty, can be made more readily dispersible with lecithin (e.g. in a ball mill at room temperature).

For traditional reasons ordinary, very cheap lecithin was initially used for this purpose, at a usage level of about 1% of the whole system. This may produce quite satisfactory results, depending on the application and formulation. However, better multifunctionality can be achieved with lecithin types that disperse spontaneously in water. These are the chemically modified types, especially acetylated and hydroxylated lecithins. Blends with non-ionogenic wetting agents have proved satisfactory, too. Purified and fractionated lecithin types offer advantages in respect of their applications. Usage levels are around 1%, and they are added before grinding. Kronstein made a very close study of this problem.

Water-based paints and varnishes (water paints) have undergone meteoric development. The reasons for this accelerated development are mainly ecological rather than qualitative.

Water paints contain binding agents that can be diluted with water. These include:

- Polycarboxylic acids. These are neutralized with ammonia or amines that evaporate with the water during drying.
- Maleic acid adducts to unsaturated fatty acid esters.
- Acrylic or methacrylic acid copolymers.
- Basic epoxy resins combined with polyamines, neutralized with low-molecular acids and phenolic resins.

In many cases the water is not added until immediately before use. Water paints have acquired major industrial significance for the electrophoretic coating of automobiles. In this anodic or cathodic process, the object is suspended in the paint bath; when an electric field of 50–400 V is applied, the paint is deposited on the surface of the object.

According to a Japanese patent, the spray chambers are treated with a wax formulation containing lecithin.

The addition of lecithin to the paint bath makes it possible to circulate the water. Paint residues can be eliminated from the water and blockages prevented.

Water paints are very susceptible to disturbance by foam. Assistance is also needed with pigment dispersion. Lecithin is able to assist in both aspects. Deoiled or modified types (acetylated, hydroxylated, hydrogenated) are recommended at different usage levels, as appropriate. As a non-foaming wetting agent, additions of 0.1–0.5% are sufficient, possibly together with other non-ionogenic wetting agents.

Radiation-hardened paints need an oligomer, e.g. acrylate, as a binder and a monomer, e.g. tripropylene glycol diacrylate, as a solvent; the latter polymerizes together with the oligomers when exposed to UV radiation, under the influence of photoinitiators (benzophenone). The biggest market for radiation-hardened paints is in the timber industry, for painting and varnishing furniture. In Germany, the classic nitrocellulose lacquers used for this in the past have now been largely replaced.

So far, radiation-hardened lacquers have always been unpigmented. But new photoinitiators are now opening up UV hardening for pigmented paints, too.

For this, dispersants and stabilizers are needed. Lecithin combined with non-ionogenic wetting agents is an obvious choice. Water-dilutable, radiation-hardened UV paints are only just emerging, but lecithin certainly promises interesting results in the development of new formulations. And not least as natural products, lecithins should be the additives of choice for this group of environment-friendly paints.

Printing inks may be liquid or in powder or paste form. They are applied by means of printing presses. In conventional printing inks the binding agents and colourants or pigments are dispersed or dissolved in organic solvents. As with the oil-based paints, aqueous printing inks are becoming more and more prominent.

Lecithins can give printing inks important practical characteristics:

- Better oil retention capacity.
- Adjustment of the rheological properties.
- Control of the flow properties through viscosity regulation (4).
- Anti-blocking properties.
- Greater colour intensity.

These useful characteristics can be attributed to lecithin as a:

- Wetting agent.
- Releasing agent.
- Lubricant.
- Dispersant.
- Emulsifier.
- Photosensitive agent.

An interesting investigation into the effect of lecithins on the rheological properties of carbon dispersions in mineral oils for the production of printing inks was carried out by Bratkowska (5). The degree of oxidation and the fatty acid composition were found to be crucial for the viscosity.

The possible applications of lecithin in printing inks were already described in the 1940s (6), in particular its influence on viscosity. The suggested usage level was 0.5 – 2.0%, based on the pigment.

In a classic printing ink consisting of lampblack and a binding agent, as little as 0.1% lecithin greatly facilitates and improves dispersion of the lampblack and noticeably enhances the intensity of the colour.

In printing inks containing petrolatum, paraffin oil, montan wax etc., lecithin serves as a flow aid and dispersant and improves the oil retention capacity and anti-blocking properties. The suggested usage level is 3–10% lecithin, based on the wax content (6). Being a natural substance, lecithin is the ideal additive for edible printing inks.

Thermographic printing (heat-sensitive transfer inks)
This modern technique is based on the thermal transfer of printing inks. It complements the technique of transfer under pressure (pressure-sensitive transfer inks).

Thermographic printing results in dot-for-dot precise, clear and bright prints without dirty marks and smears. Printing inks of this kind are heat-sensitive and consist of colourants, solvents or plasticizers, fast-melting binding agents with a hot melt base (waxes or thermoplastic organic compounds) and lecithin. They can be applied from a flexible strip of film.

Lecithin acts as a dispersant, but above all it functions as a lubricant and separating agent, preventing adhesion of the printing ink to the thermal head and ensuring smooth running. The usage level is 10–30% lecithin, based on the binding agent.

8.3.3.3 Agrochemicals and fertilizers

The diverse uses of lecithins are especially welcomed by the agricultural sector. Besides the usual outstanding capabilities of the lecithins for wetting, dispersing and emulsifying, a number of other effects come into the foreground as demonstrated by the variety of applications, patents and publications.

- Lecithin as an active fungicide, herbicide etc.
- Lecithin as a synergist for other fungicides, herbicides, nematicides, bactericides, insecticides and pesticides.
- Lecithin as an application-facilitating ingredient for active substances, especially as a spreading and penetrating agent, film former (binder) and adhesion promoter.

It is especially interesting that the addition of lecithin has made it possible to considerably reduce the necessary amounts of the herbicides, fungicides, pesticides and insecticides that are now regarded as so objectionable. This is sometimes achieved more by synergistic effects, sometimes more through action as a wetting or binding agent, and sometimes through both effects together.

Lecithin is suitable for all the forms in which agrochemicals are applied – as liquids, powders or sprays; for all types of plant protection and insect control; for fertilizers, and in products for cleaning.

Being a purely natural substance, lecithin fits well into the ecological philosophy of our time. There are no restrictions on lecithin in the food regulations. In Europe it is approved under the number E322, and in the USA, too, lecithin is not among the agrochemical substances for which tolerances for residues after harvesting must not be exceeded. Lecithin is expressly exempted from these.

Lecithin as an active ingredient

Lecithins have fungicidal, herbicidal and even nematicidal and insect-repellent properties.

Wittcoff reported on the biocidal effects of lecithin as long ago as 1951 (7). For a long time this was doubted and attributed to synergistic effects. The wetting effect of the lecithins doubtless also contributes to an increase in the efficacy of biocidal agents; this has been proved time and again.

It is true that biocidal effects cannot necessarily be deduced from the configuration of lecithin, but they are now substantiated by numerous patents, publications and practical applications.

When the fungicidal efficacy of lecithin is compared to that of fumaric acid and nicotinamide in combating mildew on cucumbers and strawberries, the relative values are 74 for lecithin, 52 for fumaric acid and 50 for nicotinamide.

Even in comparison with captan (related to the phthalimide fungicides), phosphatidic acid showed unusually good results as a fungicide with petunias.

Neudorf GmbH produces a fungicide to counteract mildew in roses and other ornamental plants, and also for cucumbers. It contains 25% lecithin as an active ingredient. This may well be one of the first fully natural fungicides.

The fungicidal and bactericidal properties of lecithin for preventing mildew on cucumbers, citrus melanose disease and rice leaf blight are used in the form of a 2% aqueous lecithin dispersion.

Again and again, lecithin has shown its value in combating mites, and in general as a fungicide.

Wills observed (8) that an aqueous emulsion of lecithin and various oils and fats reduces the shrivelling of apples during cold storage (-1°C).

Lecithin shows nematicidal activity and helps to prevent worm infestation in apples, an interesting aspect that deserves closer investigation. Nowadays especially, when there is widespread opposition to spraying chemicals and at the same time a likely increase in worm infestation, this could be a promising new application for lecithin as a natural, renewable raw material.

As an additive for agrochemicals, lecithin has a considerable influence on the spreading and distribution of active substances, for instance on the surface of plants. It increases contact activity between the substance and the plant and promotes absorption. It enables the active ingredients to penetrate the cells of the plants.

These advantages can be used in all types of agrochemicals such as herbicides, pesticides, fungicides, insecticides and virucides.

There is a demand for self-emulsifying concentrates consisting of lecithin with food acids, alcohols and water, which can be diluted with the appropriate amounts of water on-site with the addition of the active ingredients.

Formulations such as 100 parts lecithin
100 parts ethanol (96%)
100 parts ammonia solution (0.5%)
create a very stable stock dispersion that can be adjusted to a pH of around 7 with hydrochloric acid; this can be mixed with the active ingredient and sprayed as a 1–3% aqueous dispersion.

Lecithin for insect control
Lecithin itself has insecticidal properties. It acts as a synergist, boosting the effect of other insecticides, and as an additive it facilitates application, chiefly because of its wetting properties.

As little as 0.01–0.5% by weight, based on the dry grain, is sufficient for insect control. The lecithin is sprayed as an approximately 10% aqueous solution.

Van Nieuwenhuyzen observed that lecithin functions as an active insect repellent when used against mosquitos and midges (9).

Insect control with lecithin is very simple, cheap and effective in waters where mosquitos breed. The lecithin spreads over the surface of the water, forming a monomolecular layer; the mosquito pupae and larvae are unable to penetrate this layer and therefore die. Lecithin is dissolved in a solvent that is not miscible with water in the ratio of 1 part lecithin to 1–3 parts solvent. Suitable solvents are isoparaffins with high spreadability. Water-insoluble carriers can also be used for this lecithin application.

Lecithin in fertilizers

An interesting application currently opening up for lecithin is in foliar fertilizers, taken in through the leaves. Lecithin ensures an even distribution of the fertilizer over the surface of the leaves, increases the contact activity and speeds up penetration. The main benefit is that lecithin forms a film, which binds the nutrients firmly to the surface of the plants and prevents them from being washed or blown away by rain and wind.

The ability of lecithin to form stable films together with the lipid substances of the leaves is not only important for foliar fertilization. Films of this kind can also be used as protection for a wide range of fruits and vegetables (10).

8.3.3.4 Lecithin in textile chemicals

A highly interesting field from the scientific point of view, and with a wide range of practical applications, is the use of lecithin in textile additives.

Only rarely do we find a field that presents such numerous and diverse challenges as textile additives. The production, processing and finishing of textiles requires additives with all manner of different chemical and physical properties, and the composition of such additives covers the whole range of chemistry.

Depending on the application, the requirements have to do with:

- Friction reduction; releasing and separating action; softening; wetting; dispersion; emulsification/co-emulsification; detergent activity; anti-static protection; biocidal properties; skin compatibility; greasing, and fibre and dye affinity.

- In most cases these requirements overlap. For example, besides reducing friction the preparations must soften the fibres, give them anti-static properties or facilitate dyeing.

The position is similar with coning oils or greasing agents. Besides friction reduction, softening and anti-static protection, the additives must ensure a lubricant effect in order to make the fibres more tear-resistant at the different stages of processing, i.e. spinning, winding, weaving, knitting or milling, or stretching.

Additives with very good wetting properties are needed for washing, mercerizing, bleaching, kier boiling, deoiling and desizing.

Even more diverse are the requirements for subsequent treatment with the finishing agents, that have to give the fabrics lasting properties. These include brightness, colour, printing or gloss, stiffness, weight, crease resistance, anti-static properties, suppleness, body, softness or water resistance. In this case the main attributes required are softening wetting, special emulsifying properties, co-emulsification and anti-static protection. Skin compatibility is important, too. In specific cases, of course, other properties have to be taken into account in the composition of textile additives; these include resistance to oxidation, colour stability, low-temperature resistance, moisture resistance and viscosity.

Lecithin is an ideal basic and auxiliary substance that covers nearly all the required properties.

- Thanks to its molecular structure (phosphoric acid esters and glycerol esters in one molecule), lecithin is a reliable softener.
- Because of its long fatty acid chains, lecithin is a highly effective separating agent and lubricant.
- Due to its polar head groups (amine components), lecithin is a typical wetting agent.
- Lecithins have all the configurations typical of anti-static agents – quaternary ammonia groups in phosphatidylcholine, amines in phosphatidylethanolamine and phosphoric acid esters in phosphatidylinositol in particular.
- As a glycerol ester with long-chain fatty acids, lecithin has pronounced lipid characteristics.
- The hydrophilic head groups and the oleophilic, long fatty acid groups give lecithin powerful emulsifying properties that can cover a wide HLB range through specific modification.

- Being a surface-active substance, lecithin has pronounced levelling properties.

Lecithin is especially useful as a co-emulsifier for stabilizing different emulsions.

It is the chemical diversity of the numerous "active groups" in lecithin that gives rise to the practical benefits of this unique multifunctionality. So like no other additive, lecithin offers possibilities of simplifying formulations. With lecithin, only one or two constituents are often necessary instead of three or four.

8.3.3.5 Leather and fur

Lecithin is an important additive in the production of leather and furs.

Fatliquoring, softening, wetting and emulsifying are the properties of lecithin most important for the processing of hides and furs. The transformation into leather and furs takes place in four stages:

- Beamhouse.
- Tannage.
- Fatliquoring.
- Dressing (usually done at the tannage stage).

Of the numerous processing steps that take place in the beamhouse, the properties of lecithin are most useful for 'soaking'. Soaking is carried out more or less at the start of beamhouse work and serves to free the raw hide from dirt, preservatives, water-soluble proteins and fat. These are leached out, emulsified and saponified. Soaking is facilitated by enzymes. Soaking is an alkaline process. It restores the hide's original degree of swelling.

Wetting agents and emulsifiers intensify the soaking process and speed it up. Wetting agents support the effect of the enzymes, while emulsifiers help to remove proteins, fats and dirt from the hides. As a wetting and emulsifying agent or co-emulsifier, lecithin is a valuable synergist in this process.

In chrome tannage, especially, the input of the tanning materials into the hides or furs is accelerated by the mechanical movement of dressing. The additives used are dressing oils or greases.

The result expected from these additives is thorough wetting combined with softening and fatting. Lecithin promotes the wetting effect of synthetic ethoxylates and makes it possible to reduce the amount of these in the formulation. Classic alum tannage uses egg yolk as a surfactant compound. Similar results can be achieved with vegetable phosphatidylcholine (lecithin).

Chamois leather is produced by thorough and repeated rubbing of certain hides (doeskin, lamb, kid) with greasy fish oils until they are saturated. Lecithin facilitates and accelerates the absorption of the oils by the hides, and as a softener it provides the suppleness chamois leather needs for making gloves, handbags, wash leathers, and also leather shorts and inserts for riding breeches. The products used for this are acetylated or hydroxylated lecithins.

Fatliquoring
The purpose of fatliquoring is to make the leather soft and supple, also to increase the water-repellent properties of the surface and reduce the absorbency of the leather fabric, thus increasing its strength and elasticity. Natural vegetable or animal oils such as fish oils, neat's-foot oil, lanolin, castor oil, olive oil or lecithins can be used as liquoring agents. The fats are applied in the form of an emulsion. Considerable amounts of lecithin are required for fatliquoring.

Lecithin meets several requirements:

- It is itself a lipid.
- It is an emulsifier.
- It is a plasticizer.
- It contains functional groups capable of forming complexes with metals.

Leather liquored with lecithin becomes soft and supple, and less emulsifier is needed.

Lecithin as a sole fatliquoring agent
The lecithin itself must be readily emulsifiable in water. This requirement is achieved by acetylation, hydroxylation or hydrolysation (7).

100 parts of lysolecithin are stirred thoroughly with 200 parts of water at room temperature until the dispersion is homogeneous. Then another 200 parts of water are added and stirred in. The emulsion is resistant to chromium and aluminium salts. Lysolecithin is very economical. It brings good results in nearly all types of leather, for example at quantities of:

- 3–5% for hard box calf.
- 1–3% for soft box calf.
- 3–4% for nappa leather.
- 3–4% for furniture leather.
- 1.5–3% for suede (velour) leather.

The percentages are based on the liquoring bath as a whole.

Lecithin as a constituent of fat liquors

As a constituent of fat liquors, lecithin shows its effect as a co-emulsifier for other emulsifiers such as sulphonated compounds or ethoxylates. Lecithin improves the stability of the emulsion and increases the active content of the liquor bath. Modified lecithins are the most suitable. As an established wetting agent, lecithin is an ideal liquoring agent for shoe sole leather. The latter loses its tannic acid and becomes brittle fairly quickly in use due to mechanical and chemical stress, which means that it has to be liquored very thoroughly several times.

8.3.3.6 Detergents and cleaners

Lecithin must have the following properties:

- Softening.
- More powerful and faster wetting of the dirt through co-wetting properties.
- Better emulsification of the dirt.
- Better dirt-carrying capacity as a co-emulsifier.
- Anti-corrosive properties.
- Refatting action.
- Complex formation.

Intensification of the wetting and emulsifying properties by lecithin is an advantage in practically all types of detergents and cleaners.

The anti-corrosive properties of lecithin are important for certain cleaners (e.g. metal and engine cleaners).

The complex-forming properties of lecithin facilitate stain removal, especially in the case of stains caused by metals.

Refatting properties are desirable in dishwashing liquids, but the softening properties of lecithin are the most important of all. They give fabrics washed with detergents containing lecithin a soft, fleecy feel. This is especially desirable with wool, but not only; cotton, mixed fabrics and synthetics benefit, too.

In conventional products, all these properties of lecithins have to be contributed by entirely synthetic substances such as non-ionic wetting agents and non-ionogenic sulphonates, synthetic complex formers like nitrilotriacetic acid and ethylenediaminetetraacetic acid, epoxides (corrosion protection), fatty acid alkylolamides (refatting) or quaternary bases (softeners).

Lecithin is a purely natural substance. It makes it possible to reduce or sometimes even eliminate the chemical substances that have come to be regarded as dubious. Lecithin offers open-minded experts the opportunity of meeting their customers' demand for ecologically sound products.

The use of lecithin as a natural additive is practised already; it is found in detergents in powder, liquid and paste form, detergents for coloured and delicate fabrics, and special fabric conditioners.

Lecithin gives dishwashing liquids practical benefits:

- It increases the efficacy of anionic and non-ionogenic detergents in removing stubborn residues that have dried or even burnt on (11).
- Lecithin helps the water to drain off. It prevents the formation of water marks and improves the drying performance on glass.
- Lecithin has a refatting effect on the skin and makes it possible to replace the usual superfatting agents (alkylolamide salts) with a truly natural product.

Dishwasher detergents

Dishwasher detergents must work without foaming. For this reason they are usually based on strongly alkaline cleaners and contain only extremely small amounts of emulsifiers and wetting agents, or none at all. Of course wetting and emulsifying properties are a desirable advantage for this application, too. Lecithin, as a foamless wetting agent, would seem the ideal product for this purpose – especially since it helps the water to drain off and counteracts the formation of water marks.

In dishwashers, the appearance of glass can be impaired by the leaching out of minerals and the deposition of silicate material on the surface.

Naturally, the function of lecithin as a co-wetting agent and co-emulsifier can also be used in many other cleaners for speeding up the removal of dirt. Examples include: bathroom cleaners, carpet cleaners, floor cleaners, cleaners for plastics, engine cleaners, furniture polishes, metal cleaners, washing agents for fruits, upholstery cleaners, cleaners for the hobs of cooking stoves.

The intensifying effect of lecithin in detergents and cleaners is substantiated by numerous documentations and patents. This applies in particular to special cleaners such as those used for precision instruments or alloys containing aluminium or zinc.

In non-aqueous cleaners containing solids, lecithin is recommended as a dispersant and viscosity regulator.

Lecithin is a valuable additive for cleaning the hobs of cooking stoves. As a wetting agent it helps to release residues, and being an ambivalent product, it leaves a separating layer that counteracts renewed soiling.

8.3.3.7 Crude oil production – lubricating oils and greases – heating oils – fuels

In crude oil production and processing, the multifunctional properties of lecithin can be used for numerous purposes.

Drilling fluids bring abraded material (drilling mud) to the surface, lubricate the drill pipe and seal the wellbore.

Lecithin greatly reduces the loss of drilling fluid.

As an emulsifier, lecithin in drilling fluids ensures continuous cleansing of the wellbore; it removes organic deposits such as paraffin-like substances as well as drilling mud and small rock chips (12).

Further valuable attributes of lecithin in this connection are its anti-corrosive properties, since it helps to prevent the harmful effects of water.

Hydraulic crude oil extraction
Lecithin is an interesting co-wetting agent for the tenside flooding process. The term hydraulic fracturing ('fracking') is also used, for example for extracting oil

from reservoir rock. Water is forced in under high pressure, and the oil is driven to the injection wells, where it emerges again under pressure.

The efficacy of this process is increased by reducing the interfacial tension between the water and the oil. Tensides (surfactants) are therefore added. The effect of the wetting tensides – for example alkylbenzene sulphonate, ethylhexyl sulphate or soaps – is considerably impaired by fluctuations in salinity (as little as 0.1% alkaline-earth salt can have a disruptive effect).

Because of its low HLB value, native lecithin is a powerful W/O emulsifier. This property is not affected by calcium or magnesium ions – in fact at certain concentrations it is even promoted. Additions of 2 – 5% lecithin, based on sulphates and sulphonates, are recommended.

Oilwell drilling

During oilwell drilling, very stable emulsions often form which greatly disturb the subsequent process. As a W/O emulsifier, lecithin therefore serves to demulsify O/W emulsions arising from oil production.

Lubricating oils and greases

Sludge formation in lubricating oils can be prevented by adding 0.08 – 0.8% lecithin and 0.08 – 0.8% thiodiphenyamine.

The use of lecithin in extreme-pressure lubricating oils is an interesting topic. Lecithin itself has remarkable extreme-pressure properties. The addition of a mixture of

- 100 parts native lecithin (60 – 70% phospholipids) and
- 12 parts refined soybean oil

to the mineral oil at a ratio of 1:1 increases the pressure resistance of the oil from 300 lbs to 3,900 lbs.

The efficacy of lecithin in extreme-pressure lubricants is substantiated by tests, for example in sedimentation-resistant formulations.

In extreme-pressure greases based on lithium stearate, lecithin increases the heat stability.

Shkola (13) describes a reduction of the friction coefficient of lubricating oils by brominated phospholipids with a bromine content of around 10%. The product used was a phosphatidylcholine (PC)-enriched deoiled lecithin. Hydrated phospholipids produce similar results.

Emulsified lubricating oils have a tendency towards corrosion. Lecithin ensures that the systems are largely corrosion-resistant (14).

Lecithin is used as an anti-corrosive lubricant in spindle oils. Its rust-inhibiting properties can easily be demonstrated by the salt spray test (15).

In addition to the effects in lubricants described above, further aspects are important in engine oils, and lecithin can be of benefit in these, too.

- Resistance to corrosion, oxidation, gumming and sludge formation.
- Avoidance of viscosity problems.
- Lubricating effect.
- Acidity.

Lecithins stabilize engine oils, preventing sludge formation, sediments and gumming and making them resistant to corrosion. Dirt particles and combustion residues remain suspended. The crowns and top edges of the pistons stay clean (16).

Sludge formation can also be caused by moulds or bacteria, but this is suppressed effectively by the fungicidal and bactericidal properties of lecithin.

The addition of as little as 0.1–1.0% lecithin results in high resistance to oxidation and low acidity.

Lecithin lowers the viscosity of mixed mineral oils and intensifies the effect of anti-foaming agents containing silicone.

Improvements to the quality of heating oil include stabilizing suspended matter. Blockages can be prevented by adding as little as 0.05% lecithin. No slag forms on the burner if a dispersion containing lecithin is mixed with the heating oil.

Heavy fuel oil (bunker oil) settles in storage tanks and forms sludge on the bottom. Lecithin prevents this process.

One crucial aspect nowadays is combustion efficiency and nitrous oxide emissions. Here, too, lecithin can contribute to a solution. Lecithin has shown itself capable of reducing the nitrous oxide emissions of heating oil by up to 25%.

8.3.3.8 Lecithin as an additive in the manufacture and processing of paper

In paper manufacturing, lecithin is useful as a

- Releasing agent.
- Defoamer.
- Wetting agent.
- Softener.

Lecithin has wetting power without foam formation. It is added together with long-chain fatty acids before the foam is expected to develop (17).

Lecithin can fulfil an important function as a releasing agent in the paper pulp, where it counteracts the tendency of the pulp to adhere to the hot dryers (18).

Even small amounts of lecithin added as a wetting agent to the mercerizing bath in the production of alkali cellulose serve to boost reactivity. The usage level is between 0.25 and 0.5%, based on the quantity of alkali cellulose. Lecithin is the substance of choice for this purpose because it does not foam (like sodium stearate, for example) and nevertheless has good wetting power.

In paper coating masses lecithin has specific tasks to perform, but it also performs multiple functions as a

- Dispersant/emulsifier.
- Blending aid.
- Viscosity reducer (better processability, higher solid content).
- Releasing agent (counteracting adhesion to the coating rollers).
- Lubricant (faster running).
- Anti-block agent (normal duplicating paper).
- Anti-static agent (normal duplicating paper).

In coatings, lecithin alone or combined with other surfactant substances is a considerable help in speeding up the coating process (19). The usage level is between 0.2 and 5%, based on the formulation as a whole. Naturally, the lubricating properties of the lecithin are also partly responsible for the higher coating speed.

The viscosity reduction brought about by the lecithin not only facilitates processing of the coating paste; it also makes it possible to increase the solids content.

Because of its releasing properties, lecithin is used in coatings with a high-gloss finish in order to prevent adhesion to the rollers. The releasing effect of the lecithin can be increased by combination with waxes, higher fatty acids or the metal salts of these.

Calcium stearate dispersions for paper coatings precipitate and thicken. When used as a co-emulsifier, lecithin can help to produce stable liquid dispersions with 50% calcium stearate which do not subsequently thicken. The quantity required is 10 – 20% lecithin, based on the emulsifier mixture used. Hydrophilic lecithins are the most suitable.

8.3.3.9 Other technical applications

Plastics

Practically all forms of plastics processing require the use of lubricants. For articles that may come into contact with foods (packaging films, water pipes), suitable lubricants such as lecithin must be used (20).

Lecithin is compatible with PVC. It can already be worked in during the production of PVC copolymers and results in a thermostable resin. Since lecithin itself has no stabilizing effect in PVC, the effect can only be explained by intensive contact of the stabilizers with the resin – facilitated by the lecithin. At the same time the lecithin prevents, or at least reduces, electrostatic charging.

In synthetic resins, lecithin accumulates on the surface and acts as a releasing agent. Examples are urethane polymers and polyester resins.

In expanded polystyrene, lecithin enhances the non-stick properties.

One very specific application of lecithin is in polyvinyl butyral films used as interlayers in automobile windscreens (laminated safety glass).

The film must adhere firmly to the glass, but on the other hand it must have a certain degree of slippage vis-à-vis the glass. This can be achieved with very small amounts of lecithin (0.005 – 0.01%).

The rheological properties of plastisols are greatly improved by the addition of 0.25% lecithin.

Rubber

Lecithin is effective in most types of natural and synthetic rubber, including isoprene and butadiene rubber and also silicone rubber based on organopolysiloxanes.

Being a natural, organic material, lecithin is the ideal additive for articles that come into contact with food.

Lecithin is used as an emulsifier in butadiene rubber and latex; as a wetting agent and dispersant it ensures excellent distribution of the fillers. It acts as a lubricant and separating agent in the rounding of tyres. Rubber gloves are less sticky if lecithin is added. It is a constituent of rubber solutions from which seals are made (21); as a plasticizer it facilitates manufacturing processes (22) and also the recycling of waste rubber.

Lecithin has constituents that assist vulcanization; it acts as an accelerator in the vulcanization process and increases hardness and resistance to tearing.

The usage levels lie between 2 and 4% based on the rubber or 10% based on the filler.

Building materials, construction chemicals

The main applications are in:

- Cold bitumen emulsion.
- Roofing felts.
- Formwork lubes, demoulding agents.
- Compressive strength and water resistance of concrete.
- Refractory (fireproof) compositions.
- Timber.

Lecithin is especially effective as an adhesive or binding agent for strengthening the adhesion of the bitumen to the filling materials and the substrate. Road surfaces with optimal adhesion can be produced by adding 2% lecithin dissolved in vegetable oil at a ratio of 1:1. Excellent adhesion is achieved even at low usage levels. In cold bitumen emulsion based on coal tar, lecithin is used as a binding agent.

The benefits for the production of roofing felts are similar. It is also interesting that lecithin increases the effect of flame retardants by facilitating thorough and uniform distribution of the metal hydroxides, borates and antimony trioxide in the roofing felt compound.

Concrete, mortar, plaster and other building materials are poured into moulds made of wood, metal or plastic to harden. After hardening, the material must be separated from the mould. In order to prevent adhesion of the material and ensure clean, smooth removal, separating agents – so-called formwork lubes or demoulding agents – are needed. In this case lecithin fulfils the functions of an anti-corrosion and separating agent and acts as a wetting and spreading agent. Usage levels may be as high as 10%.

Rust-inhibiting separating agents for concrete moulds made of steel contain 4 to 8% lecithin. The compressive strength of concrete can be increased by the addition of 0.1–3% lecithin (23).

Aqueous silicone compounds are often used as hydrophobizing agents for concrete. The addition of lecithin as a dispersant is recommended.

Unsaturated, long-chain fatty acids and magnesium oxide are added to special building materials, for example those used for lining stoves and furnaces. 0.5% lecithin ensures good wetting and distribution of the additives.

Timber is much in demand as a building material, but it has to be protected against mould with fungicides. Lecithin itself has fungicidal and insecticidal properties. As a synergist it also strengthens the effect of other biocides (24). Lecithin helps to reduce the amount of (toxic) wood preservatives by 40–60%.

Metalworking

Drilling and cutting fluids and rolling and wire-drawing oils (25,26) are aqueous emulsions with a bifunctional effect:

- The aqueous constituent acts as a coolant to remove the frictional heat.
- The oil contained acts as the actual lubricant.

Lecithin lowers the friction coefficient and strengthens the lubricating properties, thus reducing wear. This prolongs the service life of the tools. Lecithin increases the pressure resistance of the metalworking fluids.

Adhesives

Although not in itself an adhesive, lecithin can give adhesives useful practical properties.

The addition of 0.01 – 0.03% lecithin to adhesive layers of cross-linked polyurethane resins or urea-formaldehyde and melamine-formaldehyde resins gives these better rheological properties and greater flexibility and toughness.

Self-adhesive labels, envelopes, wallpapers, films, medicinal sticking plasters or "Scotch"-type pressure-sensitive tapes require adhesive substances that develop the necessary attributes under pressure after removal of a protective covering. The relevant adhesive components – 'tackifiers' – are mainly hydrocarbon resins, e.g. d-limonene resins.

The ideal layer should have only low affinity with the substrate on one side (pressure side) and high affinity with the substrate on the other side (adhesive side). This can be achieved with 0.5 – 3% interface-active substances. Besides fatty acid esters with polyvalent alcohols, such as glycerol monostearate, sorbitan monostearate or diethylene glycol monooleate, lecithin has proved highly effective.

For the production of starch paste, starch is first hydrolysed enzymatically. The lecithin is added before spray-drying. Lecithin promotes wetting of the substrate and spreading.

Environmental technology

As a biocompatible, natural substance, lecithin opens up more environmentally interesting possibilities than any other wetting agent or any other emulsifier. Lecithin spares its users the necessity of making compromises in formulations.

Fields of environmental technology with relevance for lecithin are:

- Recycling of paper (deinking).
- Splitting of polymers.
- Biocompatible packaging materials.
- Microbiological purification of water and soil.
- Dust binding.

The repeated use of printed waste paper darkens the colour more and more, so a system was introduced for removing the printing ink before reuse of the paper. The most suitable emulsifiers for deinking are alkali soaps; their efficacy is heightened by bleaches and complex formers like EDTA. Lecithin in the amount of 2–5%, based on the soap content, strengthens the wetting and emulsifying effect.

Polymers disposed of in landfills rot faster when water-soluble additives are supported by lecithin. Their decomposition in the soil starts after only 1.5 months.

Lecithin is the ideal plasticizer and perfect wetting agent for biocompatible packaging materials. The addition of up to 5% lecithin ensures good flow properties and viscosity control during operations, in some cases together with triglycerides.

Lecithins have established themselves as highly effective separating agents in food-safe coatings, for example for sausage skins based on cellulose, storage nets for ham, and the insides of cans. Lecithin can also be used as a dispersant for food colourants.

Soil and water purification

Lecithin stimulates the effect of microorganisms that degrade hydrocarbons. In this process it acts as a nitrogen and phosphorus donor; it is non-toxic and has a wetting and emulsifying effect. Hydrocarbons in soils and water can therefore be converted into biodegradable emulsions with the aid of lecithins.

It is interesting that lecithin is recommended as a nutrient for mould growth (27). The biodegradability of native oils and fats can be speeded up, too. Blockages in sewage treatment plants can be avoided by faster breakdown of the triglycerides.

Wastewater from metalworking can be degraded more quickly when lecithin is used as a wetting agent.

Crude oil that has spilled into fresh or salt waters (oil pollution) is adsorbed and contained by aluminium oxide coated with PC-enriched lecithin. The mass sinks down and the adhering oil is degraded microbiologically. But even without an adsorbent, crude oil in fresh or salt water is broken down by the microorganisms if these have a sufficient supply of lecithin as a phosphorus and nitrogen donor.

Dust-binding
Being a biocompatible, natural product, lecithin is an ideal dust binder. Its chemical structure enables it to meet the requirements of oleophilicity and hydrophilicity. Its good wetting capability for all manner of different dusts, and its anti-static properties, increase its affinity for dust.

Agrochemicals in powder form can be made dust-free with lecithin. Explosions in flour mills, feed production plants and storage facilities are well-known examples of the risk involved with dust. Spraying with 0.2 g of a mixture of soybean oil and lecithin binds the dust (28,29) and greatly reduces the risk of an explosion.

8.3.4 Conclusion

From the physical and chemical point of view, lecithins are an extremely interesting group of substances. Not only do they have a wide range of uses; in some processes they may be of crucial importance.

The surface and interfacial activity of the lecithins can differ greatly and covers numerous systems, depending on the application:

- Wetting of surfaces.
- Stabilization of systems.
- Dispersion of solids in liquids.
- Rendering surfaces repellent.
- Emulsifying non-miscible fluids.
- Viscosity adjustment.
- Plasticizing effect.

All bodies have surfaces: solid, liquid or gaseous. However, these surfaces can differ greatly. Viewed under a microscope, an apparently smooth surface may show itself to be extremely irregular (with craters, cracks, pores or moisture).

This chapter aims to show that lecithin makes it possible to transform surfaces and interfaces. That is not, in fact, a recent discovery. Surface and interface-active substances have been known to exist for centuries, and soap – one of the oldest representative examples – is part of our daily life.

References

1. Kronstein, M., 1978. Mod. Paint Coating. 68, 59-63.
2. Kronstein, M., Eichberg, J., 1982. Production of paint containing lecithin – treated metal oxide. USP 4 520 153.
3. Eichberg, J., MacDonald, K.S., 1992. Lecithin corrosion inhibitor. USP 5 120 357.
4. Ehrhardt, G. H., 1993. Flow agent – dispersant. Composition. USP 5 188 765.
5. Bratkowska, I., Zwierzykowski, W., 1987. Effect of raw materials on rheological properties of carbon black dispersion in mineral oil used for printing inks production. Przem. Chem. 66 (8), 393-395.
6. Rosen., 1946. American Ink Maker (12), 29-33.
7. Wittcoff, H., 1951. The Phosphatides. J. Am. Pharm. Assoc. 41 (3), 168.
8. Wills, R.B.H., Hopkirk, G., Scott, K.J., 1980. Use of fatty acid methyl esters and edible fats and oils to reduce soft scald of apples. J. Sci. Food Agr. 31 (7), 663-666.
9. van Nieuwenhuyzen, W., 1981. The industrial uses of special lecithins. J. Am. Oil Chem.' Soc. 58, 886-888.
10. Shillington, W.L., Liggett J.J., 1970. Compositions and methods for coating fruits and vegetables with lecithin. USP 3 533 810.
11. Lowell, W.B., 1981. Process for removing hard-to-remove soils from hardware. USP 4 297 251.
12. Lummus, J.L., 1960. Oil based drilling fluid with a combination of calcium chloride and lecithin. USP 2943051.
13. Shkola et.al., 1992. Chem. Tech. 44 (9), 311-315.
14. Dorer, C.J., 1966 Substituted succinic acid-boron-alkylene amine phosphatide derived additive and lubricating oil containing same. USP 3 284 409.
15. Pardun, H. 1981. SOFW, 107, 409-411.
16. Elliott, F.S., 1955. Lubricant compositions to improve marine engine oils. USP 2 718 502.
17. Jurick, M.J., 1981. Antifoam composition with lecithin. USP 2 923 687.
18. Martinez, E.D., Duncan, T.F., 1990. Epoxidized polyamide wet strength resin containing lecithin. USP 4 970250.
19. Nikoloff, K., Peltier, R.J., 1988. Phospholipid lubricant for coating moving webs. USP 4 766 015.
20. Riethmayer, S., 1972. Gleitmittel für PVC. SÖFW 7,8,9,12.
21. Cahill, G.F., Arons, I.J., 1967. Gasket-forming non-settling solvent based rubber compositions containing lecithin. USP 3 344 099.
22. Markley K.S., 1950. Soybeans and Soybean products. Vol.2, 593-647.
23. Proell, W., A., 1970. Additive for cementitious mixtures. USP 3 537 869.
24. Sietz., 1981. SÖFW 107, 629.
25. Nass - Schmierstoffe, Peter Greven, Fettchemie.
26. Riethmayer, S., 1970. SÖFW 96, (9), 298-300.
27. Holtz, Barry, R., 1983. Nutrient for mushroom growth and for producing same. USP 4 370 159.
28. Bürgi, H., Schwarz, K., Froesch, E.R. 1969. Arbeiten DLG 125.
29. N.N., 1986. Mühle und Mischfuttertechnik 123, 703-705.

About the editors

Waldemar Buxmann

Dr Waldemar Buxmann graduated in food technology at the University of Bonn and subsequently prepared his doctoral thesis on the techno-functional properties of egg phospholipids at the German Institute of Food Technologies (DIL e.V.) Quakenbrück, Germany in collaboration with the Technical University of Berlin. During his time as a research associate in the research platform on Food Structure and Functionality at the DIL, he headed several publicly funded research projects focussing on lecithin and phospholipids.

In 2012 he took the step into the food industry by joining Sternchemie, the lecithin producing company within the Stern-Wywiol Gruppe in Hamburg. He is currently head of the technical department at Sternchemie. His interests are lecithin and phospholipids, product and process development and applications of lecithin in food.

Rüdiger Ziegelitz

After leaving school in 1948, Rüdiger Ziegelitz initially trained as a chemical laboratory assistant from 1948 to 1951. From 1955 to 1959 he attended the Evening Technical College in Hamburg, where he qualified as a chemical technologist. From his initial training until 1995, Rüdiger Ziegelitz was employed by Dr Meyer-Castens & Co. GmbH, Hamburg (subsequently Lucas Meyer), where he later held general commercial power of attorney. From 1980 he was a managing director responsible for product control, applications of lecithin products (including customer advice), development of new products, process technology for the production of oil-free phospholipids, fractionation of phospholipids and enzymatic modification, and also processing of soy proteins and their applications in food and feed. He organized phospholipid symposiums in 1986,1989 and 1993 and the first and second Hamburg Soy Conferences in 1986 and 1991. From 1995 until 2015 he acted as a free-lance consultant to the Stern-Wywiol-Gruppe.

Parris Kidd

Dr Parris M. Kidd was born in Jamaica, earned his Bachelor degree in zoology at the University of the West Indies, then acquired his doctorate in zoology – cell biology at the University of California in Berkeley. He was a research investigator at the University of California San Francisco Medical Center until entering integrative medicine with a textbook on nutritional antioxidants. He was the leading scientist in establishing the phospholipid phosphatidylserine (PS) as a brain nutrient on five continents. He is recognized internationally for his expertise on the nutritional applications of lecithins, phospholipids, omega-3 fatty acids, carotenoids, and other nutrient classes. Dr Kidd now focuses on the use of nutritional medicine to improve memory and other mental performance. He is currently Chief Science Officer for BrainMD of Irvine, California, USA.

Volkmar Wywiol

Volkmar Wywiol, born in 1935, has been closely associated with the products lecithin and phospholipids for over 65 years. As a long-standing employee and subsequent managing director of the Hamburg company "Lucas Meyer – The Lecithin People" he was among the pioneers of the worldwide development and marketing of lecithin.

In 1980 Volkmar Wywiol ventured into business on his own with Sternchemie. Together with the oil mill group and the leading American lecithin producer Central Soya (acquired by DuPont, today IFF) he later established the joint venture Stern-Lecithin & Soja. In the 1990s, this company grew to become Europe's leading supplier of lecithin, with a broad basis of applications research, acknowledged lecithin experts and a modern tank farm and processing facility in Rotterdam.

From 2000, Volkmar Wywiol extended the product range of the Stern-Wywiol Gruppe by purchasing and establishing further companies in the field of food ingredients and feed additives. These included Berg+Schmidt, Hydrosol, Planteneers, Mühlenchemie, DeutscheBack, SternVitamin, SternLife, SternEnzym, Herza Schokolade, SternMaid and OlbrichtArom. With 12 specialist ingredient firms in Germany and 18 affiliates abroad, the owner-managed Group is now one of the world's most competent developers and manufacturers of customized functional ingredients for food and feed.

In 2020, Volkmar Wywiol received the "Lifetime Achievement Award" from the Mayor of Hamburg.

List of Contributers

Dana Middendorf

Dr Dana Middendorf studied food chemistry at the Technical University of Brunswick, Germany. After graduating as a state-certified food chemist in 2011, she moved to the German Institute of Food Technologies – DIL e.V. – in Quakenbrück, Germany. She is currently working as a research associate in DIL's Research Platform on Food Structure and Functionality and the business unit Center of Food Physics. In 2015 she acquired her doctorate on the subject of adsorption of emulsifiers to sucrose particle surfaces in cocoa-butter-based suspensions, using atomic force microscopy. Her main research interests are the formation and characterization of disperse food systems, the design of food matrices and interfaces. Dr Middendorf combines a knowledge of food chemistry and chemical analysis of food with fundamental aspects of food physics and food process engineering.

Eckhard Flöter

Prof. Dr Eckhard Flöter started his academic career as a student of chemical engineering at the Technical University of Berlin. He moved to the Netherlands in 1992 to work for a doctorate in the field of thermodynamics of reservoir fluids. In 1996 he switched his interest to food technology by joining Unilever R&D in Vlaardingen, where his first business card described him as a 'Fat Technologist'. During his 15 years with Unilever Dr Flöter held several posts, kept in close contact with science and was an active inventor, generating a substantial patent portfolio bearing his name. In 2011 he took the step back into the academic world. Since then he has held the chair of Food Process Engineering at the Technical University of Berlin. His laboratory work focuses on phase transitions, in particular crystallization, and the resulting product structures. The main substrates studied are fats and oils, starch grains and sugar, since the chair also hosts the Berlin Sugar Institute.

Gerhard Helmerich

Dr Gerhard Helmerich is a certified food chemist with 15 years' experience in the pharmaceutical industry. He obtained his doctorate in Food Chemistry at the German Research Centre for Food Chemistry in Munich. His doctoral thesis on Functional Properties of Phospholipids in Emulsifiers earned him an award from the Federation of German Large Bakeries in 2004. In the same year he was granted certification at the Bavarian State Office for Health and Food Safety. Since 2004 he has worked in the pharmaceutical industry, holding various posts with increasing responsibility in the areas of project and external business management as well as strategic and operational roles. He is an expert in product development, manufacturing and the commercial supply chain for small molecules and biotechnology products.

Gunther Arnold

Dr Gunther Arnold initially graduated in food processing and thereafter worked as a research assistant at the Institute of Food and Bioprocess Engineering of the Technical University of Dresden, where he acquired his doctorate with a focus on lecithins and phospholipids in oil-based suspensions. He has been working for Anton Paar Germany GmbH since 2014 and is currently employed as an international product manager for rheometers.

Hans-Ulrich Waiblinger

Hans-Ulrich Waiblinger studied food chemistry at the University of Stuttgart. Since 1991 he has been employed with the Food Safety Inspectorate in Baden-Württemberg, the focus of his work being on examination for genetic modifications and allergens and molecular-biological species differentiation. For many years he has served as an expert in European and national study groups on molecular-biological and biochemical food analysis; these include CEN, DIN, BVL and GDCh. He is currently head of the "Origin and Integrity" department of the State Institute for Chemical and Veterinary Analysis (CVUA) in Freiburg.

Harald Rohm

Prof. Dr Harald Rohm studied at the University of Natural Resources and Life Sciences (BOKU) in Vienna, Austria, where he also obtained his doctorate and subsequently his postdoctoral lecturer qualification. He worked in the Department of Dairy Science and Bacteriology at BOKU from 1984 to 2002, then accepted a call from the Technical University of Dresden for the chair of Food Engineering. Professor Rohm works in several fields of food technology and food process engineering and is internationally recognized for his expertise in chocolate manufacturing.

Irina Ostroumova

Prof. Dr Irina Ostroumova holds a doctorate in biological sciences and is a professor and chief researcher at the State Research Institute for Lake and River Fisheries, named after L.S. Berg (GosNIORKh), of the St. Petersburg branch of FGBNU VNIRO. She is a leading scientist specializing in the physiology and feeding of fish in aquaculture and the author of more than 250 scientific publications and works on scientific methodology. These include the widely known monograph "Biological Basics of Fish Feeding," 2001, republished and supplemented in 2012. For many years she has concentrated on the use of phospholipids in the nutrition of fish of different species and ages.

Jean-Pierre Arnaud

Dr Jean-Pierre Arnaud graduated in chemical engineering and water treatment at the National Institute of Applied Sciences (INSA) of Toulouse, France, then obtained his doctorate in food science and technology at the Laval University in Quebec City, Canada. He was Technical Director of Lucas Meyer Cosmetics for 20 years and is recognized internationally for his expertise in phospholipid applications in the cosmetics industry. He currently acts as a consultant to several international cosmetic ingredient manufacturers.

Mabel Tomás

Dr Mabel C. Tomás is a research scientist at Consejo Nacional de Investigaciones Científicas y Técnicas (CONICET), a professor of Food Toxicology at the Facultad de Ciencias Exactas (UNLP) and in the post-graduate Master in Food Technology and Hygiene (UNLP), Argentina. She received her doctorate in chemical sciences from the Universidad Nacional de La Plata (UNLP), Argentina. She has been President of the International Society for Fat Research (ISF) and holds the chair of the Phospholipid Division of the American Oil Chemists' Society (AOCS), USA, recognized by the International Lecithin and Phospholipid Society (ILPS) with the Willem van Nieuwenhuyzen AWARD. The focus of her research is on functional lipids, (omega-3 fatty acids, phospholipids, lecithin) associated with the development, description and application of the delivery systems of various bioactive compounds in foods that constitute a benefit to human health.

Estefania Guiotto

Dr Estefania N. Guiotto is a research scientist with CONICET at Centro de Investigación y Desarrollo en Criotecnología de Alimentos (CIDCA), Argentina. She received her doctorate from Facultad de Ciencias Exactas (UNLP), Argentina, for research into the use of chia (Salvia hispanica L.) and sunflower by-products (Helianthus annuus L.), e.g. sunflower lecithin, in foods.

Michael Schneider

The chemist Dr Michael Schneider acquired his doctorate at Hamburg University and subsequently worked on tumour diagnosis at the University of Berlin from 1976 to 1980. From 1980 to 1999 he was head of R&D, QC and applications services and also a board member with Lucas Meyer. From 1999 to 2000, Dr Schneider worked for Degussa following its takeover of Lucas Meyer, and from 2000 to 2002 he was a board member of Lipoid. Since 2003 he has been the founder and owner of the company Lecithos Consulting (lecithin and phospholipids, product and process development, nutritional applications).

Roland Adelmann

Dr Roland Adelmann studied agricultural sciences at the Justus Liebig University of Giessen, specializing in animal production. He then became a research assistant at the Institute of Animal Nutrition of the University of Giessen, where he acquired his doctorate. Dr Adelmann has worked in the feed industry since 1992, and has been employed with Berg+Schmidt, Hamburg since 2000. He is currently head of R&D at Berg+Schmidt.

Tabea Masiak

Tabea Masiak obtained her Bachelor and Master's degrees in food technology at the Technical University of Berlin. During her studies she worked on the applications of lecithin and wrote her Master's dissertation on the use of commercial lecithins for oleogelation. She is currently employed in R&D for the dairy industry at a German subsidiary of Andros.

Ute Bindrich

Dr Ute Bindrich studied food process engineering at the Technical University of Dresden; she was then engaged in research and prepared her doctoral and university teaching thesis (Dr.-Ing. habil.). Since 1993 she has been working in the field of food physics at the German Institute of Food Technologies e.V. (DIL) Quakenbrück. She is currently head of the research platform Structure and Functionality and Center for Food Physics. Her research fields are generation and characterization of disperse multiphase systems and the design and characterization of interfaces.

Yvonne Schneider

Dr Yvonne Schneider graduated in processing technology and process engineering at the Technical University of Dresden, specializing in food engineering. She then engaged in scientific work at the Institute of Food and Bioprocess Engineering (now Natural Materials Technology) and prepared her doctoral thesis in the field of ultrasonic cutting of foods at the Technical University of Dresden. From 2008 to 2018 she was senior scientist and head of the research group "Disperse Food Systems". Since 2018 she has headed the food industry division of the Saxony Economic Development Corporation.

Jingbo Li

Dr Jingbo Li initially studied bioengineering at the Hefei University of Technology, then went on to acquire a Master's degree in biochemical engineering at the South China University of Technology and a doctorate in lipid technology at Aarhus University in Denmark. He started working at Food Oil, DLG Group, Denmark after his doctorate study for 5 months and then moved to Massachusetts Institute of Technology, the USA as a Postdoctoral Researcher. He is now working at Ginkgo Bioworks, Boston, USA. He is recognized internationally for his expertise in lipid biotechnology and lipid processing and received three awards from the American Oil Chemists' Society in 2017.

Zheng Guo

Prof. Dr Zheng Guo acquired a degree in applied chemistry and a doctorate in biochemical engineering at Tianjin University. He has published over 170 peer-reviewed papers and dozens of book chapters on lipid processing, biotransformation, enzyme discovery & engineering and green chemistry. Dr Zheng Guo is a leading scientist in lipid and phospholipid biorefining and green processing. He is currently a professor of Agro-Biotechnology Science at Aarhus University, Denmark, serving also as the associate editor of JAOCS and on the editorial board of Biochemical Engineering Journal, Chinese Journal of Chemical Engineering and Bioresources and Bioprocessing.

Gerd Konrad

Dr Gerd Konrad studied dairy technology at der Humboldt University of Berlin, where he obtained his doctorate in 1976. From 1976 to 1990 he worked as a research assistant at the Institute of Dairy Technology in Oranienburg, developing numerous methods for the chemical and enzymatic hydrolysis of proteins in the production of meat flavourings. Up to his retirement in 2013 he continued his research work at Anhalt University of Applied Sciences in Köthen. His special field was the integral utilization of whey ingredients. Dr Konrad is an internationally acknowledged expert in the application of membrane filtration and the isolation and modification of individual milk proteins, which he has made the subject of numerous international publications. In 2012 he was honoured with the Best Paper Award of the Phospholipid Division of the American Oil Chemists' Society.

Erik Løvaas

After initial studies at the University of Bergen, Norway, from 1967 to 1973, where he earned his cand. real. degree Dr Erik Løvaas worked as a project associate (NIH) at the Enzyme Institute, Madison, Wisconsin, USA, from 1974 to 1976. From 1977 to 1979 he was a NATO Visiting Professor and EMBO Fellow based in the Bari area of Italy. Following this, Erik Løvaas became a consultant with the Wellcome Foundation in Oslo (Norway), a position he held from 1979 to 1981. Since then he has held various posts as a researcher and professor at the University of Tromsø, Norway, and has founded, chaired or managed 5 spinoff companies from this university. He is a recognized expert in the field of "exploitation of marine biological resources" with the special focus on lipids, omega-3 phospholipids and antioxidants. A further field of interest is basic research into MPL liposomes.

Willem van Nieuwenhuyzen

Willem van Nieuwenhuyzen held various international posts with Unilever, Eridania Beghin-Say, Central Soya and Solae-DuPont. He was also one of the founders of Lecipro Consulting in Limmen, Netherlands. Furthermore, he was a founder member and executive director of the International Lecithin and Phospholipid Society (ILPS) until 2017. He played an active role in the Phospholipids Division of AOCS, served on the Membership Development Committee and was an associate editor of Inform. As such, he contributed many reports on meetings and articles to the magazine.

Abbreviations

Abbreviation	Meaning
AA	arachidonic acid
AAMI	age-associated memory impairment
ABC transporter	ATP-binding cassette transporter
ACh	acetylcholine
ACTH	adrenocorticotropic hormone
ADHD	attention deficit/ hyperactivity disorder
ADL	activities of daily living
ADM	Archer-Daniels-Midland in Chicago
ADPH	adipophilin
AFM	atomic force microscopy
AI	acetone-insoluble
AL	acetone-solubles
AL	algal lecithin
ALC	American Lecithin Company
ALD	alcoholic liver disease
AMF	anhydrous milk fat
AOSC	American Oil Chemists' Society
APCI	atmospheric pressure chemical ionization
ARA	omega-6 arachidonic acid
ARCD	age related cognitive decline
ATP	adenosine triphosphate
AV	acid value
Å	Angstrom
BC	brain cortex
BHT	butylated hydroxytoluene
BMI	body mass index
BSE	bovine spongiform encephalopathy
CAC	calcium alginate-chitosan
CaMV	cauliflower mosaic virus
CAS	Chemical Abstracts Service
CAVI	cardio-ankle vascular index
CC	coagulation and clotting
CER	ceramide
CFR	Code of Federal Regulations
CIR	Cosmetic Ingredient Review
CIS	Commonwealth of Independent States
CL	cardiolipin
CLA	conjugated linoleic acid
CLEAR	Canadian Low Erucic Acid Rapeseed
CLSM	confocal laser scanning microscopy
CM	chia mucilage
CM	chylomicrons
CMC	chemistry, manufacturing and control
CNS	central nervous system
COPD	chronic obstructive lung disease
CPL	phospholipid concentration
CRP	C-reactive protein
CTAB	hexadecyltrimethylammonium bromide
CTP	chloroplast transit peptide
DAG	diacylglycerol
DATEM	diacetyl tartaric acid ester of mono- and diglycerides
DB	double blind
DEAE	diethylaminoethyl
DED	dry eye disease
DF	diafiltration
DGF	German Society for Fat Science
DHA	omega-3 docosahexaenoic acid
DI	dispersibility index
DL	deoiled sunflower lecithin
DM	dry matter
DNA	desoxyribonucleic acid
DOPC	1,2-dioleoyl-sn-glycero-3-phosphocholine
DP	dispersive part
DPA	docosapentaenoic acid
DPPC	dipalmitoylphosphatidylcholine
DPPG	1,2-dipalmitoyl-sn-glycero-3-phosphorylglycerol
DSM	Diagnostic and Statistical Manual of Mental Disorders
DSPE	1,2-distearoyl-sn-glycero-3-phosphorylethanolamine
EC	emulsifying capacity
ECS	endocannabinoid system
EDTA	ethylenediaminetetraacetic acid
EEG	electroencephalography
ELISA	enzyme-linked immunosorbent assay
ELS	evaporative light-scattering
ELSD	evaporative light-scattering detector
EMA	European Medicines Agenda
ENGL	European Network of GMO Laboratories
EPA	eicosapentaenoic acid
EPL	essential phospholipids
EPSPS	enol-pyruvyl-shikimate-3-phosphate-synthase
ER	endoplasmic reticulum
ESI	electrospray ionization
EU	European Union
FA	fatty acid
FAB	fast atom bombardment
FABP	fatty-acid binding protein
FAME	fatty acid methyl esters
FAO	Food and Agriculture Organization of the United Nations
FCC	Food Chemical Codex
FDA	US Food and Drug Administration
FFA	free fatty acids
FI	flame ionization
FLD	fatty liver disease
FO	fish oil
FU	Farinograph unit

GANG	gangliosides
GC	gel chromatography
GD3	ganglioside GD3
GDP	gross domestic product
GDP	guanosine diphosphate
GI	gastrointestinal
GLA	gamma-linolenic acid
GM	genetically modified
GM3	ganglioside GM3
GMO	genetically modified organisms
GMP	genetically modified plants
GMS	green manures
GOED	Global Organization for EPA and DHA
GOST	governmental standard
GPC	glycerophosphocholine
GPCR	G protein coupled receptors
GPL	glycero-phospholipids
GRAS	Generally Recognized As Safe
GS	gangliosides
GTP	guanosine triphosphate
HA	hyaluronic acid
HACCP	hazard analysis and critical control points
HCY	homocysteine
HDL	high-density lipoproteins
HDS	Hasegawa Dementia Scale
HI	hexane insoluble
HLB	hydrophilic-lipophilic balance
HP	high protein
HPLC	high performance liquid chromatography
HPTLC	high-performance thin-layer chromatography
IC	instant criterion
IDF	International Dairy Federation
IDL	intermediate-density lipoproteins
IL	ionic liquid
IMP	integral membrane proteins
INS	International Numbering System for Food Additives
IP	inositol phosphate
IP-programmes	identity preservation programmes
IR	infrared
IUCN	International Union for the Conservation of Nature and Natural Resources
JPE	Japanese Pharmaceutical Excipients
KL	krill lecithin
LC	long-chain
LC/MS	liquid chromatography/mass spectrometry
LCAT	lecithin: cholesterol acyltransferase
LC-PUFA	long-chain polyunsaturated fatty acids
LDL	low-density lipoproteins
LOD	limit of detection
LOQ	limit of quantification
LP	lysophosphatidic
LPA	lysophosphatidic acid
LPC	lysophosphatidylcholine
LPE	lysophosphatidylethanolamine
LPI	lysophosphatidylinositol
LPL	lysophospholipids
LPS	lysophosphatidylserine
LRT	lipid replacement therapy
LSD	lysergic acid diethylamide
LVT	low viscosity torque
MCT	medium-chain triglycerides
MFGM	milk fat globule membrane
MMSE	mini-mental state examination
MPEG	methoxy linear PEG
MPEG	monodisperse PEG
MPL	marine phospholipids
MS	mass spectrometry
MSC	Marine Stewardship Council
MSG	monosodium glutamate
MTHF	5-methyltetrahydrofolate
MUC	mucin
NAC	N-acetylcysteine
NAFLD	non-alcoholic fatty liver disease
NANA	N-acetyl-neuraminic acid
NAPE	N- acyl-phosphatidylethanolamine
NAR	nicotinic acetylcholine receptor
NFE	nitrogen-free extract
NGF	nerve growth factor
NHP	non-hydratable phospholipids
NK	natural killer
NKA	Na K–ATPase
NKO	Neptune Krill Oil
NL	neutral lipids
NMR	nuclear magnetic resonance
NOS	nopaline synthase
NRC	U.S. National Research Council
NSAIDs	non-steroidal anti-inflammatory drugs
NTA	nitrilotriacetic acid
NVG herring	Norwegian spring-spawning herring
O/W	oil-in-water
OA	osteoarthritis
OM	outer membrane
PA	phosphatidic acid
PAF	platelet activating factor
PAGE	polyacrylamide gel electrophoresis
PAS complex	phosphatidic acid/phosphatidylserine complex

PAT	phosphinothricin-acetyltransferase
PC	phosphatidylcholine
PCF	phosphatidylcholine enriched fraction
PCR	polymerase chain reaction
PE	phosphatidylethanolamine
PEG	polyethylene glycol
PET	positron emission tomography
PG	phosphatidylglycerol
PGPR	polyglycerol polyricinoleate
PI	phosphatidylinositol
PKC	protein kinase c
PL	phospholipid
PLA	phospholipase A
PLA1	phospholipase A1
PLA2	phospholipase A2
PLB	phospholipase B
PLC	phospholipase C
PLD	phospholipase D
PMS	premenstrual syndrome
^{31}P NMR	phosphorus-31 nuclear magnetic resonance
POMS	Profile of Mood States
POV	peroxide value
POX	peroxygenase esterified phospholipid headgroup
PP	polar part
PPC	polyenyl phosphatidylcholine
PPI	poly-phosphoinositides
PS	phosphatidylserine
PS-DHA	PS omega-3 lecithin
PUFA	polyunsaturated fatty acid
PVC	polyvinyl chloride
QA	quality assurance
QHC	qualified health claims
RAP	Rapslecithin
RBC	red blood cells
RCT	randomized controlled trial
RCT	reverse cholesterol transport
RDMP	roller-dried milk powder
RDS	respiratory distress syndrome
RES	reticuloendothelial system
rFMMM	revised fluid mosaic membrane model
RI	refractive index
RRS	Roundup Ready® soybean
RSV	respiratory syncytial virus
S1P	sphingosine-1-phosphate
SAMe	S-adenosyl-methionine
SD	standard deviation
SDMP	spray-dried milk powder
SDS	sodium dodecyl sulfate
SL	sphingolipid
SL	sulfolipid
SM	sphingomyelin
SOD	superoxide dismutase
SPH	sphingolipids
SPM	specialized pro-resolving mediators
SPM	sphingomyelin
ST	signal transduction
SUN	sunflower lecithin
Syt1	synaptotagmin-1
TAG	triacylglyceride
TBARS	thiobarbituric acid reactive substances
TEM	transmission electron microscope
TEWL	transepidermal water loss
TG	triglycerides
TI	toluene-insoluble
TLC	thin layer chromatography
TMA	trimethylamine
TMAO	trimethylamine oxide
TNF	tumor necrosis factor
TOVA	test of variables of attention
TSST	Trier social stress test
UC	ulcerative colitis
UF	ultrafiltration
UTP	uniform transmembrane pressure
UV	ultraviolet
VLDL	very low-density lipoproteins
W/O	water-in-oil
WHO	World Health Organization
WI	wettability index
WMS	Wechsler memory scale
WPC	whey protein concentrate
WPI	whey protein isolate

Page numbers with addition "f" refer to figures, those with addition "t" to tables.